Advances in Sustainable Polymeric Materials II

Advances in Sustainable Polymeric Materials II

Editors

Cristina Cazan
Mihai Alin Pop

Basel • Beijing • Wuhan • Barcelona • Belgrade • Novi Sad • Cluj • Manchester

Editors

Cristina Cazan
Product Design, Mechatronics
and Environment Department
Transilvania University
of Brasov
Brasov
Romania

Mihai Alin Pop
Department of Materials Science
Transylvania University
of Brasov
Brasov
Romania

Editorial Office
MDPI
St. Alban-Anlage 66
4052 Basel, Switzerland

This is a reprint of articles from the Special Issue published online in the open access journal *Polymers* (ISSN 2073-4360) (available at: www.mdpi.com/journal/polymers/special_issues/DN455O4838).

For citation purposes, cite each article independently as indicated on the article page online and as indicated below:

Lastname, A.A.; Lastname, B.B. Article Title. *Journal Name* **Year**, *Volume Number*, Page Range.

ISBN 978-3-7258-1326-1 (Hbk)
ISBN 978-3-7258-1325-4 (PDF)
doi.org/10.3390/books978-3-7258-1325-4

© 2024 by the authors. Articles in this book are Open Access and distributed under the Creative Commons Attribution (CC BY) license. The book as a whole is distributed by MDPI under the terms and conditions of the Creative Commons Attribution-NonCommercial-NoDerivs (CC BY-NC-ND) license.

Contents

Preface vii

Catarina E. S. Ferreira, Isabel Santos-Vieira, Carlos R. Gomes, Salete S. Balula and Luís Cunha-Silva
Porous Coordination Polymer MOF-808 as an Effective Catalyst to Enhance Sustainable Chemical Processes
Reprinted from: *Polymers* 2024, 16, 968, doi:10.3390/polym16070968 1

Erika Pajares, Josu Fernández Maestu, Irati Fernandez-de-Mendiola, Unai Silvan, Pedro Costa and Iker Agirrezabal-Telleria et al.
Strategies for Improving Sustainability in the Development of High-Performance Styrenic Block Copolymers by Developing Blends with Cellulose Derivatives
Reprinted from: *Polymers* 2024, 16, 856, doi:10.3390/polym16060856 17

Joana M. Rocha, Rui P. C. L. Sousa, Raul Fangueiro and Diana P. Ferreira
The Potential of Electrospun Membranes in the Treatment of Textile Wastewater: A Review
Reprinted from: *Polymers* 2024, 16, 801, doi:10.3390/polym16060801 34

Vincenzo Algieri, Loredana Maiuolo, Debora Procopio, Paola Costanzo, Fiore Pasquale Nicoletta and Sonia Trombino et al.
Reactive Deep Eutectic Solvent for an Eco-Friendly Synthesis of Cellulose Carbamate
Reprinted from: *Polymers* 2024, 16, 757, doi:10.3390/polym16060757 68

Eleonora Ruffini, Andrea Bianchi Oltolini, Mirko Magni, Giangiacomo Beretta, Marco Cavallaro and Raffaella Suriano et al.
Crosslinked Polyesters as Fully Biobased Coatings with Cutin Monomer from Tomato Peel Wastes
Reprinted from: *Polymers* 2024, 16, 682, doi:10.3390/polym16050682 84

Ivana A. Boškov, Ivan M. Savić, Naa . Grozdanić Stanisavljević, Tatjana D. Kundaković-Vasović, Jelena S. Radović Selgrad and Ivana M. Savić Gajić
Stabilization of Black Locust Flower Extract via Encapsulation Using Alginate and Alginate–Chitosan Microparticles
Reprinted from: *Polymers* 2024, 16, 688, doi:10.3390/polym16050688 102

David Choque-Quispe, Carlos A. Ligarda-Samanez, Yudith Choque-Quispe, Sandro Froehner, Aydeé M. Solano-Reynoso and Elibet Moscoso-Moscoso et al.
Stability in Aqueous Solution of a New Spray-Dried Hydrocolloid of High Andean Algae *Nostoc sphaericum*
Reprinted from: *Polymers* 2024, 16, 537, doi:10.3390/polym16040537 122

Kazuma Matsumoto, Naoto Iwata and Seiichi Furumi
Cholesteric Liquid Crystals with Thermally Stable Reflection Color from Mixtures of Completely Etherified Ethyl Cellulose Derivative and Methacrylic Acid
Reprinted from: *Polymers* 2024, 16, 401, doi:10.3390/polym16030401 140

Endarto Yudo Wardhono, Nufus Kanani, Mekro Permana Pinem, Dwinanto Sukamto, Yenny Meliana and Khashayar Saleh et al.
Fluid Mechanics of Droplet Spreading of Chitosan/PVA-Based Spray Coating Solution on Banana Peels with Different Wettability
Reprinted from: *Polymers* 2023, 15, 4277, doi:10.3390/polym15214277 156

Carlos A. Ligarda-Samanez, Elibet Moscoso-Moscoso, David Choque-Quispe, Betsy S. Ramos-Pacheco, José C. Arévalo-Quijano and Germán De la Cruz et al.
Native Potato Starch and Tara Gum as Polymeric Matrices to Obtain Iron-Loaded Microcapsules from Ovine and Bovine Erythrocytes
Reprinted from: *Polymers* **2023**, *15*, 3985, doi:10.3390/polym15193985 . . . . **169**

Moises Job Galindo-Pérez, Lizbeth Martínez-Acevedo, Gustavo Vidal-Romero, Luis Eduardo Serrano-Mora and María de la Luz Zambrano-Zaragoza
Preservation of Fresh-Cut 'Maradol' Papaya with Polymeric Nanocapsules of Lemon Essential Oil or Curcumin
Reprinted from: *Polymers* **2023**, *15*, 3515, doi:10.3390/polym15173515 . . . . **188**

Zhibing Ni, Jianan Shi, Mengya Li, Wen Lei and Wangwang Yu
FDM 3D Printing and Soil-Burial-Degradation Behaviors of Residue of Astragalus Particles/Thermoplastic Starch/Poly(lactic acid) Biocomposites
Reprinted from: *Polymers* **2023**, *15*, 2382, doi:10.3390/polym15102382 . . . . **212**

Joaquín Hernández-Fernández, Esneyder Puello-Polo and Edgar Márquez
Furan as Impurity in Green Ethylene and Its Effects on the Productivity of Random Ethylene–Propylene Copolymer Synthesis and Its Thermal and Mechanical Properties
Reprinted from: *Polymers* **2023**, *15*, 2264, doi:10.3390/polym15102264 . . . . **228**

Tunde Borbath, Nicoleta Nicula, Traian Zaharescu, Istvan Borbath and Tiberiu Francisc Boros
The Contribution of $BaTiO_3$ to the Stability Improvement of Ethylene–Propylene–Diene Rubber: Part I—Pristine Filler
Reprinted from: *Polymers* **2023**, *15*, 2190, doi:10.3390/polym15092190 . . . . **241**

Bunsita Wongvasana, Bencha Thongnuanchan, Abdulhakim Masa, Hiromu Saito, Tadamoto Sakai and Natinee Lopattananon
Reinforcement Behavior of Chemically Unmodified Cellulose Nanofiber in Natural Rubber Nanocomposites
Reprinted from: *Polymers* **2023**, *15*, 1274, doi:10.3390/polym15051274 . . . . **259**

Fuyun Pei, Lijuan Liu, Huie Zhu and Haixin Guo
Recent Advances in Lignocellulose-Based Monomers and Their Polymerization
Reprinted from: *Polymers* **2023**, *15*, 829, doi:10.3390/polym15040829 . . . . **276**

Preface

Sustainability and environmental protection challenges are becoming increasingly significant. Therefore, chemistry and materials engineering play an important role in the development of innovative solutions to reduce the negative impact on our planet.

This reprint, "Advances in Sustainable Polymeric Materials II", contains 16 studies and further cutting-edge research that illustrate how these challenges can be addressed through the use of advanced materials, sustainable polymers, and green technologies.

This volume presents various aspects of materials research and development, with particular emphasis on efficient catalysts, bio-based materials, and green chemical processes. Each chapter provides detailed insight into how natural resources can be harnessed to create high-performance products while maintaining a strong commitment to environmental protection.

The presented research demonstrates the innovative use of porous coordination polymers as catalysts, providing sustainable solutions for chemical processes. Strategies to improve sustainability in the development of performance materials emphasize the importance of integrating cellulose derivatives and other natural materials into polymer compositions. These innovations not only improve the performance of materials but also contribute to reducing the ecological footprint of industrial processes.

Another key aspect addressed in this reprint is the use of electrospun membranes and reactive eutectic solvents to develop greener methods of wastewater treatment and materials synthesis. These emerging technologies demonstrate the enormous potential of innovative approaches to solving environmental problems.

Some studies also address the use of agri-food waste and biomaterials for the development of fully bio-based polymer coatings and materials. These studies highlight how often waste can be transformed into valuable and sustainable products, thus contributing to the circular economy and reducing waste.

Another area of interest presented in this reprint is the stabilization and preservation of natural extracts and active ingredients, demonstrating the importance of encapsulation and the use of nanotechnology in preserving the beneficial properties of these substances.

We hope this reprint will inspire and guide researchers, engineers, and practitioners in their efforts to transform environmental challenges into opportunities for innovation and sustainable progress.

Cristina Cazan and Mihai Alin Pop
Editors

Article

Porous Coordination Polymer MOF-808 as an Effective Catalyst to Enhance Sustainable Chemical Processes

Catarina E. S. Ferreira [1], Isabel Santos-Vieira [2], Carlos R. Gomes [3], Salete S. Balula [1,*] and Luís Cunha-Silva [1,*]

1. LAQV/REQUIMTE & Department of Chemistry and Biochemistry, Faculty of Sciences, University of Porto, 4169-007 Porto, Portugal; up201804944@fc.up.pt
2. CICECO—Aveiro Institute of Materials, Department of Chemistry, University of Aveiro, 3810-193 Aveiro, Portugal; ivieira@ua.pt
3. CIMAR/CIIMAR—Centro Interdisciplinar de Investigação Marinha e Ambiental & Faculdade de Ciências, Universidade do Porto, 4169-007 Porto, Portugal; crgomes@fc.up.pt
* Correspondence: sbalula@fc.up.pt (S.S.B.); l.cunha.silva@fc.up.pt (L.C.-S.)

Abstract: The improvement of sustainable chemical processes plays a pivotal role in safe environmental and societal development, for example, by reducing the use of hazardous substances, preventing chemical waste, and improving the efficiency of chemical reactions to obtain added-value compounds. In this context, the porous coordination polymer MOF-808 (MOF, metal–organic framework) was prepared by a straightforward method in water, at room temperature, and was unequivocally characterized by powder X-ray diffraction, vibrational spectroscopy, thermogravimetric analysis, and scanning electron microscopy. MOF-808 material was applied for the first time as catalysts in ring-opening aminolysis reactions of epoxides. It demonstrated high activity and selectivity for reactions of styrene oxide and cyclohexene oxide with aniline, using a very low amount of an eco-sustainable solvent (0.5 mL of EtOH), at 70 °C. Moreover, MOF-808 demonstrated high stability in the catalytic reaction conditions applied, and a notable reuse capacity of up to 20 consecutive reaction cycles, without significant variation in its catalytic performance. In fact, this Zr-based porous coordination polymer prepared by environment-friendly conditions proved to be a novel efficient heterogeneous catalyst, promoting the ring-opening reaction of epoxides under more sustainable conditions, and using a very low amount of catalyst.

Keywords: porous coordination polymers; metal–organic framework; MOF-808; heterogeneous catalysts; epoxides ring opening; sustainable processes

Citation: Ferreira, C.E.S.; Santos-Vieira, I.; Gomes, C.R.; Balula, S.S.; Cunha-Silva, L. Porous Coordination Polymer MOF-808 as an Effective Catalyst to Enhance Sustainable Chemical Processes. *Polymers* 2024, *16*, 968. https://doi.org/10.3390/polym16070968

Academic Editor: Shin-Ichi Yusa

Received: 1 March 2024
Revised: 26 March 2024
Accepted: 29 March 2024
Published: 2 April 2024

Copyright: © 2024 by the authors. Licensee MDPI, Basel, Switzerland. This article is an open access article distributed under the terms and conditions of the Creative Commons Attribution (CC BY) license (https://creativecommons.org/licenses/by/4.0/).

1. Introduction

Metal–organic frameworks (MOFs) are a class of crystalline materials consisting of metal centers (nodes) interconnected by organic ligands (linkers) via coordination bonds, ultimately forming porous three-dimensional (3D) porous coordination polymers [1]. In the last few years, numerous studies have shown that MOFs can overcome the potential of other known porous solid materials such as zeolites and porous carbon-based materials in a variety of applications [2]. In fact, the scientific interest in these materials is due to the huge variety of organic and inorganic components used for the preparation of MOFs originating large structural variety and the easy introduction of functional groups in the ligand, enhancing MOFs for several applications and areas of interest such as gas storage, adsorption, luminescence, sensors, catalysis, therapy for various diseases, such as cancer, water/air purification, and others [1,3–7].

The use of MOFs as heterogeneous catalysts was one of the most promising applications, along with its rapid development in the last twenty years [8]. Because of the enormous possibility of combining various active catalytic sites (active metal centers and/or active functional organic linkers) in the same material, porous MOFs have attracted significant scientific interest as heterogeneous catalysts in a wide range of chemical reactions [9–14]. In general, the structures of MOFs reveal active sites uniformly dispersed throughout the framework, and the characteristic porosity of MOFs tends to facilitate the transport of substrates and products and to facile access of the catalytic active sites. Frequently, MOFs present highly active and stable structures, which can possibly be recycled or reused in up to several reaction cycles. In fact, some MOFs can demonstrate high catalytic performances as homogeneous catalysts and also the capacity for reutilization that is characteristic of heterogeneous catalysts [8,15,16]. In particular, Zr(IV)-based MOFs with carboxylate-type ligands, for example, MOF-808, tend to be formed by the Zr–O cluster interconnected by the several mentioned ligands. These peculiar structural features provide Zr-based MOFs with interesting characteristics, such as high (nano)porosity, excellent thermal and hydrolytic stability, high specific surface area, and the possibility to accommodate framework defects, giving them enormous catalytic potential as heterogeneous catalysts [17–19]. Specifically, MOF-808 is formed by clusters $[Zr_6O_4(OH)_4]^{12+}$ like those present in the well-known UiO-66, but each cluster is connected only by six trimesate ligands, and the other coordination positions of the Zr cations are saturated by ion-shaped molecules.

The ring-opening reactions of an epoxide by amines have relevant importance since the products of this reaction, such as β-amino alcohols, have an extraordinary interest in the preparation of biologically active and synthetic compounds, as well as other neutral products, such as chiral pharmaceutical molecules, insecticides, and others [20–23] (Figure 1). β-amino alcohols are generally synthesized by a direct reaction of epoxide aminolysis. Epoxides are one of the most versatile intermediates used in organic synthesis and react with a wide variety of reagents such as electrophiles, nucleophiles, acids, bases, reducing agents, and some oxidizing agents [24]. These present an easy-to-reproduce synthesis due to ring deformation and still react with different nucleophiles with high regioselectivity, originating products that have an open ring [25]. The fact that there is a high reactivity with several nucleophiles produces highly regioselective and trans-stereospecific ring-opening products. Therefore, there is currently a great interest in the study of opening reactions of the epoxy ring [26,27]. However, this reaction most often involves a large excess of amines, and the use of other extreme reaction conditions, such as elevated temperatures, drastically reduces its industrial attractiveness. The cost-efficiency and sustainability of this type of process can be improved using efficient, selective, and recyclable solid catalysts.

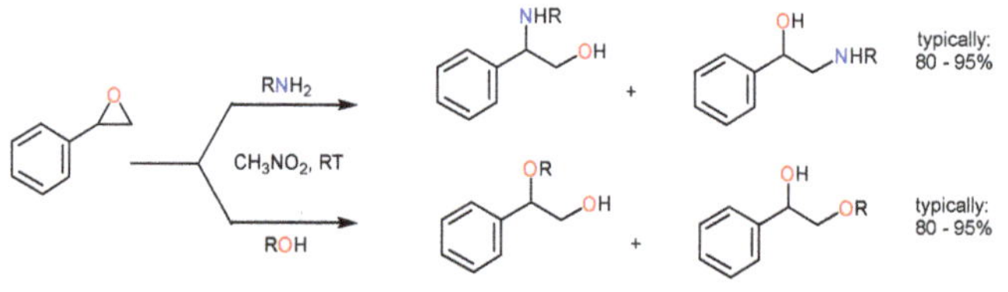

Figure 1. Scheme of the epoxide ring-opening reaction, both by amines and alcohols, using nitromethane as a solvent at room temperature; the products represented are usually the majority (with yields usually ranging from 80% to 95%).

Most of the catalysts reported for these types of reactions are homogeneous and thus, are practically impossible to reuse, and they are often expensive, inefficient, toxic, or unstable. For this reason, it is important and fundamental to evaluate unprecedented materials as sustainable heterogeneous catalysts [28,29]. Following the research interest of our research group in the development of MOF-based materials towards different sustainable catalytic processes [30–36], MOF-808 was prepared by a greener procedure, properly characterized, and evaluated as a catalyst in two ring-opening catalytic reactions, one of styrene oxide and the other of cyclohexene oxide (another Zr-based porous coordination polymer, UiO-66-NH$_2$, was also prepared and applied for comparative proposes; Figure 2). MOF-808 proved to be an efficient heterogeneous catalyst for this type of reaction.

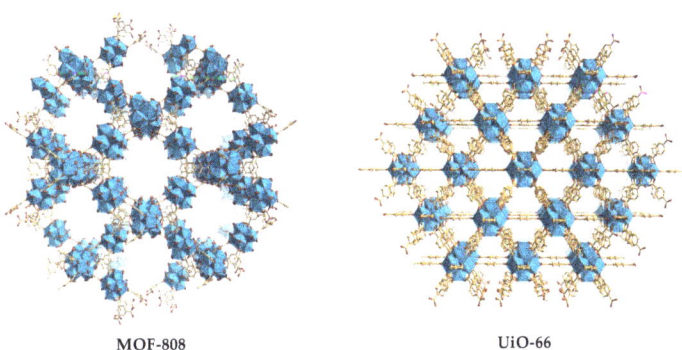

Figure 2. Crystalline structures of the porous MOF materials, MOF-808 and UiO-66 (identical to UiO-66-NH$_2$), revealing the same metal cluster center; the images were prepared from CIF files obtained from the Cambridge Structural Database (CSD, reference codes BOHWUS [37] and AZALUL [38], respectively).

2. Materials and Methods

2.1. Reagents and Products

Styrene oxide (C_8H_8O, 97%), cyclohexene oxide ($C_6H_{10}O$, 98%), aniline (C_6H_7N, 99%), and trimesic acid ($C_9H_6O_6$, 95%) were acquired from Aldrich. The phosphomolybdic acid hydrate ($H_3PMo_{12}O_{40} \cdot xH_2O$) was acquired from Sigma-Aldrich. Formic acid (CH_2O_2, 90%) and dimethylformamide (C_3H_7NO, ≥99.5%) were acquired from Fisher. Acetonitrile (CH_3CN, 99.5%, MeCN), acetone (C_3H_6O, 99.8%), and methanol (CH_4O, ≥99.9%) were purchased from Carlo Erba. Zirconium chloride ($ZrCl_4$, 99.5%) was acquired from Alfa Aesar. Ethanol (C_2H_6O, ≥99.8%) and acetic acid ($C_2H_4O_2$, ≥99.8%) were purchased from Honeywell. Isopropanol (C_3H_8O) was acquired from Merck Millipore. These chemical compounds used in this research work were used as received, without any type of purification treatment prior to use.

2.2. Characterizations Methods

Powder X-ray diffraction (PXRD) patterns were collected at room temperature on an Empyrean PANalytical diffractometer (CuKα1/2 radiation, λ1 = 1.540598 Å, and λ2 = 1.544426 Å), equipped with a PIXcel 1D detector and a flat-plate sample holder in a Bragg–Brentano para-focusing optics configuration (45 kV, 40 mA); intensity data were obtained by the step-counting method (step 0.02°), in continuous mode in the approximate range of $3.0° \leq 2\theta \leq 50°$. Infrared spectra were recorded on a PerkinElmer FT-IR System Spectrum BX spectrometer (in the range of wavenumbers: 400 to 4000 cm^{-1} and 64 scans). FT-Raman spectra were recorded in a Bruker spectrometer RFS 100, using as excitation source a Nd: YAG laser (λ = 1064 nm) at room temperature, in the frequency range of 3600–50 cm^{-1}, with a resolution of 4 cm^{-1} (the excitation power and number of steps were selected according to the sample). Scanning Electronic Microscopy (SEM)/Energy-Dispersive X-Ray Spectroscopy (EDS) analyses were performed at the "Centro de Materiais da Universidade do Porto" (CEMUP) using an Environmental Scanning Electron Microscope, high resolution (Schottky), with X-ray Microanalysis and Analysis of Electron Diffraction Patterns: FEI Quanta 400FEG ESEM/EDAX Genesis X4M; the samples were coated with Au/Pd thin film by cathodic pulverization (sputtering) using the SPI Module Sputter Coater equipment. Thermogravimetric analyses (TGAs) were performed using a thermobalance Thermal Analysis system STA 300 Hitachi, using a heating speed of 5 C/min and an N_2 atmosphere. Inductively coupled plasma optical emission spectrometry (ICP-OES) tests were performed at the University of Santiago de Compostela, Spain, on PerkinElmer Optima 4300 DV equipment. The quantification of the products of the two ring-opening reactions along the catalytic ring-opening reactions was performed in a

Scion 8300 GC gas chromatograph, using hydrogen as drag gas, with a linear velocity of 55 $cm^3.s^{-1}$ and using a capillary column of SPB-5 Supelco (30 m long, 250 µm inner diameter and 25 µm film thickness); the total time per analysis was 18 min. Gas chromatography coupled to mass spectrometry (GC-MS) was performed in a Thermo Scientific Trace 1300 chromatograph coupled to a Thermo Scientific ISQ Single Quadruplo MS device. In both cases, TG-5MS columns (30 m; 0.25 mm (i.d.); 0.25 µm) were used.

2.3. Preparation of the Materials

The preparation of the oxoclusters and, posteriorly, the synthesis of MOF-808 with three different particle sizes was performed following procedures previously reported in the literature [39]. The synthesis at room temperature of this MOF was carried out using pre-formed octahedral oxoclusters Zr_6, which is the most common SBU in the family of MOFs based on Zr (IV) with the carboxylate functional ligands.

Oxoclusters. A total of 25.17 g (0.11 mol) of $ZrCl_4$ was mixed with 37.5 mL of acetic acid and 62.5 mL of isopropanol, and the mixture was stirred until a homogeneous solution was obtained. The solution was heated in a paraffin bath at 120 °C for 1 h, and the white suspension obtained was then centrifuged and washed twice with acetone. Finally, the material was dried by heating (60 °C) at reduced pressure (200 mbar), and the white powder obtained was characterized by FTIR-ATR.

MOF-808. A total of 0.6 g of oxoclusters was added to 1.5 mL of formic acid, and the solution was stirred at room temperature. Then, 2.5 mL of H_2O was added, and the solution changed from whitish to transparent, as reported in the literature [39]. Finally, 150 mg (0.7 mmol) of trimesic acid was added, and the reaction mixture was kept in agitation overnight. The material obtained was isolated by centrifugation, washed twice with H_2O and ethanol, and dried by heating (60 °C) at reduced pressure (200 mbar). Yield (%): 73.8.

UiO-66-NH_2. The synthesis procedure followed the solvothermal method adapted from the literature [40]. Briefly, 2.17 g (0.012 mol) of 2-aminoterafthallic acid and 3.8 g (0.016 mol) of $ZrCl_4$ were mixed in 36 mL of DMF. The mixture was stirred for 30 min and inserted in autoclaves that were placed in the oven at 120 °C for 24 h. A yellow solid was isolated by centrifugation, washed with DMF and methanol, and dried by heating at 70 °C and reduced pressure (under vacuum). Yield (%): 82.5.

2.4. Catalytic Studies

The ring-opening reactions of styrene oxide (1 mmol) and cyclohexene oxide (1 mmol) with aniline (0.9 mmol) were carried out in a borosilicate 5 mL reaction vessel with a magnetic stirrer and immersed in a thermostat oil bath under air (atmospheric pressure), using an amount of catalyst (10 mg) containing 1 µmol of Zr and 1.5 mL of acetonitrile (MeCN) or ethanol (EtOH) as a solvent, at 70 °C (Figure 3). The recycling study of the catalyst was performed by using the same portion of the solid catalyst during various consecutive cycles, dispersed in 0.5 mL of EtOH, and after each catalytic cycle, the catalyst was recovered by centrifugation, washed with EtOH (twice), and dried. The recovered MOF was weighed and reused in a new catalytic cycle maintaining the experimental conditions. Further, the reutilization ability of the catalyst was also studied, and in this case, the catalyst was not washed or dried after each catalytic cycle. The catalyst was used without treatment between cycles, maintaining all the experimental conditions. All the reactions were monitored by periodic GC analysis taking an aliquot from the reaction mixture at regular intervals until the product yields remained constant during at least 2 h of reaction. The products were analyzed by GC/GC–MS techniques. Moreover, the recovered catalysts were characterized by powder XRD and FT-IR spectroscopy after catalytic use.

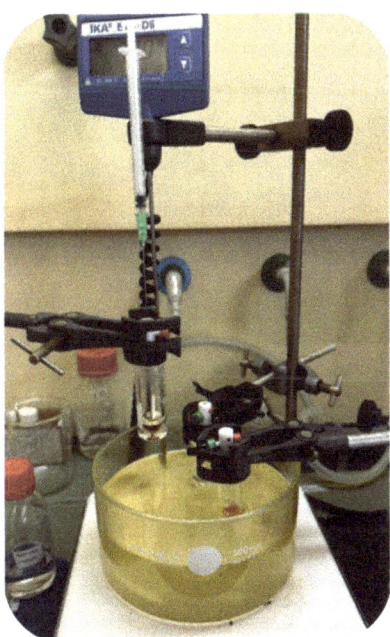

Figure 3. Experimental apparatus used to perform the catalytic experiments.

3. Results and Discussion

3.1. Preparation and Characterization of the Catalysts

The syntheses of Zr-based MOFs are usually performed by the solvothermal procedure using organic solvents, such as DMF, with some negative environmental impact. However, nowadays, it is crucial to obtain these MOF structures using more sustainable synthesis routes, especially under aqueous conditions and/or at room temperature. The research in the context of new conditions of green synthesis is an important focus in the area of MOF materials, especially motivated by the transition to an industrial-scale synthesis that would be practically impossible with the use of hazardous chemicals under adverse reaction conditions [11]. In this work, while the MOF UiO-66-NH_2 was prepared by the conventional solvothermal procedure (high temperature and DMF as a solvent), MOF-808 was prepared by a more sustainable method (room temperature and water as a solvent; Figure 4). Both porous MOFs, i.e., MOF-808 and UiO-66-NH_2, were further characterized by a myriad of techniques including vibration spectroscopies (FTIR and FT-Raman), powder XRD, TGA, elemental analysis (ICP), scanning electron microscopy (SEM), and energy-dispersive X-ray spectroscopy (EDS), confirming the preparation of the expected solid-state pure phases of the two porous MOF materials. The porous nature of these coordination polymers is well documented in the literature (MOF-808: BET area ~1600 $mg^2 cm^{-1}$, specific volume 0.69 $cm^3 g^{-1}$, pore size 18.4 Å; UiO-66-NH_2: BET area ~870 $mg^2 cm^{-1}$, specific volume 0.38 $cm^3 g^{-1}$, pore size 9.5 Å).

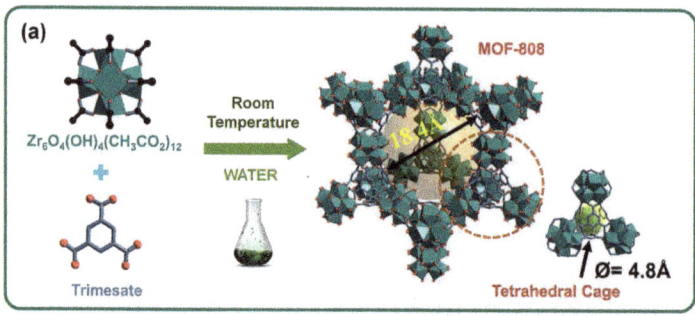
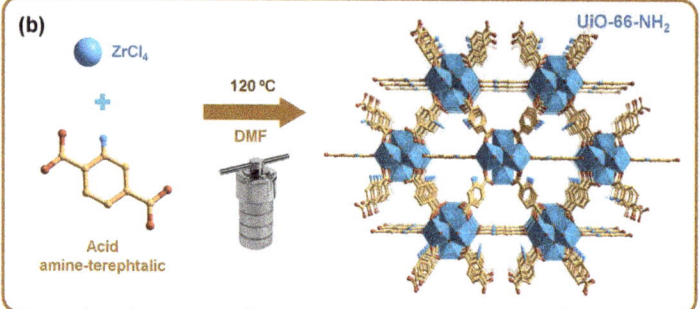

Figure 4. (a) Schematic representation of the sustainable synthesis of MOF-808 (in water and at room temperature) and (b) the scheme of the preparation of UiO-66-NH$_2$, showing some features of the crystalline structure of the porous coordination polymers. Adapted from reference [39].

The FT-IR and FT-Raman spectra obtained for both the Zr-based MOF-based materials reveal the main vibrational band characteristics as expected from the two MOF structures (Figure 5a,b; spectra shown in the 1900–400 cm^{-1} wavenumber region) [42,43]. Briefly, the medium and strong intensity bands to vibrational modes of the carboxylate groups can be assigned from 1600 to 1390 cm^{-1}; a vibrational band around 1450 cm^{-1} ascribed to aromatic (C–C) bonds of the organic ligands; as well as a group of vibrational bands associated with Zr–(μ_3-O) framework bonds in the range of 800–600 cm^{-1}, and a band (FT-IR) around 450 cm^{-1} assigned to Zr–(OC) bonds. These vibrational spectra are comparable with those previously reported, being the first clear indication of the preparation of the desired MOF materials, MOF-808 and UiO-66-NH$_2$. This fact is completely confirmed by the powder XRD patterns of the isolated materials that reveal the characteristic reflections of the MOF-808 and UiO-66-NH$_2$ crystalline phase, both in location and relative intensities (Figure 5c): in the diffractogram of MOF-808, the main 2θ diffraction peaks at the 4.3°, 8.3°, 8.7°, 10.0°, 11.0°, and 13.0° plans are assigned to the reflection plans (111), (311), (222), (400), (331), and (511), respectively; for the UiO-66-NH$_2$ diffraction pattern, the main reflections at 7.4°, 8.5°, 11.9°, 14.6°, 16.8°, and 25.6° are attributed to the plans (111), (200), (220), (222), (400), and (600), respectively. Furthermore, the powder XRD analysis also allows for discarding the coexistence of any secondary crystalline phases in the two isolated materials by comparison with the diffractogram obtained from the crystallographic data. In fact, the experimental diffractograms prove the preparation of both the MOF-808 and UiO-66-NH$_2$ materials with the expected solid-state structure [39,44]. As expected, the thermogravimetric analysis profiles of the two materials are like those previously reported and confirm the thermal stability of MOF-808 up to about 500 °C and up to 400 °C for UiO-66-NH$_2$ (Figure 5d) [45,46].

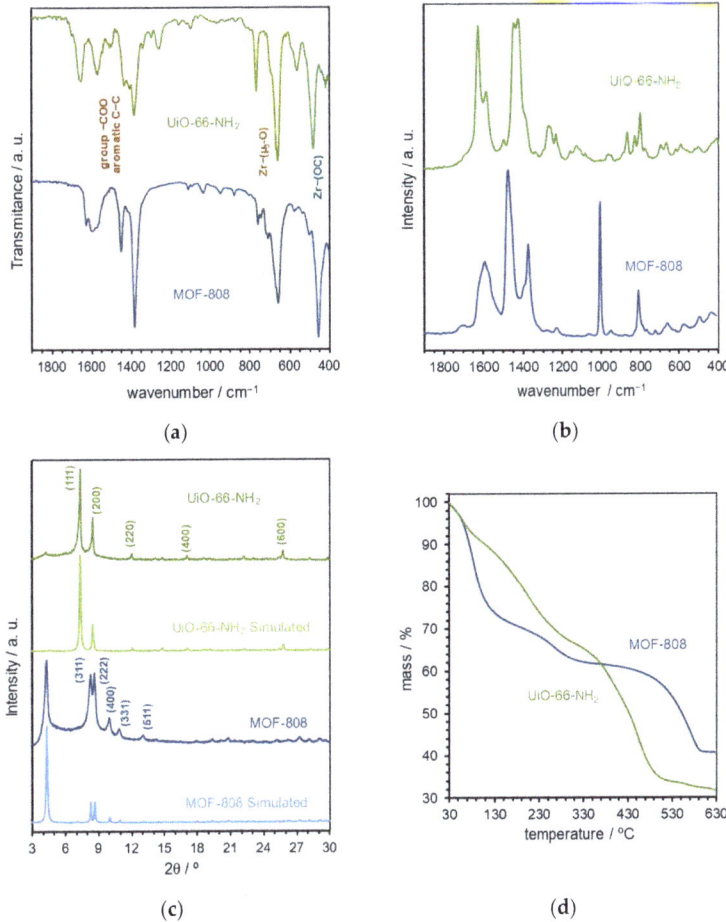

Figure 5. Selected characterization data for the prepared MOF materials, MOF-808 (blue color) and UiO-66-NH$_2$ (green color): (**a**) FT-IR and (**b**) FT-Raman showed in the wavenumber range from 1900 to 400 cm^{-1}; (**c**) simulated and experimental powder XRD patterns (the simulated diffractograms were obtained from the respective crystallographic data deposited in the Cambridge Structural Database); and (**d**) TGA profiles showed in temperature range of 30–630 °C.

The micrographs obtained by SEM for the two materials in random zones of the samples reveal agglomerated particles, apparently with superior crystalline regularity in the UiO-66-NH$_2$ sample, most probably because of its solvothermal preparation instead of the room temperature preparation of MOF-808 (Figure 6). In addition, the EDS analyses clearly demonstrate the presence of the following characteristic elements of the MOFs: Zr, O, and C (also N for UiO-66-NH$_2$). The EDS elemental mappings confirm and homogenous distribution of the main elements of the material structure. The combination of all the data obtained by several characterization techniques confirms the successful preparation of the two Zr-based porous MOF materials, MOF-808 and UiO-66-NH$_2$.

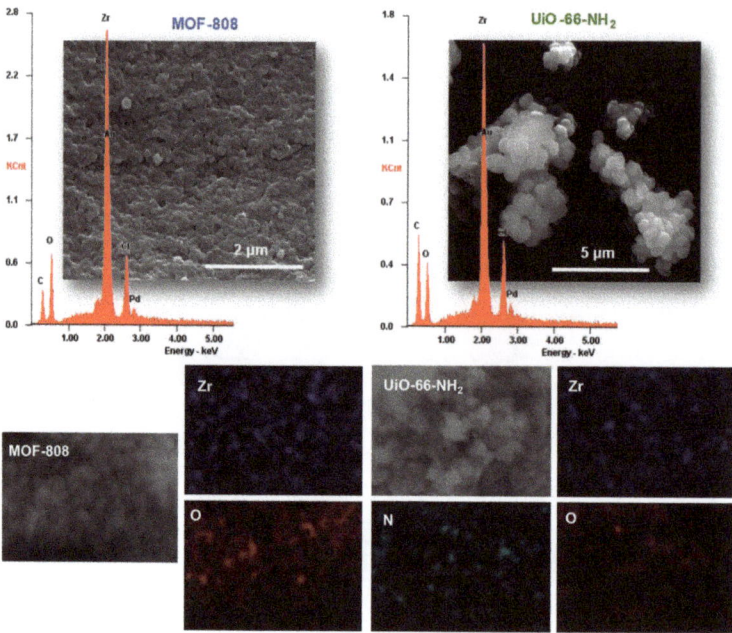

Figure 6. SEM images, EDS spectra, and elemental mappings for the prepared porous coordination polymers, MOF-808 and UiO-66-NH$_2$.

3.2. Catalytic Studies

The prepared Zr-based materials were evaluated as heterogeneous catalysts for ring-opening reactions of epoxides, in particular, styrene oxide and cyclohexene oxide, by amines (aniline). The influence of some experimental reaction conditions was initially investigated, namely, the nature and the volume of the used solvent. This study intends to use more sustainable solvents at the lowest amount needed that can guarantee the highest catalytic activity of the catalytic materials. Blank experiments were conducted without the presence of the MOF catalysts using both epoxide substrates (Table 1). The results indicate that practically no conversion of substrates was observed in the absence of the catalyst, given that the highest conversion under these conditions was less than 16% for styrene oxide, after 24 h using 0.5 mL of ethanol (EtOH).

Table 1. Conversion of the aminolysis reaction of epoxides (styrene oxide and cyclohexene oxide) with aniline in the absence of a catalyst after 24 h of reaction (blank experiments).

	Solvent	Conversion (%)
	1.5 mL EtOH	10%
Styrene oxide	0.5 mL EtOH	16%
	1.5 mL MeCN	14%
	1.5 mL EtOH	10%
Cyclohexane oxide	0.5 mL EtOH	11%
	1.5 mL MeCN	13%

The Zr-based MOF catalysts, MOF-808 and UiO-66-NH$_2$, were then used as heterogeneous catalysts. The initial experimental conditions used were based on the previously published work from our research group using the following Fe-based MOFs: MIL-101(Fe) and MIL-101(Fe)-NH$_2$, i.e., 1 mmol substrate, 0.9 mmol of aniline, 1.5 mL of MeCN, and 70 °C temperature [28]. The results obtained for the conversion of styrene oxide are presented in Figure 7. It is possible to observe that during the first 6 h of the reaction, MOF-808 showed to be more active than UiO-66-NH$_2$, mainly using EtOH as solvent instead of MeCN. After 6 h of reaction, 82.9% of styrene oxidation was obtained using MOF-808 as a catalyst and a very small amount of the sustainable solvent EtOH (0.5 mL), instead of 61.7% obtained with a higher EtOH amount (1.5 mL), and 37.6% of conversion using MeCN (1.5 mL). After 24 h, the conversion increased to 86.1% using MOF-808 and 0.5 mL of EtOH. When UiO-66-NH$_2$ was used, a similar conversion was achieved (88.3%) using 1.5 mL of EtOH. From these results, it was possible to conclude that a higher conversion was possible to be obtained with a lower amount of solvent used. Therefore, future experiments were performed with EtOH instead of MeCN.

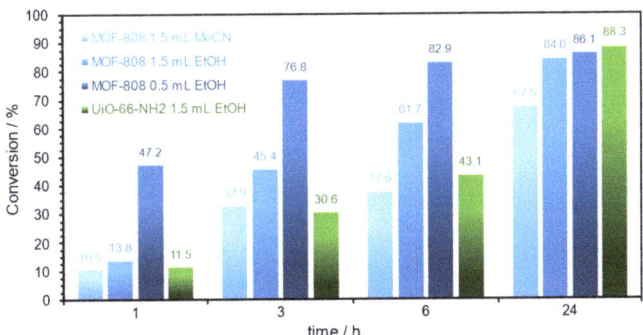

Figure 7. Conversion data for the ring-opening reaction of styrene oxide (1 mmol) in the presence of aniline (0.9 mmol), using MOF-808 as a catalyst (1.5 mL of MeCN and 1.5 mL and 0.5 mL of EtOH as solvents), and UiO-66-NH$_2$ as a catalyst (1.5 mL of EtOH as a solvent), at 70 °C.

In addition to the interesting catalytic activity of the two Zr-based MOFs for the conversion of styrene oxide with aniline, both the MOF-808 and UiO-66-NH$_2$ catalysts also showed high selectivity, obtaining amino derivative main products (Figure 8). Under the studied conditions, both MOF catalysts promoted, essentially, the production of a distinct main product (77% of aminodiphenylmethane with MOF-808 and 85% of (1S,2R)-(+)-2-amino-1,2-diphenylethanol with UiO-66-NH$_2$) and two minor identical products (diecetal benzaldehyde and 3-aniline-3-pheny propionitrile), as presented in Figure 8b. The various products obtained in the catalytic reactions were identified by GC-MS.

The use of Zr-based MOFs as catalysts did not originate β-amino alcohols as products. The mechanism to obtain these is well-known in the literature and consists of two main steps. The reaction may be divided into the following two steps: (i) the epoxide ring-opening by the interaction of the nitrogen from the aniline with one of the two carbons from the epoxide and (ii) a transfer from a proton from aniline to the alkoxide oxygen atom to yield the β-amino alcohol product [22]. The studied Zr-based MOFs can have easily coordinated vacancies to the Zr metal center, which can interact easily with the epoxide, modifying the nature of the obtained products.

The potential of MOF-808 as heterogeneous catalyst in the ring-opening reaction of epoxides with amines was further validated with a second distinct molecule, cyclohexene oxide. The experimental conditions of the reaction were the same as those used for styrene oxide (1 mmol of epoxide substrate, 0.9 mmol of aniline, catalyst, at 70 °C). The nature and the amount of the solvent were also here studied; therefore, 1.5 mL of MeCN and 1.5 mL and 0.5 mL of EtOH were used. The results are presented in Figure 9a, which reveal that the conversion of cyclohexene oxide after 24 h was 78.4% (1.5 mL MeCN), 70.4% (1.5 mL EtOH), and 88.3% (0.5 mL EtOH). As in the previous study with styrene oxide, also using cyclohexene oxide as substrate, the use of EtOH as solvent originated the highest conversion. This was achieved using low volume of this solvent, i.e., 0.5 mL of EtOH), where the the conversion obtained at lower reaction times (1, 3, and 6 h) were slightly superior than the reaction system with MeCN.

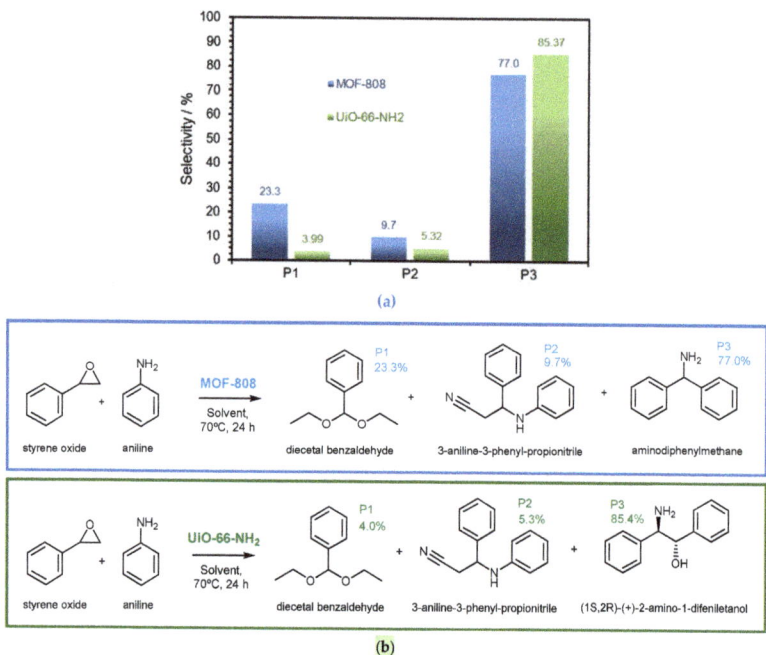

Figure 8. (**a**) Selectivity obtained for the ring-opening reaction of styrene oxide with aniline in the presence of MOF-808 (blue) and UiO-66-NH$_2$ (green) as catalysts showing the 3 main products (P1, P2, and P3) (24 h; a temperature of 70 °C and EtOH as a solvent). (**b**) Reaction schemes using the two MOFs as catalysts showing the respective obtained products (P1, P2, and P3).

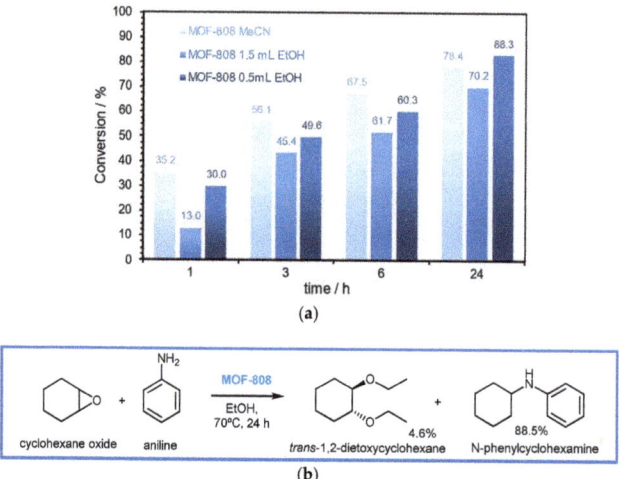

Figure 9. (**a**) Result obtained for the catalytic reaction of the ring opening of cyclohexene oxide (1 mmol) with aniline (0.9 mmol), using the MOF-808 catalyst and 1.5 mL of MeCN, and 1.5 mL and 0.5 mL of EtOH as solvents, at 70 °C. (**b**) Scheme of the reaction of the ring opening of cyclohexene oxide with aniline using the MOF-808 catalyst and 0.5 mL of EtOH.

In addition to the high activity demonstrated by MOF-808 in this ring-opening reaction of cyclohexene oxide with aniline, it revealed interesting selectivity, originating N-phenylcyclohexamine (88.5%) as the main product and as a secondary product, *trans*-1,2-dietoxycyclohexene (4.6%) (Figure 9b). In fact, the results obtained for the ring-opening reactions of styrene oxide and cyclohexene oxide with aniline confirm the good efficiency of MOF-808 as a catalyst in this type of reaction, both in terms of activity and selectivity.

3.3. Catalyst Reutilization and Stability

The capacity of recycling and reutilizing MOF-808 in several catalytic cycles of ring-opening reactions of epoxides was evaluated using styrene oxide (1 mmol), aniline (0.9 mmol), and 0.5 mL EtOH at 70 °C. Figure 10 presents the results obtained after performing the various consecutive reactions for 24 h. In the recycling procedure, after each reaction cycle, MOF-808 was recovered by centrifugation, washed carefully, and dried. To perform a consecutive reaction cycle, the recovered and treated catalyst was weighed, and all the experimental conditions were adjusted and maintained, i.e., the quantities of the reagents and volume of EtOH. For the recycling study, it was possible to use the MOF-808 catalyst for ten consecutive cycles without any relevant loss of catalytic efficiency (the conversion was maintained around 90% with a slight decrease after the seventh cycle) (Figure 10a). It was not possible to evaluate more recycle cycles because of experimental limitations, namely, the amount of catalyst that was slightly reduced after each cycle because of some losses in the separation and cleaning processes. To overcome these experimental limitations, the reutilization ability of MOF-808 was further evaluated. In this procedure, at each new reuse cycle, the solid catalyst was maintained in the reaction vessel (without any additional treatment), and the same amount of styrene oxide and aniline and the same volume of solvent was added to proceed with the reaction under the same experimental conditions for all the reuse cycles. Notably, MOF-808 also revealed high catalytic efficiency during 20 consecutive reutilization cycles for the reaction of the ring opening of styrene oxide with aniline (Figure 10b). The conversion was maintained at around 90% without a significant loss of activity during the 20 reaction cycles. Based on the recycling and reuse behavior of the MOF-808 catalyst during the high number of cycles, it is possible to predict the high structural stability of this material under the aminolysis reaction. To investigate the structural stability of the catalyst after catalytic use, this was characterized by several techniques. The FT-IR spectroscopy and powder XRD analysis were performed of the catalyst after 20 reusing cycles (Figure 11). The powder XRD pattern of the recovered catalyst revealed the same crystalline structure of the initial MOF 808, eventually showing a slight reduction in its crystallinity (Figure 11a). This evidence is supported by the FT-IR spectrum being identical to that of the pristine MOF-808 (Figure 11b). This complementary characterization confirms that the structure of the MOF material remains unchanged after reuse. In fact, these results demonstrated that MOF-808 is an effective heterogeneous catalyst, with high efficiency and notable reusability in the ring-opening reaction of epoxides with amines.

3.4. Comparison with Reported Works

Only a limited number of examples can be found in the literature reporting the application of MOF structures as catalysts for the aminolysis of epoxides with aniline (Table 2). The first work was presented by Jiang et al. in 2008, who used Cu-MOF as a heterogeneous catalyst at room temperature under a solvent-free system [47]. Under these conditions, only 32% of cyclohexene oxide was converted after 4 h, using a high excess of aniline (21 mmol instead of 0.9 mmol as used in this work). The authors state that the low conversion is mainly due to diffusion limitations in the micropores of this Cu-MOF. In 2010, Garcia et al. used Fe-BTC MOF as a catalyst for the ring opening of styrene oxide using acetonitrile as solvent (10 mL). After 24 h, a moderate yield was obtained (72%) with a substrate/catalyst ratio of 410 [48]. Eight years later, Anbu et al. used the same Fe-BTC MOF and copper $Cu_3(BTC)_2$ catalysts for styrene oxide aminolysis, under a solvent-free system, but they still used a higher catalyst excess than Gracia et al. used in 2010 [49]. The conversion obtained under these conditions was not improved compared to the previous work of Garcia et al. (Table 1). Previously, in 2010, Kumar studied the same reactions using [Co^{3+}-Ln^{3+}] heterobimetallic one-dimensional zigzag coordination polymers as heterogeneous catalysts [50]. High yields to produce β-amino alcohols were obtained using a near equimolar ratio of substrates and aniline, at room temperature and under solvent-free conditions. Further, in this study, the solid catalyst was reused for only three catalytic cycles and some loss of activity could be observed mainly for the aminolysis reaction with styrene oxide. This is a considerable advantage of this system, mainly when expensive lanthanide metals are used in a high amount of catalyst. In 2017, our research group investigated the catalytic activity of MIL-101(Fe) for the aminolysis of cyclohexene oxide [28]. At this time, MeCN was used as a solvent, and a high amount of catalyst was needed to achieve high conversion. In this work, it was also verified that the presence of amine functional groups in the MOF structure did not affect its catalytic performance. To analyze the importance of Fe in the styrene oxide aminolysis reaction, MIL-101(Fe)-NH_2 was used as a support to incorporate iron-monosubstituted polyoxometalate [$PW_{11}Fe(H_2O)O_{39}$]$^{5-}$ (abbreviated as $PW_{11}Fe$) [36]. In this case, the composite catalyst presented an appreciable increase in catalytic activity with the near complete aminolysis of styrene oxide after 1 h of reaction, instead of 22% obtained by the isolated support MIL-101(Fe)-NH_2. Comparing the results obtained previously (Table 1) with those obtained in the present work, it is possible to observe that the enormous advantage of using MOF-808 and UiO-66-NH_2 is the almost negligible amount of solid MOF catalyst necessary to achieve a near complete aminolysis reaction using a system with a low amount of a sustainable solvent (0.5 mL EtOH). Another advance present in this work was the high number of heterogenous catalyst reusing cycles (no extra solvents or thermic treatments were needed) that were possible to perform without loss of catalyst activity.

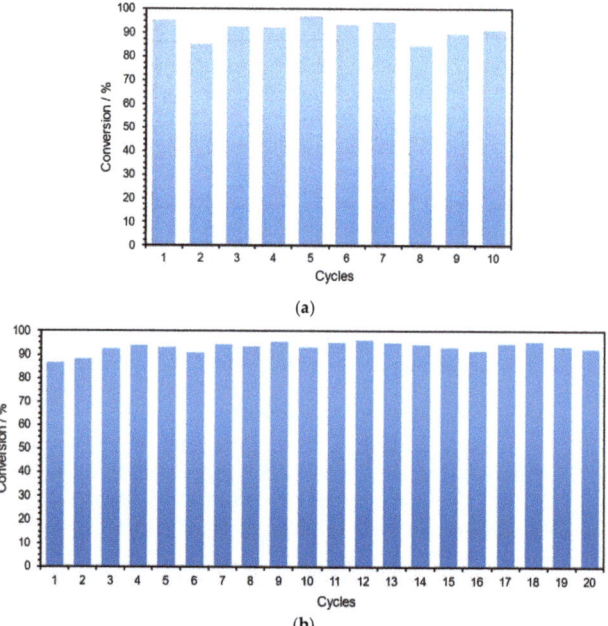

Figure 10. Results obtained in the recycling study (**a**) and for the reusing tests (**b**) of the catalyst MOF-808, obtained after 24 h of the ring-opening reaction of styrene oxide (1 mmol) with aniline (0.9 mmol), using 0.5 mL of EtOH as solvent, at 70 °C.

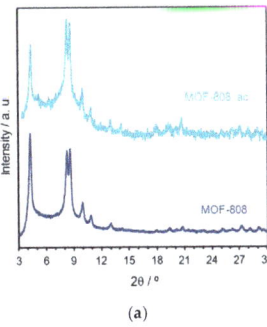 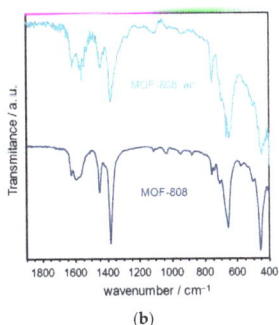

Figure 11. (a) Powder XRD patterns and (b) FT-IR spectra of MOF-808 before and after catalytic application (sample recovered after 20 consecutive cycles of reutilization).

Table 2. Reported works for the aminolysis reaction of epoxides (styrene oxide and cyclohexene oxide) with aniline catalyzed by MOF structures.

Catalyst	Epoxide	Epoxide/Aniline (mmol)	Catalyst Amount	Solvent	T (°C)	Time (h)	Conv. (%)	Refs.
Cu-MOF	CyclohexOx	1/21	0.11 mmol	no	rt	4	32	[47]
Fe-BTC	StyrOx	41.6/41.6	150 mg	MeCN	80	24	72	[48]
[Co^{3+}-Ln^{3+}] [a]	CyclohexOx	0.98/1.18	49 µmol	No	rt	4	98	[50]
[Co^{3+}-Ln^{3+}] [a]	StyrOx	0.87/1	44 µmol	No	rt	4	84	[50]
MIL-101(Fe)	CyclohexOx	1/0.9	55 µmol	MeCN	80	24	87	[28]
MIL-101(Fe)-NH_2	CyclohexOx	1/0.9	55 µmol	MeCN	80	24	86	[28]
PW_{11}Fe@MIL-101(Fe)	StyrOx	1/0.9	50 µmol	MeCN	80	1	97	[36]
MIL-101(Cr)	StyrOx	1/0.9	50 µmol	MeCN	80	5	0	[36]
$Cu_3(BTC)_2$	StyrOx	0.3/0.25	25 mg	No	60	24	75	[49]
Fe-BTC	StyrOx	0.3/0.25	25 mg	no	60	24	77	[49]
MOF-808	StyrOx	1/0.9	1 µmol	EtOH	70	6	83	here
UiO-66-NH_2	StyrOx	1/0.9	1 µmol	EtOH	70	24	88	here
MOF-808	CyclohexOx	1/0.9	1 µmol	EtOH	70	24	88	here

[a] the catalyst is [Co^{3+}-Ln^{3+}] coordination polymer with Ln = Eu or Tb; rt = room temperature.

4. Concluding Remarks

The incessant demand to develop chemical processes as sustainable as possible, aiming their application at a large scale without harming the environment, is a current research issue of extreme importance. In particular, the use of greener solvents to replace toxic solvents, which are pollutants and harmful, in the preparation of chemical materials such as MOFs, and the development of these materials as heterogeneous catalysts that are active and reusable, are important factors that have been considered along with the research work reported in this manuscript.

The Zr-based material MOF-808, a crystalline porous coordination polymer, was prepared by a straightforward room temperature procedure using water as a solvent, in contrast to the following synthesis method initially developed and widely used to prepare this MOF: the solvothermal method with DMF at a temperature of 120 °C and a time reaction of at least 24 h. This MOF-808 prepared by a much more sustainable method revealed identical structural features reported in the literature for the parent material isolated by the typical solvothermal method, as confirmed by powder XRD, vibrational (FT-IR and FT-Raman) spectroscopies, TGA, and SEM/EDS. Furthermore, this material was evaluated as a heterogeneous catalyst for ring-opening reactions of epoxides with amines for the first time. In particular, the reactions of styrene oxide or cyclohexene oxide with aniline in the presence of MOF-808 have been studied using MeCN or EtOH solvents. This MOF material revealed high activity and selectivity for both the ring-opening reactions of styrene oxide and cyclohexene oxide using a very small amount of solvent (0.5 mL of EtOH) and catalyst (an equivalent amount containing 1 µmol of Zr). Furthermore, it demonstrated notable recycling and reuse ability since MOF-808 was used for 10 consecutive recycling cycles and 20 successive reusing cycles in the reaction of styrene oxide (in EtOH) without significant loss of efficiency and maintaining its structural stability.

The investigation reported in this manuscript is a clear demonstration of the possibility of enhancing the sustainability of the chemical processes either in the preparation of functional MOFs or in their application as effective heterogeneous catalysts. Now, other MOF materials are being prepared by alternative green methods, showing their potential application as catalysts, as well as in gas separation/capture and harmful gas sensing.

Author Contributions: Conceptualization, S.S.B. and L.C.-S.; methodology, S.S.B. and L.C.-S.; validation, I.S.-V., S.S.B. and L.C.-S.; formal analysis, S.S.B. and L.C.-S.; investigation, C.E.S.F. and I.S.-V.; resources, I.S.-V., S.S.B. and L.C.-S.; data curation, L.C.-S.; writing—original draft preparation, C.E.S.F.; writing—review and editing, S.S.B. and L.C.-S.; visualization, C.R.G. and S.S.B.; supervision, C.R.G. and L.C.-S.; project administration, L.C.-S.; funding acquisition, S.S.B. and I.S.-V. All authors have read and agreed to the published version of the manuscript.

Funding: This work received financial support from Fundação para a Ciência e a Tecnologia/Ministério da Ciência, Tecnologia e Ensino Superior (FCT/MCTES) by national funds, through LAQV/REQUIMTE (Ref. UIDP/50006/2020 DOI 10.54499/UIDP/50006/2020; LA/P/0008/2020 DOI 10.54499/LA/P/0008/2020; UIDB/50006/2020 DOI 10.54499/UIDB/50006/2020) and the CICECO-Aveiro Institute of Materials (UIDB/50011/2020, UIDP/50011/2020 and LA/P/0006/2020).

Institutional Review Board Statement: Not applicable.

Data Availability Statement: The data are not publicly available due to privacy.

Acknowledgments: L.C.S. and S.S.B. thank FCT/MCTES for supporting their contract positions via the Individual Call to Scientific Employment Stimulus (Ref. CEECIND/00793/2018 and Ref. CEECIND/03877/2018, respectively). The position held by I.S.-V. (Ref. 197_97_ARH-2018) was supported by national funds (OE), through FCT, I. P., in the scope of the framework contract foreseen in the numbers 4, 5, and 6 of article 23 of the Decree-Law 57/2016 of 29 August, changed by Law 57/2017 of 19 July.

Conflicts of Interest: The authors declare no conflicts of interest.

References

1. Zhou, H.C.; Long, J.R.; Yaghi, O.M. Introduction to Metal-Organic Frameworks. *Chem. Rev.* **2012**, *112*, 673–674. [CrossRef] [PubMed]
2. Bai, Y.; Dou, Y.B.; Xie, L.H.; Rutledge, W.; Li, J.R.; Zhou, H.C. Zr-based metal-organic frameworks: Design, synthesis, structure, and applications. *Chem. Soc. Rev.* **2016**, *45*, 2327–2367. [CrossRef] [PubMed]
3. Lin, Z.J.; Zheng, H.Q.; Chen, J.; Zhuang, W.E.; Lin, Y.X.; Su, J.W.; Huang, Y.B.; Cao, R. Encapsulation of Phosphotungstic Acid into Metal-Organic Frameworks with Tunable Window Sizes: Screening of PTA@MOF Catalysts for Efficient Oxidative Desulfurization. *Inorg. Chem.* **2018**, *57*, 13009–13019. [CrossRef] [PubMed]
4. Păun, C.; Motelică, L.; Ficai, D.; Ficai, A.; Andronescu, E. Metal-organic frameworks: Versatile platforms for biomedical innovations. *Materials* **2023**, *16*, 6143. [CrossRef] [PubMed]
5. Mousavi, S.M.; Hashemi, S.A.; Fallahi Nezhad, F.; Binazadeh, M.; Dehdashtijahromi, M.; Omidifar, N.; Ghahramani, Y.; Lai, C.W.; Chiang, W.-H.; Gholami, A. Innovative metal-organic frameworks for targeted oral cancer therapy: A review. *Materials* **2023**, *16*, 4685. [CrossRef] [PubMed]
6. Xie, Y.; Lyu, S.; Zhang, Y.; Cai, C. Adsorption and Degradation of Volatile Organic Compounds by Metal–Organic Frameworks (MOFs): A Review. *Materials* **2022**, *15*, 7727. [CrossRef] [PubMed]
7. Saeb, M.R.; Rabiee, N.; Mozafari, M.; Mostafavi, E. Metal-Organic Frameworks (MOFs)-based nanomaterials for drug delivery. *Materials* **2021**, *14*, 3652. [CrossRef] [PubMed]
8. Jiao, L.; Wang, Y.; Jiang, H.-L.; Xu, Q. Metal–Organic Frameworks as Platforms for Catalytic Applications. *Adv. Mater.* **2018**, *30*, 1703663. [CrossRef] [PubMed]
9. Chughtai, A.H.; Ahmad, N.; Younus, H.A.; Laypkov, A.; Verpoort, F. Metal-organic frameworks: Versatile heterogeneous catalysts for efficient catalytic organic transformations. *Chem. Soc. Rev.* **2015**, *44*, 6804–6849. [CrossRef] [PubMed]
10. Liu, J.W.; Chen, L.F.; Cui, H.; Zhang, J.Y.; Zhang, L.; Su, C.Y. Applications of metal-organic frameworks in heterogeneous supramolecular catalysis. *Chem. Soc. Rev.* **2014**, *43*, 6011–6061. [CrossRef] [PubMed]

11. Dhakshinamoorthy, A.; Opanasenko, M.; Cejka, J.; Garcia, H. Metal organic frameworks as heterogeneous catalysts for the production of fine chemicals. *Catal. Sci. Technol.* **2013**, *3*, 2509–2540. [CrossRef]
12. Yoon, M.; Srirambalaji, R.; Kim, K. Homochiral Metal-Organic Frameworks for Asymmetric Heterogeneous Catalysis. *Chem. Rev.* **2012**, *112*, 1196–1231. [CrossRef] [PubMed]
13. Corma, A.; Garcia, H.; Xamena, F. Engineering metal-organic frameworks for heterogeneous catalysis. *Chem. Rev.* **2010**, *110*, 4606–4655. [CrossRef] [PubMed]
14. Liu, Y.; Xuan, W.M.; Cui, Y. Engineering Homochiral Metal-Organic Frameworks for Heterogeneous Asymmetric Catalysis and Enantioselective Separation. *Adv. Mater.* **2010**, *22*, 4112–4135. [CrossRef] [PubMed]
15. Tang, F.S.; Zhao, J.C.G.; Chen, B.L. Porous Coordination Polymers for Heterogeneous Catalysis. *Curr. Org. Chem.* **2018**, *22*, 1773–1791. [CrossRef]
16. Fan, F.Q.; Zhao, L.; Zeng, Q.Q.; Zhang, L.Y.; Zhang, X.M.; Wang, T.Q.; Fu, Y. Self-Catalysis Transformation of Metal-Organic Coordination Polymers. *ACS Appl. Mater. Interfaces* **2023**, *15*, 37086–37092. [CrossRef]
17. Gogoi, C.; Nagarjun, N.; Roy, S.; Mostakim, S.K.; Volkmer, D.; Dhakshinamoorthy, A.; Biswas, S. A Zr-based metal-organic framework with a DUT-52 structure containing a trifluoroacetamido-functionalized linker for aqueous phase fluorescence sensing of the cyanide ion and aerobic oxidation of cyclohexane. *Inorg. Chem.* **2021**, *60*, 4539–4550. [CrossRef] [PubMed]
18. Wang, Y.; Li, L.; Yan, L.; Cao, L.; Dai, P.; Gu, X.; Zhao, X. Continuous synthesis for zirconium metal-organic frameworks with high quality and productivity via microdroplet flow reaction. *Chin. Chem. Lett.* **2018**, *29*, 849–853. [CrossRef]
19. Gu, Y.; Ye, G.; Xu, W.; Zhou, W.; Sun, Y. Creation of Active Sites in MOF-808(Zr) by a Facile Route for Oxidative Desulfurization of Model Diesel Oil. *ChemistrySelect* **2020**, *5*, 244–251. [CrossRef]
20. Ruediger, E.; Martel, A.; Meanwell, N.; Solomon, C.; Turmel, B. Novel 3′-deoxy analogs of the anti-HBV agent entecavir: Synthesis of enantiomers from a single chiral epoxide. *Tetrahedron Lett.* **2004**, *45*, 739–742. [CrossRef]
21. Aramesh, N.; Yadollahi, B.; Mirkhani, V. Fe(III) substituted Wells–Dawson type polyoxometalate: An efficient catalyst for ring opening of epoxides with aromatic amines. *Inorg. Chem. Commun.* **2013**, *28*, 37–40. [CrossRef]
22. Liu, J.W.; Tang, W.; Wang, C. Nickel-Catalyzed Regio- and Enantioselective Ring Opening of 3,4-Epoxy Amides and Esters with Aromatic Amines. *Chem. A Eur. J.* **2023**, *29*, e202300704. [CrossRef] [PubMed]
23. Deshpande, N.; Parulkar, A.; Joshi, R.; Diep, B.; Kulkarni, A.; Brunelli, N.A. Epoxide ring opening with alcohols using heterogeneous Lewis acid catalysts: Regioselectivity and mechanism. *J. Catal.* **2019**, *370*, 46–54. [CrossRef]
24. Bonini, C.; Righi, G. Regio- and chemoselective synthesis of halohydrins by cleavage of oxiranes with metal halides. *Synthesis* **1994**, *1994*, 225–238. [CrossRef]
25. Mancilla, G.; Femenía-Ríos, M.; Macías-Sánchez, A.J.; Collado, I.G. Sn(OTf)$_2$ catalysed regioselective styrene oxide ring opening with aromatic amines. *Tetrahedron* **2008**, *64*, 11732–11737. [CrossRef]
26. Danafar, H.; Yadollahi, B. (TBA)$_4$PFeW$_{11}$O$_{39}$·3H$_2$O catalyzed efficient and facile ring opening reaction of epoxides with aromatic amines. *Cat. Commun.* **2009**, *10*, 842–847. [CrossRef]
27. Mohseni, S.; Bakavoli, M.; Morsali, A. Theoretical and Experimental Studies on the Regioselectivity of Epoxide Ring Opening by Nucleophiles in Nitromethane without any Catalyst: Nucleophilic-Chain Attack Mechanism. *Prog. React. Kinet. Mech.* **2014**, *39*, 89–102. [CrossRef]
28. Barbosa, A.D.S.; Julião, D.; Fernandes, D.M.; Peixoto, A.F.; Freire, C.; de Castro, B.; Granadeiro, C.M.; Balula, S.S.; Cunha-Silva, L. Catalytic performance and electrochemical behaviour of metal–organic frameworks: MIL-101(Fe) versus NH2-MIL-101(Fe). *Polyhedron* **2017**, *127*, 464–470. [CrossRef]
29. Xu, S.; Gao, Q.; Zhou, C.; Li, J.; Shen, L.; Lin, H. Improved thermal stability and heat-aging resistance of silicone rubber via incorporation of UiO-66-NH2. *Mater. Chem. Phys.* **2021**, *274*, 125182. [CrossRef]
30. Mirante, F.; Leo, P.; Dias, C.N.; Cunha-Silva, L.; Balula, S.S. MOF-808 as an Efficient Catalyst for Valorization of Biodiesel Waste Production: Glycerol Acetalization. *Materials* **2023**, *16*, 7023. [CrossRef] [PubMed]
31. Silva, D.F.; Faria, R.G.; Santos-Vieira, I.; Cunha-Silva, L.; Granadeiro, C.M.; Balula, S.S. Simultaneous sulfur and nitrogen removal from fuel combining activated porous MIL-100(Fe) catalyst and sustainable solvents. *Cat. Today* **2023**, *423*, 114250. [CrossRef]
32. Gao, Y.; Granadeiro, C.M.; Cunha-Silva, L.; Zhao, J.; Balula, S.S. Peroxomolybdate@MOFs as effective catalysts for oxidative desulfurization of fuels: Correlation between MOF structure and catalytic activity. *Catal. Sci. Technol.* **2023**, *13*, 4785–4801. [CrossRef]
33. Lara-Serrano, M.; Morales-delaRosa, S.; Campos-Martin, J.M.; Abdelkader-Fernández, V.K.; Cunha-Silva, L.; Balula, S.S. One-Pot Conversion of Glucose into 5-Hydroxymethylfurfural using MOFs and Brønsted-Acid Tandem Catalysts. *Adv. Sustain. Syst.* **2022**, *6*, 2100444. [CrossRef]
34. Lara-Serrano, M.; Morales-DelaRosa, S.; Campos-Martin, J.M.; Abdelkader-Fernández, V.K.; Cunha-Silva, L.; Balula, S.S. Isomerization of glucose to fructose catalyzed by metal-organic frameworks. *Sustain. Energy Fuels* **2021**, *5*, 3847–3857. [CrossRef]
35. Viana, A.M.; Julião, D.; Mirante, F.; Faria, R.G.; de Castro, B.; Balula, S.S.; Cunha-Silva, L. Straightforward activation of metal-organic framework UiO-66 for oxidative desulfurization processes. *Cat. Today* **2021**, *362*, 28–34. [CrossRef]
36. Julião, D.; Barbosa, A.D.S.; Peixoto, A.F.; Freire, C.; De Castro, B.; Balula, S.S.; Cunha-Silva, L. Improved catalytic performance of porous metal-organic frameworks for the ring opening of styrene oxide. *CrystEngComm* **2017**, *19*, 4219–4226. [CrossRef]
37. Furukawa, H.; Gándara, F.; Zhang, Y.-B.; Jiang, J.; Queen, W.L.; Hudson, M.R.; Yaghi, O.M. Water Adsorption in Porous Metal–Organic Frameworks and Related Materials. *J. Am. Chem. Soc.* **2014**, *136*, 4369–4381. [CrossRef] [PubMed]

38. Ma, Y.; Han, X.; Xu, S.; Wang, Z.; Li, W.; da Silva, I.; Chansai, S.; Lee, D.; Zou, Y.; Nikiel, M.; et al. Atomically Dispersed Copper Sites in a Metal–Organic Framework for Reduction of Nitrogen Dioxide. *J. Am. Chem. Soc.* **2021**, *143*, 10977–10985. [CrossRef] [PubMed]
39. Dai, S.; Simms, C.; Dovgaliuk, I.; Patriarche, G.; Tissot, A.; Parac-Vogt, T.N.; Serre, C. Monodispersed MOF-808 nanocrystals synthesized via a scalable room-temperature approach for efficient heterogeneous peptide bond hydrolysis. *Chem. Mater.* **2021**, *33*, 7057–7066. [CrossRef]
40. Luu, C.L.; Van Nguyen, T.T.; Nguyen, T.; Hoang, T.C. Synthesis, characterization and adsorption ability of UiO-66-NH2. *Adv. Nat. Sci. Nanosci. Nanotechnol.* **2015**, *6*, 025004. [CrossRef]
41. Dreischarf, A.C.; Lammert, M.; Stock, N.; Reinsch, H. Green Synthesis of Zr-CAU-28: Structure and Properties of the First Zr-MOF Based on 2,5-Furandicarboxylic Acid. *Inorg. Chem.* **2017**, *56*, 2270–2277. [CrossRef] [PubMed]
42. Romero-Muñiz, I.; Romero-Muñiz, C.; del Castillo-Velilla, I.; Marini, C.; Calero, S.; Zamora, F.; Platero-Prats, A.E. Revisiting Vibrational Spectroscopy to Tackle the Chemistry of Zr6O8 Metal-Organic Framework Nodes. *ACS Appl. Mater. Interfaces* **2022**, *14*, 27040–27047. [CrossRef] [PubMed]
43. Timofeev, K.L.; Kulinich, S.A.; Kharlamova, T.S. NH2-Modified UiO-66: Structural Characteristics and Functional Properties. *Molecules* **2023**, *28*, 3916. [CrossRef] [PubMed]
44. Nguyen, K.D.; Vo, N.T.; Le, K.T.M.; Ho, K.V.; Phan, N.T.S.; Ho, P.H.; Le, H.V. Defect-engineered metal–organic frameworks (MOF-808) towards the improved adsorptive removal of organic dyes and chromium (vi) species from water. *New J. Chem.* **2023**, *47*, 6433–6447. [CrossRef]
45. Ye, G.; Wan, L.; Zhang, Q.; Liu, H.; Zhou, J.; Wu, L.; Zeng, X.; Wang, H.; Chen, X.; Wang, J. Boosting Catalytic Performance of MOF-808(Zr) by Direct Generation of Rich Defective Zr Nodes via a Solvent-Free Approach. *Inorg. Chem.* **2023**, *62*, 4248–4259. [CrossRef] [PubMed]
46. Farrando-Pérez, J.; Martinez-Navarrete, G.; Gandara-Loe, J.; Reljic, S.; Garcia-Ripoll, A.; Fernandez, E.; Silvestre-Albero, J. Controlling the Adsorption and Release of Ocular Drugs in Metal–Organic Frameworks: Effect of Polar Functional Groups. *Inorg. Chem.* **2022**, *61*, 18861–18872. [CrossRef] [PubMed]
47. Jiang, D.; Mallat, T.; Krumeich, F.; Baiker, A. Copper-based metal-organic framework for the facile ring-opening of epoxides. *J. Catal.* **2008**, *257*, 390–395. [CrossRef]
48. Dhakshinamoorthy, A.; Alvaro, M.; Garcia, H. Metal–organic frameworks as efficient heterogeneous catalysts for the regioselective ring opening of epoxides. *Chem. Eur. J.* **2010**, *16*, 8530–8536. [CrossRef]
49. Anbu, N.; Dhakshinamoorthy, A. Regioselective ring opening of styrene oxide by carbon nucleophiles catalyzed by metal–organic frameworks under solvent-free conditions. *J. Ind. Eng. Chem.* **2018**, *58*, 9–17. [CrossRef]
50. Kumar, G.; Singh, A.P.; Gupta, R. Synthesis, structures, and heterogeneous catalytic applications of {Co^{3+}–Eu^{3+}} and {Co^{3+}–Tb^{3+}} heterodimetallic coordination polymers. *Eur. J. Inorg. Chem.* **2010**, *2010*, 5103–5112. [CrossRef]

Disclaimer/Publisher's Note: The statements, opinions and data contained in all publications are solely those of the individual author(s) and contributor(s) and not of MDPI and/or the editor(s). MDPI and/or the editor(s) disclaim responsibility for any injury to people or property resulting from any ideas, methods, instructions or products referred to in the content.

Strategies for Improving Sustainability in the Development of High-Performance Styrenic Block Copolymers by Developing Blends with Cellulose Derivatives

Erika Pajares [1,2], Josu Fernández Maestu [1], Irati Fernandez-de-Mendiola [1,3], Unai Silvan [1,4], Pedro Costa [5], Iker Agirrezabal-Telleria [2], Carmen R. Tubio [1,*], Sergio Corona-Galván [6] and Senentxu Lanceros-Mendez [1,4,5]

1. BCMaterials, Basque Center for Materials, Applications and Nanostructures, UPV/EHU Science Park, 48940 Leioa, Spain; erikapajares@hotmail.com (E.P.); josu.fernandez@bcmaterials.net (J.F.M.); irati.fernandezdemendiola@ehu.eus (I.F.-d.-M.); unai.silvan@bcmaterials.net (U.S.); senentxu.lanceros@bcmaterials.net (S.L.-M.)
2. Sustainable Process Engineering Group, Department of Chemical and Environmental Engineering, University of the Basque Country (UPV/EHU), 48013 Bilbao, Spain; iker.aguirrezabal@ehu.eus
3. Department of Cell Biology and Histology, Faculty of Medicine and Nursing, University of the Basque Country (UPV/EHU), 48940 Leioa, Spain
4. IKERBASQUE, Basque Foundation for Science, 48009 Bilbao, Spain
5. Physics Center of Minho and Porto Universities (CF-UM-UP) and LaPMET—Laboratory of Physics for Materials and Emergent Technologies, University of Minho, 4710-057 Braga, Portugal; pcosta@fisica.uminho.pt
6. Dynasol, Titán 15, 9th Floor, 28045 Madrid, Spain; scoronag@repsol.com
* Correspondence: carmen.rial@bcmaterials.net

Abstract: Next-generation high-performance polymers require consideration as sustainable solutions. Here, to satisfy these criteria, we propose to combine high-performance styrenic block copolymers, a class of thermoplastic elastomer, with cellulose derivatives as a reinforcing agent with the aim of maintaining and/or improving structural and surface properties. A great advantage of the proposed blends is, besides their biocompatibility, a decrease in environmental impact due to blending with a natural polymer. Particularly, we focus on identifying the effect of different blending compounds and blend ratios on the morphological, structural, thermal, mechanical, electrical and cytotoxic characteristics of materials. This research provides, together with novel material formulations, practical guidelines for the design and fabrication of next-generation sustainable high-performance polymers.

Keywords: high-performance polymers; blends; styrenic block copolymer; thermoplastic elastomer; cellulose derivatives

1. Introduction

High-performance polymers (HPP) are essential candidates for a broad range of applications, including in the aerospace, automotive, electronic, medical, oil, gas and military fields [1–4]. Unlike conventional polymers, HPP are known to offer a superior range of properties, particularly mechanical, thermal and chemical, with the ability to tolerate and resist harsh environments and conditions, such as a corrosive environment and high pressure and temperature, among others [5,6]. Significant efforts have been successfully made to develop and enhance their properties, in particular by the synthesis of new formulations, blends and composites [7] as well as by developing advanced processing methods, including additive manufacturing technologies such as fused deposition modelling (FDM) and direct ink writing (DIW) [6,8,9].

Different kinds of HPP, including thermoplastics and thermosets have been developed. They include liquid crystalline polymers, fluoropolymers, epoxy resins, polyurethanes, and siloxanes, among others [10]. In particular, thermoplastic elastomers (TPEs) have received considerable attention because of their mechanical properties and processability.

They consist of two phases (elastomeric or soft phase, and thermoplastic or hard phase), each one providing different features [11]. One of the most relevant TPEs is the styrene-b-(ethylene-co-butylene)-b-styrene (SEBS) copolymer [12,13]. Petroleum-based SEBS is obtained through the hydrogenation of the styrene-butylene-styrene (SBS) copolymer to remove the unsaturation present in polybutadiene; the latter, in turn, being obtained through the polymerization of two containing monomers (styrene and butadiene). SEBS is biocompatible, stable under thermal and oxidizing conditions, resistant to UV radiation, elastic and easy to process [12]. Such versatile features make SEBS very attractive for use in the automotive industry, footwear, adhesives, and sensors, as well as in medical devices [14–17].

Recent efforts have significantly improved specific performance parameters and functional properties, in particular by the development of SEBS-based composites and blends. As representative examples, the introduction of inorganic fillers ranging from clay [18], graphite [19], and carbon black [20] to multiwalled carbon nanotubes [16] allow the tuning of mechanical and electrical properties, depending on filler type and content. SEBS-based blends with organic compounds of synthetic origin, such as polyamide 6 (PA) [21,22] and polystyrene matrices [23], result in the improvement of impact strength, tensile strength and Young's modulus, respectively. Further, SEBS has been also used as a compatibilizing agent in polymer blends. This is the case for polystyrene/high-density polyethylene [24], and polypropylene/polyamide 6 [25] blends, in which mechanical properties such as ductility were generally improved.

Despite considerable progress, an increase in environmental concerns and regulations has resulted in an urgent need for research and development of more sustainable polymer and, in particular, SEBS compounds. In particular, the production of SEBS is based on polymerization techniques, which are related to considerable energy consumption and greenhouse gas emissions, with negative environmental impacts. The growing demand for more sustainable products based on the principles of a circular economy requires materials to have a low environmental impact without compromising on the physicochemical properties needed for applications. In this regard, the use of organic agents of natural origin to meet sustainable demands is increasingly being considered. In the particular case of SEBS as a polymer matrix, it can be blended with biomass derivatives ranging from cellulose nanofibers [26] and cork [27] to pineapple leaf fibers [28]. However, more research is necessary to improve functionality and properties, and to broaden the potential range of applications.

Lignocellulosic biomass is particularly attractive as sustainable feedstock due to its renewability and abundance [29–31]. It consists of three main components: lignin, hemicellulose and cellulose. The latter is the most abundant component, with great potential as a biodegradable biopolymer. Cellulose is a polymer formed by glucose units with alternating amorphous and crystalline regions and containing hydrogen bonds [32]. In fact, the combination of SEBS and cellulose as a biopolymer has been addressed [26,33], with cellulose derivatives contributing to a mechanical reinforcing of the composites. In order to properly tune materials' properties, the affinity between both matrices in the blends is critical, which is reliant on the chemical structure of the polymers.

In this work, we make use of a high-performance SEBS thermoplastic elastomer as well as SEBS functionalized with maleic anhydride (SEBS-g-MA) as the main components of a polymer blend in order to form a compatible and homogeneous or heterogeneous matrix with cellulose as the second component. The purpose of this study is to investigate the compatibility of cellulose derivatives with SEBS and SEBS-g-MA, and to evaluate the physical–chemical properties of the blends, in the scope of their potential applicability. Compatibility in the blend is based on exploring three cellulose derivatives: ethyl cellulose (EC), cellulose acetate (CA) and microcrystalline cellulose (MCC) [32,34,35]. These cellulose derivatives have been selected to reflect different hydrophobic/hydrophilic character and polarity (Figure 1). Notably, they have differences related to the number of hydroxyl groups reduced by etherification (EC), and esterification and transesterification (CA), whereas the degree of crystallinity is increased by purification and partial depolymerization (MCC). Using a solvent casting method, we successfully prepared several polymer blends with different ratios, and the morphological, structural, thermal, mechanical, electrical and cytotoxic properties were evaluated. The analyzed properties were used to investigate the effect of the addition of different types of cellulose into the non-functionalized and functionalized SEBS matrix, as well as to assess the effects of blending ratios. The results obtained from this work are expected to shed light on the design of more renewable polymer blends, leading to a transition to more sustainable high-performance polymers.

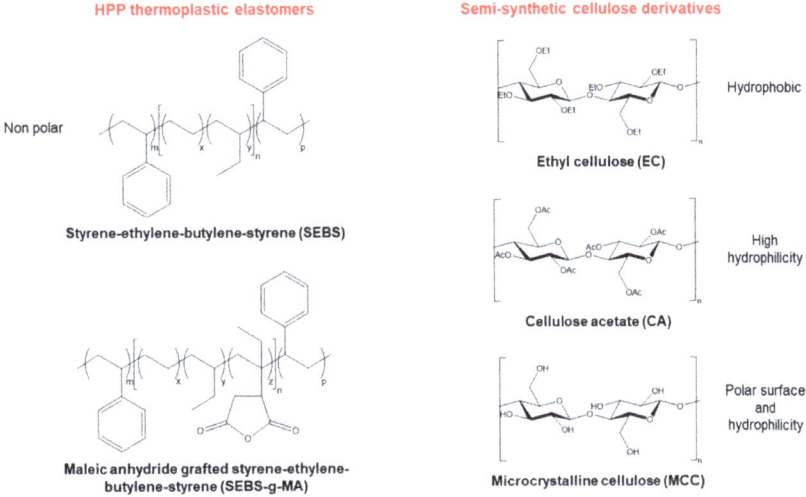

Figure 1. Chemical structures of SEBS, SEBS-g-MA, and cellulose derivatives.

2. Materials and Methods

2.1. Materials

SEBS (15 wt% styrene content, 62% vinyl content, melt flow = 37.5 g/10 min (230 °C/2.16 kg)) and SEBS-g-MA (degree of functionalization 1.6 wt%, derived from a SEBS having 15 wt% styrene, 62% vinyl and melt flow = 10 g/10 min (230 °C/2.16 kg)) were provided by Dynasol Elastomers, Madrid, Spain. Ethyl cellulose powder (48–49.5% (W/W) ethyl basis), cellulose acetate powder (Mn ~30,000 g/mol, 39.8 wt% acetyl), and cellulose microcrystalline powder (particle size = 51 µm) were supplied by Sigma-Aldrich (St. Louis, MI, USA). Tetrahydrofuran (THF, GPC grade, Scharlab) was selected as a solvent. All the reagents and the solvent were used as received.

2.2. Sample Preparation

The films were produced by solvent casting, where blends of SEBS:cellulose derivatives, and blends of SEBS-g-MA:cellulose derivatives were produced in the blend ratios (by weight) of 100:0, 90:10, 80:20, and 70:30 (Table 1). The polymer to solvent ratio was 1:5 (wt:v). Specific solutions were prepared by dissolving the specific polymer in the THF solvent at ambient temperature and under magnetic stirring until complete dissolution. Then, the solutions were mixed under magnetic stirring at ambient temperature for 2 h. The mixture solutions were finally deposited using the doctor blade technique onto glass substrates, dried at room temperature for 12 h and films were peeled from the glass substrates. The obtained films show a thickness of around 50–70 µm, and a size of 2×5 cm^2. The blends with MCC were prepared following the same procedure, with a previous ultrasonic step in an ultrasonic bath (Model ATM3L, ATU, Valencia, Spain) for around 2 h before mixing to ensure good dispersion of the MCC particles in the THF solvent.

Table 1. Blend formulation according to weight percentage (wt%) of each component.

Identification Sample	SEBS (wt%)	SEBS-g-MA (wt%)	EC (wt%)	CA (wt%)	MCC (wt%)
SEBS100	100	0	0	0	0
SEBS90:EC10	90	0	10	0	0
SEBS80:EC20	80	0	20	0	0
SEBS70:EC30	70	0	30	0	0
SEBS90:MCC10	90	0	0	0	10
SEBS80:MCC20	80	0	0	0	20
SEBS70:MCC30	70	0	0	0	30
SEBS-g-MA100	0	100	0	0	0
SEBS-g-MA90:EC10	0	90	10	0	0
SEBS-g-MA80:EC20	0	80	20	0	0
SEBS-g-MA70:EC30	0	70	30	0	0
SEBS-g-MA90:CA10	0	90	0	10	0
SEBS-g-MA80:CA20	0	80	0	20	0
SEBS-g-MA70:CA30	0	70	0	30	0

2.3. Characterization

Cross-section scanning electronic microscopy (SEM) images were obtained using a Hitachi S-4800 microscope (Tokyo, Japan), using an accelerating voltage of 15 kV. Samples were coated with a 20 nm gold layer via sputtering deposition with a Polaron SC 502 sputter coater (Laughton, UK).

Fourier transform infrared spectroscopy (FTIR) spectra were recorded by using a Jasco FT/IR-4100 (Easton, MD, USA). in the attenuated total reflectance (ATR) mode from 600 to 4000 cm^{-1} with a resolution of 4 cm^{-1}.

Contact angle measurements were carried out using a contact angle goniometer (OCA 15EC, Neurtek, Guipuzkoa, Spain).

Mechanical properties were evaluated by analyzing the tensile stress–strain curves of the samples. These tests were carried out using a Shimadzu AGS-J universal testing set up (Kyoto, Japan), with a load cell of 500 N until fracture at a test velocity of 3 mm/min. Stress–strain hysteresis cycles (up to 500 loading-unloading cycles) were recorded by applying three different maximum strains: 5%, 10%, and 30% of the initial sample length.

Dielectric measurements were carried out at room temperature using a Quadtech 1920 Precision LCR Meter (Sussex, WI, USA) in a 1 kHz–1 MHz frequency range with an applied voltage of 1 V. Five-millimeter diameter gold electrodes were sputtered with a 25 nm thin gold layer, using a Polaron SC502 (Quorum, Laughton, UK) sputter coater under nitrogen atmosphere. The capacity (C) and dielectric losses (tan δ) of the polymer blends prepared in the parallel plate configuration were obtained as a function of frequency. The real and imaginary part of the dielectric constant (ε' and ε'') and the real component of the alternating current (AC) electrical conductivity (σ'_{AC}) were obtained according to the following equations:

$$\varepsilon' = \frac{C \cdot d}{\varepsilon_0 \cdot A} \quad (1)$$

$$\tan \delta = \frac{\varepsilon''}{\varepsilon'} \quad (2)$$

$$\sigma'_{AC} = \varepsilon_0 \cdot \omega \cdot \varepsilon'' \quad (3)$$

where A indicates the area, d is the sample thickness, ε_0 (8.85×10^{-12} F/m) is the permittivity of free space, and $w = 2 \cdot \pi \cdot f$ is the angular frequency.

The cytotoxicity of the samples was evaluated using an extract exposure test. Briefly, mouse embryonic fibroblasts (MEFs) were seeded in 24-well plates and cultured in complete culture medium composed of Dulbecco's Modified Eagle Medium (DMEM) (Gibco), 10% fetal bovine serum and antibiotics (Pen/Strep) at 37 °C and 5% CO_2. Samples of 1 cm^2

of the SEBS and SEBS-g-MA-based blends were sterilized by UV irradiation for 30 min and incubated in 1 mL of complete culture medium for 24 h at 37 °C. Next, 600 µL of the extracts were added to the MEF cultures and allowed to exert their effect for an additional 24 h, before carrying out a Live/Dead cell viability assay (Thermofisher, Waltham, MA, USA). For the positive control, MEFs cultured in complete culture medium were used, and for the negative control, cells were permeabilized using cold 100% methanol. Fluorescence intensity for Calcein (live cells) and Ethidium homodimer (dead cells) was measured using a plate reader (Tecan, Switzerland). Each material was analyzed in triplicate. MEF images were taken using a Nikon Eclipse Ti-S/L100 inverted fluorescence microscope (Melville, NY, USA).

3. Results and Discussion

3.1. Morphological and Chemical and Surface Properties

The performance of polymer blends is highly dependent on the miscibility of polymers, requiring a suitable morphological analysis. Factors such as the nature of both materials and blend composition are essential to determine miscibility. We tested combinations of styrenic block copolymers (SEBS and SEBS-g-MA) and cellulose derivatives (CA, CE, MCC) at 100:0, 90:10, 80:20 and 70:30 weight ratios. The mechanical consistency of films is suitable for most of the combinations, but the SEBS:CA and SEBS-g-MA:MCC samples presented several problems related to ink mixing and peeling films. This is attributed to the significant differences in the polarity of the materials [36,37]. Representative cross-sectional SEM images of the obtained SEBS and SEBS-g-MA-based blends with different weight ratios are shown in Figures 2 and 3, respectively. The neat SEBS and SEBS-g-MA matrices (Figure S1) have a smooth and compact morphology. Moreover, the incorporation of cellulose derivatives resulted in a phase separation process, where two distinct phases are clearly identified: the styrene-based thermoplastic elastomer (continuous polymer matrix) and the cellulose derivatives (dispersed phase). This is due to the immiscibility of both components mainly caused by their difference in polarity. Interestingly, oval-shaped EC agglomerates are found uniformly dispersed through the SEBS (Figure 2A–C) and SEBS-g-MA (Figure 3A–C) matrices, which are oriented with the plane in the direction in which the films are produced through the doctor blade technique. Voids (green arrows) also appear at the interface of the materials' components as a result of phase separation processes during solvent evaporation in the casting technique and the interfacial interaction of the material components [38]. By comparison, EC shows a better adhesion to SEBS (Figure 2A–C) than to SEBS-g-MA (Figure 3A–C). On the other hand, the lack of homogeneity and the extended agglomerate presence in the blends formed by SEBS and MCC (Figure 2D–F) is a clear indication of their poor compatibility. CA presents a similar adhesion than EC in the SEBS-g-MA matrix (Figure 3), but with regular circular-shaped agglomerates. In particular, compatibility is a function of the relative polarity between the two components and is thus critical for obtaining a homogeneous hybrid phase. SEBS and the chemically modified cellulose derivatives by etherification (EC) and esterification (CA) allow tuning of the interfacial compatibility when compared to SEBS with MCC cellulose derivative.

On the other hand, as seen from the optical microscope images in Figure S2A, color changes were observed after the cellulose derivative addition in the neat SEBS and SEBS-g-MA samples. In particular, the SEBS and SEBS-g-MA samples exhibit a transparent appearance. Moreover, the optical microscopy images reveal that the surface of the SEBS70:EC30 blend remains transparent. However, it can be seen that the SEBS70:MCC30 blend has aggregates in the form of islands over the sample surface. In contrast, it was observed that the color of the SEBS-g-MA blends changed from transparent to white after the cellulose derivative addition.

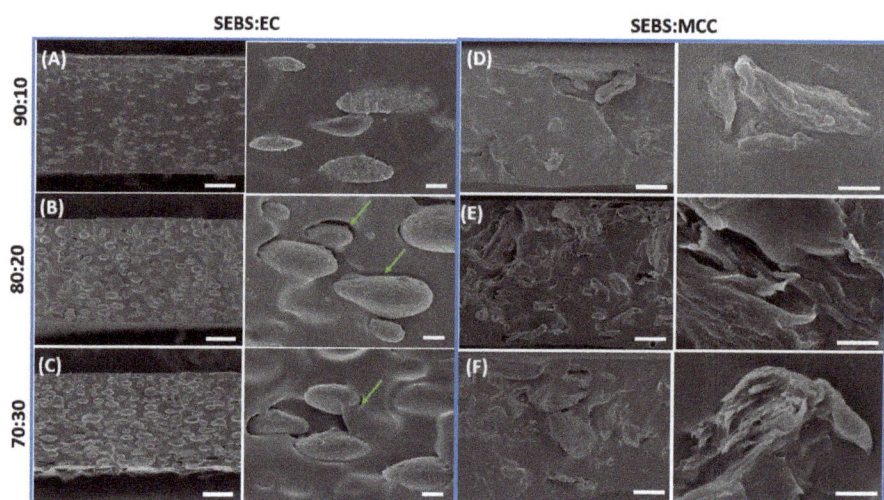

Figure 2. Cross-sectional SEM images of the SEBS-based blends at low magnification (left, 30 μm scale bar) and high magnification (right, 1 μm scale bar): (**A**) SEBS90:EC10, (**B**) SEBS80:EC20, (**C**) SEBS70:EC30, (**D**) SEBS90:MCC10, (**E**) SEBS80:MCC20, and (**F**) SEBS70:MCC30. The green arrows indicate voids at the different materials' interfaces.

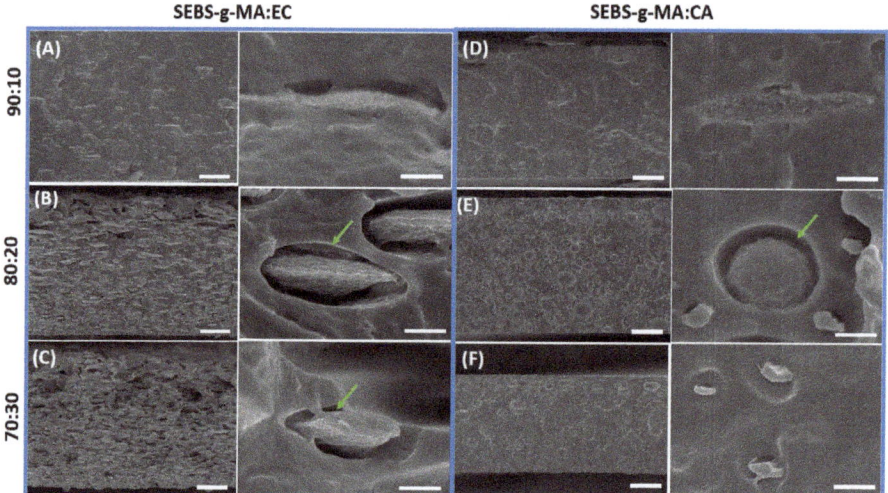

Figure 3. Cross-sectional SEM images of the SEBS-g-MA-based blends at low magnification (left, 30 μm scale bar) and high magnification (right, 1 μm scale bar): (**A**) SEBS-g-MA90:EC10, (**B**) SEBS-g-MA80:EC20, (**C**) SEBS-g-MA70:EC30, (**D**) SEBS-g-MA90:CA10, (**E**) SEBS-g-MA80:CA20, and (**F**) SEBS-g-MA70:CA30. The green arrows indicate voids at the different materials' interfaces.

The surface of the hybrid materials has been also assessed by FTIR spectroscopy, as represented in Figure 4 for the samples with the highest cellulose contents. The FTIR spectra for SEBS and SEBS-g-MA differ due to the stretching vibrations bands of the carbonyl groups at 1769 and 1715 cm^{-1}, respectively, as well as for the C–O–C stretching vibration bands at 1254 cm^{-1}, arising from the maleic anhydride grafted onto the SEBS [39]. Styrene is related to the stretching vibration bands of the unsaturated bonds of the aliphatic groups at wavenumbers above 3000 cm^{-1}, the C=C aromatic stretching at 1455–1600 cm^{-1}

and the out-of-plane bending at 754 and 697 cm^{-1}. Butadiene is reflected in the symmetric bending vibration band of the methyl group at 1380 cm^{-1}, whereas ethylene is related to the C–H rocking vibration band at 718 cm^{-1}. Both butadiene and ethylene contribute to CH$_2$ group bending vibrations (1455 cm^{-1}) [40–42]. The spectra of the neat cellulose derivatives (EC, MCC and CA) share several bands, the main differences being the hydroxyl group stretching at 3300 cm^{-1} observed in all samples, although with different intensities, because ethylation and acetylation in EC and CA strongly reduce this band [43]. Likewise, the common bands for EC and CA are related to the methylene and methyl group at 1455 and 1375 cm^{-1} [44,45], whereas the one corresponding just to CA is the carbonyl vibration at 1738 cm^{-1} [45].

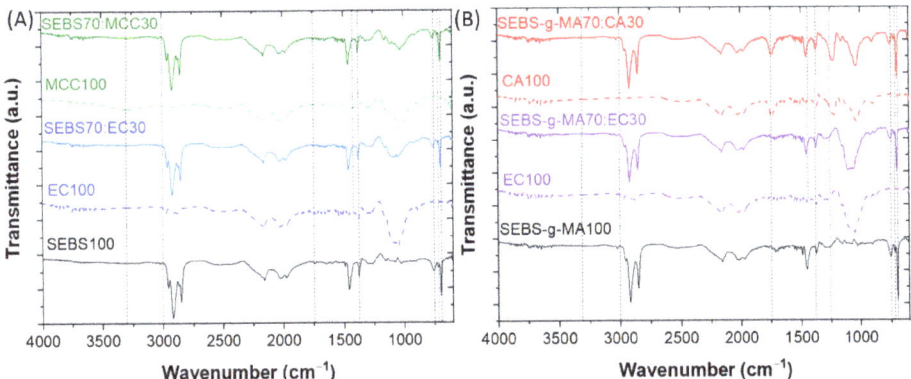

Figure 4. FTIR-ATR spectra of (**A**) SEBS and (**B**) SEBS-g-MA-based blends.

By comparing the spectra of the neat compounds with those of the blends, the FTIR bands presented in Figure 4 demonstrate that there is no chemical interaction between the two materials as no new bands appear indicating the possible formation of new hydrogen or primary bonds.

On the other hand, we also tested the hydrophobic/hydrophilic surface characteristics of the samples (Figure 5). Neat SEBS films are hydrophobic with a contact angle of around 102°, whereas SEBS-g-MA shows a contact angle of 100°, highlighting the hydrophobic character of these polymers [46,47]. Regarding the effect of EC content within the polymer blend, no relevant variations have been observed. On the other hand, the inclusion of MCC leads to a reduction in the contact angle down to 77° for the sample with the higher filler contents, due to its hydrophilic character. Similarly, CA also leads to filler content surface variations of SEBS-g-MA-based blends, as reflected by the water contact angle around 80° obtained for the sample with 30 wt% CA content [48].

3.2. Mechanical Properties

The stress–strain mechanical curves of the SEBS and SEBS-g-MA-based blends with varying cellulose derivative contents are shown in Figure 6A,B, respectively. Note that the curves could not be compared with the SEBS-g-MA80:EC20 and SEBS-g-MA70:EC30 samples because tensile testing could not be conducted due to the fragility of the samples. Figure S2B shows a photograph of the SEBS-g-MA70:EC30 sample after being stretched. From the testing results, a strong variation in the mechanical characteristics of the blend with respect to the ones of the neat polymers can be observed. In particular, the elongation at break decreases significantly due the presence of cellulose derivatives in the blend. Thus, the addition of cellulose derivatives leads to a decrease in the elasticity and flexibility of the TPE. Figure 6C–E shows the dependence of the Young´s modulus and breaking tensile and strain of the blends as a function of cellulose contents. For the neat SEBS-g-MA, the Young´s modulus and the breaking tensile values are 21.48 MPa and 4.7 MPa, respectively, which are also much higher than that of neat SEBS with values of 2.44 MPa and 1.56 MPa, respectively. On the other hand, with the addition of cellulose derivatives, the Young´s modulus increases, whereas the tensile values decrease. Interestingly, the use of 30 wt% CA in combination with SEBS-g-MA results in the highest value for Young´s modulus (44.47 MPa). By comparison, among the different cellulose derivatives, the addition of EC in the SEBS and SEBS-g-MA-based blends caused just a slight variation in the Young´s modulus and tensile values.

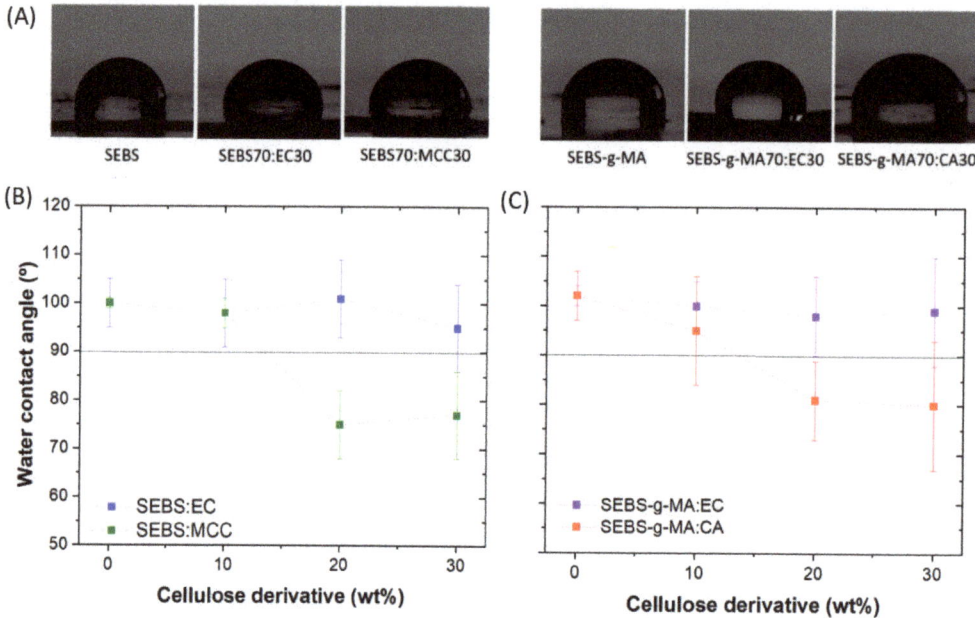

Figure 5. (**A**) Optical images of water droplets on the different blends. Water contact angle measurement of (**B**) SEBS and (**C**) SEBS-g-MA-based blends. The measurement was performed in three samples and the error bar represents the standard deviation.

In general, the variation in the results between the different compositions is caused by factors such as the interfacial interaction between the polymer matrix and the filler agent, or the degree of dispersion of the filler agent in the matrix due to low compatibility as a result of polarity differences. In this way, despite the non-polarity of the SEBS matrix, there is an increase in its polarity due to the modification through the maleic anhydride graft. Therefore, neat TPE polymers possess different properties, as reflected in the mechanical results. However, it is also worth noting that the polarity of the cellulose derivative causes changes in the blends, as can be seen in the SEM results. This is due to the modification of the cellulose and the replacement of hydroxyl groups by other functions, facilitating better adhesion and uniformity in hybrids of similar polarity. This can be observed in the previous SEM images of SEBS:EC and SEBS-g-MA:EC blends, which are more uniform than the blends with MCC and CA fillers. For this reason, the SEBS:EC blends maintain similar and better mechanical properties, in spite of the decrease in ductility caused by the reduction in elongation.

The data suggest that the presence of cellulose fillers results in less stress transfer between the matrix and the dispersed phase. In particular, the agglomerates act as a breaking point due to the defects at the interface caused by the reduced adhesion between the two components. Depending on the material application, it is evident that the blends of thermoplastic elastomer and renewable filler agent at low concentrations can serve as a replacement for the use of neat thermoplastic elastomer.

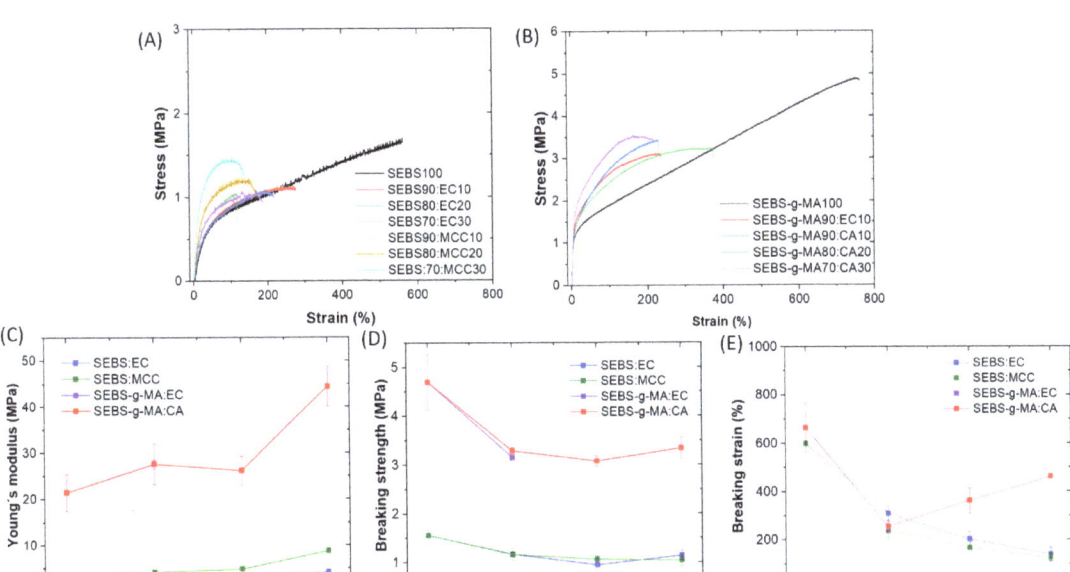

Figure 6. Stress–strain curves of (**A**) SEBS, and (**B**) SEBS-g-MA-based blends. Variation of (**C**) Young's modulus, (**D**) breaking strength, and (**E**) breaking strain values with the cellulose content in the blends.

A loading–unloading mechanical cyclic tensile test was then carried out to investigate the mechanical stability of the different composites. The cyclic tests were conducted under 5%, 10% and 30% applied strains for 500 cycles. Figure 7 shows the cyclic behavior of the neat styrenic block copolymers, and of the corresponding blends with the highest contents of cellulose derivatives. All samples show hysteretic behavior, indicative of the energy dissipation, and exhibit the Mullins effect, which is characterized by a decrease in the stress upon unloading compared to the stress upon loading at the same strain [49]. In addition, the maximum stress for a given strain of each cycle is higher for SEBS-g-MA blends than for SEBS blends, indicating a more flexible interface due to the presence of maleic polar groups. The addition of EC and MCC to the SEBS matrix (Figure 7A) shows a significant impact on the hysteresis loop, resulting in an increase in the loop, i.e., in the dissipated energy, the effect being more pronounced with the addition of MCC. Meanwhile, the incorporation of EC in SEBS-g-MA blends (Figure 7B) does not produce significant response differences, whereas the addition of CA causes an increase in the hysteresis loop for all evaluated strains.

3.3. Thermal Properties

Figure 8 shows the TGA curves for the SEBS and SEBS-g-MA-based blends. Also, the corresponding derivative thermogravimetry (DTG) curves are shown, where each peak determined the maximum rate of the degradation processes.

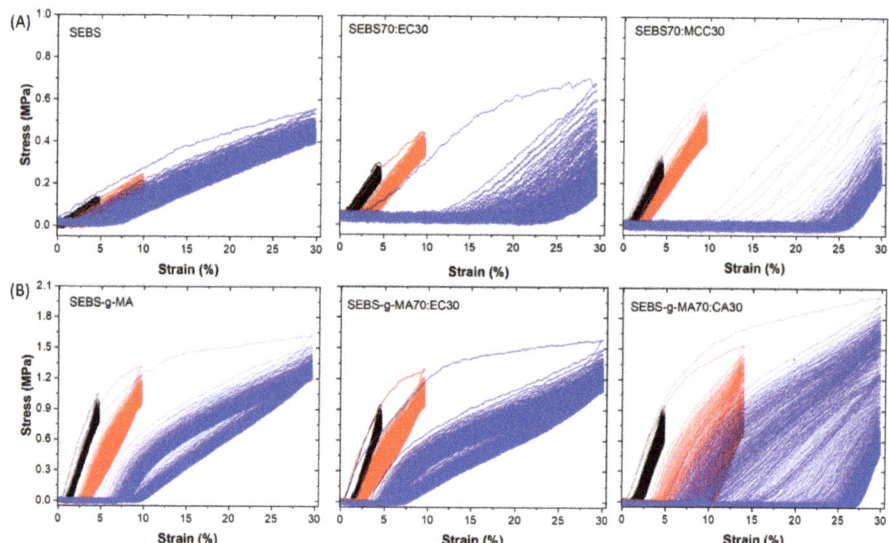

Figure 7. Loading–unloading mechanical tensile curves for (**A**) SEBS, and (**B**) SEBS-g-MA-based blends under 5% strain (black line), 10% strain (red line) and 30% strain (blue line).

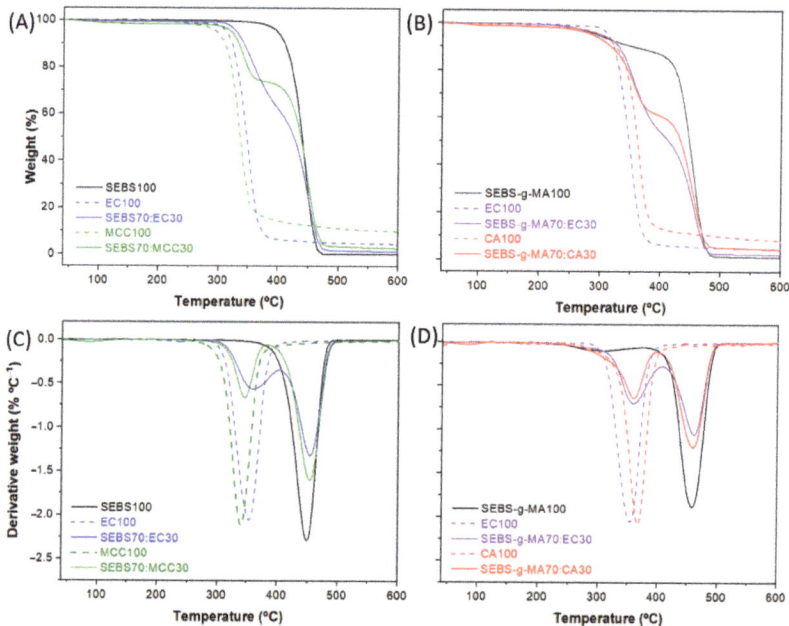

Figure 8. TGA and DTG thermograms of the (**A**,**B**) SEBS and (**C**,**D**) SEBS-g-MA-based blends.

When comparing neat SEBS with the neat, functionalized SEBS-g-MA, the degradation of samples starts at higher temperatures, at a 10 wt% loss temperature (T10%) of 409 °C in the case of SEBS and at lower temperature (at 349 °C) in the case of SEBS-g-MA. Thus, the presence of maleic anhydride grafted onto the copolymer decreases the degradation temperature and thus the thermal stability. This effect is observed in many other polymers when they are grafted with MA, such as poly(hydroxybutyrate-co-hydroxyvalerate)

(PHBV) [50] and poly(acrylamide/gelatin) hydrogels [51], among others. Moreover, it can be distinguished that SEBS is characterized by a single degradation stage in the range from 350 to 480 °C, whereas two stages are observed in SEBS-g-MA, one between 225 and 380 °C and the other between 380 and 500 °C. In the case of the SEBS-g-MA sample, the first step could relate to anhydride units of MA [52].

On the other hand, in the case of neat cellulose derivatives, a single degradation stage is observed. This stage and weight loss is attributed to the breakdown of the anhydroglucose polymeric chain, i.e., the breaking of the β-1,4-glycosidic bonds that hold the glucose units together, followed by the primary decomposition of volatile and dehydrated compounds. It corresponds to the processes of dehydration, depolymerization and decomposition of glysosyl units [53].

With respect to the composite samples, several stages of degradation can be distinguished, related to the composite composition. With regard to the SEBS:cellulose samples (Figure 8A,B), the first stage of degradation is attributed to cellulose (EC or MCC) and appears in the range from 300 to 370 °C, and the second stage is related to the degradation of the block copolymer and appears in the range from 370 to 490 °C. For this reason, the SEBS blends with cellulose derivatives present a decrease in thermal stability compared to the neat SEBS. This decrease is caused by the cellulose because, as described previously, the decomposition of the anhydroglucose polymer chain in the cellulose occurs at a lower temperature than in the copolymer. Several processes are involved, such as the breaking of the β-1,4-glycosidic bonds holding the glucose units together and then the primary decomposition of volatile and dehydrated compounds.

On the other hand, in the SEBS-g-MA-based blends (Figure 8C,D) prior to the decomposition of the anhydroglucose chain, a slight SEBS degradation occurs at 230 °C for SEBS-g-MA blends with cellulose derivatives. This is caused by the presence of anhydride units of MA, affecting the weight loss to a minor extent at such low content. The presence of the maleic anhydride graft in SEBS, however, decreases the thermal stability of the copolymer. Subsequently, the two main stages mentioned above (cellulose and copolymer degradation) occur after 230 °C. In addition, the substitution of hydroxyl groups in the cellulose by ethyl or acetate groups leads to a slight increase in thermal stability. Therefore, the SEBS:EC blend is thermally more stable than the SEBS:MCC, and the SEBS-g-MA:EC blend is somewhat more stable than the SEBS-g-MA:CA. The onset of CA degradation occurs earlier than that of EC, although the maximum degradation temperature is slightly higher in CA. By comparing the different polymer blends, all the maximum degradation temperatures lay ±10 °C compared to neat materials, confirming the immiscibility of the polymers. After the thermogravimetric analysis of the samples, the content of the final residue depends on the amount of added filler and its crystallinity.

3.4. Electrical Properties

In order to study the potential of blends for electronic applications, the dielectric properties of the blends have been evaluated. The frequency dependence of the real dielectric permittivity (ε') for the different SEBS (Figure S3A) and SEBS-g-MA (Figure S4A) blends shows a relative stable behavior over frequency variation. Meanwhile, the corresponding dependence of the dielectric loss (tan δ) on the frequency for SEBS (Figure S3B) and SEBS-g-MA (Figure S4B) blends shows similar behavior in both samples. The tan δ values are maintained below 0.15 over the measured frequency range. For comparison, the ε' value at 1 kHz for SEBS and SEBS-g-MA-based blends with cellulose derivatives is shown in Figure 9A,B, respectively. Significantly, dielectric permittivity increases by a factor of two for the blends with the 30 wt% EC, when compared to the neat SEBS polymer ($\varepsilon' = 1.75$) and neat SEBS-g-MA polymer ($\varepsilon' = 2.15$). Finally, the ac electrical conductivity was calculated for the different samples (Figure 9C,D). An increase of nearly 2 orders of magnitude is obtained in the SEBS blend (Figure 9C) with the highest content of EC and MCC. It is in agreement with previous results on the effect of MCC that improves dielectric properties [54]. On the other hand, it is evidenced that the AC conductivity in the SEBS-g-MA-based blends (Figure 9D) remains stable with the EC and CA addition.

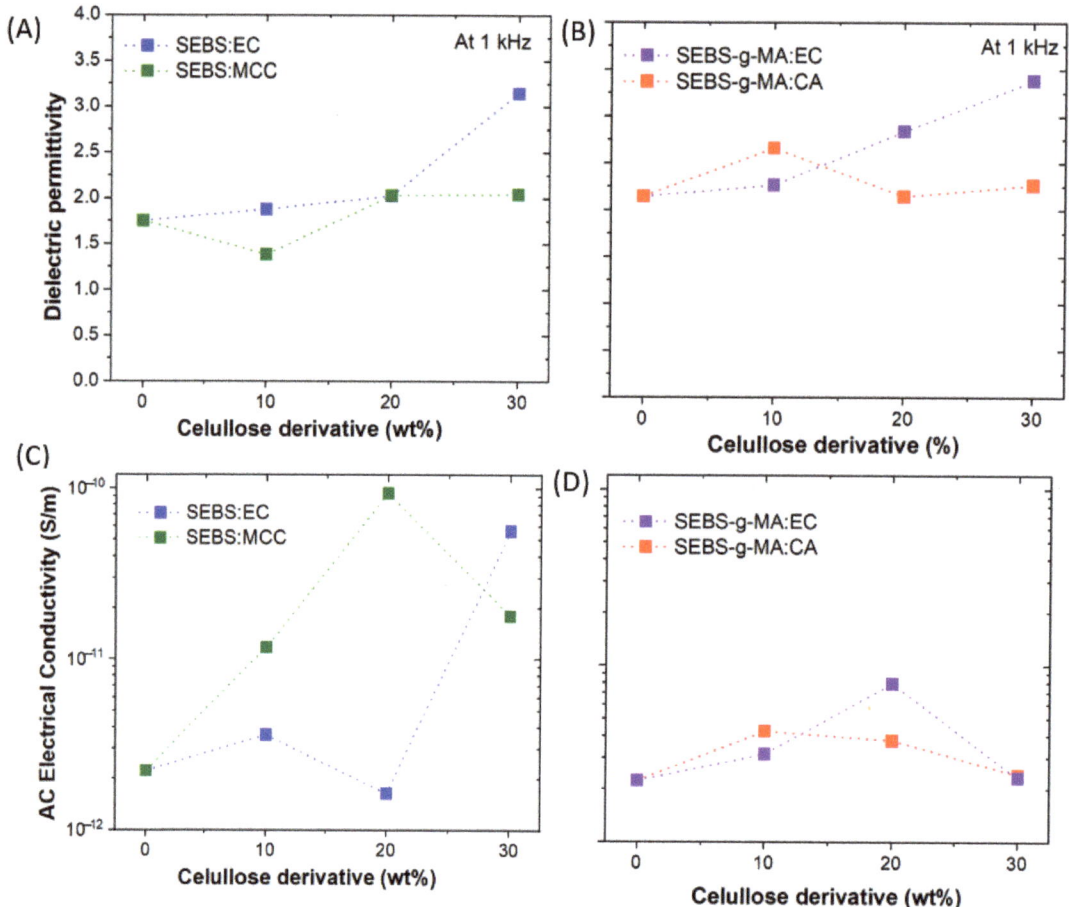

Figure 9. Dielectric permittivity values at 1 kHz for (**A**) SEBS and (**B**) SEBS-g-MA-based blends. Evolution of aAC electrical conductivity as a function of cellulose derivatives content for (**C**) SEBS and (**D**) SEBS-g-MA-based blends.

3.5. Indirect Cytotoxicity Analysis

The biocompatibility of the SEBS and SEBS-g-MA-based blends was evaluated after an extract exposure test. For this, we incubated the materials in complete culture medium for 24 h and subsequently used it to grow mouse embryonic fibroblasts (MEFs). The obtained results show high biocompatibility of all analyzed materials with similar survival values to those of the positive control (Figure 10A,B). Thus, the results are consistent with the literature that indicate negligible cytotoxicity for SEBS samples [55], as well as cellulose derivatives [56].

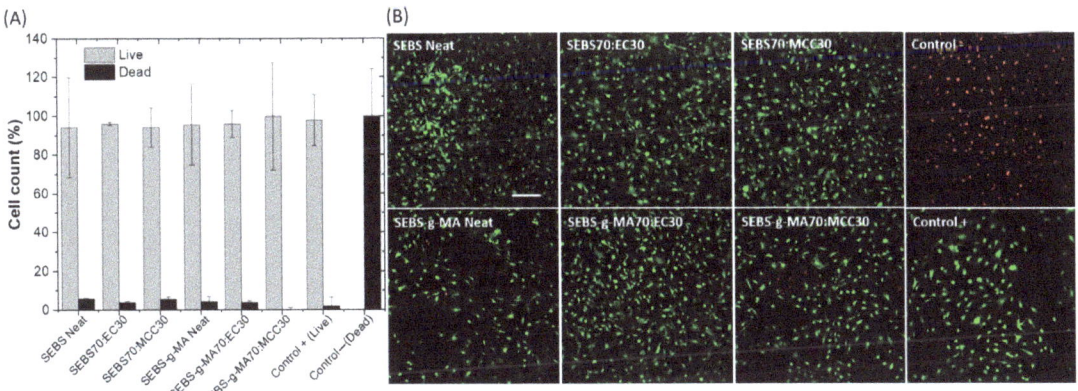

Figure 10. (**A**) All blend extracts display high biocompatibility as estimated using a Live/Dead assay. Data shown as average +/− SD. (**B**) Fluorescence images show a large number of live cells (green) in the cultures in contact with the material extracts with a residual number of dead cells (red). Scale bar represents 200 μm.

4. Discussion

Cellulose derivatives up to 30 wt% have been explored as potential additives to improve the sustainability of high-performance styrenic block thermoplastic elastomer. The strategy implemented here explores styrenic block copolymers SEBS (no polar) and SEBS-g-MA (polar) and cellulose derivatives with a different polar surface area (PSA) from weaker polar EC (PSA = 134.53 Å), and MCC (PSA = 167.53 Å) to moderately polar CA (PSA = 238.09 Å). SEBS, SEBS-g-MA and EC are hydrophobic, differing from the hydrophilic CA and MCC. Blends are produced by solvent casting, leading to composites with different mechanical and electrical characteristics (Table 2), which can be taken advantage of for different applications. The use of EC in SEBS and SEBS-g-MA blends preserves the processability and wettability of styrene copolymers and increases the dielectric response. Also, the incorporation of EC leads to a decrease in the mechanical properties, such as the tensile strength, modulus and elongation at break, without impacting significantly on the dissipative properties of neat styrene block copolymers. This is an interesting result, considering the potential applications of the styrenic block copolymers, where the mechanical properties are essential. Therefore, the role of the hydrophobic character of these three polymers is a critical factor in controlling compatibility.

Table 2. Comparison between the main characteristics of the different blends.

Blends	Morphology	Mechanical	Wettability	Dielectric	Citotoxicity
SEBS:EC	Regular oval-shaped aggregates		Hydrophobic	Enhanced	Not cytotoxic
SEBS:CA	Not applicable	Not applicable	Not applicable	Not applicable	Not applicable
SEBS:MCC	Irregularly-shaped aggregates		Hydrophilic	Preserved	Not cytotoxic
SEBS-g-MA:EC	Regular oval-shaped aggregates	Preserved tensile energy dissipation	Hydrophobic	Enhanced	Not cytotoxic

Table 2. *Cont.*

Blends	Morphology	Mechanical	Wettability	Dielectric	Citotoxicity
SEBS-g-MA:CA	Regular circular-shaped aggregates		Hydrophilic	Preserved	Not cytotoxic
SEBS-g-MA:MCC	Not applicable	Not applicable	Not applicable	Not applicable	Not applicable

On the other hand, the use of hydrophilic MCC and CA polymers results in more negative effects on the physicochemical properties of the blends. Specifically, crack-free films can only be obtained when using polymers with similar polarity (SEBS:MCC, SEBS-g-MA:CA). However, the other combinations of SEBS:CA and SEBS-g-MA:MCC result in broken films, which reveals the importance of wettability and polarity. In addition, these cellulose polymers do not provide significant changes in dielectric properties, when compared to neat styrenic block copolymers. Instead, the wettability is modified, with respect to the pristine polymer, in all compositions.

5. Conclusions

In summary, we report the use of cellulose derivatives in the formation of styrenic block copolymer-based blends to improve the sustainability of high-performance polymers. A series of experimental tests were conducted to assess the influence of different cellulose derivatives and contents on the physicochemical and structural characteristics of styrene-based blends. The results show that the morphology of blends is highly dependent on the cellulose derivative type and content. The SEM results revealed that the presence of EC causes the formation of regular oval-shape aggregates, where the interfacial adhesion with SEBS is higher than with SEBS-g-MA, also depending on the polarity of the specific cellulose derivative. On the other hand, blending with cellulose derivatives leads to a decrease in tensile properties, such as tensile strength, and elongation at break, compared to the neat styrenic block copolymers. However, the presence of EC showed a positive effect on the tensile energy dissipation, without significant changes. Furthermore, the analysis of FTIR showed no interaction among the components of blends, whereas thermal testing shows that the cellulose derivatives decrease thermal stability compared to the neat styrene copolymers. Meanwhile, the dielectric permittivity of both SEBS and SEBS-g-MA were enhanced by a factor of two upon addition of 30 wt% EC.

Finally, all SEBS and SEBS-g-MA-based blends showed no cytotoxicity after 24 h for MEFs cells. Based on the above results, it can be concluded that EC at different concentrations can effectively be used in blends with SEBS type materials in order to tune mechanical, thermal and electrical properties, representing a promising pathway for the preparation of more sustainable high-performance styrenic block copolymers-based blends.

Supplementary Materials: The following supporting information can be downloaded at: https://www.mdpi.com/article/10.3390/polym16060856/s1, Figure S1: Cross-sectional SEM images of SEBS and SEBS-g-MA samples at low magnification (left, 30 µm scale bar) and high magnification (right, 5 µm scale bar) Figure S2: (A) Optical microscopy images for the SEBS and SEBS-g-MA based blend films with different cellulose derivatives. Scale bar: 500 µm. (B) Photograph of stretchable SEBS-g-MA70:EC30 blend film. Figure S3: (A) Dielectric permittivity, and (B) dielectric loss as a function of frequency for SEBS based blends with EC and MCC fillers. Figure S4: (A) Dielectric permittivity, and (B) dielectric loss as a function of frequency for SEBS-g-MA based blends with EC and CA fillers.

Author Contributions: Conceptualization, I.A.-T., C.R.T. and S.L.-M.; methodology, I.A.-T., C.R.T. and S.L.-M.; validation, I.A.-T., C.R.T. and S.L.-M.; formal analysis, E.P., J.F.M., I.F.-d.-M., U.S., P.C., I.A.-T., C.R.T., S.C.-G. and S.L.-M.; investigation, E.P., J.F.M., I.F.-d.-M., U.S., P.C., I.A.-T., C.R.T., S.C.-G. and S.L.-M.; resources, U.S., P.C., I.A.-T., S.C.-G. and S.L.-M.; data curation, E.P., J.F.M., I.F.-d.-M. and

P.C.; writing—original draft preparation, E.P. and C.R.T.; writing—review and editing, I.A.-T., C.R.T. and S.L.-M.; visualization, E.P., C.R.T., P.C. and U.S.; supervision, I.A.-T., C.R.T. and S.L.-M.; project administration, S.L.-M.; funding acquisition, I.A.-T. and S.L.-M. All authors have read and agreed to the published version of the manuscript.

Funding: The authors acknowledge the financial support provided to the Sustainable Process Engineering (SUPREN, IT1554-22) group at the University of the Basque Country (UPV/EHU, Spain). Funding is from the Basque Government Industry Department under the ELKARTEK program.

Institutional Review Board Statement: Not applicable.

Data Availability Statement: Data available on request due to restrictions of privacy. The data presented in this study are available on request from the corresponding author. The data are not publicly available due to restrictions of privacy.

Acknowledgments: Technical and human support provided by SGIker (UPV/EHU, MICINN, GV/EJ, EGEF and ESF) are gratefully acknowledged. The authors acknowledge Dynasol Elastomers for providing SEBS and SEBS-g-MA samples.

Conflicts of Interest: Author Sergio Corona-Galván was employed by the company Dynasol. The remaining authors declare that the research was conducted in the absence of any commercial or financial relationships that could be construed as a potential conflict of interest.

References

1. de Leon, A.C.C.; da Silva, Í.G.M.; Pangilinan, K.D.; Chen, Q.; Caldona, E.B.; Advincula, R.C. High Performance Polymers for Oil and Gas Applications. *React. Funct. Polym.* **2021**, *162*, 104878. [CrossRef]
2. Wiesli, M.G.; Özcan, M. High-Performance Polymers and Their Potential Application as Medical and Oral Implant Materials: A Review. *Implant Dent.* **2015**, *24*, 448–457. [CrossRef]
3. Yen, H.J.; Liou, G.S. Solution-Processable Triarylamine-Based High-Performance Polymers for Resistive Switching Memory Devices. *Polym. J.* **2016**, *48*, 117–138. [CrossRef]
4. Yang, X.; Loos, J.; Veenstra, S.C.; Verhees, W.J.H.; Wienk, M.M.; Kroon, J.M.; Michels, M.A.J.; Janssen, R.A.J. Nanoscale Morphology of High-Performance Polymer Solar Cells. *Nano Lett.* **2005**, *5*, 579–583. [CrossRef] [PubMed]
5. Kurdi, A.; Chang, L. Recent Advances in High Performance Polymers-Tribological Aspects. *Lubricants* **2018**, *7*, 2. [CrossRef]
6. De Leon, A.C.; Chen, Q.; Palaganas, N.B.; Palaganas, J.O.; Manapat, J.; Advincula, R.C. High Performance Polymer Nanocomposites for Additive Manufacturing Applications. *React. Funct. Polym.* **2016**, *103*, 141–155. [CrossRef]
7. DeMeuse, M.T. *Polymer Blends Handbook*; Utracki, L., Wilkie, C., Eds.; Springer: Dordrecht, The Netherlands, 2014.
8. Weyhrich, C.W.; Long, T.E. Additive Manufacturing of High-Performance Engineering Polymers: Present and Future. *Polym. Int.* **2022**, *71*, 532–536. [CrossRef]
9. Boydston, A.J.; Cui, J.; Lee, C.U.; Lynde, B.E.; Schilling, C.A. 100th Anniversary of Macromolecular Science Viewpoint: Integrating Chemistry and Engineering to Enable Additive Manufacturing with High-Performance Polymers. *ACS Macro Lett.* **2020**, *9*, 1119–1129. [CrossRef]
10. Yevgen Mamunya, M.I. *Advances in Progressive Thermoplastic and Thermosetting Polymers, Perspectives and Applications*; Technopress Editura: Lasi, Romania, 2012.
11. Spontak, R.J.; Patel, N.P. Thermoplastic Elastomers: Fundamentals and Applications. *Curr. Opin. Colloid Interface Sci.* **2000**, *5*, 333–340. [CrossRef]
12. Maji, P.; Naskar, K. Styrenic Block Copolymer-Based Thermoplastic Elastomers in Smart Applications: Advances in Synthesis, Microstructure, and Structure–property Relationships—A Review. *J. Appl. Polym. Sci.* **2022**, *139*, 5–7. [CrossRef]
13. Lin, F.; Wu, C.; Cui, D. Synthesis and Characterization of Crystalline Styrene-b-(Ethylene-Co-Butylene)-b-Styrene Triblock Copolymers. *J. Polym. Sci. Part A Polym. Chem.* **2017**, *55*, 1243–1249. [CrossRef]
14. Downey, A.; Pisello, A.L.; Fortunati, E.; Fabiani, C.; Luzi, F.; Torre, L.; Ubertini, F.; Laflamme, S. Durability and Weatherability of a Styrene-Ethylene-Butylene-Styrene (SEBS) Block Copolymer-Based Sensing Skin for Civil Infrastructure Applications. *Sens. Actuators A Phys.* **2019**, *293*, 269–280. [CrossRef]
15. Wu, T.; Hu, Y.; Rong, H.; Wang, C. SEBS-Based Composite Phase Change Material with Thermal Shape Memory for Thermal Management Applications. *Energy* **2021**, *221*, 119900. [CrossRef]
16. Castro, H.F.; Correia, V.; Pereira, N.; Costab, P.; Oliveiraa, J.; Lanceros-Méndez, S. Printed Wheatstone Bridge with Embedded Polymer Based Piezoresistive Sensors for Strain Sensing Applications. *Addit. Manuf.* **2018**, *20*, 119–125. [CrossRef]
17. Zhang, Z.X.; Dai, X.R.; Zou, L.; Wen, S.B.; Sinha, T.K.; Li, H. A Developed, Eco-Friendly, and Flexible Thermoplastic Elastomeric Foam from Sebs for Footwear Application. *Express Polym. Lett.* **2019**, *13*, 948–958. [CrossRef]
18. Chang, Y.W.; Shin, J.Y.; Ryu, S.H. Preparation and Properties of Styrene-Ethylene/Butylene-Styrene (SEBS)-Clay Hybrids. *Polym. Int.* **2004**, *53*, 1047–1051. [CrossRef]

19. Grigorescu, R.M.; Ciuprina, F.; Ghioca, P.; Ghiurea, M.; Iancu, L.; Spurcaciu, B.; Panaitescu, D.M. Mechanical and Dielectric Properties of SEBS Modified by Graphite Inclusion and Composite Interface. *J. Phys. Chem. Solids* **2016**, *89*, 97–106. [CrossRef]
20. Kuester, S.; Merlini, C.; Barra, G.M.O.; Ferreira, J.C.; Lucas, A.; De Souza, A.C.; Soares, B.G. Processing and Characterization of Conductive Composites Based on Poly(Styrene-b-Ethylene-Ran-Butylene-b-Styrene) (SEBS) and Carbon Additives: A Comparative Study of Expanded Graphite and Carbon Black. *Compos. Part B Eng.* **2016**, *84*, 236–247. [CrossRef]
21. Gao, W.; Yu, B.; Li, S.; Chen, S.; Zhu, Y.; Zhang, B.; Zhang, Y.; Cai, H.; Han, B. Preparation and Properties of Reinforced SEBS-Based Thermoplastic Elastomers Modified by PA6. *Polym. Eng. Sci.* **2022**, *62*, 1052–1060. [CrossRef]
22. Yin, L.; Yin, J.; Shi, D.; Luan, S. Effects of SEBS-g-BTAI on the Morphology, Structure and Mechanical Properties of PA6/SEBS Blends. *Eur. Polym. J.* **2009**, *45*, 1554–1560. [CrossRef]
23. Sharudin, R.W.B.; Ohshima, M. Preparation of Microcellular Thermoplastic Elastomer Foams from Polystyrene-b-Ethylene-Butylene-b-Polystyrene (SEBS) and Their Blends with Polystyrene. *J. Appl. Polym. Sci.* **2013**, *128*, 2245–2254. [CrossRef]
24. Abis, L.; Abbondanza, L.; Braglia, R.; Castellani, L.; Giannotta, G.; Po, R. Syndiotactic Polystyrene/High-Density Polyethylene Blends Compatibilized with SEBS Copolymer: Thermal, Morphological, Tensile, Dynamic-Mechanical, and Ultrasonic Characterization. *Macromol. Chem. Phys.* **2000**, *201*, 1732–1741. [CrossRef]
25. Roeder, J.; Oliveira, R.V.B.; Gonçalves, M.C.; Soldi, V.; Pires, A.T.N. Polypropylene/Polyamide-6 Blends: Influence of Compatibilizing Agent on Interface Domains. *Polym. Test.* **2002**, *21*, 815–821. [CrossRef]
26. Yadav, C.; Saini, A.; Maji, P.K. Cellulose Nanofibres as Biomaterial for Nano-Reinforcement of Poly[Styrene-(Ethylene-Co-Butylene)-Styrene] Triblock Copolymer. *Cellulose* **2018**, *25*, 449–461. [CrossRef]
27. Gama, N.; Ferreira, A.; Evtuguin, D.; Barros-Timmons, A. Modified Cork/SEBS Composites for 3D Printed Elastomers. *Polym. Adv. Technol.* **2022**, *33*, 1881–1891. [CrossRef]
28. Saikrasun, S.; Yuakkul, D.; Amornsakchai, T. Thermo-Oxidative Stability and Remarkable Improvement in Mechanical Performance for Styrenic-Based Elastomer Composites Contributed from Silane-Treated Pineapple Leaf Fiber and Compatibilizer. *Int. J. Plast. Technol.* **2017**, *21*, 252–277. [CrossRef]
29. Haldar, D.; Purkait, M.K. Micro and Nanocrystalline Cellulose Derivatives of Lignocellulosic Biomass: A Review on Synthesis, Applications and Advancements. *Carbohydr. Polym.* **2020**, *250*, 116937. [CrossRef] [PubMed]
30. Lee, H.V.; Hamid, S.B.A.; Zain, S.K. Conversion of Lignocellulosic Biomass to Nanocellulose: Structure and Chemical Process. *Sci. World J.* **2014**, *2014*, 631013. [CrossRef]
31. Vasile, C.; Baican, M. Lignins as Promising Renewable Biopolymers and Bioactive Compounds for High-Performance Materials. *Polymers* **2023**, *15*, 3177. [CrossRef]
32. Heinze, T.; El Seoud, O.A.; Koschella, A. *Principles of Cellulose Derivatization BT—Cellulose Derivatives: Synthesis, Structure, and Properties*; Springer Series on Polymer and Composite Materials; Springer Nature: Basel, Switzerland, 2018.
33. Dinesh; Kumar, B.; Kim, J. Mechanical and Dynamic Mechanical Behavior of the Lignocellulosic Pine Needle Fiber-Reinforced SEBS Composites. *Polymers* **2023**, *15*, 1225. [CrossRef]
34. Murtaza, G. Ethylcellulose Microparticles: A Review. *Acta Pol. Pharm. Drug Res.* **2012**, *69*, 11–22.
35. Fischer, S.; Thümmler, K.; Volkert, B.; Hettrich, K.; Schmidt, I.; Fischer, K. Properties and Applications of Cellulose Acetate. *Macromol. Symp.* **2008**, *262*, 89–96. [CrossRef]
36. Tejada-Oliveros, R.; Balart, R.; Ivorra-Martinez, J.; Gomez-Caturla, J.; Montanes, N.; Quiles-Carrillo, L. Improvement of Impact Strength of Polylactide Blends with a Thermoplastic Elastomer Compatibilized with Biobased Maleinized Linseed Oil for Applications in Rigid Packaging. *Molecules* **2021**, *26*, 240. [CrossRef]
37. Gan, H.; Xu, J.; He, W. Study on the Properties of Cellulose Acetate Composite Synergistically Toughened by Glycerol Triacetate and Modified Nano Titania. *J. Phys. Conf. Ser.* **2023**, *2539*, 012065. [CrossRef]
38. Garhwal, A.; Maiti, S.N. Influence of Styrene–ethylene–butylene–styrene (SEBS) Copolymer on the Short-Term Static Mechanical and Fracture Performance of Polycarbonate (PC)/SEBS Blends. *Polym. Bull.* **2016**, *73*, 1719–1740. [CrossRef]
39. Ganguly, A.; Bhowmick, A.K. Effect of Polar Modification on Morphology and Properties of Styrene-(Ethylene-Co-Butylene)-Styrene Triblock Copolymer and Its Montmorillonite Clay-Based Nanocomposites. *J. Mater. Sci.* **2009**, *44*, 903–918. [CrossRef]
40. Fernandes, L.C.; Correia, D.M.; Pereira, N.; Tubio, C.R.; Lanceros-Méndez, S. Highly Sensitive Transparent Piezoionic Materials and Their Applicability as Printable Pressure Sensors. *Compos. Sci. Technol.* **2021**, *214*, 108976. [CrossRef]
41. Gupta, P.; Bera, M.; Maji, P.K. Nanotailoring of Sepiolite Clay with Poly [Styrene-b-(Ethylene-Co-Butylene)-b-Styrene]: Structure–property Correlation. *Polym. Adv. Technol.* **2017**, *28*, 1428–1437. [CrossRef]
42. Zhou, T.; Wu, Z.; Li, Y.; Luo, J.; Chen, Z.; Xia, J.; Liang, H.; Zhang, A. Order-Order, Lattice Disordering, and Order-Disorder Transition in SEBS Studied by Two-Dimensional Correlation Infrared Spectroscopy. *Polymer* **2010**, *51*, 4249–4258. [CrossRef]
43. Senusi, N.A.; Mohd Shohaimi, N.A.; Halim, A.Z.A.; Shukr, N.M.; Abdul Razab, M.K.A.; Mohamed, M.; Mohd Amin, M.A.; Mokhtar, W.N.A.W.; Ismardi, A.; Abdullah, N.H.; et al. Preparation & Characterization of Microcrystalline Cellulose from Agriculture Waste. *IOP Conf. Ser. Earth Environ. Sci.* **2020**, *596*, 012035.
44. Gunduz, O.; Ahmad, Z.; Stride, E.; Edirisinghe, M. Continuous Generation of Ethyl Cellulose Drug Delivery Nanocarriers from Microbubbles. *Pharm. Res.* **2013**, *30*, 225–237. [CrossRef]
45. Ghorani, B.; Russell, S.J.; Goswami, P. Controlled Morphology and Mechanical Characterisation of Electrospun Cellulose Acetate Fibre Webs. *Int. J. Polym. Sci.* **2013**, *2013*, 256161. [CrossRef]

46. Kurusu, R.S.; Demarquette, N.R. Blending and Morphology Control to Turn Hydrophobic SEBS Electrospun Mats Superhydrophilic. *Langmuir* **2015**, *31*, 5495–5503. [CrossRef] [PubMed]
47. Kurusu, R.S.; Demarquette, N.R. Surface Properties Evolution in Electrospun Polymer Blends by Segregation of Hydrophilic or Amphiphilic Molecules. *Eur. Polym. J.* **2017**, *89*, 129–137. [CrossRef]
48. Qi, J.; Chen, Y.; Zhang, W.T.; Li, L.; Huang, H.D.; Lin, H.; Zhong, G.J.; Li, Z.M. Imparting Cellulose Acetate Films with Hydrophobicity, High Transparency, and Self-Cleaning Function by Constructing a Slippery Liquid-Infused Porous Surface. *Ind. Eng. Chem. Res.* **2022**, *61*, 7962–7970. [CrossRef]
49. Cantournet, S.; Desmorat, R.; Besson, J. Mullins Effect and Cyclic Stress Softening of Filled Elastomers by Internal Sliding and Friction Thermodynamics Model. *Int. J. Solids Struct.* **2009**, *46*, 2255–2264. [CrossRef]
50. Lemes, A.P.; Talim, R.; Gomes, R.C.; Gomes, R.C. Preparation and Characterization of Maleic Anhydride Grafted Poly(Hydroxybutirate-CO-Hydroxyvalerate)—PHBV-g-MA. *Mater. Res.* **2016**, *19*, 229–235.
51. Eid, M.; Abdel-Ghaffar, M.A.; Dessouki, A.M. Effect of Maleic Acid Content on the Thermal Stability, Swelling Behaviour and Network Structure of Gelatin-Based Hydrogels Prepared by Gamma Irradiation. *Nucl. Instrum. Methods Phys. Res. Sect. B Beam Interact. Mater. Atoms* **2009**, *267*, 91–98. [CrossRef]
52. Bhuyan, K.; Dass, N.N. Thermal Studies of Copolymers of Styrene and Maleic Anhydride. *J. Therm. Anal.* **1989**, *35*, 2529–2533. [CrossRef]
53. De Oliveira, R.L.; Da Silva Barud, H.; De Assunção, R.M.N.; Da Silva Meireles, C.; Carvalho, G.O.; Filho, G.R.; Messaddeq, Y.; Ribeiro, S.J.L. Synthesis and Characterization of Microcrystalline Cellulose Produced from Bacterial Cellulose. *J. Therm. Anal. Calorim.* **2011**, *106*, 703–709. [CrossRef]
54. Zyane, A.; Belfkira, A.; Brouillette, F.; Lucas, R.; Marchet, P. Microcrystalline Cellulose as a Green Way for Substituting $BaTiO_3$ in Dielectric Composites and Improving Their Dielectric Properties. *Cellul. Chem. Technol.* **2015**, *49*, 783–787.
55. Ribeiro, S.; Costa, P.; Ribeiro, C.; Sencadas, V.; Botelho, G.; Lanceros-Méndez, S. Electrospun Styrene-Butadiene-Styrene Elastomer Copolymers for Tissue Engineering Applications: Effect of Butadiene/Styrene Ratio, Block Structure, Hydrogenation and Carbon Nanotube Loading on Physical Properties and Cytotoxicity. *Compos. Part B Eng.* **2014**, *67*, 30–38. [CrossRef]
56. Sainorudin, M.H.; Abdullah, N.A.; Asmal Rani, M.S.; Mohammad, M.; Mahizan, M.; Shadan, N.; Abd Kadir, N.H.; Yaakob, Z.; El-Denglawey, A.; Alam, M. Structural Characterization of Microcrystalline and Nanocrystalline Cellulose from *Ananas comosus* L. Leaves: Cytocompatibility and Molecular Docking Studies. *Nanotechnol. Rev.* **2021**, *10*, 793–806. [CrossRef]

Disclaimer/Publisher's Note: The statements, opinions and data contained in all publications are solely those of the individual author(s) and contributor(s) and not of MDPI and/or the editor(s). MDPI and/or the editor(s) disclaim responsibility for any injury to people or property resulting from any ideas, methods, instructions or products referred to in the content.

Review

The Potential of Electrospun Membranes in the Treatment of Textile Wastewater: A Review

Joana M. Rocha, Rui P. C. L. Sousa , Raul Fangueiro and Diana P. Ferreira *

Centre for Textile Science and Technology (2C2T), University of Minho, 4800 Guimarães, Portugal; b13622@2c2t.uminho.pt (J.M.R.); rp.cls@hotmail.com (R.P.C.L.S.); rfangueiro@det.uminho.pt (R.F.)
* Correspondence: diana.ferreira@det.uminho.pt

Abstract: Water security and industrial wastewater treatment are significant global concerns. One of the main issues with environmental contamination has been the discharge of dye wastewater from the textile and dye industries, contributing to an ever-growing problem with water pollution, poisoning water supplies, and harming the ecosystem. The traditional approach to wastewater treatment has been found to be inefficient, and biosorption techniques and mechanisms have been proven to be a successful replacement for conventional methods. Recent developments have led to the recognition of fibrous materials as an environmentally friendly option with broad application in several industries, including wastewater treatment. This review explores the potential of fibrous materials produced by the electrospinning technique as adsorbents for wastewater treatment, while at the same time, for the removal of adsorbates such as oil, dyes, heavy metals, and other substances, as reported in the literature. Textile wastewater filtering structures, produced by electrospinning, are summarized and the use of synthetic and natural polymers for this purpose is discussed. The limitations of electrospun textile wastewater filtering structures are also mentioned. Electrospun nanofibrous membranes appear to be a very promising route to filter textile wastewater and therefore contribute to water reuse and to reducing the contamination of water courses.

Keywords: adsorbents; electrospinning; fibrous materials; industries; wastewater treatment

Citation: Rocha, J.M.; Sousa, R.P.C.L.; Fangueiro, R.; Ferreira, D.P. The Potential of Electrospun Membranes in the Treatment of Textile Wastewater: A Review. *Polymers* **2024**, *16*, 801. https://doi.org/10.3390/polym16060801

Academic Editors: Cristina Cazan and Mihai Alin Pop

Received: 29 January 2024
Revised: 1 March 2024
Accepted: 11 March 2024
Published: 13 March 2024

Copyright: © 2024 by the authors. Licensee MDPI, Basel, Switzerland. This article is an open access article distributed under the terms and conditions of the Creative Commons Attribution (CC BY) license (https://creativecommons.org/licenses/by/4.0/).

1. Introduction

Water pollution is one of the many environmental issues that have recently surfaced as a result of overcrowding, urbanization, an increase in various industrial and human activities, the growth of landfills, mining activities, and urban wastewater [1,2]. Huge quantities of different toxic pollutants are discharged into water resources on a daily basis, causing increasingly serious water pollution problems, contaminating water bodies, and damaging the environment [2,3]. Therefore, water supplies are contaminated with a variety of pollutants, including oil, microbes, viruses, pharmaceutical waste, organic and inorganic pollutants, and nanoparticles [4]. Dyes and heavy metal ions are considered the main water pollutants [4]. Water contamination is one of the most important worldwide issues at the moment because there are not enough water sources. By 2025, around 3.5 billion people worldwide will face a shortage of fresh water [5]. As a result, researchers are working to create sustainable, affordable, and alternative solutions for treating or recycling wastewater [6].

The particular case of the textile industry is alarming. This industry is one of the major contributors to effluent wastewater due to the huge consumption of water in different parts of the process. It is estimated that the textile industry consumes about 200 L of water per kilogram of fabric processed, with these processes generating around 17 to 20% of total industrial wastewater [7,8]. In the wet fabric processing industry, processing operations include steps such as scouring, bleaching, dying, printing, and finishing stages, which generate different pollutants in the wastewater. Effluent wastewater may contain chemicals like alkalis, reactive dyes and other organic compounds, hydrogen peroxide, surfactants, or metal ions [9]. It has been estimated that each year, approximately 100 tons of dyes are released into drinking water [10]. Therefore, a variety of treatment methods, including physical (adsorption and filtration), oxidation (ozone, H_2O_2, or Fenton's process), biological (fungi, algae, or bacteria), and hybrid processes, have been developed to effectively treat textile wastewater to avoid the discharge of the mentioned pollutants [11]. However, many innovative technologies are emerging to optimize and improve the current methods in terms of recycling, reducing costs, and increasing sustainability.

With a focus on producing nanomaterials with particular characteristics for use in a range of water purification applications, nanotechnologies and nanomaterial sciences have received a lot of attention recently. Of all of the nanostructured materials currently available, nanofibers have proven to have unique qualities and characteristics that allow them to overcome the constraints of traditional fibrous structures [12]. One of the most significant methods for producing these nanomaterials is electrospinning. It is a sophisticated method for creating continuous fibers that range from micro- to nanoscale in size. Furthermore, it is an adaptable method that works with a variety of spinnable materials, is accurate and controllable, and has an excellent cost-effectiveness ratio [13]. Generally, an electric field strong enough to overcome the surface tension of the polymer material is created by applying an electrostatic force over a predetermined distance between a collector and a syringe filled with a polymer solution. As a result, the polymer solution is transformed into a fiber structure that is deposited on the collector, creating a non-woven mat with good mechanical qualities, a high surface area, a low weight connected porous structure, and high porosity. In order to tailor nanofiber mats for particular applications [14], functional additives such as surface coating, interfacial polymerization (IP), or active species can be added to the electrospinning solution [15–17]. These modification techniques provide electrospun nanofibers with strong advantages for different applications, such as water purification.

The utilization of electrospinning techniques to create nanofibrous structures has demonstrated significant promise in wastewater treatment because of its advantageous properties for the elimination of both organic and inorganic contaminants from aqueous systems [2,5,10,18,19]. The abovementioned superior characteristics of these structures and the tunability of the electrospinning process turn electrospun nanostructures into a promising route for wastewater treatment. Additionally, through employing such cutting-edge methods and scalable production processes, society can become less dependent on conventional purifying technologies and cut down on the resources and energy that these technologies consume [18].

Despite some reports on water purification approaches with various materials and techniques, the application of different types of nanofibers obtained by the electrospinning technique in water purification systems is quite limited. In this work, the potential of filtering electrospun structures for textile wastewater treatment is discussed. An overview of the electrospinning process and its processing parameters is given. The reports on filtering structures obtained by electrospinning are reviewed and the limitations of this field are summarized.

2. Textile Wastewater: Composition and Purification Requirements

The composition of textile wastewater is complex and diverse, with substantial limitations to successful pollutant treatment and removal. Textile wastewater comprises several types of contaminants, with textile dyes being particularly problematic given their xenobiotic properties [20]. Furthermore, organic chemicals including aldehydes, alcohols, ketones, and surfactants add to the complexity of textile effluent. Wastewater is generated during textile processing steps, such as dyeing, printing, and finishing (Figure 1) [21]. The discharge of wastewater from the textile industry without adequate treatment can result in the discharge of pollutants that include carcinogens, mutagens, or heavy metals, endangering human health and aquatic environments [22].

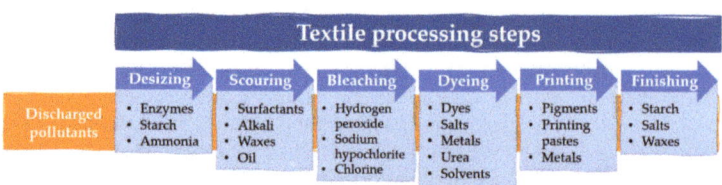

Figure 1. Textile processing steps and pollutants the may be discharged in each step.

Textile wastewater's physicochemical features, such as high amounts of suspended substances, a high chemical oxygen demand, acidity, heat, and color, emphasize the wide range of contaminants found in its composition [21]. While oxidation and biological methods, such as ozonation, Fenton's process, or bacterial treatment, have been widely explored for the removal of textile wastewater constituents [23], physical methods such as the use of porous membranes with high selectivity and efficiency may be the solution for the adsorption/filtration of dyes, other organic compounds, and heavy metals.

Membrane separation processes include microfiltration, ultrafiltration, nanofiltration, and reverse osmosis [24]. Other membrane technologies have been recently used for wastewater treatment, such as forward osmosis, membrane distillation, electrodialysis, or membrane bioreactors. A comparative analysis of the different membrane technologies was carried out by Ma et al. with the help of the software Circos (version 0.69–6) (Figure 2) [25]. Fibrous membranes exhibit tunable pore sizes that can act in some of the mentioned processes, as the effective separation of components smaller than the pore size is essential for efficient filtration [26]. Indeed, membrane separation can not only efficiently remove the mentioned contaminants, but also adsorb and recycle dyes and other organic chemicals for reuse in the textile processing.

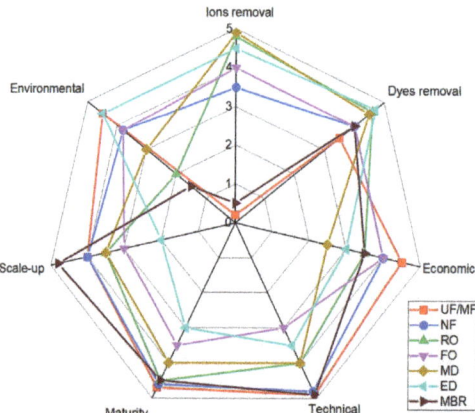

Figure 2. Comparative analysis of the membrane technologies in textile wastewater treatment, with an evaluation of the following parameters: ion removal efficiency, dye removal efficiency, economic analysis, technical feasibility, maturity, scalability, and environmental analysis. The scale indicates the rating from lowest to highest, with a minimum of 1 and a maximum of 5 points. Reprinted with permission from reference [25].

The effectiveness of membranes for wastewater adsorption or filtration relies on specific characteristics. Membranes' tunable nature, porosity, surface area-to-volume ratio, morphology, tensile strength and elongation, and wettability play crucial roles in determining their suitability for filtration applications. Electrospun membranes show structural and mechanical features which render them potentially advantageous for application in the filtration of textile wastewater, such as their high porosity, large surface area-to-volume ratio, and mechanical stability.

3. Electrospinning—Process Technology and Operation

Electrospinning is a simple, fast, and versatile technique that uses electrostatic forces to produce fibers with controllable diameters from polymer solutions [27]. Fibers produced by electrospinning are characterized by their small diameters, which can vary from nanometers to micrometers, their high surface-to-volume ratio, high porosity, easy surface functionalization, and high flexibility to tune the shape and size [28,29]. Furthermore, this technique allows for the use of different polymeric solutions, made up of natural and/or synthetic polymers, which can also incorporate reinforcing agents and bioactive agents, allowing for the production of structures with advanced properties. These characteristics mean that nanofibers produced by electrospinning are of potential interest for applications in various areas, namely in the textile industry for military protective clothing, in environmental engineering for environmental monitoring, in tissue engineering for the development of structures or supports, and in biomedical applications, for dressings with the release of active ingredients [30,31]. Control of the morphology and final dimensions of the produced fibers depends on the different processing parameters, which include the parameters of the polymer solution, the process/equipment, and also the environmental parameters [32].

The fundamental arrangement of electrospinning apparatus is shown schematically in Figure 3. It has three main components: a grounded metal collector, a capillary tube with a smalldiameter metal needle coupled to a high voltage source, and a syringe containing the polymer solution [33,34].

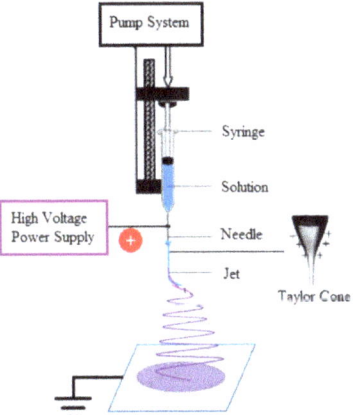

Figure 3. Schematic representation of electrospinning equipment. Reproduced with permission from reference [35].

In the electrospinning process, the polymer solution is ejected from the needle into the metal collector by applying high voltage between the needle and the collector. In detail, the production of fibers by using electrospinning begins with the transfer of the polymer solution, which is stored in the syringe, through the capillary tube to the needle. Subsequently, the applied electric field induces charges in the polymer solution, which are evenly distributed over the entire surface of the drop at the tip of the needle. As the intensity of the electric field increases, the polymer droplet elongates, acquiring a conical shape commonly referred to as the Taylor cone. When the applied potential reaches the critical value necessary to overcome the surface tension of the polymer droplet, a jet of electrically charged fluid is ejected from the tip of the Taylor cone. As the polymer jet travels from the needle to the collector, the solvent gradually evaporates, and the polymer fibers are deposited on the collector [36,37].

3.1. Processing Parameters

Several processing parameters have a major impact on the electrospinning process and can change the morphology, diameter, and shape of the fibers that are produced. There are three main groups for this set of parameters: environmental, process, and solution parameters [38].

3.1.1. Solution Parameters

Solution parameters such as polymer molecular weight, concentration, viscosity, and electrical conductivity directly influence the morphology, geometry, and size of the fibers produced by electrospinning. These parameters are related to the physicochemical properties of the polymers and solvents, as well as the interactions between them [39,40].

- Concentration and viscosity

The concentration of the polymer in the solution and the viscosity of the solution play an important role in the formation of fibers during the electrospinning process. Low concentrations usually cause the polymer's molecular chains to break down before reaching the collector, leading to the appearance of defects, namely droplets, along with the deposited fibers. On the other hand, the use of high concentrations, combined with solutions with greater viscosity, promotes the production of fibers with fewer defects, as there is greater conjugation of the polymer molecular chains in the solution, favoring the production of a continuous jet. In addition, solutions with a higher viscosity are associated with the production of more uniform fibers with larger diameters [29,34].

Mirtic et al. evaluated the influence of the total concentration of the polymer (PEO and alginate) in the solution on the morphology of the nanofibers developed by electrospinning. To this end, they tested three percentages, 2.5%, 3.5%, and 4.5%, and found that with higher concentrations, nanofibers with proportionally larger diameters were obtained and defects were eliminated (Figure 4) [41]. Similar results were obtained by Dhandayuthapani et al., who evaluated the effect of the concentration of CS and gel on the production of CS nanofibers and gel nanofibers. The authors observed that the use of low concentrations of both polymers, associated with low viscosity, resulted in the production of non-uniform nanofibers with numerous defects. As the concentration of the polymers in the solution increased, it was possible to obtain uniform nanofibers without defects [42]. Finally, Nezarati et al. evaluated the effect of the concentration and viscosity of the polycarbonate urethane (PCU) polymer solution on the morphology of the fibers. Firstly, they found that increasing the concentration of PCU in the solution led to an increase in its viscosity. They also showed that low viscosity values (7.2 ± 1.7 Pa.s), resulting from lower concentration solutions, led to fibers with defects in their structure. Intermediate viscosity values (10.1 ± 0.5 Pa.s) allowed for the development of defect-free fibers with homogeneous diameters. Finally, for high viscosity values (22.5 ± 1.4 Pa.s), there was a substantial increase in the average diameter of the fibers produced, increasing from 1.2 µm to 3.5 µm [43].

Thus, considering the studies presented above, it can be concluded that low polymer concentrations, and consequently reduced viscosity, promote the appearance of defects in the nanofibers produced, while increasing the concentration favors the development of more uniform nanofibers with fewer defects and a larger diameter. However, excessively high concentrations may prevent the solution from flowing through the needle tip and may inhibit fiber formation due to an excessive increase in the solution's viscosity. It is therefore necessary to find the ideal polymer concentration at which the solution reaches viscosity values suitable for fiber formation in the electrospinning process [29,44].

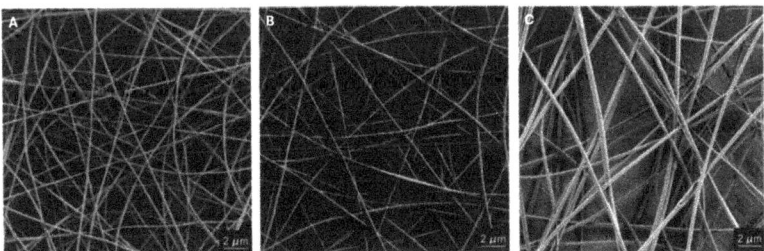

Figure 4. FESEM images of PEO and alginate nanofibers with a total polymer percentage of (**A**) 2.5% (m/m); (**B**) 3.5% (m/m); and (**C**) 4.5% (m/m), using a PEO with a molecular weight of 2 MDa and a proportion of 15% (m/m). Reprinted with permission from reference [41].

- Molecular weight

The morphology of fibers created by electrospinning is also significantly influenced by the molecular weight of the polymer. The conjugation of the polymer chains reflects the molecular weight, which in turn affects the solution's viscosity [45]. Sohi et al. studied the effect of the molecular weight of chitosan (CS) on the production of CS/PEO fibers, where three molecular weights were evaluated: high ($(3.10–3.75) \times 10^5$ g/mol), medium ($(1.90–3.1) \times 10^5$ g/mol), and low molecular weight ($(0.5–1.9) \times 10^5$ g/mol). Initially, using CS solutions of different concentrations and different molecular weights, the authors found that when using 1.5% (w/v) and 2% (w/v) CS, it was not possible to produce fibers using medium and low molecular weight CS. In addition, with high molecular weight CS, it was not possible to produce fibers with polymer concentrations in the solution lower than 1.5% (m/v). Therefore, PEO was incorporated in different proportions in order to improve the properties of the solution for use in electrospinning. From this point, the authors only evaluated the effect of low and medium molecular weight CS. Thus, for a polymer percentage of 2% (m/v) and a CS/PEO ratio of 1:3, the authors obtained defect-free fibers using low molecular weight CS. With medium molecular weight CS, some defects were observed along with the fibers produced [46]. Similarly, Roldán et al. evaluated the morphology of PCL fibers obtained using different polymer molecular weights, 14,000, 45,000, and 80,000 g/mol. For a polymer percentage of 15% in a solution of acetone and acetic acid (3:7), the authors found that with low molecular weight PCL, no fibers were formed by the electrospinning process. On the other hand, with PCL of 45,000 g/mol, fibers were formed along with the appearance of some droplets. Increasing the molecular weight to 80,000 g/mol allowed for the development of uniform fibers without defects in their structure, although with larger diameters [47]. The effect of the polymer polyvinyl alcohol (PVA) with different molecular weights was also evaluated by Akduman et al., where identical results were observed. Three molecular weights were used, 89,000–98,000 g/mol, 125,000 g/mol, and 146,000–186,000 g/mol. The most uniform fibers were obtained when high molecular weight PVA was used, where no formation of defects in the fiber was observed, unlike for low molecular weight PVA [48].

Generally, high molecular weight polymers are used in the electrospinning process, as they provide the desired viscosity for fiber formation. However, the use of low molecular weight polymers may be sufficient to provide adequate viscosities and thus guarantee the formation of a stable and uniform jet during the electrospinning process. Furthermore, the conjugation of two or more polymers can also facilitate the electrospinning process and thus enable the use of polymers with lower molecular weights [49].

- Conductivity

The conductivity of the solution is mainly determined by the type of polymer, solvent, and the availability of ionizable salts. A solution with high electrical conductivity will have a greater capacity to transport charges than one with low conductivity. Thus, raising the solution's conductivity to a crucial value will raise the charge on the polymer droplet's surface and promote the Taylor cone's development. Several authors have verified the existence of an inverse proportional relationship between the electrical conductivity and the diameter of the fibers produced by electrospinning. Thus, higher conductivity values result in a smaller fiber diameter, while low conductivity leads to the formation of fibers with larger diameters due to insufficient stretching of the polymer jet [29,50]. A study carried out by Zuo et al. using the polymer poly(3-hydroxybutyrate-co-3-hydroxyvalerate) (PHBV) showed that the conductivity of the solution also has a major influence on the formation of defects in nanofibers produced by electrospinning. In addition, the authors found that the conductivity of the solution varies with the type of solvent used. In fact, the addition of alcohol to the PHBV solution in chloroform promoted an increase in the solution's conductivity from 0.3190 to 13.50 μS/cm, resulting in the formation of more uniform fibers with no defects in their structure. On the other hand, the incorporation of carbon tetrachloride into the PHBV/chloroform solution decreased the solution's conductivity values (0.01700 μS/cm) and consequently favored the appearance of various defects along the deposited fibers. Finally, the addition of dimethylformamide to the PHBV/chloroform solution significantly increased the conductivity (22.50 μS/cm), and smooth fibers with fewer defects were produced (Figure 5) [51].

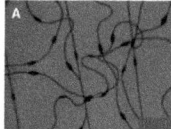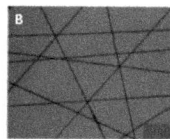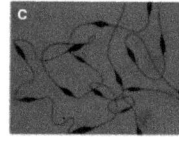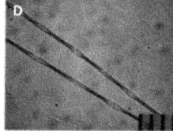

Figure 5. FESEM images of PHBV fibers with different solvents, (**A**) chloroform; (**B**) chloroform/alcohol (3:1); (**C**) chloroform/carbon tetrachloride (3:1); (**D**) chloroform/dimethylformamide (3:1), using a voltage of 22 kV and a feed rate of 4 mL/h. Reproduced with permission from reference [51].

Amariei et al. also showed that the electrical conductivity of the solutions is a determining factor in the diameter of the fibers produced by electrospinning. In this study, 10% PVA was dissolved in different solvents, water, ethanol (20% and 50%), and acetic acid (20% and 50%). The highest electrical conductivity value, 1.39 mS/cm, was obtained with 20% acetic acid, resulting in fibers with diameters of 82 ± 16 nm. The conductivity values decreased with the use of 50% acetic acid, water, and 20% ethanol, respectively. Finally, the solution with the lowest conductivity value (0.54 mS/cm), obtained with 50% ethanol, produced less uniform fibers with larger diameters (955 ± 123 nm) [52]. It can be seen that increasing the electrical conductivity of the solution leads to greater uniformity and a decrease in the diameter of the fibers, while at the same time reducing the number of associated defects [53].

3.1.2. Process Parameters

Included in the process parameters are all of the variables related to the electrospinning apparatus, such as the voltage applied, the distance between the needle and the collector, the feed rate, the needle diameter, and the kind of collector utilized. The user chooses these parameters based on the type of solution to be used, and each one directly affects the diameter and shape of the fibers that are developed [29,54].

- Applied voltage

The applied tension is a parameter that can be manipulated in order to control the quality and diameter of the fibers produced during the electrospinning process. In this process, sufficiently strong tension is required to overcome the surface tension of the polymer droplet and thus cause a jet to form through the Taylor cone [38]. In this way, very low-tension values may not be sufficient for the correct formation of the Taylor cone, leading to the formation of droplets hanging from the tip of the needle [45].

Haghju et al. evaluated the effect of different voltage values, 20 and 30 kV, on the production of CS/PVA fibers by electrospinning. For a constant polymer concentration in the solution, applying a voltage of 20 kV produced fibers with diameters of 522.55 nm and 365.64 nm for feed flow rates of 0.1 and 0.5 mL/h, respectively. As the voltage increased to 30 kV, there was a decrease in fiber diameter to 290.38–405.62 nm and 162.19–318.97 nm, depending on the feed rate. On the other hand, lower values of applied voltage resulted in more homogeneous fibers with fewer defects in their structure [55]. The influence of the applied tension on the morphology and diameter of Eudragite® L100 fibers produced by electrospinning was also evaluated by Reda et al. Four different voltages were tested, 10, 15, 20, and 25 kV. At a voltage of 10 kV, fibers with average diameters of 123.91 ± 48.35 nm were produced. As the voltage value increased to 15, 20, and 25 kV, a progressive increase in fiber diameter was observed to 125.58 ± 33.19, 182.75 ± 62.60, and 186.43 ± 51.01 nm, respectively [56].

The studies described above show that the relationship between the tension applied and the diameter of the fibers produced by electrospinning is ambiguous. In this sense, the tension applied must be adjusted to appropriate values according to the composition of the solution under study in order to guarantee a continuous jet.

- Distance between the needle and collector

The distance between the needle and the collector significantly influences the shape and size of the fibers created by electrospinning. This parameter has a direct influence on the deposition time, the solvent evaporation rate, and the stability of the jet, so a minimum distance must be found that allows the fibers to dry properly before they reach the collector. In addition, the distance will always be related to the concentration of the polymer solution and the applied voltage [57].

Fallah et al. studied the influence of the distance between the needle and the collector in PCL and gelatin fibers [58]. Three different distances were tested (10, 18, and 21 cm), with the feed rate and applied tension remaining constant. At a distance of 10 cm, fibers with diameters of around 130 nm were obtained. As the distance increased to 18 cm, smaller diameters were obtained (95 nm). However, when the distance was increased to 21 cm, the diameter increased again. This result is due to the fact that when the distance increases from 10 cm to 18 cm, the solvent has more time to evaporate before reaching the collector and, consequently, the final diameter of the fibers is reduced. However, when the distance increases to 21 cm, there is a reduction in the electrostatic force on the ejected jet, preventing the jet and the resulting nanofibers from being sufficiently elongated, resulting in larger diameters. Similar results were obtained by Doshi et al. in their study of another polymeric system [59].

Thus, using the results presented by the previous studies, it is possible to conclude that there is a minimum and maximum limit for the value of the distance between the needle and the collector in order to reduce the diameter of the fiber to a minimum value.

- Flow rate

The flow rate of the polymer solution is an important parameter in the electrospinning process, as it influences the speed of the jet and the transfer rate of the polymer solution. To achieve proper evaporation of the solvent and obtain smooth (defect-free) nanofibers, lower feed rates are generally preferable. On the other hand, excessively low flow rates can lead to the unavailability of the solution at the needle tip. However, because the solvent does not completely evaporate before reaching the collector, using extremely high flow rates might lead to the generation of fibers with greater diameters and structural flaws [31,60].

Kyselica et al. studied the influence of applying different flow rates to a PEO solution. In this study, the authors found that low flow rates, but that were sufficient to supply the polymer solution to form the Taylor cone, were the ones that allowed for the formation of nanofibers with the fewest defects. On the other hand, feed rates that were too low led to interruptions in the jet due to the unavailability of the solution at the tip of the needle, while high flow rates led to dripping and the production of thicker fibers [32].

Similar outcomes were obtained by Zargham et al. in their study which varied the feed rate between 0.1 and 1.5 mL/h in Nylon 6 solutions. When a flow rate of 0.5 mL/h was applied at a voltage of 29 kV, with a polymer content of 20% in the solution and a distance of 15 cm between the needle and the collector, uniform, flawless fibers with diameters of about 237 nm were obtained. Values below (0.1 mL/h) and above (1 mL/h) this flow rate favored the development of fibers with more heterogeneous diameters and more defects in their structure and, in the case of greater tension, higher average diameters. In addition, the application of an excessively high flow rate (1.5 mL/h) led to the formation of an unstable jet, with a tendency to form fibers with larger and more heterogeneous diameters, while at the same time resulting in a lower density of deposited fiber area (Figure 6) [61].

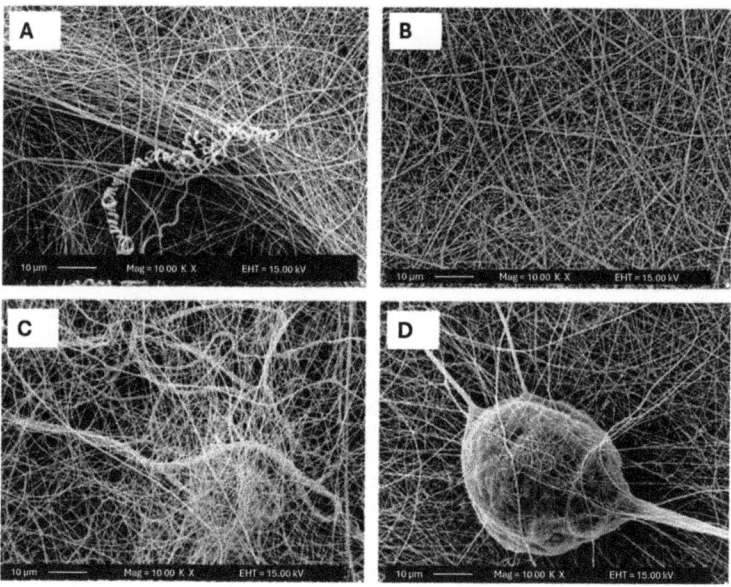

Figure 6. FESEM images of Nylon 6 fibers produced by electrospinning, applying different feed flow rates: (**A**) 0.1 mL/h; (**B**) 0.5 mL/h; (**C**) 1 mL/h; and (**D**) 1.5 mL/h. Adapted with permission from reference [61].

It can therefore be concluded that the value of the flow rate must be defined in such a way as to ensure the necessary time for the solvent to evaporate properly, while at the same time being sufficient to guarantee the correct availability of the solution at the tip of the needle [45].

- Needle diameter

Although it is not a parameter widely evaluated by researchers, the diameter of the needle also has a significant influence on the morphology of the fibers produced by electrospinning [62].

Pisani et al. evaluated the effect of using needles with different diameters, 18 G (1200 μm) and 27 G (400 μm), on PCL/Polylactic Acid (PLA) fibers. In this study, for a total polymer percentage of 10%, the use of the smaller diameter needle led to the development of fibers with numerous defects in their structure. In contrast, the 27 G needle led to the development of defect-free fibers with good morphology and diameters of around 200 nm. For the higher percentages of polymer used (15, 20, and 25%), fibers with good morphology were developed for the different needle diameters used. However, the smaller diameter needle showed fibers with smaller diameters and greater membrane porosity when compared to the larger diameter needle [38]. Similarly, Kuchi et al. tested needles with different diameters, 330, 500, 570, and 720 μm, for the production of polyvinylpyrrolidone (PVP) fibers with titanium dioxide (TiO$_2$). Fibers with an average diameter of 150 ± 20 nm were created using the needle with the smallest diameter. The average diameter of the fibers increased to 200 ± 20, 250 ± 20, and 350 ± 20 nm, respectively, when the needle's diameter was increased to 500, 570, and 720 μm (Figure 7) [63].

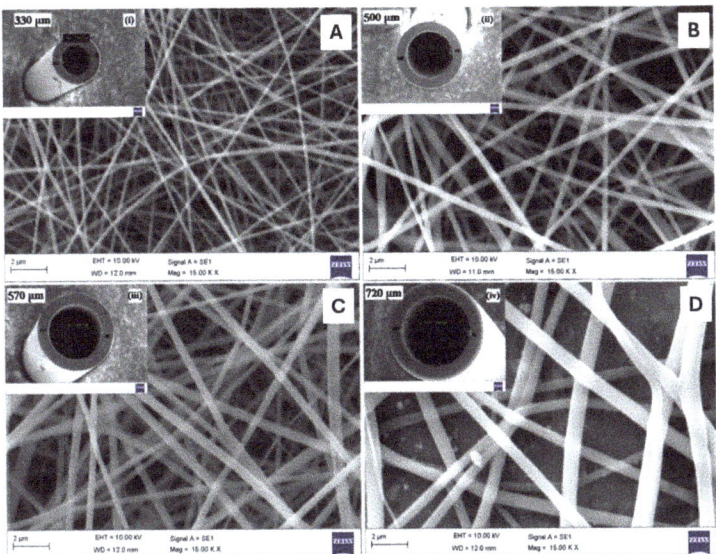

Figure 7. FESEM images of PVP and TiO$_2$ fibers using needles with different diameters: (**A**) 330 μm; (**B**) 500 μm; (**C**) 570 μm; and (**D**) 720 μm. Inset: diameter measurements: (**i**) 330 μm; (**ii**) 500 μm; (**iii**) 570 μm; and (**iv**) 720 μm. Adapted with permission from reference [63].

Thus, smaller needle diameters not only favor the development of fibers with smaller diameters and without defects, but also increase the porosity of the membranes developed when compared to fibers obtained with larger needle diameters [64].

- Collector

The type of collector utilized in the electrospinning process is a critical factor. The collector serves as a conductive substrate for the deposition of fibers in this method. Typically, the collector is covered with aluminum foil, although some authors have used alternative materials including conductive paper, wire mesh, or metal [65,66]. The collector can differ in shape, being a static collector, in the form of a flat metal plate, or a dynamic collector, such as a rotating metal cylinder, depending on the type of fiber to be obtained. The use of a flat plate makes it possible to obtain thicker membranes (using the same deposition time), but it has a small collection area and does not allow the fibers to be aligned. On the other hand, using the rotating cylinder makes it possible to obtain aligned fibers and, due to its larger collection area, to obtain membranes with larger fibers. However, there is some difficulty in guaranteeing the alignment of all of the fibers deposited, as well as the possibility of the fibers breaking if the rotation is too high [49]. In fact, several parameters influence the orientation of the fibers, such as the design and rotation of the collector. Several studies have evaluated these parameters and found that higher rotations are associated with greater fiber orientation [67,68].

3.1.3. Environmental Parameters

The influence of environmental parameters is generally underestimated by many researchers, as only a few studies have monitored their effect on the electrospinning process.

However, the structure and morphology of fibers are also affected by environmental conditions, namely temperature and humidity [69].

- Temperature

Temperature has a greater influence on the viscosity of the solution, since an increase in temperature leads to a decrease in the viscosity value and, consequently, a reduction in the average diameter of the nanofibers obtained. On the other hand, low temperatures tend to increase the viscosity of the solution, promoting the formation of fibers with larger diameters [69].

In a study by Mituppatham et al., the synthesis of polyamide fibers with thinner diameters was favored by an increase in temperature that was accompanied by a decrease in viscosity. The resulting nanofibers had a diameter of 98 nm at 30 °C and 90 nm at 60 °C, respectively [70]. Similar results were obtained by Yang et al. [71]. These authors studied the effect of temperature on polyacrylonitrile (PAN) and polyvinylpyrrolidone (PVP) fibers (Figure 8) and showed that in both cases, there was a gradual decrease in the diameter of the nanofibers from 530 to 280 nm for PAN fibers and from 830 to 540 nm for PVP fibers as the temperature increased from 20 °C to 60 °C and from 20 °C to 50 °C, respectively. However, for excessively high temperatures, it was found that the solvent evaporated too quickly and, consequently, the diameter of the fibers reached higher values.

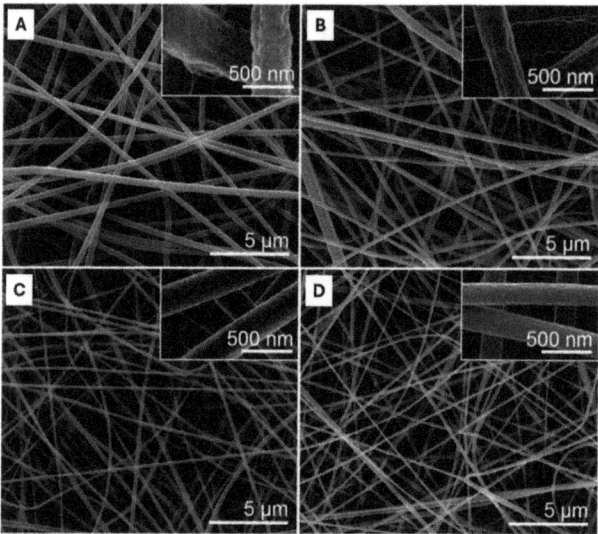

Figure 8. FESEM images of PAN/PVP nanofibers obtained at different ambient temperatures: (**A**) 20 °C; (**B**) 40 °C; (**C**) 60 °C; (**D**) 80 °C. Reprinted with permission from reference [71].

In general, an increase in temperature favors the formation of fiber membranes with smaller average diameters. However, excessively high temperature values show the opposite effect, increasing the average diameter of the fibers [29].

- Humidity

Humidity is an environmental parameter that influences the final morphology of fibers produced by electrospinning. At low humidity values, the fibers dry out quickly due to an increase in the evaporation rate of the solvent. In addition, there is a chance that the fluid will dry up at the needle tip, which could cause issues with the electrospinning procedure. On the other hand, high humidity can promote the development of pores in the nanofibers and lead to an increase in their diameters [29,69].

Ghobeira et al. evaluated the effect of relative humidity on the diameter of PCL fibers, where they varied the humidity between 35% and 65%, keeping the other parameters constant. For a humidity value of 35%, the fibers had average diameters of 249 nm. As the humidity increased to 65%, there was a progressive increase in the diameter of the PCL fibers to 841 nm (Figure 9). This result can be explained by the presence of more water molecules between the needle and the collector, which reduces the load on the polymer jet. Consequently, there is a decrease in the intensity of the electric field, which will limit the proper elongation of the jet, resulting in fibers with larger diameters [68]. The effect of the humidity parameter was also evaluated by Casper et al. [72] and by Medeiros et al. for polystyrene fibers [73], where they found that increasing humidity promoted the production of nanofibers with larger diameters.

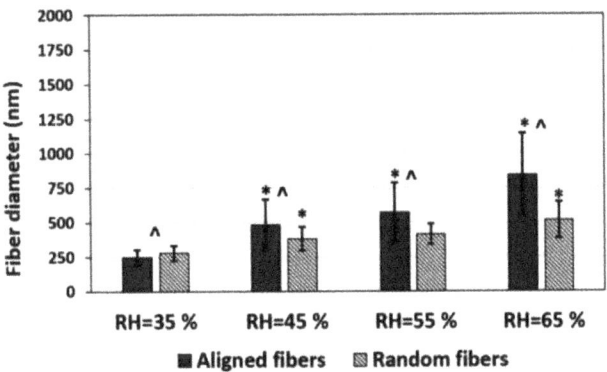

Figure 9. Effect of ambient humidity on the diameter of PCL fibers produced by electrospinning. (*: statistically significant difference between fibers electrospun with 2 different and consecutive parameters within the same aligned or random fiber group; ^: statistically significant difference between aligned and random fibers electrospun with the same parameter; $p < 0.05$). Reproduced with permission from reference [68].

Thus, by examining previous studies, it can be concluded that in the electrospinning process, an increase in humidity favors the development of fibers with larger diameters, associated with an increase in water molecules in the atmosphere, and consequently results in changes in the elongation of the polymer jet. On the other hand, very low humidity values mean that the solvent evaporates very quickly, preventing the correct formation of the jet by drying the solution on the needle tip [29,74].

4. Electrospun Membranes for Water Treatment

The choice of polymers used in the electrospinning solution influences the solution's viscosity, electrical conductivity, and surface tension, among other properties [75]. The surface tension of the spinning solution must be overcome by the supplied voltage, and lowering the surface tension promotes the creation of fibers free of granules [76].

Differentiated nanofibers and membranes are required for different applications, and the tunability of these materials may be limited by the use of a single polymer. By using the technology of electrospinning, polymers with various characteristics can be combined and spun together in a container to enhance membrane performance [77]. Because functionality from individual components can be integrated, nanofibers generated from a mixture of polymers can also lead to novel applications [76].

4.1. Synthetic Polymers

Among the various materials used to manufacture nanofibrous membranes by electrospinning, synthetic polymers including polyacrylonitrile (PAN), polyvinyl acetate (PVA), polyvinylpyrrolidone (PVP), polyetherimide (PEI), or polycaprolactone (PCL) have been the focus of much attention due to their physicochemical stability, film-forming properties, and spinning capacity. Table 1 shows some examples of the most common materials used to produce nanofibrous membranes with possible application in textile wastewater purification.

Table 1. Examples of electrospun membranes for water treatment.

Membrane Composition	Results	Ref.
Polyacrylonitrile (PAN)	Retention of 99.99% of bacteriophages and 99.9999% of bacteria.	[78]
PAN Haloisite nanotubes (HNTs)	The incorporation of HNTs, especially 1% w/w, improved the mechanical and thermal properties of the membranes. Rejection rate of 99.5% of oil/water for membranes with 3% w/w HNTs, and removal efficiency of 31.1% of heavy metal ions.	[79]
PAN TIME Jute cellulose nanowhiskers	Good mechanical properties, efficient nanoparticle filtration capacity, and good oil/water separation (with a rejection rate of over 99%).	[80]
PAN Diethylenetriamine (DETA)	Maximum adsorption capacities: methylene blue—184.84 mg/g; rhodamine B—367.65 mg/g; safratin T—195.7 mg/g. The membrane showed higher maximum adsorption capacities when compared to conventional adsorbents.	[81]
PAN Structure of zeolitic imidazole-67 (ZIF-67)	Maximum adsorption capacities of malachite green: ZIF-67 membranes—2545 mg/g; ZIF-67/PAN membrane—1305 mg/g. After four regeneration cycles, the ZIF-67/PAN membrane showed more than 92% of its original capacity. It also showed good adsorption abilities for Congo red (849 mg/g) and fuchsin (730 mg/g). The membrane can be reused by washing it with ethanol.	[82]
PAN Graphene oxide (GO) Titanium dioxide (TiO_2) β-cyclodextrin (β-CD)	In 5 h, the degradation efficiency for methyl orange and methylene blue was around 93.52% and 90.92%, respectively. The membranes' MB and MO degradation efficiency was 80% for the first three cycles, but dropped to around 68.42% and 65.13% in the fifth cycle, respectively. Antibacterial properties against *E. coli* and *S. aureus*.	[83]
PAN Carbon nanotubes (CNTs)	Almost complete degradation after 120 min and 60 min for methylene blue and indigo carmine, respectively. Improvements of 38% and 84% in tensile strength and elastic modulus, respectively, with just 0.05 wt% CNTs.	[84]
Polyacrylonitrile-co-maleic acid (PANCMA) GO TiO_2	Under optimized conditions, by the E-spun RGO/TiO_2/PANCMA NFs, 90.6% of malachite green and 93.7% of leucomalachite green were adsorbed in 2 min, and subsequently 91.4% and 95.2% adsorbed were degraded in 60 min under UV irradiation, respectively. Good recyclability. Before the 14th cycle, the removal efficiencies of malachite green and leucomalachite were over 91%.	[85]
PAN GO	High rejection performance (almost 100% rejection of Congo red, 56.7% for Na_2SO_4, and 9.8% for NaCl). The water flow under extremely low external pressure (1.0 bar) increased significantly due to the structure of the graphene oxide layer and the nanofibrous support.	[86]

Table 1. *Cont.*

Membrane Composition	Results	Ref.
PAN Cellulose acetate (CA)	The optimum solution pH values for the adsorption of Fe(III), Cu(II), and Cd(II) ions were 2, 5, and 6, respectively, and the adsorption equilibria were obtained in 5, 20, and 60 min. The amount of saturation adsorption of the nanofibrous membranes (at 25 °C) for Fe(III), Cu(II), and Cd(II) was 7.47, 4.26, and 1.13 mmol/g, respectively. After five consecutive adsorption and desorption tests, the desorption rate of the metal ions maintained more than 80% of their first desorption rate. The AOPAN/RC nanofibers showed excellent regeneration capacity.	[87]
PAN ZIF-8	With relatively larger surface areas (of 871.0 m^2/g) and adequate pore sizes (from around 0.6 to 0.8 nm), the nanofibers exhibited greater Cr(VI) adsorption capacity (with q_{max} of 39.68 mg/g) and good recyclability.	[88]
PAN Polyaniline (PANI)	The maximum adsorption capacities for lead and $Cr_2O_7^{2-}$ on the PANI-coated membranes were 290.12 and 1202.53 mg/g, respectively. Greater removal of lead ions (99%) compared to chromium (VI) ions (90%) at 5 mg/L. The PAN/PANI membrane retained almost 58% and 60% of its initial adsorption capacity after four cycles for $Cr_2O_7^{2-}$ and Pb(II).	[89]
Polyvinyl acetate (PVA) 1,2,3,4 butanetetracarboxylic acid (BTAC) crosslinked	Good performance in adsorbing the dye Reactive red 141. The maximum adsorption capacity reached 88.31 mg/g. If the temperature is increased from 110 °C to 130 °C, the adsorption capacity decreases.	[90]
PVA Silica (SiO$_2$) Chitosan	The addition of 1.0% wt SiO$_2$ resulted in a significant improvement in dye rejection and water permeability. Under 0.4 bar transmembrane pressure, the improved nanocomposite membrane yielded 98% Direct Red 23 rejection with a water flux value as high as 1711 L/m^{2}h. It was discovered that the membranes were reusable and antifouling.	[91]
PVA SiO$_2$ Cyclodextrin	The maximum adsorption capacity for the indigo carmine dye reached 495 mg/g and adsorption equilibrium was reached in less than 40 min. Recycled through acidification.	[92]
PVA Chitosan	The maximum adsorption capacity was 266.12 mg/g (Pb(II)) and 148.79 mg/g (Cd(II)). Detailed adsorption studies were carried out at pH 8 and 6 for Cd(II) and Pb(II), respectively. It is a simpler and more sustainable process than conventional methods.	[93]
PVA Konjac glucomannan (KGM) Zinc oxide nanoparticles (ZnO NPs)	Filtration efficiency for ultrafine particles (300 nm) of over 99.99%, superior to commercial HEPA filters. Methyl orange removal efficiency of over 98%, with an initial concentration of 20 mg/L, during 120 min of solar irradiation. Antibacterial activity (*E. coli* and *Bacillus subtilis*).	[94]
Polyvinylpyrrolidone (PVP) Copper (II) acetate hydrate Zinc (II) acetate	A total of 100% degradation of mixed dyes (methylene blue, rhodamine B, and methyl orange, 10 ppm each) in 90 min. Good reusability (94.1% after five cycles).	[95]
PVP Graphitic carbon nitride (g-C$_3$N$_4$) Niobium pentoxide (Nb$_2$O$_5$)	After 120 min in visible light, 98.1% degradation was recorded for rhodamine B and phenol (10 mg/L each). No obvious change in the performance of the nanofibers was recorded after four cycles (remained $\approx$ 98%).	[96]
PVP Zinc oxide (ZnO) Bismuth oxide (Bi$_2$O$_3$)	The compound with a molar ratio of 23:1 (ZnO/Bi$_2$O$_3$) showed the best activity under both excitations (UV and visible light). Approximately 95% degradation of rhodamine B (1.0 $\times$ 10^{-5} M, 60 mL) was reported after 90 min.	[97]

Table 1. Cont.

Membrane Composition	Results	Ref.
PVP Zinc tungstate (ZnWO$_4$)	The degradation efficiency of rhodamine B (10 mg/L) was over 90% in about 90 min of irradiation. There was no decline in photocatalytic activity after five photodegradation cycles.	[98]
Lacase Polyetherimide (PEI) Polycaprolactone (PCL)	After ten cycles, PCL/PEI/TTL and PCL/PEI/TVL had residual activities of 33.2 ± 0.2% and 26.0 ± 0.9%, respectively. At 50 °C and pH 5, PCL/PEI/TTL demonstrated the highest decolorization efficiency of orange II and malachite green, reaching over 86% and 46%, respectively, after eight continuous uses. PCL/PEI/TTL and PCL/PEI/TVL had maximum removal efficiencies of 64.5 ± 7.6% and 52.6 ± 0.1%, respectively, and successfully decomposed 2,6-dichlorophenol. Environmentally friendly, sustainable materials.	[99]
PEI TiO$_2$	The PEI membrane modified with 0.2% TiO$_2$ achieved a significant removal rate of E. coli (99%) and humic acid (≈80%). Degradation of 85% of methylene blue during the photocatalytic process.	[100]
Polystyrene (PS) GO	PSGO films had a removal capacity ≈ 2.3 times higher than that of pure PS membranes. After 120 min, the equilibrium value of the adsorption capacity (qe = 114 mg/g) was reached for all of the methylene blue concentrations that were examined. After the first cycle, the removal capacity was reduced to ≈65%, a value that became constant during subsequent cycles (up to a maximum of five cycles).	[101]
PEN Bisphenol A (BPA) Hydroquinone methanesulfonic acid potassium salt (HQS) 2,6-difluorobenzonitrile (DFBN)	Methylene blue exhibited a high adsorption capacity of 796.25 mg/g. Even after eight separation–regeneration cycles, the optimized membrane achieved a 99% selective removal efficiency of cationic dyes. Good recyclability and stability at high temperatures.	[102]

The electrospinning approach has effectively converted many synthetic polymers into nanofibers [103], with polyacrylonitrile (PAN) being of particular interest for the adsorption of heavy metal ions. This is due to the fact that amidoxime groups (-C(NH$_2$)=NOH) are easily formed from nitrile groups (-C≡N) in PAN by reacting with hydroxylamine in an aqueous solution at room temperature [104–106]. The resulting amidoxime groups have a strong ability to adsorb heavy metal ions through the formation of coordination/chelation bonds [107–109]. PAN is usually combined with other materials to enhance its characteristics. In the study by Feng et al., an innovative nanomaterial was prepared and used for the removal of heavy metal ions from wastewater. Polyacrylonitrile/cellulose acetate (PAN/CA) nanofibrous membranes were created by electrospinning, and subsequently, amidoxime polyacrylonitrile/regenerated cellulose (AOPAN/RC) membranes were synthesized through a combination of hydrolysis and amidoximation modification. After characterizing the membranes, they found that the optimum solution pH values for the adsorption of Fe(III), Cu(II), and Cd(II) ions were 2, 5, and 6, respectively. At 25 °C, the nanofibrous membranes' saturation adsorption amounts for Fe(III), Cu(II), and Cd(II) were 7.47, 4.26, and 1.13 mmol/g, respectively, with adsorption equilibrium being reached in 5, 20, and 60 min (Figure 10). After five consecutive adsorption and desorption tests, the desorption rate of the metal ions remained more than 80% of their first desorption rate [87].

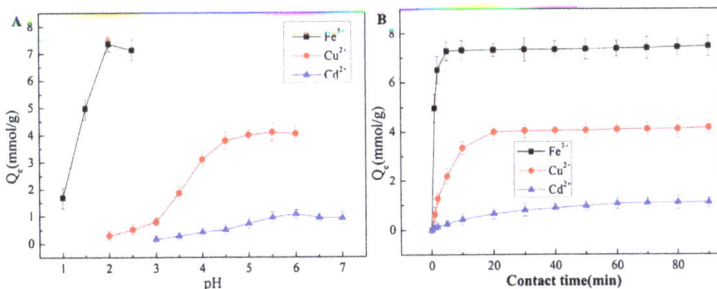

Figure 10. Effect of solution pH value (**A**) and contact time (**B**) on adsorption capacity of different heavy metal ions. During the experiments, the initial concentrations of Fe(III), Cu(II), and Cd(II) ions were 100 mmol/L. Each of the reported data is the average value of 3 replicas, and an error bar represents one standard deviation. Reprinted with permission from reference [87].

Given its exceptional mechanical, thermal, and chemical resistance, polyacrylonitrile (PAN) has become a material that is frequently utilized in the water treatment industry. Since electrospun PAN nanofibers range in size from tens to hundreds of nanometers, they have excellent ion adsorption capability as well as tremendous potential for use in the adsorption of organic dyes due to their high surface area and better surface-to-volume ratio [110]. In a study by Du et al., RGO/TiO$_2$/PANCMA composite nanofibers were obtained by electrospinning the dispersive solution of PANCMA, GO, and TiO$_2$, followed by post-chemical reduction. Under optimized conditions, by E-spun RGO/TiO$_2$/PANCMA nanofibers, 90.6% of malachite green and 93.7% of leucomalachite green were adsorbed in 2 min, and subsequently, 91.4% and 95.2% of adsorbed malachite green and leucomalachite green were degraded in 60 min under UV irradiation, respectively. In addition, the nanofibers showed good recyclability and can be reused in several cycles of operations for adsorption and photocatalytic degradation. In fact, before the 14th cycle, the removal efficiencies of malachite green and leucomalachite were over 91% [85]. Zhang et al. used the cationic dye methylene blue (MB) and the anionic dye methyl orange (MO) as model pollutants to investigate the photocatalytic activity of PAN/β-cyclodextrin (β-CD)/TiO$_2$/GO nanofibrous membranes, which were produced by electrospinning in conjunction with ultrasound-assisted electrospinning. The membrane that demonstrated the highest dye degradation efficiency was the one with a minimum diameter of 84.66 ± 40.58 nm and an 8:2 mass ratio of TiO$_2$ to GO. In 5 h, the degradation efficiencies for methyl orange and methylene blue were approximately 90.92% and 93.52%, respectively. The membranes maintained an 80% degradation efficiency of MB and MO over the course of three cycles, declining to around 68.42% and 65.13%, respectively, in the fifth cycle. Additionally, the PAN/β-CD/TiO$_2$/GO membrane demonstrated antibacterial properties against *Staphylococcus aureus* and *Escherichia coli* [83].

The potential of metal–organic frameworks (MOFs) for dye adsorption is also very significant. However, the actual use of MOFs is hindered by the difficulty of removing MOF powder from an aqueous solution. The incorporation of MOFs into polymeric fiber membranes offers a novel approach to attain outcomes that preserve MOFs' dye adsorption capability while facilitating the facile extraction of the MOF/dye complex from aqueous solutions. Ji et al. recently used the electrospinning method to successfully construct a zeolitic fibrous membrane of imidazole-67/polyacrylonitrile (ZIF-67/PAN) with a ZIF-67 loading ratio of 54%. The ZIF-67 and ZIF-67/PAN membranes have maximal malachite green adsorption capacities of 2545 and 1305 mg/g, respectively. The ZIF-67/PAN membrane retained more than 92% of its original fibers after four regeneration cycles. In addition, the ZIF-67/PAN fibers also showed good adsorption abilities for Congo red (849 mg/g) and fuchsin (730 mg/g). After being cleaned with ethanol, the membrane could be utilized again. Therefore, because of their straightforward manufacturing process, superior adsorption qualities, ease of separation, and advantageous reusability, ZIF-67/PAN fibers seem to be attractive adsorbents for the removal of dyes in industrial applications [82].

Polyvinyl acetate (PVA), a biocompatible polymer with good mechanical properties and biodegradability, has been widely used to form homogeneous miscible systems with strong chemical stability, film-forming ability, and high hydrophilicity. Furthermore, PVA's side chain contains a greater number of hydroxyl groups, which increases its solubility in water [111]. Therefore, various materials can be hybridized with PVA to form nanofiber composites to be used as an economical adsorbent for ions and dyes from wastewater. As stated in an article by Karim et al., lead (Pb(II)) and cadmium (Cd(II)) ions were selectively and highly adsorbed onto nanofiber membranes made of polyvinyl acetate/chitosan (PVA/Chi) using the electrospinning technique, depending on how acidic the solution was (Figure 11).

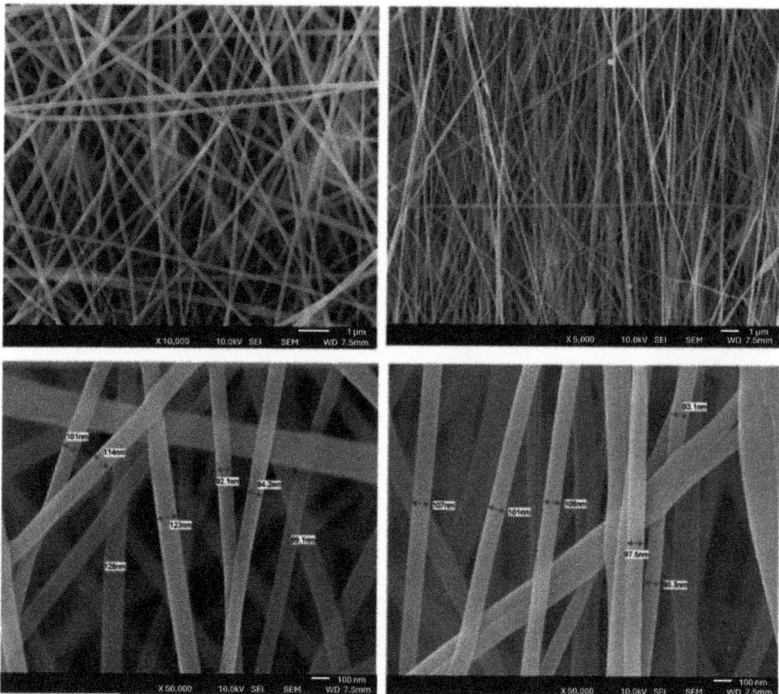

Figure 11. FESEM images of PVA/Chi nanofibers membranes. Adapted with permission from reference [93].

Under ideal conditions, as seen in Figure 12, the maximum adsorption capacity was 266.12 mg/g (Pb(II)) and 148.79 mg/g (Cd(II)). For Cd(II) and Pb(II), in-depth adsorption investigations were conducted at pH 8 and 6, respectively. The PVA/Chi membranes' ability to adsorb Pb(II) or Cd(II) ions was unaffected by foreign ions. Thus, PVA/Chi nanofiber membranes (70:30 PVA/Chi ratio) produced by electrospinning are considered effective and promising for removing Pb(II) and Cd(II) ions from wastewater with high efficiency. Furthermore, this method is a simpler and more sustainable process than conventional methods [93].

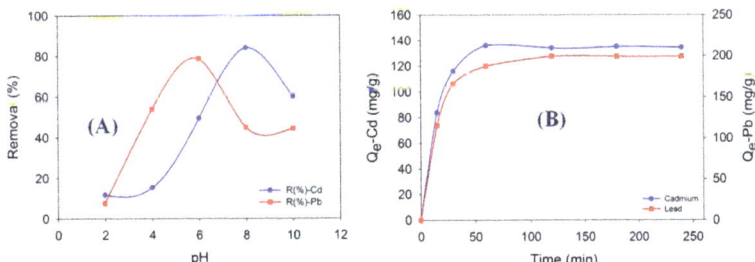

Figure 12. (**A**) Percentage adsorption of Cd(II) and Pb(II) ions at different pH ranges. (Conditions: 25 mg adsorbent, 10 mL of 400 mg/L solution of Pb(II)/Cd(II) ions, and contact time = 60 min.) (**B**) Adsorption of Cd(II) and Pb(II) ions on PVA/Chi NFs membrane. (Conditions: 0.025 g adsorbent, 10 mL of 400 mg/L solution of heavy-metal ions, and contact time = 5–240 min.) Reproduced with permission from reference [93].

Hosseini et al. developed innovative electrospun nanofibrous membranes with PVA, chitosan, and SiO_2 nanocomposite materials to enhance their mechanical strength and permeability capabilities. The investigation focused on the morphology, fiber diameter, porosity, thermomechanical characteristics, and permeability of the synthesized membranes in relation to different concentrations of SiO_2 in the spinning solution (0—M1, 0.5—M2, 1.0—M3, and 2.0—M4 wt%). The affinity membranes that were produced were used to extract dye from wastewater. It was discovered that adding SiO_2 as a reinforcing ingredient increased the nanocomposite membranes' resilience. The produced membranes' Young's modulus nearly doubled, from 0.74 MPa for PVA/chitosan to 1.69 MPa for nanocomposite membranes with the addition of 0.5% wt SiO_2. The main discovery was that the addition of 1.0% wt SiO_2 resulted in a significant improvement in dye rejection and water permeability, as seen in Figure 13. Under 0.4 bar transmembrane pressure, the improved nanocomposite membrane yielded 98% Direct Red 23 rejection with a water flux value as high as 1711 L/m^{2}h. It was also discovered that the membranes were reusable and antifouling [91].

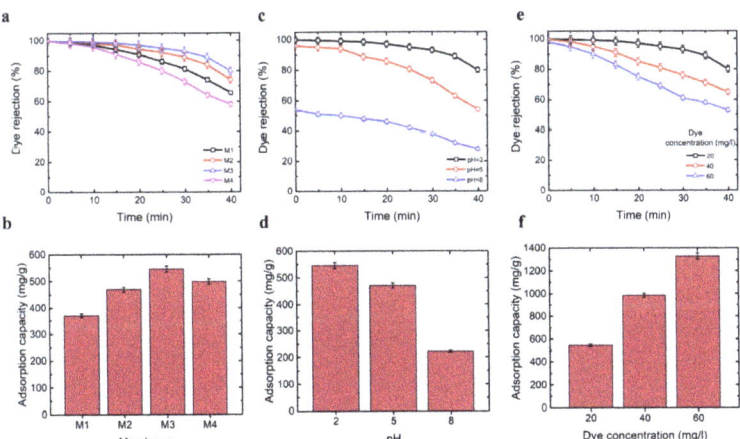

Figure 13. Effect of SiO_2 concentration on (**a**) dye removal and (**b**) adsorption capacity (pH = 2 and dye concentration = 20 mg/L). Effect of pH on (**c**) dye removal and (**d**) adsorption capacity (dye concentration = 20 mg/L). Effect of dye concentration on (**e**) dye removal and (**f**) adsorption capacity (pH = 2). Reprinted with permission from reference [91].

PVA/SiO_2 composite nanofibers including cyclodextrin-functionalized mesostructures were synthesized, as reported by Teng et al. The preparation procedure combined the use of the electrospinning technique with the sol–gel process. The indigo carmine dye was effectively adsorbed by the PVA/SiO$_2$ nanofiber membranes. In less than 40 min, the adsorption equilibrium was reached, and the maximum adsorption capacity was 495 mg/g. Furthermore, for practical usage, the membranes offer good recycling capabilities. It is also convenient to fix and recover the mesoporous membranes. As a result, this novel substance may offer a new method for removing dye molecules [92]. Lv and colleagues synthesized zinc oxide nanoparticle-loaded nanofiber membranes using PVA and konjac glucomannan (KGM) by means of environmentally friendly thermal crosslinking and green electrospinning. ZnO@PVA/KGM membranes exhibited a filtration performance of over 99.99% for ultrafine particles (300 nm), surpassing that of commercial HEPA filters. After 120 min of solar light, methyl orange was effectively decolorized with a clearance efficiency of more than 98% at an initial concentration of 20 mg/L. The resultant fibrous membranes exhibited enhanced photocatalytic and antibacterial activity against Escherichia coli and Bacillus subtilis, in addition to their efficiency in air filtration [94].

Polyvinylpyrrolidone (PVP), a very water-soluble polymer, is used to produce systems that allow for the degradation of dyes from aqueous solutions, serving as a base for other components. Z-type-CuS/ZnS@PVP nanofibers were created by Sitinjak et al. using electrospinning with the intention of converting 4-nitrophenol to 4-aminophenol and degrading the mixtures of methyl orange, rhodamine B, and methylene blue when exposed to sun light. In 2 h, 4-nitrophenol was reduced to 4-aminophenol, and in 90 min, the combined dyes (10 ppm each) were completely degraded (Figure 14). The nanofibers showed excellent stability, with good reuse (94.1% after five cycles) of the catalyst [95].

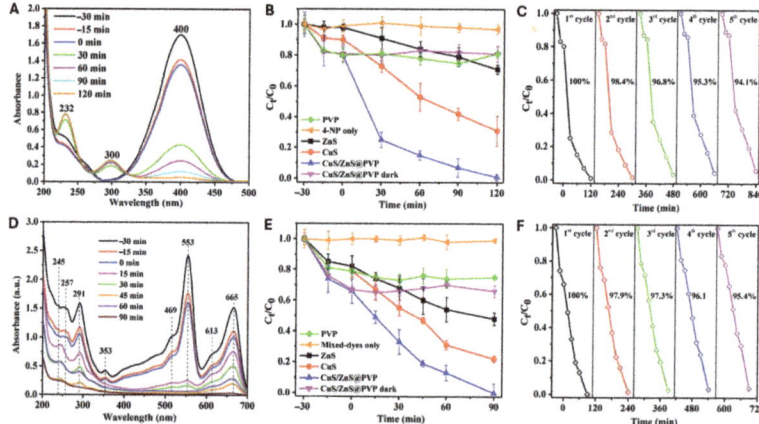

Figure 14. UV–vis spectra showing the conversion of (**A**) 4-nitrophenol and (**D**) mixed dyes. Plots of C_t/C_0 vs. reaction time comparing PVP only, organic pollutants only, and CuS, ZnS, and CuS/ZnS@PVP nanofibers to convert (**B**) 4-nitrophenol and (**E**) mixed dyes. The reusability performances after five cycles of CuS/ZnS@PVP nanofibers on (**C**) 4-nitrophenol and (**F**) mixed dyes. Reprinted with permission from reference [95].

In the work of Lu et al., zinc tungstate fibers (ZnWO$_4$) with PVP were successfully produced by electrospinning. A total of 100 mL of aqueous rhodamine B solution (10 mg/L) was used to test the photocatalytic activity of the fibers under solar irradiation. The degradation efficiency was over 90% within about 90 min of irradiation. In addition, there was no decline in photocatalytic activity after five photodegradation cycles [98].

Not widely used in this area, polyarylene ether nitrile (PEN) is a type of high-performance thermoplastic with adaptable molecular structures, excellent mechanical properties, chemical and high-temperature resistance, good spinnability, and biocompatibility [112]. In the study by Li et al., nanofibrous membranes named PEN(B3S7), PEN(B5S5), and PEN(B7S3) were produced by electrospinning PEN, bisphenol (BPA), hydroquinone monosulfonic acid potassium salt (HQS), and 2,6-difluorobenzonitrile (DFBN). When it came to the cationic organic dye methylene blue, the optimized nanofibers demonstrated a high adsorption capacity of 796.25 mg/g. Even after eight separation–regeneration cycles, the cationic dyes' removal efficiency reached 99% thanks to the improved polymer membrane, which enabled the quick and selective removal of the dyes from mixtures containing other dye molecules [102]. Thus, the optimized membranes made of PEN showed good performance in removing dyes under adverse conditions and an easy regeneration characteristic, as well as the possibility of reuse. Based on these characteristics, it is plausible to say that this observation provides new information for the development of advanced nanofibrous adsorbents and separation membranes for the purification of wastewater contaminated with organic dyes.

4.2. Natural Polymers

Bio-based polymers have been used as effective substitutes for synthetic polymers to reduce their negative effects on the environment because of their benign and ecological properties as well as their potential for commercial use [113]. With a unique chemical composition, biocompatibility, biodegradability, and functional chemical groups, such as hydroxyl, amine, and carboxyl groups, biopolymers have great potential for wastewater treatment by removing dyes and metal ions through various mechanisms, such as chelation, electrostatic attraction, and ion exchange [113,114].

4.2.1. Cellulose Acetate

Cellulose acetate (CA) is the most researched derivative of cellulose since it may be deacetylated to regenerate pure cellulose. [115,116]. The process of acetylation, which involves substituting glucose units for cellulose hydroxyl groups to a degree of between two and four on average, produces cellulose acetate. For more than 70 years, CA has been effectively used in membrane filtration [117–119]. It has been extensively employed in nanofiltration and ultrafiltration as a selective layer. Considered among the most effective metal adsorbents, it has distinct functional groups that are modifiable [117,120]. The functionalization of CA with -COOH, -SO_3H, and NH_2 offers an opportunity for the application of CA in the complexation of heavy metals [121–124]. To improve a cellulose acetate membrane's ability to adsorb metals, certain nanofillers can be added [117,121,122].

Taha et al. successfully produced cellulose acetate/silica nanofibrous membranes functionalized with NH_2 through the sol–gel process combined with electrospinning technology. The membranes were used to remove Cr(VI) from aqueous solutions. The Langmuir adsorption model provides a good description of Cr(VI)'s adsorption behavior. At pH = 1, Cr(VI)'s maximum adsorption capacity was calculated to be 19.46 mg/g. After 60 min, for a Cr(VI) solution concentration of 10 ppm, the membrane's adsorption efficiency was 98%. The membrane can also be regenerated through alkalinization [122]. In a project by Zayadi et al., a palm nanoleaf titanium fiber membrane (Nano-PFTF) was fabricated using cellulose acetate (CA) derived from palm leaf oil (OPF). The nanofiber membrane combined the adsorption and photocatalytic degradation of pollutants by nitrogen-doped titanium dioxide (N-TiO_2). Under visible light and UV radiation, the Nano-PTFT membrane (CA/N-TiO_2) achieved a 96.72% rejection of 10 ppm MB (methylene blue) and a 99% rejection of 10 ppm Cr(VI) in 120 min, respectively. In addition to separating the pollutants from the water, the membrane simultaneously reduced the pollutants with the presence of the photocatalyst N-TiO_2 [125].

Cheng et al. successfully developed a uniform coating layer of polydopamine (PDA) on the surface of deacetylated cellulose acetate (DA) nanofibers using electrospinning. In order to extract methylene blue (MB) from an aqueous solution, the membrane was used as a highly effective adsorbent. After 30 h of adsorption at 25 °C and pH 6.5, the DA@PDA nanofiber membrane's capacity to adsorb the MB dye was 88.2 mg/g, around 8.6 times higher than that of DA nanofibers, as seen in Figure 15 [126].

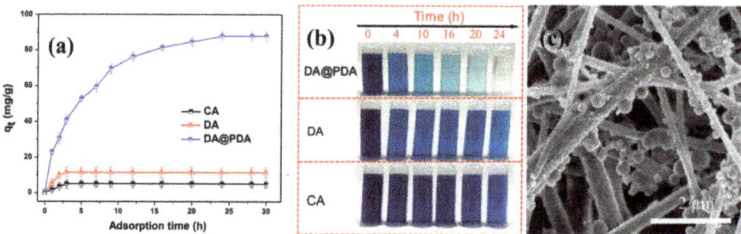

Figure 15. (**a**) Adsorption capacity of CA, DA, and DA@PDA nanofiber membranes with increasing adsorption time for the adsorption of MB dye. (**b**) Digital photographs of the MB solution after being immersed in the representative CA, DA, and DA@PDA nanofiber membranes. (**c**) SEM image of the DA@PDA composite nanofiber after MB adsorption for 24 h. (Adsorption conditions: original MB concentration was 50 mg/L, weight of adsorbent was 10 mg, temperature was 298 K, and pH was 6.5). Reproduced with permission from reference [126].

In a study by Akduman et al., cellulose acetate nanofiber membranes incorporated with diatomite (DE) were produced by electrospinning. A range of concentrations of 10, 20, and 30% DE (w/w polymer) was employed to investigate how different quantities of DE affected the removal of the dye Reactive Red 141. As a control, pure AC nanofibers were prepared by electrospinning. The equilibrium adsorption capacity of the AC, AC-10DE, AC-20DE, and AC-30DE nanofiber membranes after 24 h was found to be 66.26 ± 3.57, 67.83 ± 3.62, 70.71 ± 3.13, and 72.27 ± 2.90 mg/g, respectively, for an initial dye concentration of 85 mg/L [127].

Furthermore, San and colleagues discovered that the electrospun web of cellulose acetate nanofibers (AC-WNF) effectively immobilized bacterial cells. Three different species of bacteria (*Aeromonas eucrenophila*, *Clavibacter michiganensis*, and *Pseudomonas aeruginosa*) immobilized on AC-WNF were used to decolorize methylene blue (MB) dye in aqueous conditions. Effective decolorization of the MB dye was achieved in 24 h and the removal rate was 95%. The web showed good reusability; approximately 45% of the dye's decolorization capacity was obtained at the end of the fourth cycle. More precisely, for *A. eucrenophila*, *C. michiganensis*, and *P. aeruginosa*, MB decolorization decreased to 45.7%, 43.1%, and 48.04%, respectively [128].

4.2.2. Chitosan

Following cellulose, chitin is the second most prevalent natural polymer on the planet. Chitin is either deacetylated in an alkaline environment or hydrolyzed enzymatically in the presence of chitin deacetylase (CDA) to produce chitosan. It is an aminopolysaccharide with particular characteristics and functions that have a variety of uses, including industrial and biomedical. Chitosan is a copolymer composed of 2-acetamido-2-deoxy-β-D-glucopyranose and 2-amino-2-deoxy-β-glucopyranose connected by (1-4)-β-glycosidic linkages. For a variety of uses, many forms of chitin and chitosan have been created, including gels, membranes, spheres, microparticles, nanoparticles, and nanofibers [129,130]. Chitosan is well known for its unique antimicrobial activity and its capacity to adsorb metals [131,132].

Li et al. produced pure chitosan membranes through electrospinning with average fiber diameters of 86 ± 18, 114 ± 17, and 164 ± 28 nm. Batch adsorption experiments were carried out using the chitosan nanofiber membranes as an adsorbent to remove acid blue-113. The chitosan membrane, with an average fiber diameter of 86 nm, had an adsorption capacity of 1377 mg/g, greater than that of the microscale chitosan sample, which had an adsorption capacity of 412 mg/g. Furthermore, following four cycles, the membranes demonstrated good regeneration. Adsorption capacity following the fourth regeneration cycle was 596.6 mg/g [133]. Likewise, Haider et al. produced chitosan membranes by electrospinning with nanofibers with a diameter of approximately 235 nm, but for the adsorption of ions from an aqueous solution. Cu(II) and Pb(II) had equilibrium adsorption capacities of 185.44 mg/g (2.85 mmol/g) and 263.15 mg/g (0.79 mmol/g), respectively, higher than the 45.20 mg/g value for the removal of Cu(II) ions from an aqueous solution using chitosan pellets [134]. Also, Razzaz et al. developed chitosan membranes with TiO$_2$ nanoparticles (NPs) incorporated by electrospinning. In a batch setup, the produced nanofibers' potential for Pb(II) and Cu(II) ion removal was examined. Cu(II) and Pb(II) ions had maximal adsorption capacities of 710.3 and 579.1 mg/g at 45 °C and 30 min, respectively, at equilibrium. The chitosan/TiO$_2$ nanofibers could be reused frequently without significant loss of adsorption performance during five adsorption/desorption cycles (>80%), as shown in Figure 16 [135].

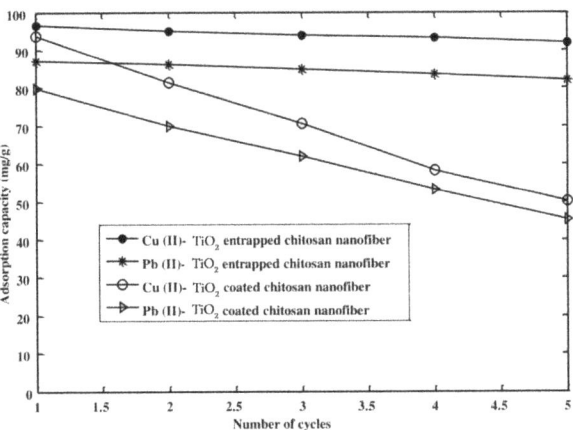

Figure 16. Five cycles of adsorption–desorption of Pb (II) and Cu (II) ions using chitosan/TiO$_2$ nanofibrous adsorbents. Reprinted with permission from reference [135].

Mokhena et al. obtained a thin film of chitosan, with and without silver nanoparticles (AgNPs), on alginate nanofibers produced by electrospinning. Both membranes, with and without AgNPs, showed a similar flux and a high rejection of nanoparticles (>98%) and oil (>93%). Over the course of the five filtration cycles examined, the inclusion of AgNPs increased the dye Congo red's rejection, achieving more than 95% rejection. The membranes demonstrated strong antibacterial action against Escherichia coli and Staphylococcus aureus as a result of the NPs [136].

4.2.3. Alginate

Originating from seaweed, alginate is a widely recognized polyelectrolyte binary copolymer [137–140]. It contributes to the flexibility and resistance of marine algae against the adverse forces of water. It is typically found in the intercellular substance (mucilage) and cell wall matrix. Alginate is a linear polysaccharide consisting of 1–4 glycosidic linkages connecting D-mannuronic (M) and L-guluronic (G) units. These units depend on the source or species, growing conditions, season, and extraction depth, and they show up along the polymer chain in different sequences and proportions (M/G). The physical characteristics and reactivity of alginate are determined by the change in M and G along the alginate chain. It has been discovered that the affinity of alginates for heavy metals is significantly influenced by functional groups, particularly carbonate ions, and the molecule shape [140]. When alginate comes into contact with divalent or polyvalent metal ions, it can gel at ambient temperature. For specific applications, this phenomenon has been used to prepare various morphologies and structures. Spheres [141], films [142,143], and hydrogels [144], as well as porous membranes and nanofibers [138], have been manufactured for different applications, such as wound dressings and metal adsorption. Due to its antibacterial efficacy and structural resemblance to glycosaminoglycan (GAG), one of the important constituents of the natural extracellular matrix (EMC) present in mammalian tissues, alginate has attracted a great deal of interest for usage in biomedical applications. Furthermore, alginate's availability, non-toxicity, biodegradability, and strong cell compatibility have made it possible to use it in a variety of applications, including metal recovery [140,141].

However, it remains controversial why it is not possible to carry out alginate electrospinning on its own. This has been attributed to the alginate solution's strong conductivity, lack of entanglement, and gelation at low concentrations (below the concentration required for entanglement formation). This is further influenced by the strong surface tension, worm-like molecular structure, and stiffness of the chains. However, Mokhena et al. were able to produce alginate membranes by electrospinning for the biosorption of heavy metals from aqueous solutions. Since the electrospinnability of alginate from its aqueous solution is a problem, polyethylene glycol (PEO) was used to facilitate its spinnability. The membranes showed a maximum monolayer adsorption capacity (Q_0) of 15.6 mg/g at a pH of 4 for Cu(II) ions. In a competitive experiment, the removal of metal ions in the mixture followed the order Cu > Ni > Cd > Co. The removal percentages were 39.2, 37.1, 25.3, and 21.8%, respectively. After five reuse cycles, the adsorption percentage had only decreased by 2% from the initial value [145].

Sodium alginate (SA) is a natural copolymer widely found in brown algae. This polymer's outstanding biocompatibility, biodegradability, and non-toxicity make it useful in biological applications like tissue engineering, drug manufacturing, and other industrial uses. The production of adsorbents for use in adsorption operations is another usage for SA [146]. Unfortunately, because of its rigid structure, poor solubility, and high solution viscosity, sodium alginate presents several challenges when used as an adsorbent in the creation of nanofibers. Combining this polymer with others that have appropriate chain rotation and low viscosity, like polyvinyl acetate (PVA), is one way to solve this issue [147]. The study by Ebrahimi et al. aimed to manufacture nanofibers using polyvinyl acetate (PVA) and sodium alginate (SA) to remove cadmium metal ions from aqueous solutions. To this end, PVA/SA nanofibers (volume ratio 40:60) were produced by electrospinning. The maximum equilibrium adsorption amount for cadmium, under optimal experimental conditions, was 67.05 mg/g [148]. In addition, because sodium alginate is highly soluble in water, SA nanofiber membranes have inadequate stability in aqueous conditions. Thus, proper crosslinking is also a key factor in achieving the practical application of SA membranes in dye adsorption. Currently, the most used crosslinking technique for materials based on alginate is the use of calcium chloride ($CaCl_2$). Wang and colleagues electrospun sodium alginate membranes, which were subsequently crosslinked using calcium chloride ($CaCl_2$). The values of the SA electrospun nanofiber membranes' surface area and pore volume before and after three different crosslinking techniques are displayed in Table 2.

Table 2. BET analysis and mechanical property results of non-crosslinked and differentially crosslinked SA nanofiber membranes. Table reprinted with permission from reference [149].

Sample	Surface Area (m^2/g)	Pore Volume (cm^3/g)	Tensile Strength (MPa)	Elongation at Break (%)
Non-crosslinked	13.97	0.0256	3.8	9.8
$CaCl_2$ crosslinked	13.56	0.0450	10.4	9.9
GA vapor crosslinked	11.86	0.0185	3.7	11.2
TFA crosslinked	15.26	0.0455	3.6	12.3

All of the membranes showed excellent integrated adsorption performance for methylene blue (MB), with a maximum effective adsorption capacity of 2230 mg/g and an adsorption equilibrium time of 50 min (Figure 17). The methylene blue/methyl orange (MB/MO) mixture solution can be separated by the nanofiber membranes based on the selective adsorption of SA, and they can continue to separate the solution with a high separation efficiency even after five cycles (>90%) [149].

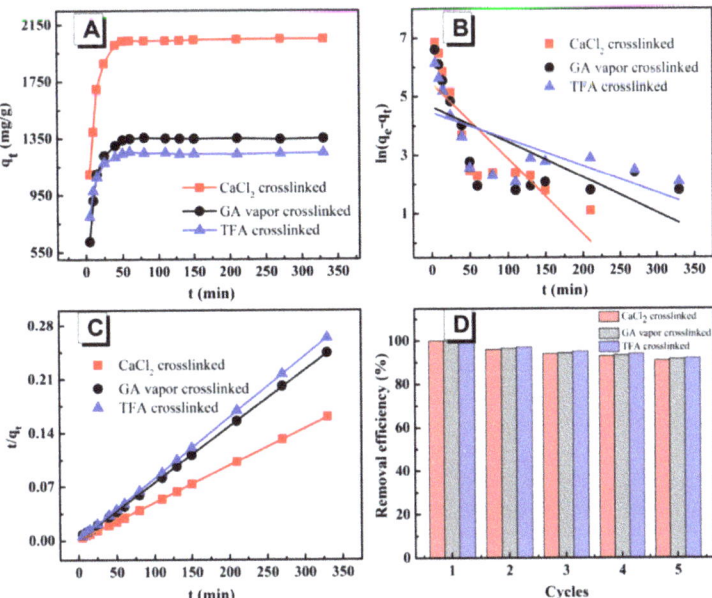

Figure 17. Adsorption kinetics (**A**) and the corresponding pseudo-first-order kinetic plots (**B**); pseudo-second-order kinetic plots (**C**); and adsorption–desorption cycles (**D**) for MB adsorption onto differentially crosslinked SA nanofiber membranes at 298 K. Reproduced with permission from reference [149].

4.2.4. β-Cyclodextrin

β-cyclodextrin (β-CD) is an important polysaccharide [150,151] that has been widely used as an adsorbent [152,153], since its macrocyclic structure can encapsulate a variety of hydrophobic molecules or parts of molecules inside the cavity through non-covalent interactions to form host–guest inclusion complexes [154,155]. It is not possible to directly remove target molecules from water using β-CD molecules since they are soluble in water. Therefore, grafting the β-CD molecules onto an existing substrate is required for water decontamination purposes [156,157] or to produce β-CD polymers that are insoluble [158–160].

Cyclodextrin fibers produced by electrospinning (CD-F) are very attractive materials for encapsulating bacteria for bioremediation purposes. For instance, the encapsulated bacteria utilize cyclodextrin fibers as a food source, in addition to acting as a transport matrix. Keskin et al. demonstrated an easy approach to encapsulate bacteria in a cyclodextrin fiber matrix (CD-F), Figure 18, for application in wastewater treatment. *Lysinibacillus sphaericus*, the bacteria, were encased in cyclodextrin nanofibers made by means of electrospinning in order to treat textile dye (Reactive Black 5, RB5) and heavy metals (nickel (II) and chromium (VI)). The bacteria/CD biocomposite showed Ni(II), Cr(VI), and Reactive Black 5 dye removal efficiencies of 70 ± 0.2%, 58 ± 1.4%, and 82 ± 0.8%, respectively [161].

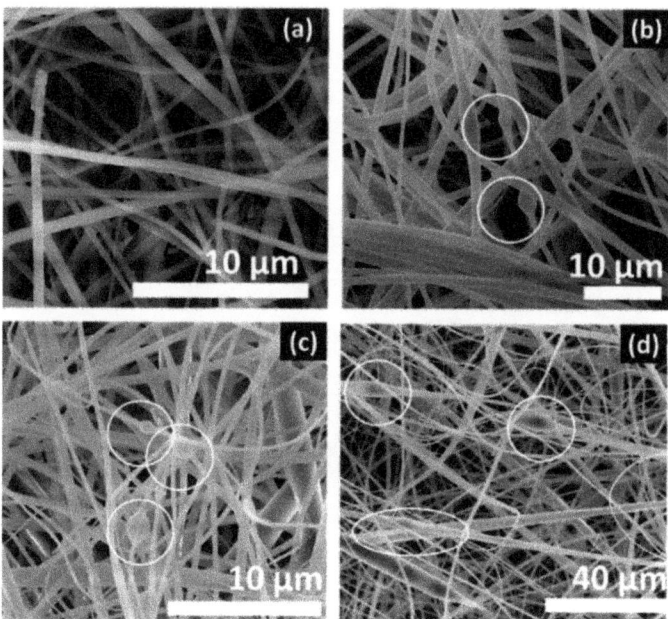

Figure 18. Representative SEM images of cyclodextrin fiber (CD-F) encapsulating (**a**) 0.25% (w/w), (**b**) 0.5% (w/w), (**c**) 1% (w/w), and (**d**) 2% (w/w) of bacteria. The circles show some of the encapsulated bacteria in the electrospun fiber matrix. Reprinted with permission from reference [161].

Zhao et al. synthesized water-insoluble fibers based on β-cyclodextrin by electrospinning followed by thermal crosslinking. With good recyclability, the crosslinked fibers demonstrated a high adsorption capacity for the cationic dye methylene blue (MB). Considering the Langmuir model, the maximal adsorption capacity was 826.45 mg/g. The fibers exhibited poor adsorption toward the negatively charged anionic dye methyl orange (MO) as a result of electrostatic repulsion. The membrane could dynamically filter the MB/MO combination solution at a high flow rate of 150 mL/min based on selective adsorption. Even after five filtration–regeneration cycles (more than 90%), the fibers were still able to retain their excellent shape and great separation efficiency [162]. In their study, Liu et al. described a strategy for a film-type water purifier prepared by including a cyclodextrin oligomer (CD) in ultrathin bacterial cellulose (BC) nanofibers. The membrane showed a high adsorption capacity for four representative organic pollutants, including phenol, bisphenol A (BPA), glyphosate, and 2,4-dichlorophenol (2,4-DCP). The equilibrium adsorption capacity values were attained after 80 min for phenol, BPA, and 2,4-DCP, and 120 min for glyphosate. The values were 90.3 mg/g for phenol, 68.1 mg/g for BPA, 81.3 mg/g for glyphosate, and 222.6 mg/g for 2,4-DCP. The membrane maintained its superior adsorption ability in the presence of different anions and macromolecules and across a broad pH range. More importantly, it can be reused after treatment with methanol under ultrasonication. The removal efficiency in the 5th and 10th cycles (Figure 19) only showed a slight decrease (<5%) compared to the original removal efficiency [163]. When it comes to the number of target pollutants, including phenol, BPA, glyphosate, and 2,4-DCP, the ideal product exhibits an amazing removal capacity that surpasses the majority of adsorbents, including porous carbon-based materials, that have been previously documented.

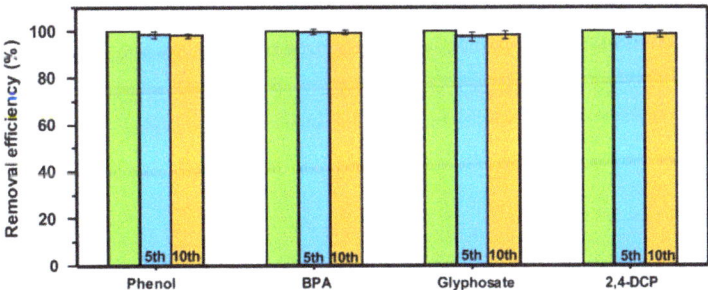

Figure 19. Removal efficiency of BC-CD-2 for each pollutant in the 5th and 10th cycle of the adsorption–desorption experiment. Reproduced with permission from reference [163].

5. Limitations of Textile Wastewater Filtering Structures by Electrospinning

The past decade has seen a significant increase in interest in advanced electrospinning technology because of its many benefits, including a wide range of material options, high porosity (>90%), and strong adaptability. Nevertheless, there are some limitations of this technique owing to the lengthy preparation cycle, the somewhat bigger average membrane pore size, the poor mechanical strength for modulation, making it difficult to produce a large mass volume of electrospun fibers, and long-term operation [6,164,165]. An electrospun nanofibrous membrane's effectiveness might be impacted by several preparation and processing factors, as seen above. Polymer solution characteristics, operating circumstances, and environmental variables are the three primary categories. For instance, the characteristics and functionality of the membranes are greatly influenced by the pore size and thickness of the nanofibers. In general, when the fiber diameter is reduced, the membrane's pore size shrinks as well, reducing water flux and increasing the salt interception rate [166–168]. Additionally, the high cost and accessibility of some of the used materials, the equipment (including maintenance), and the expert operators are among some major drawbacks of this technique [6,164].

Electrospun filtering structures may also show other limitations such as low pressure and temperature sensitivity, wettability, difficult cleaning, microorganism contaminations, and membrane fouling [169–172]. Membrane stability is also quite poorly addressed (there is a lack of long-term experiments in this regard) [164,173,174]. Another drawback of these materials is the use of non-environmentally friendly materials. However, this work already provides some insights into the use of eco-friendly natural polymers.

To increase the efficiency of electrospun membranes, the polymer materials and electrospinning procedures shall be further improved to overcome some of the mentioned limitations. The ecological impact of the membranes, such as their non-toxicity, long-term durability, and degradability properties, must be continuously optimized.

Besides the limitations of the developed structures, there are some literature gaps that shall be addressed. First, other types of polymers, both synthetic and natural, are quite unexplored for wastewater treatment applications. The lack of full-scale or even pilot-scale cases reveals a scalability issue that has to be addressed [25]. The small amount of pilot-scale reports lacks economic and environmental analysis [175–177]. Emerging technologies, such as electrodialysis and membrane bioreactors, are also under-addressed and should be considered in future works.

6. Conclusions and Outlook

For a very long time, access to safe drinking water has been a serious issue across the world. Currently, one of the main sources of pollution is wastewater from textile dyeing. Various combinations of biological, physical, and chemical approaches are used in traditional technologies to treat textile effluent; however, these methods come with substantial capital and operating expenses. Membrane-based technologies provide one of the best alternatives available for large-scale environmentally sustainable treatment processes. In this regard, nanofibers have given the problem of water filtration a new dimension. These nanofibers offer advantageous qualities such as large surface areas, strength, and an ideal pore size. An easier and more practical way of creating nanofibers is electrospinning, which allows us to adjust the pore size of the nanofibers. In the last few decades, there have been innumerable advances in the field of electrospun nanofibrous membranes. Nanofibrous membranes for the treatment of water and wastewater are incredibly promising based on the present view of development and application. This technique offers access to a vast variety of nanomaterials with unique features, such as their eco-friendliness, degradability, renewability, non-toxicity, and mechanical and thermal properties, as well as their availability, long-term durability, and cost-effectiveness.

This paper summarizes the process variables that influence the electrospun nanofibers' physical characteristics and functional capabilities, including large surface area, roughness, porosity, and surface chemistry (hydrophilicity/hydrophobicity), to offer ways to improve their efficacy in water and wastewater purification. We have discussed the most recent developments in the study of water purification using electrospun nanofibers. The most effective composite nanofibers for filtration applications include electrospun PAN, PVA, PVP, and CA nanofibers. The use of other natural polymers, such as chitosan or alginate, that can be extracted from natural sources is a growing field. Electrospun nanofibrous membranes based on these polymers have already been shown as powerful adsorbents for several metal ions and dyes, being a promising alternative to synthetic polymers for the development of textile wastewater filtering structures.

Despite their advantages, electrospun nanofibrous membranes are not without limitations. These include problems with inadequate nanoscale selectivity, mechanical weakness, high cost of and accessibility to the used materials, the equipment (including maintenance), and the expert operators. It is also important to note that an electrospun nanofibrous membrane's effectiveness is impacted by several preparation and processing factors, such as room temperature and humidity. Another constraint is the lack of large-scale studies. Due to the variability in the composition of textile wastewater, the application of electrospun membrane systems in the processing of textile wastewater is difficult, which limits full-scale studies. The economic analysis of the scarce pilot-scale cases reported is insufficient, and the application of this type of material in real cases must be further studied. In order to develop affordable, industrial-scale modules and effective electrospun nanofibrous membranes for water treatment, research organizations and industrial businesses must work together to overcome many obstacles. Technical constraints during the fabrication and operation of nanofiber electrospun membranes, such as membrane stability and mechanical weakness, should be the subject of future research. Future works shall address the scalability issue, and joint efforts between scholars and the industry must be made. Economic and technical analyses are needed in pilot-scale reports, and the environmental impact of the methods and technologies developed should be considered. The use of natural fibers plays a crucial part in the development of more sustainable methodologies for water filtration. The recycling of not only textile residues but also natural (marine, agricultural, etc.) waste can contribute to the development of water treatment technologies in a circular way.

Overall, the application of the electrospinning technique to water treatment technologies provides a great number of opportunities for effectively purifying textile wastewater, promising a cleaner and more sustainable future for the industry. Electrospun membranes do, in fact, have a bright future and are anticipated to play a significant role in the treatment of refractory contaminated water.

Author Contributions: Conceptualization, D.P.F. and R.F.; investigation, D.P.F., J.M.R. and R.P.C.L.S.; writing—original draft preparation, J.M.R. and R.P.C.L.S.; writing—review and editing, D.P.F., R.F., J.M.R. and R.P.C.L.S.; supervision, D.P.F. and R.F. All authors have read and agreed to the published version of the manuscript.

Funding: The authors acknowledge the financial support from the integrated project GIATEX—Gestão Inteligente da Água na ITV (Investment RE-C05-i01.01—Mobilizing Agendas/Alliances for Business Innovation), promoted by the Recovery and Resilience Plan (RRP), Next Generation EU, for the period 2022-2025, and from the European Regional Development Fund through the Operational Competitiveness Program and the National Foundation for Science and Technology of Portugal (FCT) under the projects UID/CTM/00264/2020 of the Centre for Textile Science and Technology (2C2T) on its components base (https://doi.org/10.54499/UIDB/00261/2020, accessed on 10 March 2024) and program (https://doi.org/10.54499/UIDP/00264/2020, accessed on 10 March 2024). The authors are also thankful to project DRI/India/0447/2020 (https://doi.org/10.54499/DRI/India/0447/2020, accessed on 10 March 2024). Diana P. Ferreira is thankful to CEECIND/02803/2017, funded by National Funds through FCT/MCTES (https://doi.org/10.54499/CEECIND/02803/2017/CP1458/CT0003, accessed on 10 March 2024).

Institutional Review Board Statement: Not applicable.

Data Availability Statement: No new data were created or analyzed in this study. Data sharing is not applicable to this article.

Conflicts of Interest: The authors declare no conflicts of interest.

References

1. Sanaeepur, H.; Ebadi Amooghin, A.; Shirazi, M.M.A.; Pishnamazi, M.; Shirazian, S. Water desalination and ion removal using mixed matrix electrospun nanofibrous membranes: A critical review. *Desalination* **2022**, *521*, 115350. [CrossRef]
2. Peydayesh, M.; Mezzenga, R. Protein nanofibrils for next generation sustainable water purification. *Nat. Commun.* **2021**, *12*, 3248. [PubMed]
3. Adam, M.R.; Othman, M.H.D.; Kurniawan, T.A.; Puteh, M.H.; Ismail, A.F.; Khongnakorn, W.; Rahman, M.A.; Jaafar, J. Advances in adsorptive membrane technology for water treatment and resource recovery applications: A critical review. *J. Environ. Chem. Eng.* **2022**, *10*, 107633.
4. Sultana, M.; Rownok, M.H.; Sabrin, M.; Rahaman, M.H.; Alam, S.M.N. A review on experimental chemically modified activated carbon to enhance dye and heavy metals adsorption. *Clean. Eng. Technol.* **2022**, *6*, 100382. [CrossRef]
5. Wang, X.; Hsiao, B.S. Electrospun nanofiber membranes. *Curr. Opin. Chem. Eng.* **2016**, *12*, 62–81. [CrossRef]
6. Kugarajah, V.; Ojha, A.K.; Ranjan, S.; Dasgupta, N.; Ganesapillai, M.; Dharmalingam, S.; Elmoll, A.; Hosseini, S.A.; Muthulakshmi, L.; Vijayakumar, S.; et al. Future applications of electrospun nanofibers in pressure driven water treatment: A brief review and research update. *J. Environ. Chem. Eng.* **2021**, *9*, 105107. [CrossRef]
7. Wang, Z.; Xue, M.; Huang, K.; Liu, Z. Textile Dyeing Wastewater Treatment. In *Advances in Treating Textile Effluent*; InTech Open: London, UK, 2011; Chapter 5; pp. 91–116. [CrossRef]
8. Kant, R. Textile dyeing industry an environmental hazard. *Nat. Sci.* **2012**, *4*, 22–26. [CrossRef]
9. Holkar, C.R.; Jadhav, A.J.; Pinjari, D.V.; Mahamuni, N.M.; Pandit, A.B. A critical review on textile wastewater treatments: Possible approaches. *J. Environ. Manag.* **2016**, *182*, 351–366. [CrossRef]
10. Al-Ahmed, Z.A.; Al-Radadi, N.S.; Ahmed, M.K.; Shoueir, K.; El-Kemary, M. Dye removal, antibacterial properties, and morphological behavior of hydroxyapatite doped with Pd ions. *Arab. J. Chem.* **2020**, *13*, 8626–8637. [CrossRef]
11. Siddique, K.; Rizwan, M.; Shahid, M.J.; Ali, S.; Ahmad, R.; Rizvi, H. Textile Wastewater Treatment Options: A Critical Review. *Enhancing Cleanup Environ. Pollut.* **2017**, *2*, 183–207.
12. Li, Y.; Zhu, J.; Cheng, H.; Li, G.; Cho, H.; Jiang, M.; Gao, Q.; Zhang, X. Developments of Advanced Electrospinning Techniques: A Critical Review. *Adv. Mater. Technol.* **2021**, *6*, 2100410. [CrossRef]
13. Patel, K.D.; Kim, H.W.; Knowles, J.C.; Poma, A. Molecularly Imprinted Polymers and Electrospinning: Manufacturing Convergence for Next-Level Applications. *Adv. Funct. Mater.* **2020**, *30*, 2001955. [CrossRef]
14. Nasreen, S.A.A.N.; Sundarrajan, S.; Nizar, S.A.S.; Ramakrishna, S. Nanomaterials: Solutions to water-concomitant challenges. *Membranes* **2019**, *9*, 40. [CrossRef]
15. Subrahmanya, T.M.; Arshad, A.B.; Lin, P.T.; Widakdo, J.; Makari, H.K.; Austria, H.F.M.; Hu, C.C.; Lai, J.Y.; Hung, W.S. A review of recent progress in polymeric electrospun nanofiber membranes in addressing safe water global issues. *RSC Adv.* **2021**, *11*, 9638–9663.
16. Zia, Q.; Tabassum, M.; Lu, Z.; Khawar, M.T.; Song, J.; Gong, H.; Meng, J.; Li, Z.; Li, J. Porous poly(L–lactic acid)/chitosan nanofibres for copper ion adsorption. *Carbohydr. Polym.* **2020**, *227*, 115343. [CrossRef]
17. Zhang, K.; Li, Z.; Deng, N.; Ju, J.; Li, Y.; Cheng, B.; Kang, W.; Yan, J. Tree-like cellulose nanofiber membranes modified by citric acid for heavy metal ion (Cu^{2+}) removal. *Cellulose* **2019**, *26*, 945–958. [CrossRef]
18. Nalbandian, M.J.; Greenstein, K.E.; Shuai, D.; Zhang, M.; Choa, Y.H.; Parkin, G.F.; Myung, N.V.; Cwiertny, D.M. Tailored synthesis of photoactive TiO_2 nanofibers and Au/TiO_2 nanofiber composites: Structure and reactivity optimization for water treatment applications. *Environ. Sci. Technol.* **2015**, *49*, 1654–1663. [CrossRef]

19. Thyavihalli Girijappa, Y.G.; Mavinkere Rangappa, S.; Parameswaranpillai, J.; Siengchin, S. Natural Fibers as Sustainable and Renewable Resource for Development of Eco-Friendly Composites: A Comprehensive Review. *Front. Mater.* **2019**, *6*, 226. [CrossRef]
20. Yaseen, D.A.; Scholz, M. *Textile Dye Wastewater Characteristics and Constituents of Synthetic Effluents: A Critical Review*; Springer: Berlin/Heidelberg, Germany, 2019; Volume 16.
21. Wang, X.; Jiang, J.; Gao, W. Reviewing textile wastewater produced by industries: Characteristics, environmental impacts, and treatment strategies. *Water Sci. Technol.* **2022**, *85*, 2076–2096. [CrossRef]
22. Malik, A.; Hussain, M.; Uddin, F.; Raza, W.; Hussain, S.; Habiba, U.E.; Malik, T.; Ajmal, Z. Investigation of textile dyeing effluent using activated sludge system to assess the removal efficiency. *Water Environ. Res.* **2021**, *93*, 2931–2940. [CrossRef] [PubMed]
23. Dojčinović, B.P.; Obradović, B.M.; Kuraica, M.M.; Pergal, M.V.; Dolić, S.D.; Indić, D.R.; Tosti, T.B.; Manojlović, D.D. Application of non-thermal plasma reactor for degradation and detoxification of high concentrations of dye Reactive Black 5 in water. *J. Serbian Chem. Soc.* **2016**, *81*, 829–845. [CrossRef]
24. Tijing, L.D.; Yao, M.; Ren, J.; Park, C.H.; Kim, C.S.; Shon, H.K. *Nanofibers for Water and Wastewater Treatment: Recent Advances and Developments*; Springer Nature: Singapore, 2019.
25. Ma, Z.; Chang, H.; Liang, Y.; Meng, Y.; Ren, L.; Liang, H. Research progress and trends on state-of-the-art membrane technologies in textile wastewater treatment. *Sep. Purif. Technol.* **2024**, *333*, 125853. [CrossRef]
26. Müller, A.K.; Xu, Z.K.; Greiner, A. Filtration of Paint-Contaminated Water by Electrospun Membranes. *Macromol. Mater. Eng.* **2022**, *307*, 2200238. [CrossRef]
27. Nayl, A.A.; Abd-Elhamid, A.I.; Awwad, N.S.; Abdelgawad, M.A.; Wu, J.; Mo, X.; Gomha, S.M.; Aly, A.A.; Bräse, S. Review of the Recent Advances in Electrospun Nanofibers Applications in Water Purification. *Polymers* **2022**, *14*, 1594. [CrossRef]
28. Mingjun, C.; Youchen, Z.; Haoyi, L.; Xiangnan, L.; Yumei, D.; Bubakir, M.M.; Weimin, Y. An example of industrialization of melt electrospinning: Polymer melt differential electrospinning. *Adv. Ind. Eng. Polym. Res.* **2019**, *2*, 110–115. [CrossRef]
29. Haider, A.; Haider, S.; Kang, I.K. A comprehensive review summarizing the effect of electrospinning parameters and potential applications of nanofibers in biomedical and biotechnology. *Arab. J. Chem.* **2018**, *11*, 1165–1188. [CrossRef]
30. Dadras Chomachayi, M.; Solouk, A.; Akbari, S.; Sadeghi, D.; Mirahmadi, F.; Mirzadeh, H. Electrospun nanofibers comprising of silk fibroin/gelatin for drug delivery applications: Thyme essential oil and doxycycline monohydrate release study. *J. Biomed. Mater. Res. Part A* **2018**, *106*, 1092–1103. [CrossRef]
31. Kalantari, K.; Afifi, A.M.; Jahangirian, H.; Webster, T.J. Biomedical applications of chitosan electrospun nanofibers as a green polymer—Review. *Carbohydr. Polym.* **2019**, *207*, 588–600. [CrossRef]
32. Kyselica, R.; Enikov, E.T.; Polyvas, P.; Anton, R. Electrostatic focusing of electrospun Polymer(PEO) nanofibers. *J. Electrostat.* **2018**, *94*, 21–29. [CrossRef]
33. Chen, X.; Xu, C.; He, H. Electrospinning of silica nanoparticles-entrapped nanofibers for sustained gentamicin release. *Biochem. Biophys. Res. Commun.* **2019**, *516*, 1085–1089. [CrossRef]
34. Martins, A.; Reis, R.L.; Neves, N.M. Electrospinning: Processing technique for tissue engineering scaffolding. *Int. Mater. Rev.* **2008**, *53*, 257–274. [CrossRef]
35. Zeyrek Ongun, M.; Paralı, L.; Oğuzlar, S.; Pechousek, J. Characterization of β-PVDF-based nanogenerators along with Fe_2O_3 NPs for piezoelectric energy harvesting. *J. Mater. Sci. Mater. Electron.* **2020**, *31*, 19146–19158. [CrossRef]
36. Mirjalili, M.; Zohoori, S. Review for application of electrospinning and electrospun nanofibers technology in textile industry. *J. Nanostruct. Chem.* **2016**, *6*, 207–213. [CrossRef]
37. Aruna, S.T.; Balaji, L.S.; Kumar, S.S.; Prakash, B.S. Electrospinning in solid oxide fuel cells—A review. *Renew. Sustain. Energy Rev.* **2017**, *67*, 673–682. [CrossRef]
38. Pisani, S.; Dorati, R.; Conti, B.; Modena, T.; Bruni, G.; Genta, I. Design of copolymer PLA-PCL electrospun matrix for biomedical applications. *React. Funct. Polym.* **2018**, *124*, 77–89. [CrossRef]
39. Costa, R.G.F.; De Oliveira, J.E.; De Paula, G.F.; De Picciani, P.H.S.; De Medeiros, E.S.; Ribeiro, C.; Mattoso, L.H.C. Eletrofiação de polímeros em solução. Parte I: Fundamentação teórica. *Polimeros* **2012**, *22*, 170–177. [CrossRef]
40. Leidy, R.; Maria Ximena, Q.C. Use of electrospinning technique to produce nanofibres for food industries: A perspective from regulations to characterisations. *Trends Food Sci. Technol.* **2019**, *85*, 92–106. [CrossRef]
41. Mirtič, J.; Balažic, H.; Zupančič, Š.; Kristl, J. Effect of Solution Composition Variables on Electrospun Alginate Nanofibers: Response Surface Analysis. *Polymers* **2019**, *11*, 692. [CrossRef]
42. Dhandayuthapani, B.; Krishnan, U.M.; Sethuraman, S. Fabrication and characterization of chitosan-gelatin blend nanofibers for skin tissue engineering. *J. Biomed. Mater. Res. Part B Appl. Biomater.* **2010**, *94*, 264–272. [CrossRef]
43. Nezarati, R.M.; Eifert, M.B.; Cosgriff-Hernandez, E. Effects of Humidity and Solution Viscosity on Electrospun Fiber Morphology. *Tissue Eng. Part C Methods* **2013**, *19*, 810–819. [CrossRef]
44. Li, Z.; Wang, C. *One-Dimensional Nanostructures Electrospinning Technique and Unique Nanofibers*; Springer: Berlin/Heidelberg, Germany, 2016.
45. Cheng, Y.-L.; Lee, C.-Y.; Huang, Y.-L.; Buckner, C.A.; Lafrenie, R.M.; Dénommée, J.A.; Caswell, J.M.; Want, D.A.; Gan, G.G.; Leong, Y.C.; et al. *Electrospinning and Drug Delivery*; Intech: Vienna, Austria, 2016; Volume 11, p. 13.

46. Sohi, A.N.; Naderi-Manesh, H.; Soleimani, M.; Mirzaei, S.; Delbari, M.; Dodel, M. Influence of Chitosan Molecular Weight and Poly(ethylene oxide): Chitosan Proportion on Fabrication of Chitosan Based Electrospun Nanofibers. *Polym. Sci. Ser. A* **2018**, *60*, 471–482. [CrossRef]
47. Colmenares-Roldán, G.J.; Quintero-Martínez, Y.; Agudelo-Gómez, L.M.; Rodríguez-Vinasco, L.F.; Hoyos-Palacio, L.M. Influence of the molecular weight of polymer, solvents and operational condition in the electrospinning of polycaprolactone. *Rev. Fac. Ing.* **2017**, *2017*, 35–45. [CrossRef]
48. Akduman, Ç.; Perrin, E.; Kumabasar, A., Çay, A. Effect of Molecular Weight on the Morphology of Electrospun Poly(Vinyl Alcohol) Nanofibers. In Proceedings of the XIIIth International Izmir Textile and Apparel Symposium, Izmir, Turkey, 2–5 April 2014; pp. 127–134.
49. Bhardwaj, N.; Kundu, S.C. Electrospinning: A fascinating fiber fabrication technique. *Biotechnol. Adv.* **2010**, *28*, 325–347. [CrossRef] [PubMed]
50. Sill, T.J.; von Recum, H.A. Electrospinning: Applications in drug delivery and tissue engineering. *Biomaterials* **2008**, *29*, 1989–2006. [CrossRef]
51. Zuo, W.; Zhu, M.; Yang, W.; Yu, H.; Chen, Y.; Zhang, Y. Experimental Study on Relationship Between Jet Instability and Formation of Beaded Fibers During Electrospinning. *Polym. Eng. Sci.* **2005**, *45*, 704–709. [CrossRef]
52. Mahmud, M.M.; Perveen, A.; Matin, M.A.; Arafat, M.T. Effects of binary solvent mixtures on the electrospinning behavior of poly (vinyl alcohol). *Mater. Res. Express* **2018**, *5*, 115407. [CrossRef]
53. Gazquez, G.C.; Smulders, V.; Veldhuis, S.A.; Wieringa, P.; Moroni, L.; Boukamp, B.A.; Ten Elshof, J.E. Influence of Solution Properties and Process Parameters on the Formation and Morphology of YSZ and NiO Ceramic Nanofibers by Electrospinning. *Nanomaterials* **2017**, *7*, 16. [CrossRef]
54. Abid, S.; Hussain, T. Materials Science & Engineering C Current applications of electrospun polymeric nano fi bers in cancer therapy. *Mater. Sci. Eng. C* **2019**, *97*, 966–977.
55. Haghju, S.; Bari, M.R.; Khaled-Abad, M.A. Affecting parameters on fabrication of β-D-galactosidase immobilized chitosan/poly (vinyl alcohol) electrospun nanofibers. *Carbohydr. Polym.* **2018**, *200*, 137–143. [CrossRef] [PubMed]
56. Reda, R.I.; Wen, M.M.; El-Kamel, A.H. Ketoprofen-loaded Eudragit electrospun nanofibers for the treatment of oral mucositis. *Int. J. Nanomed.* **2017**, *12*, 2335–2351. [CrossRef] [PubMed]
57. Megelski, S.; Stephens, J.S.; Bruce Chase, D.; Rabolt, J.F. Micro- and nanostructured surface morphology on electrospun polymer fibers. *Macromolecules* **2002**, *35*, 8456–8466. [CrossRef]
58. Fallah, M.; Bahrami, S.H.; Ranjbar-Mohammadi, M. Fabrication and characterization of PCL/gelatin/curcumin nanofibers and their antibacterial properties. *J. Ind. Text.* **2016**, *46*, 562–577. [CrossRef]
59. Doshi, J.; Reneker, D.H. Electrospinning process and applications of electrospun fibers. *J. Electrostat.* **1995**, *35*, 151–160. [CrossRef]
60. Patil, J.V.; Mali, S.S.; Kamble, A.S.; Hong, C.K.; Kim, J.H.; Patil, P.S. Electrospinning: A versatile technique for making of 1D growth of nanostructured nanofibers and its applications: An experimental approach. *Appl. Surf. Sci.* **2017**, *423*, 641–674. [CrossRef]
61. Zargham, S.; Bazgir, S.; Tavakoli, A.; Rashidi, A.S.; Damerchely, R. The Effect of Flow Rate on Morphology and Deposition Area of Electrospun Nylon 6 Nanofiber. *J. Eng. Fiber. Fabr.* **2012**, *7*, 42–49. [CrossRef]
62. Hekmati, A.H.; Rashidi, A.; Ghazisaeidi, R.; Drean, J.Y. Effect of needle length, electrospinning distance, and solution concentration on morphological properties of polyamide-6 electrospun nanowebs. *Text. Res. J.* **2013**, *83*, 1452–1466. [CrossRef]
63. Kuchi, C.; Harish, G.S.; Reddy, P.S. Effect of polymer concentration, needle diameter and annealing temperature on TiO_2-PVP composite nanofibers synthesized by electrospinning technique. *Ceram. Int.* **2018**, *44*, 5266–5272. [CrossRef]
64. Abunahel, B.M.; Azman, N.Z.N.; Jamil, M. Effect of Needle Diameter on the Morphological Structure of Electrospun n-Bi_2O_3/Epoxy-PVA Nanofiber Mats. *Int. J. Chem. Mater. Eng.* **2018**, *12*, 296–299.
65. Wang, X.; Um, I.C.; Fang, D.; Okamoto, A.; Hsiao, B.S.; Chu, B. Formation of water-resistant hyaluronic acid nanofibers by blowing-assisted electro-spinning and non-toxic post treatments. *Polymer* **2005**, *46*, 4853–4867. [CrossRef]
66. Sundaray, B.; Subramanian, V.; Natarajan, T.S.; Xiang, R.Z.; Chang, C.C.; Fann, W.S. Electrospinning of continuous aligned polymer fibers. *Appl. Phys. Lett.* **2004**, *84*, 1222–1224. [CrossRef]
67. Anindyajati, A.; Boughton, P.; Ruys, A. The Effect of Rotating Collector Design on Tensile Properties and Morphology of Electrospun Polycaprolactone Fibres. *MATEC Web Conf.* **2015**, *27*. [CrossRef]
68. Ghobeira, R.; Asadian, M.; Vercruysse, C.; Declercq, H.; De Geyter, N.; Morent, R. Wide-ranging diameter scale of random and highly aligned PCL fibers electrospun using controlled working parameters. *Polymer* **2018**, *157*, 19–31. [CrossRef]
69. De Vrieze, S.; Van Camp, T.; Nelvig, A.; Hagström, B.; Westbroek, P.; De Clerck, K. The effect of temperature and humidity on electrospinning. *J. Mater. Sci.* **2009**, *44*, 1357–1362. [CrossRef]
70. Supaphol, P.; Mit-uppatham, C.; Nithitanakul, M. Ultrafine Electrospun Polyamide-6 Fibers: Effects of Solvent System and Emitting Electrode Polarity on Morphology and Average Fiber Diameter. *Macromol. Mater. Eng.* **2005**, *290*, 933–942. [CrossRef]
71. Yang, G.Z.; Li, H.P.; Yang, J.H.; Wan, J.; Yu, D.G. Influence of Working Temperature on The Formation of Electrospun Polymer Nanofibers. *Nanoscale Res. Lett.* **2017**, *12*, 55. [CrossRef] [PubMed]
72. Casper, C.L.; Stephens, J.S.; Tassi, N.G.; Chase, D.B.; Rabolt, J.F. Controlling Surface Morphology of Electrospun Polystyrene Fibers: Effect of Humidity and MolecularWweight in the Electrospinning Process. *Macromolecules* **2004**, *37*, 573–578. [CrossRef]

73. Medeiros, E.S.; Mattoso, L.H.C.; Offeman, R.D.; Wood, D.F.; Orts, W.J. Effect of relative humidity on the morphology of electrospun polymer fibers. *Can. J. Chem.* **2008**, *86*, 590–599. [CrossRef]
74. Long, Y.Z.; Yan, X.; Wang, X.X.; Zhang, J.; Yu, M. Electrospinning: The Setup and Procedure. In *Electrospinning: Nanofabrication and Applications*; Elsevier Inc.: Amsterdam, The Netherlands, 2018; pp. 21–52.
75. Herrero-Herreo, M. Role of Electrospinning Parameters on Poly(Lactic-co-Glycolic Acid) and Poly(Caprolactone-co-Glycolic Acid) Membranes. *Polymers* **2021**, *13*, 695. [CrossRef]
76. Tang, Y.; Cai, Z.; Sun, X.; Chong, C.; Yan, X.; Li, M.; Xu, J. Electrospun Nanofiber-Based Membranes for Water Treatment. *Polymers* **2022**, *14*, 2004. [CrossRef]
77. Xue, J.; Wu, T.; Dai, Y.; Xia, Y. Electrospinning and Electrospun Nanofibers: Methods, Materials, and Applications. *Chem. Rev.* **2019**, *119*, 5298–5415. [CrossRef]
78. Ma, H.; Hsiao, B.S.; Chu, B. Functionalized electrospun nanofibrous microfiltration membranes for removal of bacteria and viruses. *J. Membr. Sci.* **2014**, *452*, 446–452. [CrossRef]
79. Makaremi, M.; De Silva, R.T.; Pasbakhsh, P. Electrospun nanofibrous membranes of polyacrylonitrile/halloysite with superior water filtration ability. *J. Phys. Chem. C* **2015**, *119*, 7949–7958. [CrossRef]
80. Cao, X.; Huang, M.; Ding, B.; Yu, J.; Sun, G. Robust polyacrylonitrile nanofibrous membrane reinforced with jute cellulose nanowhiskers for water purification. *Desalination* **2013**, *316*, 120–126. [CrossRef]
81. Haider, S.; Binagag, F.F.; Haider, A.; Mahmood, A.; Al Masry, W.A.; Alhoshan, M.; Khan, S.U.D. Fabrication of the diethylenetriamine grafted polyacrylonitrile electrospun nanofibers membrane for the aqueous removal of cationic dyes. *Sci. Adv. Mater.* **2015**, *7*, 309–318. [CrossRef]
82. Jin, L.; Ye, J.; Wang, Y.; Qian, X.; Dong, M. Electrospinning Synthesis of ZIF-67/PAN Fibrous Membrane with High-capacity Adsorption for Malachite Green. *Fibers Polym.* **2019**, *20*, 2070–2077. [CrossRef]
83. Zhang, R.; Ma, Y.; Lan, W.; Sameen, D.E.; Ahmed, S.; Dai, J. Enhanced photocatalytic degradation of organic dyes by ultrasonic-assisted electrospray TiO_2/graphene oxide on polyacrylonitrile/β-cyclodextrin nanofibrous membranes. *Ultrason. Sonochem.* **2021**, *70*, 105343. [CrossRef]
84. Mohamed, A.; Yousef, S.; Ali Abdelnaby, M.; Osman, T.A.; Hamawandi, B.; Toprak, M.S.; Muhammed, M.; Uheida, A. Photocatalytic degradation of organic dyes and enhanced mechanical properties of PAN/CNTs composite nanofibers. *Sep. Purif. Technol.* **2017**, *182*, 219–223. [CrossRef]
85. Du, F.; Sun, L.; Huang, Z.; Chen, Z.; Xu, Z.; Ruan, G.; Zhao, C. Electrospun reduced graphene oxide/TiO_2/poly (acrylonitrile-co-maleic acid) composite nanofibers for efficient adsorption and photocatalytic removal of malachite green and leucomalachite green. *Chemosphere* **2020**, *239*, 124764. [CrossRef]
86. Wang, J.; Zhang, P.; Liang, B.; Liu, Y.; Xu, T.; Wang, L.; Cao, B.; Pan, K. Graphene Oxide as an Effective Barrier on a Porous Nanofibrous Membrane for Water Treatment. *ACS Appl. Mater. Interfaces* **2016**, *8*, 6211–6218. [CrossRef]
87. Feng, Q.; Wu, D.; Zhao, Y.; Wei, A.; Wei, Q.; Fong, H. Electrospun AOPAN/RC blend nanofiber membrane for efficient removal of heavy metal ions from water. *J. Hazard. Mater.* **2018**, *344*, 819–828. [CrossRef]
88. Yang, X.; Zhou, Y.; Sun, Z.; Yang, C.; Tang, D. Effective strategy to fabricate ZIF-8@ZIF-8/polyacrylonitrile nanofibers with high loading efficiency and improved removing of Cr(VI). *Colloids Surf. A Physicochem. Eng. Asp.* **2020**, *603*, 125292. [CrossRef]
89. Mohammad, N.; Atassi, Y. Enhancement of removal efficiency of heavy metal ions by polyaniline deposition on electrospun polyacrylonitrile membranes. *Water Sci. Eng.* **2021**, *14*, 129–138. [CrossRef]
90. Akduman, C.; Akçakoca Kumbasar, E.P.; Morsunbul, S. Electrospun nanofiber membranes for adsorption of dye molecules from textile wastewater. *IOP Conf. Ser. Mater. Sci. Eng.* **2017**, *254*, 102001. [CrossRef]
91. Hosseini, S.A.; Vossoughi, M.; Mahmoodi, N.M.; Sadrzadeh, M. Efficient dye removal from aqueous solution by high-performance electrospun nanofibrous membranes through incorporation of SiO_2 nanoparticles. *J. Clean. Prod.* **2018**, *183*, 1197–1206. [CrossRef]
92. Teng, M.; Li, F.; Zhang, B.; Taha, A.A. Electrospun cyclodextrin-functionalized mesoporous polyvinyl alcohol/SiO_2 nanofiber membranes as a highly efficient adsorbent for indigo carmine dye. *Colloids Surf. A Physicochem. Eng. Asp.* **2011**, *385*, 229–234. [CrossRef]
93. Karim, M.R.; Aijaz, M.O.; Alharth, N.H.; Alharbi, H.F.; Al-Mubaddel, F.S.; Awual, M.R. Composite nanofibers membranes of poly(vinyl alcohol)/chitosan for selective lead(II) and cadmium(II) ions removal from wastewater. *Ecotoxicol. Environ. Saf.* **2019**, *169*, 479–486. [CrossRef] [PubMed]
94. Lv, D.; Wang, R.; Tang, G.; Mou, Z.; Lei, J.; Han, J.; De Smedt, S.; Xiong, R.; Huang, C. Ecofriendly Electrospun Membranes Loaded with Visible-Light-Responding Nanoparticles for Multifunctional Usages: Highly Efficient Air Filtration, Dye Scavenging, and Bactericidal Activity. *ACS Appl. Mater. Interfaces* **2019**, *11*, 12880–12889. [CrossRef] [PubMed]
95. Sitinjak, E.M.; Masmur, I.; Marbun, N.V.M.D.; Hutajulu, P.E.; Gultom, G.; Sitanggang, Y. Direct Z-scheme of n-type CuS/p-type ZnS@electrospun PVP nanofiber for the highly efficient catalytic reduction of 4-nitrophenol and mixed dyes. *RSC Adv.* **2022**, *12*, 16165–16173. [CrossRef]
96. Wang, L.; Li, Y.; Han, P. Electrospinning preparation of g-C_3N_4/Nb_2O_5 nanofibers heterojunction for enhanced photocatalytic degradation of organic pollutants in water. *Sci. Rep.* **2021**, *11*, 22950. [CrossRef]
97. Xing, Y.; Que, W.; Yin, X.; He, Z.; Liu, X.; Yang, Y.; Shao, J.; Kong, L.B. Electrospun ZnO/Bi_2O_3 Nanofibers with Enhanced Photocatalytic Cctivity. *Nanomaterials* **2014**, *2014*, 130539.

98. Lu, J.; Liu, M.; Zhou, S.; Zhou, X.; Yang, Y. Electrospinning fabrication of ZnWO4 nanofibers and photocatalytic performance for organic dyes. *Dye. Pigment.* **2017**, *136*, 1–7. [CrossRef]
99. Kulak, S.; Bırhanlı, E.; Boran, F.; Bakar, B.; Ulu, A.; Yeşilada, Ö.; Ateş, B. Tailor-made novel electrospun polycaprolactone/polyethyleneimine fiber membranes for laccase immobilization: An all in one material to biodegrade textile dyes and phenolic compounds. *Chemosphere* **2023**, *313*, 137478. [CrossRef] [PubMed]
100. Al-Ghafri, B.; Lau, W.J.; Al-Abri, M.; Goh, P.S.; Ismail, A.F. Titanium dioxide-modified polyetherimide nanofiber membrane for water treatment. *J. Water Process Eng.* **2019**, *32*, 100970. [CrossRef]
101. De Farias, L.M.S.; Ghislandi, M.G.; De Aguiar, M.F.; Silva, B.R.S.; Leal, A.N.R.; Silva, F.D.A.O.; Fraga, T.J.M.; De Melo, C.P.; Kleber, G.; Alves, B. Electrospun polystyrene/graphene oxide fibers applied to the remediation of dye wastewater. *Mater. Chem. Phys.* **2022**, *276*, 125356. [CrossRef]
102. Li, X.; Yi, K.; Ran, Q.; Fan, Z.; Liu, C.; Liu, X.; Jia, K. Selective removal of cationic organic dyes via electrospun nanofibrous membranes derived from polyarylene ethers containing pendent nitriles and sulfonates. *Sep. Purif. Technol.* **2022**, *301*, 121942. [CrossRef]
103. Greiner, A.; Wendorff, J.H. Electrospinning: A fascinating method for the preparation of ultrathin fibers. *Angew. Chem. Int. Ed.* **2007**, *46*, 5670–5703. [CrossRef] [PubMed]
104. Lin, W.; Lu, Y.; Zeng, H. Studies of the preparation, structure, and properties of an acrylic chelating fiber containing amidoxime groups. *J. Appl. Polym. Sci.* **1993**, *47*, 45–52. [CrossRef]
105. Zhang, L.; Luo, J.; Menkhaus, T.J.; Varadaraju, H.; Sun, Y.; Fong, H. Antimicrobial nano-fibrous membranes developed from electrospun polyacrylonitrile nanofibers. *J. Membr. Sci.* **2011**, *369*, 499–505. [CrossRef]
106. Huang, F.; Xu, Y.; Liao, S.; Yang, D.; Hsieh, Y.L.; Wei, Q. Preparation of amidoxime polyacrylonitrile chelating nanofibers and their application for adsorption of metal ions. *Materials* **2013**, *6*, 969–980. [CrossRef]
107. Saeed, K.; Haider, S.; Oh, T.J.; Park, S.Y. Preparation of amidoxime-modified polyacrylonitrile (PAN-oxime) nanofibers and their applications to metal ions adsorption. *J. Membr. Sci.* **2008**, *322*, 400–405. [CrossRef]
108. Wu, Z.; Zhang, Y.; Wang, B.; Qian, G.; Tao, T. Synthesis of palladium dendritic nanostructures on amidoxime modified polyacrylonitrile fibers through a complexing-reducing method. *Mater. Sci. Eng. B Solid-State Mater. Adv. Technol.* **2013**, *178*, 923–929. [CrossRef]
109. Zhao, H.; Liu, X.; Yu, M.; Wang, Z.; Zhang, B.; Ma, H.; Wang, M.; Li, J. A study on the degree of amidoximation of polyacrylonitrile fibers and its effect on their capacity to adsorb uranyl ions. *Ind. Eng. Chem. Res.* **2015**, *54*, 3101–3106. [CrossRef]
110. Jatoi, A.W.; Gianchandani, P.K.; Kim, I.S.; Ni, Q.Q. Sonication induced effective approach for coloration of compact polyacrylonitrile (PAN) nanofibers. *Ultrason. Sonochem.* **2019**, *51*, 399–405. [CrossRef]
111. Hu, X.Q.; Ye, D.Z.; Tang, J.B.; Zhang, L.J.; Zhang, X. From waste to functional additives: Thermal stabilization and toughening of PVA with lignin. *RSC Adv.* **2016**, *6*, 13797–13802. [CrossRef]
112. Lin, G.; Bai, Z.; Liu, C.; Liu, S.; Han, M.; Huang, Y.; Liu, X. Mechanically robust, nonflammable and surface cross-linking composite membranes with high wettability for dendrite-proof and high-safety lithium-ion batteries. *J. Membr. Sci.* **2022**, *647*, 120262. [CrossRef]
113. Phan, D.N.; Khan, M.Q.; Nguyen, N.T.; Phan, T.T.; Ullah, A.; Khatri, M.; Kien, N.N.; Kim, I.S. A review on the fabrication of several carbohydrate polymers into nanofibrous structures using electrospinning for removal of metal ions and dyes. *Carbohydr. Polym.* **2021**, *252*, 117175. [CrossRef]
114. Mokhena, T.C.; Jacobs, V.; Luyt, A.S. A review on electrospun bio-based polymers for water treatment. *Express Polym. Lett.* **2015**, *9*, 839–880. [CrossRef]
115. Frey, M.W. Electrospinning cellulose and cellulose derivatives. *Polym. Rev.* **2008**, *48*, 378–391. [CrossRef]
116. Sato, A.; Wang, R.; Ma, H.; Hsiao, B.S.; Chu, B. Novel nanofibrous scaffolds for water filtration with bacteria and virus removal capability. *J. Electron Microsc.* **2011**, *60*, 201–209. [CrossRef]
117. Chu, B.; Brook, S.; Hsiao, B.S.; Ma, H. High Flux High Efficiency Nanofiber Membranes and Methods of Production Thereof. No. 61/103,479. 2009. Available online: https://patentimages.storage.googleapis.com/e1/3a/7e/5f49ad5f2ac2dd/WO2010042647A2.pdf (accessed on 10 March 2024).
118. Ma, Z.; Kotaki, M.; Ramakrishna, S. Electrospun cellulose nanofiber as affinity membrane. *J. Membr. Sci.* **2005**, *265*, 115–123. [CrossRef]
119. Wenten, I.G. Recent development in membrane science and its industrial applications. *J. Sci. Technol.* **2003**, *24*, 1009–1024.
120. Ma, H.; Burger, C.; Hsiao, B.S.; Chu, B. Highly permeable polymer membranes containing directed channels for water purification. *ACS Macro Lett.* **2012**, *1*, 723–726. [CrossRef]
121. Ji, F.; Li, C.; Tang, B.; Xu, J.; Lu, G.; Liu, P. Preparation of cellulose acetate/zeolite composite fiber and its adsorption behavior for heavy metal ions in aqueous solution. *Chem. Eng. J.* **2012**, *209*, 325–333. [CrossRef]
122. Taha, A.A.; Wu, Y.; Wang, H.; Li, F. Preparation and application of functionalized cellulose acetate/silica composite nanofibrous membrane via electrospinning for Cr(VI) ion removal from aqueous solution. *J. Environ. Manag.* **2012**, *112*, 10–16. [CrossRef] [PubMed]
123. Bódalo, A.; Gómez, J.L.; Gómez, E.; León, G.; Tejera, M. Ammonium removal from aqueous solutions by reverse osmosis using cellulose acetate membranes. *Desalination* **2005**, *184*, 149–155. [CrossRef]

124. Konwarh, R.; Karak, N.; Misra, M. Electrospun cellulose acetate nanofibers: The present status and gamut of biotechnological applications. *Biotechnol. Adv.* **2013**, *31*, 421–437. [CrossRef] [PubMed]
125. Ismail, I.I.N.; Zayadi, R.A.; Ho, K.C.; Soo, J.Z.; Idris, M.S.; Tay, K.Y. Electrospun Nano-Palm Frond Titania Fiber (Nano-PFTF) Membrane for Industrial Wastewater Treatment. *Solid State Sci. Technol.* **2020**, *28*, 87–102.
126. Cheng, J.; Zhan, C.; Wu, J.; Cui, Z.; Si, J.; Wang, Q.; Peng, X.; Turng, L.S. Highly Efficient Removal of Methylene Blue Dye from an Aqueous Solution Using Cellulose Acetate Nanofibrous Membranes Modified by Polydopamine. *ACS Omega* **2020**, *5*, 5389–5400. [CrossRef]
127. Akduman, Ç. Fabrication and characterization of diatomite functionalized cellulose acetate nanofibers. *AATCC J. Res.* **2019**, *6*, 28–36. [CrossRef]
128. San, N.O.; Celebioglu, A.; Tümtaş, Y.; Uyar, T.; Tekinay, T. Reusable bacteria immobilized electrospun nanofibrous webs for decolorization of methylene blue dye in wastewater treatment. *RSC Adv.* **2014**, *4*, 32249–32255. [CrossRef]
129. Schiffman, J.D.; Schauer, C.L. A review: Electrospinning of biopolymer nanofibers and their applications. *Polym. Rev.* **2008**, *48*, 317–352. [CrossRef]
130. Jayakumar, R.; Menon, D.; Manzoor, K.; Nair, S.V.; Tamura, H. Biomedical applications of chitin and chitosan based nanomaterials—A short review. *Carbohydr. Polym.* **2010**, *82*, 227–232. [CrossRef]
131. Muzzarelli, R.A.A. Potential of chitin/chitosan-bearing materials for uranium recovery: An interdisciplinary review. *Carbohydr. Polym.* **2011**, *84*, 54–63. [CrossRef]
132. Wan Ngah, W.S.; Teong, L.C.; Hanafiah, M.A.K.M. Adsorption of dyes and heavy metal ions by chitosan composites: A review. *Carbohydr. Polym.* **2011**, *83*, 1446–1456. [CrossRef]
133. Li, C.; Lou, T.; Yan, X.; Long, Y.; Cui, G.; Wang, X. Fabrication of pure chitosan nanofibrous membranes as effective absorbent for dye removal. *Int. J. Biol. Macromol.* **2018**, *106*, 768–774. [CrossRef]
134. Haider, S.; Park, S.Y. Preparation of the electrospun chitosan nanofibers and their applications to the adsorption of Cu(II) and Pb(II) ions from an aqueous solution. *J. Membr. Sci.* **2009**, *328*, 90–96. [CrossRef]
135. Razzaz, A.; Ghorban, S.; Hosayni, L.; Irani, M.; Aliabadi, M. Chitosan nanofibers functionalized by TiO_2 nanoparticles for the removal of heavy metal ions. *J. Taiwan Inst. Chem. Eng.* **2016**, *58*, 333–343. [CrossRef]
136. Mokhena, T.C.; Luyt, A.S. Development of multifunctional nano/ultrafiltration membrane based on a chitosan thin film on alginate electrospun nanofibres. *J. Clean. Prod.* **2017**, *156*, 470–479. [CrossRef]
137. Alsberg, E.; Anderson, K.W.; Albeiruti, A.; Franceschi, R.T.; Mooney, D.J. Cell-interactive alginate hydrogels for bone tissue engineering. *J. Dent. Res.* **2001**, *80*, 2025–2029. [CrossRef]
138. Dar, A.; Shachar, M.; Leor, J.; Cohen, S. Optimization of cardiac cell seeding and distribution in 3D porous alginate scaffolds. *Biotechnol. Bioeng.* **2002**, *80*, 305–312. [CrossRef]
139. Hashimoto, T.; Suzuki, Y.; Tanihara, M.; Kakimaru, Y.; Suzuki, K. Development of alginate wound dressings linked with hybrid peptides derived from laminin and elastin. *Biomaterials* **2004**, *25*, 1407–1414. [CrossRef] [PubMed]
140. Davis, T.A.; Volesky, B.; Mucci, A. A review of the biochemistry of heavy metal biosorption by brown algae. *Water Res.* **2003**, *37*, 4311–4330. [CrossRef]
141. Papageorgiou, S.K.; Katsaros, F.K.; Kouvelos, E.P.; Kanellopoulos, N.K. Prediction of binary adsorption isotherms of Cu2+, Cd2+ and Pb2+ on calcium alginate beads from single adsorption data. *J. Hazard. Mater.* **2009**, *162*, 1347–1354. [CrossRef] [PubMed]
142. Xiao, C.; Liu, H.; Lu, Y.; Zhang, L. Preparation and Physical Properties of Blend Films from Sodium Alginate and Polyacrylamide Solutions. *J. Macromol. Sci. Part A Pure Appl. Chem.* **2000**, *37*, 1663–1675. [CrossRef]
143. Çaykara, T.; Demirci, S.; Eroğlu, M.S.; Güven, O. Poly(ethylene oxide) and its blends with sodium alginate. *Polymer* **2005**, *46*, 10750–10757. [CrossRef]
144. Omidian, H.; Rocca, J.G.; Park, K. Elastic, superporous hydrogel hybrids of polyacrylamide and sodium alginate. *Macromol. Biosci.* **2006**, *6*, 703–710. [CrossRef] [PubMed]
145. Mokhena, T.C.; Jacobs, N.V.; Luyt, A.S. Electrospun alginate nanofibres as potential bio-sorption agent of heavy metals in water treatment. *Express Polym. Lett.* **2017**, *11*, 652–663. [CrossRef]
146. Fang, D.; Liu, Y.; Jiang, S.; Nie, J.; Ma, G. Effect of intermolecular interaction on electrospinning of sodium alginate. *Carbohydr. Polym.* **2011**, *85*, 276–279. [CrossRef]
147. Li, W.; Li, X.; Chen, Y.; Li, X.; Deng, H.; Wang, T.; Huang, R.; Fan, G. Poly(vinyl alcohol)/sodium alginate/layered silicate based nanofibrous mats for bacterial inhibition. *Carbohydr. Polym.* **2013**, *92*, 2232–2238. [CrossRef]
148. Ebrahimi, F.; Sadeghizadeh, A.; Neysan, F.; Heydari, M. Fabrication of nanofibers using sodium alginate and Poly(Vinyl alcohol) for the removal of Cd2+ ions from aqueous solutions: Adsorption mechanism, kinetics and thermodynamics. *Heliyon* **2019**, *5*, 1–10. [CrossRef]
149. Wang, Q.; Ju, J.; Tan, Y.; Hao, L.; Ma, Y.; Wu, Y.; Zhang, H.; Xia, Y.; Sui, K. Controlled synthesis of sodium alginate electrospun nanofiber membranes for multi-occasion adsorption and separation of methylene blue. *Carbohydr. Polym.* **2019**, *205*, 125–134. [CrossRef] [PubMed]
150. Szejtli, J. Introduction and general overview of cyclodextrin chemistry. *Chem. Rev.* **1998**, *98*, 1743–1753. [CrossRef] [PubMed]
151. Crini, G. Review: A history of cyclodextrins. *Chem. Rev.* **2014**, *114*, 10940–10975. [CrossRef]
152. Liu, H.; Cai, X.; Wang, Y.; Chen, J. Adsorption mechanism-based screening of cyclodextrin polymers for adsorption and separation of pesticides from water. *Water Res.* **2011**, *45*, 3499–3511. [CrossRef] [PubMed]

153. Liu, J.; Liu, G.; Liu, W. Preparation of water-soluble β-cyclodextrin/poly(acrylic acid)/graphene oxide nanocomposites as new adsorbents to remove cationic dyes from aqueous solutions. *Chem. Eng. J.* **2014**, *257*, 299–308. [CrossRef]
154. Connors, K.A. The stability of cyclodextrin complexes in solution. *Chem. Rev.* **1997**, *97*, 1325–1357. [CrossRef]
155. Sherje, A.P.; Dravyakar, B.R.; Kadam, D.; Jadhav, M. Cyclodextrin based nanosponges: A critical review. *Carbohydr. Polym.* **2017**, *173*, 37–49. [CrossRef]
156. Liu, F.; Sun, Y.; Gu, J.; Gao, Q.; Sun, D.; Zhang, X.; Pan, B.; Qian, J. Highly efficient photodegradation of various organic pollutants in water: Rational structural design of photocatalyst via thiol-ene click reaction. *Chem. Eng. J.* **2020**, *381*, 122631. [CrossRef]
157. Liu, Q.; Zhou, Y.; Lu, J.; Zhou, Y. Novel cyclodextrin-based adsorbents for removing pollutants from wastewater: A critical review. *Chemosphere* **2020**, *241*, 125043. [CrossRef]
158. Alsbaiee, A.; Smith, B.J.; Xiao, L.; Ling, Y.; Helbling, D.E.; Dichtel, W.R. Rapid removal of organic micropollutants from water by a porous β-cyclodextrin polymer. *Nature* **2016**, *529*, 190–194. [CrossRef]
159. Morin-Crini, N.; Winterton, P.; Fourmentin, S.; Wilson, L.D.; Fenyvesi, É.; Crini, G. Water-insoluble β-cyclodextrin–epichlorohydrin polymers for removal of pollutants from aqueous solutions by sorption processes using batch studies: A review of inclusion mechanisms. *Prog. Polym. Sci.* **2018**, *78*, 1–23. [CrossRef]
160. Jeong, D.; Joo, S.W.; Shinde, V.V.; Jung, S. Triple-crosslinked β-cyclodextrin oligomer self-healing hydrogel showing high mechanical strength, enhanced stability and pH responsiveness. *Carbohydr. Polym.* **2018**, *198*, 563–574. [CrossRef] [PubMed]
161. San Keskin, N.O.; Celebioglu, A.; Sarioglu, O.F.; Uyar, T.; Tekinay, T. Encapsulation of living bacteria in electrospun cyclodextrin ultrathin fibers for bioremediation of heavy metals and reactive dye from wastewater. *Colloids Surf. B Biointerfaces* **2018**, *161*, 169–176. [CrossRef] [PubMed]
162. Zhao, R.; Wang, Y.; Li, X.; Sun, B.; Wang, C. Synthesis of β-cyclodextrin-based electrospun nanofiber membranes for highly efficient adsorption and separation of methylene blue. *ACS Appl. Mater. Interfaces* **2015**, *7*, 26649–26657. [CrossRef] [PubMed]
163. Liu, F.; Chen, C.; Qian, J. Film-like bacterial cellulose/cyclodextrin oligomer composites with controllable structure for the removal of various persistent organic pollutants from water. *J. Hazard. Mater.* **2021**, *405*, 124122. [CrossRef] [PubMed]
164. Chen, H.; Huang, M.; Liu, Y.; Meng, L.; Ma, M. Functionalized electrospun nanofiber membranes for water treatment: A review. *Sci. Total Environ.* **2020**, *739*, 139944. [CrossRef] [PubMed]
165. Brown, T.D.; Dalton, P.D.; Hutmacher, D.W. Melt electrospinning today: An opportune time for an emerging polymer process. *Prog. Polym. Sci.* **2016**, *56*, 116–166. [CrossRef]
166. Jian, S.; Zhu, J.; Jiang, S.; Chen, S.; Fang, H.; Song, Y.; Duan, G.; Zhang, Y.; Hou, H. Nanofibers with diameter below one nanometer from electrospinning. *RSC Adv.* **2018**, *8*, 4794–4802. [CrossRef]
167. Xue, J.; Xie, J.; Liu, W.; Xia, Y. Electrospun Nanofibers: New Concepts, Materials, and Applications. *Acc. Chem. Res.* **2017**, *50*, 1976–1987. [CrossRef] [PubMed]
168. Kaur, S.; Sundarrajan, S.; Rana, D.; Matsuura, T.; Ramakrishna, S. Influence of electrospun fiber size on the separation efficiency of thin film nanofiltration composite membrane. *J. Membr. Sci.* **2012**, *392–393*, 101–111. [CrossRef]
169. Dobosz, K.M.; Kuo-Leblanc, C.A.; Martin, T.J.; Schiffman, J.D. Ultrafiltration Membranes Enhanced with Electrospun Nanofibers Exhibit Improved Flux and Fouling Resistance. *Ind. Eng. Chem. Res.* **2017**, *56*, 5724–5733. [CrossRef]
170. Park, M.J.; Gonzales, R.R.; Abdel-Wahab, A.; Phuntsho, S.; Shon, H.K. Hydrophilic polyvinyl alcohol coating on hydrophobic electrospun nanofiber membrane for high performance thin film composite forward osmosis membrane. *Desalination* **2018**, *426*, 50–59. [CrossRef]
171. Shokrollahzadeh, S.; Tajik, S. Fabrication of thin film composite forward osmosis membrane using electrospun polysulfone/polyacrylonitrile blend nanofibers as porous substrate. *Desalination* **2018**, *425*, 68–76. [CrossRef]
172. Zhou, T.; Li, J.; Guo, X.; Yao, Y.; Zhu, P.; Xiang, R. Freestanding PTFE electrospun tubular membrane for reverse osmosis brine concentration by vacuum membrane distillation. *Desalin. Water Treat.* **2019**, *165*, 63–72. [CrossRef]
173. Xu, Y.; Yang, Y.; Fan, X.; Liu, Z.; Song, Y.; Wang, Y.; Tao, P.; Song, C.; Shao, M. In-situ silica nanoparticle assembly technique to develop an omniphobic membrane for durable membrane distillation. *Desalination* **2021**, *499*, 114832. [CrossRef]
174. Gontarek-Castro, E.; Castro-Muñoz, R.; Lieder, M. New insights of nanomaterials usage toward superhydrophobic membranes for water desalination via membrane distillation: A review. *Crit. Rev. Environ. Sci. Technol.* **2022**, *52*, 2104–2149. [CrossRef]
175. Ağtaş, M.; Yılmaz, Ö.; Dilaver, M.; Alp, K.; Koyuncu, İ. Pilot-scale ceramic ultrafiltration/nanofiltration membrane system application for caustic recovery and reuse in textile sector. *Environ. Sci. Pollut. Res.* **2021**, *28*, 41029–41038. [CrossRef]
176. Sahinkaya, E.; Tuncman, S.; Koc, I.; Guner, A.R.; Ciftci, S.; Aygun, A.; Sengul, S. Performance of a pilot-scale reverse osmosis process for water recovery from biologically-treated textile wastewater. *J. Environ. Manag.* **2019**, *249*, 109382. [CrossRef]
177. Kurt, E.; Koseoglu-Imer, D.Y.; Dizge, N.; Chellam, S.; Koyuncu, I. Pilot-scale evaluation of nanofiltration and reverse osmosis for process reuse of segregated textile dyewash wastewater. *Desalination* **2012**, *302*, 24–32. [CrossRef]

Disclaimer/Publisher's Note: The statements, opinions and data contained in all publications are solely those of the individual author(s) and contributor(s) and not of MDPI and/or the editor(s). MDPI and/or the editor(s) disclaim responsibility for any injury to people or property resulting from any ideas, methods, instructions or products referred to in the content.

Article

Reactive Deep Eutectic Solvent for an Eco-Friendly Synthesis of Cellulose Carbamate

Vincenzo Algieri [1], Loredana Maiuolo [1], Debora Procopio [2], Paola Costanzo [1], Fiore Pasquale Nicoletta [2], Sonia Trombino [2], Maria Luisa Di Gioia [2,*] and Antonio De Nino [1,*]

[1] Department of Chemistry and Chemical Technologies, University of Calabria, Via P. Bucci, Cubo 12C, 87036 Rende, CS, Italy; vincenzo.algieri@unical.it (V.A.); maiuolo@unical.it (L.M.); paola.costanzo@unical.it (P.C.)

[2] Department of Pharmacy, Health and Nutritional Sciences, University of Calabria, ed. polifunzionale, 87036 Rende, CS, Italy; debora.procopio@unical.it (D.P.); fiore.nicoletta@unical.it (F.P.N.); sonia.trombino@unical.it (S.T.)

* Correspondence: ml.digioia@unical.it (M.L.D.G.); denino@unical.it (A.D.N.)

Abstract: The limited solubility of natural cellulose in water and common organic solvents hinders its diverse applications, despite being one of the most abundant and easily accessible biopolymers on Earth. Chemical derivatization, such as cellulose carbamate (CC), offers a pathway to enhance both solubility and industrial processability. In this study, CC was synthesized by exploiting a novel type IV deep eutectic solvent (DES) composed of erbium trichloride and urea. This DES was shown to be not only an environmentally friendly reaction medium/catalyst but also actively participated in the synthetic process as a reagent. The resultant cellulose carbamate samples were characterized through FT-IR and elemental analysis. A nitrogen content value of 1.59% was afforded determining a degree of substitution corresponding to a value of 0.19. One of the key scientific advancements lies in the preparation of cellulose carbamate using a straightforward and cost-effective method. This approach utilizes non-toxic compounds, aligning with the principles of green chemistry and contributing to sustainable development in cellulose derivative production.

Keywords: cellulose; carbamate; derivatization; reactive deep eutectic solvents; erbium trichloride; green chemistry

Citation: Algieri, V.; Maiuolo, L.; Procopio, D.; Costanzo, P.; Nicoletta, F.P.; Trombino, S.; Di Gioia, M.L.; De Nino, A. Reactive Deep Eutectic Solvent for an Eco-Friendly Synthesis of Cellulose Carbamate. *Polymers* **2024**, *16*, 757. https://doi.org/10.3390/polym16060757

Academic Editors: Cristina Cazan and Mihai Alin Pop

Received: 3 February 2024
Revised: 1 March 2024
Accepted: 6 March 2024
Published: 9 March 2024

Copyright: © 2024 by the authors. Licensee MDPI, Basel, Switzerland. This article is an open access article distributed under the terms and conditions of the Creative Commons Attribution (CC BY) license (https://creativecommons.org/licenses/by/4.0/).

1. Introduction

The accumulation of synthetic polymers in the environment has led to significant environmental pollution issues, which can only be mitigated by exploring biodegradable and non-toxic biopolymers [1]. In this context, industries play a crucial role in focusing on sustainable resources and increasing the use of renewable raw materials. Natural cellulose (Figure 1), due to its abundance and versatility as a biopolymer, represents a significant resource for various applications [2,3]. Its easy accessibility, biocompatibility, high mechanical stability, and unique physiochemical properties make it suitable for applications in biomedical science, environmental science, and sustainable packaging [4,5]. Despite the enormous potential of cellulose, only a small fraction is currently utilized for further processing [6]. This is attributed to its complex inter- and intramolecular hydrogen bonding network, leading to reduced solubility in water and most common solvents [7–9]. The OH groups at position 6 are the ones responsible for intermolecular bonding in cellulose. Consequently, the accessibility of these groups is the limiting factor for cellulose solubility.

Numerous solvents have been explored and/or have been introduced into industry to improve cellulose solubility, such as *N,N*-dimethylacetamide/lithium chloride (DMAc)/LiCl, dimethyl sulfoxide/tetrabutylammonium fluoride (DMSO/TBAF), *N,N*-dimethylformamide/dinitrogen tetroxide (DMF/N_2O_4), *N*-methyl-morpholine-*N*-oxide (NMMO), and ionic liquids [10–17]. However, these methods often present limitations, especially in terms of cost and environmental impact, necessitating the development of more cost-effective and environmentally friendly alternatives.

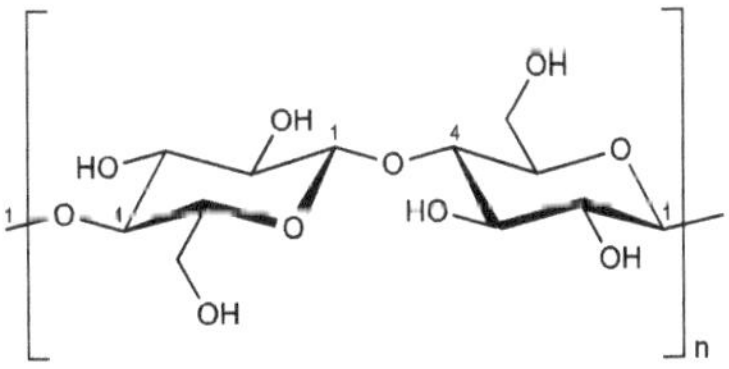

Figure 1. Chemical structure of Cellulose.

A frequently used approach to enhance cellulose solubility is pre-chemical modification, such as converting it into cellulose carbamate (CC), as shown in Scheme 1. This modification renders cellulose biodegradable and biocompatible, making it an eco-friendly material with applications in absorbent products, food packaging, chromatography, and fireproof products [18–24] The conventional "CarbaCell" process involves heating cellulose with urea, resulting in cellulose carbamates with a nitrogen content (N%) of around 1–2.5% (Scheme 1) [25]. However, this process faces challenges in industrial reproducibility due to rigorous conditions, including high temperatures, long reaction times, and the use of a catalyst [26].

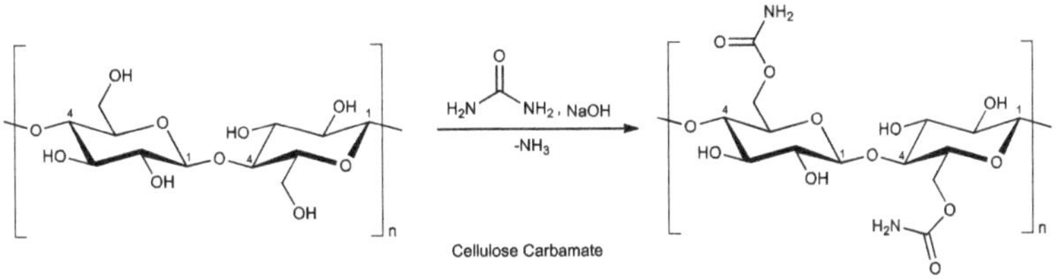

Scheme 1. Chemical modification of cellulose into cellulose carbamate (CC).

In response to the need for innovative and efficient methods for cellulose carbamate preparation, it is worth exploring environmentally sustainable solvents and methods [27–32]. Ionic liquids (ILs), which are traditionally considered environmentally friendly and very versatile non-conventional solvents for chemical transformations [33–37], have limitations in terms of biodegradability and sustainability [38–41]. Recently, Deep Eutectic Solvents (DESs) have gained attention as potential green solvents for biomass processing, offering biocompatibility and applications in bio-pharma industries [42–53]. Several studies have emerged in which DESs are used as solvents for the derivatization of cellulose and to produce various types of nanocelluloses [54–57]. The IV-type class of Deep Eutectic Solvents formed using hydrated metal salts as a cationic component have been widely used in the treatment of biomass [36,51]. Recently, a DES based on dimethylurea and zinc chloride was proposed for the preparation of cellulose methyl carbamate [58]. Nevertheless, zinc chloride is known to be a toxic salt, capable of accumulating if improperly disposed of [59–64].

Encouraged to seek out more eco-friendly solutions and to address concerns about toxic salts, we propose the possible combination of erbium trichloride (ErCl$_3$) and urea in different molar ratios to form a type IV DES. In this regard, erbium salts have emerged as a sustainable catalytic solution for a series of organic transformations that require the presence of Lewis acid catalysts [65–67]. Erbium trichloride (ErCl$_3$) was chosen for its cost-effectiveness, low toxicity, and versatility [65,68–70]. This Lewis acid was combined with urea, which is not only one of the most used components in DESs but also represents a promising chemical for cellulose modification as it is a low-cost, readily available, and non-toxic compound [71]. The developed reactive deep eutectic solvent (RDES), acting as a reaction medium, reagent, and catalyst, holds promise in improving the environmental sustainability of cellulose carbamate production.

2. Materials and Methods

2.1. Materials

Microcrystalline cellulose, Avicel PH-101 (MCC, Mw = 53,470, Mn = 24,235, DP = 350) was purchased from Sigma–Aldrich (St. Louis, MO, USA). Urea (99.7%) and Er (III) chloride hexahydrate ($ErCl_3 \cdot 6H_2O$, 99.9%) were purchased from Merck and used as received without further purification. Deionized water was used in all experiments. The heating necessary for the preparation of DESs was provided using a Rotavapor Heidolph Laborota 4000 (Heidolph Instruments Italia, Milano, Italy). Drying was carried out using a Vismara Vacuum Oven 65 stove. The product, after recovery by washing with a solution of HCl 12 M and filtration, was dried using an Edwards XDS5 diaphragm pump (Crawley, West Sussex, UK). Ultrasonic irradiation was carried out using a Hielscher VP 2005 Sonicator (Hielscher Ultrasonics GmbH, Teltow, Germany).

2.2. Preparation of Reactive Deep Eutectic Solvents (RDESs)

2.2.1. Preparation of DES-1 Using the Vacuum Evaporation (VE) Method

$ErCl_3$ (0.5078 g, 0.00185 mol) and urea (0.3719 g, 0.0062 mol) in a 3:10 molar ratio were weighted in separate vials and dissolved in water. Then the two solutions were mixed together in a round-bottom flask and dried using a rotary evaporator (at 50 °C) until a clear, homogeneous pink liquid was formed (Table 1). The DES was stored in a desiccator.

Table 1. Preparation of DESs using $ErCl_3$ (anhydrous) or $ErCl_3 \cdot 6H_2O$ (hexahydrate) and urea.

Entry	DES	Components	Molar Ratio	Preparation Method	Time (min)	Aspect
1	-	$ErCl_3$ anhydrous: urea	1:1	Vacuum evaporation	60	Pink solid
2	-	$ErCl_3$ anhydrous: urea	1:3	Vacuum evaporation	60	Pink solid
3	-	$ErCl_3 \cdot 6H_2O$: urea	1:3	Vacuum evaporation	60	Pink highly viscous liquid with particles in suspension
4	DES-1	$ErCl_3 \cdot 6H_2O$: urea (+ 20% water)	3:10	Vacuum evaporation	60	Clear pink viscous liquid
5	-	$ErCl_3 \cdot 6H_2O$: urea (+20% water)	3:10	Ultrasonication-assisted preparation	30	Clear pink viscous liquid with particles in suspension
6	DES-2	$ErCl_3 \cdot 6H_2O$: urea (+20% water)	3:10	Ultrasonication-assisted preparation	60	Clear pink viscous liquid
7	DES-3	$ErCl_3 \cdot 6H_2O$: urea	3:10	Heating and stirring	60	Clear pink viscous liquid
8	DES-4	$ErCl_3 \cdot 6H_2O$: urea	3:10	Ultrasonication-assisted preparation and vacuum evaporation	120	Clear pink viscous liquid

2.2.2. Preparation of DES-2 Using the Ultrasound-Assisted (US) Method

The two components, $ErCl_3 \cdot 6H_2O$ and urea, were mixed in a 10 mL glass flask in the molar ratios indicated in Table 1. The glass flasks were placed in a KQ-300E ultra-sonic bath (Kunshan ultrasonic instruments Co., Ltd., Kunshan, China) with an ultra-sonic input power of 300 W and a frequency of 40 kHz. The mixture formed a clear and homogeneous pink liquid at room temperature after up to 1 h of ultrasonication. The final temperatures of the ultrasonic bath could reach up to 50 °C due to the long duration of ultrasonication. DES-2 was stored in a desiccator.

2.2.3. Preparation of DES-3 Using the Heating and Stirring (HS) Method

$ErCl_3 \cdot 6H_2O$ (0.5074 g, 0.00185 mol) and urea (0.3743 g, 0.00623 mol) in a 3:10 molar ratio were placed in the same vial and water (17 µL, 0.95 µmol) was added. The vial was placed in a water bath at 60 °C and stirred magnetically until the formation of a clear pink liquid was observed (see Table 1). The DES was stored in a desiccator.

2.2.4. Preparation of DES-4 Using Both Ultrasonication and Vacuum Evaporation Methods

DES-4 was prepared by placing 1.01 g of $ErCl_3 \cdot 6H_2O$ and 0.74 g of urea (molar ratio 3:10) in a round-bottom flask and combining both ultrasonication-assisted and vacuum evaporation methods. The reaction was first assisted by ultrasound for 1 h and then it was stirred using a rotary evaporator at a temperature of 40 °C for 1 h. The pink liquid (Table 1) was finally preserved in a desiccator.

All the DES samples were kept at room temperature for an extra 24 h after their preparation to ensure the formation of the liquid. All the prepared DESs were characterized by Differential Scanning Calorimetry (DSC) analysis, Fourier Transform Infrared Spectroscopy (FTIR), and Polarized Optical Microscopy (POM).

2.3. Differential Scanning Calorimetry (DSC) Analysis of RDESs

DES-1–DES-4 were characterized by Differential Scanning Analysis (DSC) performed using a DSC instrument (DSC 200 PC Netzsch, Wittelsbacherstraße 42, Selb, Germany). The temperature ranged from −80 °C to 350 °C, at 10 °C/min, after equilibration for 5 min at −80 °C. The experiments were performed under a nitrogen atmosphere (50 mL/min), with 15 mg of the sample in aluminum pans with covering lids.

2.4. Polarized Optical Microscopy (POM) Analysis

A small droplet of the prepared DESs was deposited on a microscopic slide for observation at a magnification of 10×. The Nikon ECLIPSE LV100N polarizing microscope (Nikon Corporation, Shinagawa Intercity Tower C, Tokyo, Japan) coupled with the Nikon DS-Fi2 camera was used for recording the polarized light image. The absence of a solid crystalline structure is evidenced by a polarized light image that is totally black.

2.5. Fourier Transform Infrared Spectroscopy (FT-IR) Analysis

The prepared DESs were analyzed by means of the KBr sheet. Infrared (FTIR) spectra were recorded using an FTIR Perkin-Elmer 1720 spectrophotometer over the 4.000–400 cm^{-1} range at a rate of 0.5 cm/s. Fifty scans were recorded, averaged for each spectrum, and corrected against ambient air as the background.

2.6. Synthesis of Cellulose Carbamate (CC)

2.6.1. Experiment 1: Reaction in DES-4 Using VE

DES-4 was prepared following the procedure reported above. Subsequently, 0.21 g of microcrystalline cellulose (a 10-fold molar excess of urea compared to the cellulose anhydroglucose unit) suspended in distilled water (2.23 mL) was introduced into the flask containing DES-4. A glass rod was employed to ensure the uniform mixing of cellulose in DES. The reaction mixture was subjected to continuous stirring using a rotary evaporator at a reduced pressure at 60 °C for 1 h, and then dried in an oil bath at 150 °C for an additional hour. The resulting product was dry and adhered to the flask walls.

After cooling, the recovery of the solid material was carried out by washing with 20 mL of a 1 M HCl solution, followed by filtration and subsequent washing with distilled water to remove the unreacted residual DES components. The product was filtered and dried using a vacuum pump, resulting in a white solid material.

IR νmax(KBr): amide N-H, 3300; amide C=O, 1716.34 cm^{-1}.

2.6.2. Experiment 2: Reaction in DES-4 Using VE and US

In a 10 mL round-bottom flask, DES-4 was prepared. Then microcrystalline cellulose (0.1030 g) and distilled water (2.23 g, 0.124 mol) were introduced. Gentle mixing was performed by manually shaking the flask. The reaction was assisted by ultrasound for 1 h. Subsequently, the mixture was stirred using a rotary evaporator under reduced pressure at 50 °C for 1 h. The product was dried in an oven at 100 °C for 24 h. After cooling to room temperature, the recovery and purification of the CC derivative were carried out as described in experiment 1. The recovered product was finally dried, resulting in a gray-colored powder.

2.6.3. Experiment 3: Reaction in DES-4 under US in the Absence of Water

The synthesis was conducted in the absence of water, and therefore, microcrystalline cellulose (1.0946 g) was dried to eliminate traces of residual moisture until a constant weight was reached. In a test tube, previously dried microcrystalline cellulose (0.1015 g) was added to DES-4. The reaction was assisted by ultrasound for 1 h. The product was dried in an oven at 100 °C for 24 h. The test tube was allowed to cool to room temperature, and the final recovery and purification of the CC derivative was carried out as described in experiment 1. The obtained product was dried to afford a white-colored solid material.

2.6.4. Experiment 4: Reaction in the Absence of DES-4 under US

To a 10 mL round-bottom flask dried containing cellulose (0.10 g) suspended in water (6.70 g, 0.372 mol), urea (0.317 g, 0.00062 mol) was added. The reaction mixture was subjected to continuous stirring using the rotary evaporator at a reduced pressure at 60 °C for 1 h, and then dried in an oil bath at 150 °C for an additional hour. The product was then dried in an oven at 100 °C for 24 h and finally allowed to cool to room temperature. The final recovery and purification of the CC derivative was carried out as described in procedure 1 affording a white-colored product.

2.7. FT-IR Analysis of CC

Cellulose carbamate (CC) was characterized by means of a Fourier Transform Infrared (FT-IR) spectrometer (Jasco 4200, Cremella, LC, Italy) and compared with native cellulose. FT-IR spectra were performed using a Jasco 4200 spectrometer using the KBr disk technique. Spectra were obtained in the 400–4000 cm^{-1} range.

2.8. Elemental Analysis of CC

Elemental analysis was carried out to determine the nitrogen content (N%) by a vario MICRO CHNS V4.0.10 analyzer from Elementar (Langensebold, DE, Germany), where the samples were burned within a jet injection of oxygen. The gaseous components were purified and separated on a TPD column before quantification with a thermal conductivity detector. Helium was used as a carrier gas.

Percentages of the elemental analysis for C, H, and N elements are reported in Table 2. The degree of substitution (DS) was calculated from the nitrogen content of CC via the following Equation:

$$DS = \frac{N \times 162.15}{14 \times 100 - (N \times 43)}$$

where N is the nitrogen content, 162.15 is the molecular weight of the anhydroglucose unit (AGU) of cellulose, 14 is the atomic weight of the nitrogen, atom and 43 is the molecular weight of the carbamate group [72].

3. Results

3.1. DES Formation

At the beginning of our investigation, we considered different procedures for preparing RDESs to ensure reproducibility: the vacuum evaporation method (VE), the heating and stirring method (HS), and the ultrasound-assisted method (US). These diverse approaches may yield distinct outcomes. In fact, variables like temperature, pressure, and water content, which are influenced by these methods, can shape the final composition of the DESs, thus affecting their physicochemical properties and stability [73]. Therefore, initially, we applied the vacuum evaporation method, where all available water was removed, leaving only the water interacting with the DES components in the system [74,75]. To achieve this, erbium trichloride and urea in different molar ratios were dissolved in water and stirred using a rotary evaporator under reduced pressure at 50 °C for 1 h until a clear, homogeneous pink viscous liquid was formed. $ErCl_3$ in its anhydrous form was found unsuitable for the intended purpose. The stoichiometric ratios investigated in combination with urea (1:1 or 1:3) did not produce recognizable DES due to the presence of solid particles of the salt and rapid solidification of the mixture after cooling (Table 1, entries 1, 2). Similarly, the use of the hexahydrate form of erbium trichloride, with 1:1 and 1:3 molar ratios, resulted in the formation of pink and highly viscous liquids with particles in suspension, making them unsuitable for easy handling and reaction (Table 1, entry 3). Instead, a clear, pink, and homogeneous liquid was obtained when using erbium chloride hexahydrate and urea in a 3:10 molar ratio (Table 1, entry 4, DES-1). The resulting DES was then dried in a desiccator containing silica gel until a constant product weight was achieved.

In a further attempt, the DES was prepared using the ultrasonic bath method [76]. As widely reported in the literature, the use of ultrasound allows for obtaining a more homogeneous mixture in a shorter time due to the phenomenon of cavitation [77]. Nevertheless, the DES was obtained as a clear and homogeneous pink viscous liquid only after 60 min of ultrasonication of the components (Table 1, entry 6, DES-2).

The preparation of the DES was also attempted using the heating and stirring method. In this case, 20% of water was added to a mixture of $ErCl_3 \cdot 6H_2O$ and urea (in a 3:10 molar ratio). Subsequently, the components were placed in a round-bottom flask and subjected to magnetic stirring in a 60 °C water bath for 1 h, resulting in the formation of a clear, viscous pink liquid (Table 1, entry 7, DES-3). A final experiment (Table 1, entry 8, DES-4) was conducted in the absence of water by exploiting the efficiency of both ultrasound and vacuum evaporating. The mixture, consisting of erbium trichloride and urea, underwent ultrasonication for 1 h and was then stirred under vacuum at reduced pressure at 60 °C for an additional 1 h. Even in this case, the product appeared as a pink, viscous liquid (Figure 2b).

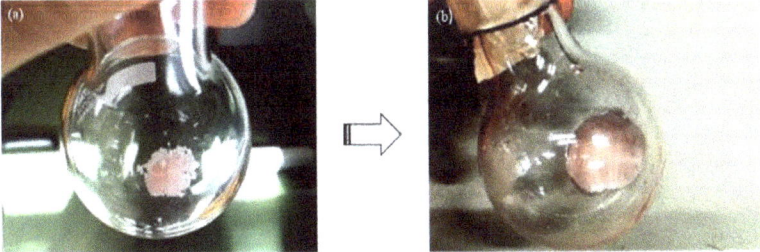

Figure 2. DES-4 prepared using both vacuum evaporation and ultrasound-assisted methods. (**a**) $ErCl_3 \cdot 6H_2O$ and urea, solid reagents in contact; (**b**) formation of the DES-4 after US and heating.

Taking into consideration the results obtained from previous tests, we inferred that erbium hexahydrate combined with urea can form a Deep Eutectic Solvent (DES) when a molar ratio of 3:10 is employed. Additionally, it is observed that the presence of additional water for DES formation is not necessary, and both heating and ultrasound simplify the product formation.

3.2. DES Characterization

The formation of the liquid mixture when the two components were combined in a 3:10 molar ratio was supported by observation and POM imaging. The POM image revealed that the DES was formed because a black image was visible, showing that the sample was completely amorphous, and no crystals remained in the system. The successful formation of the DESs was further demonstrated by measuring the melting point or glass transition temperature of the DESs by differential scanning calorimetry (DSC) analysis as shown graphically in Figure 3.

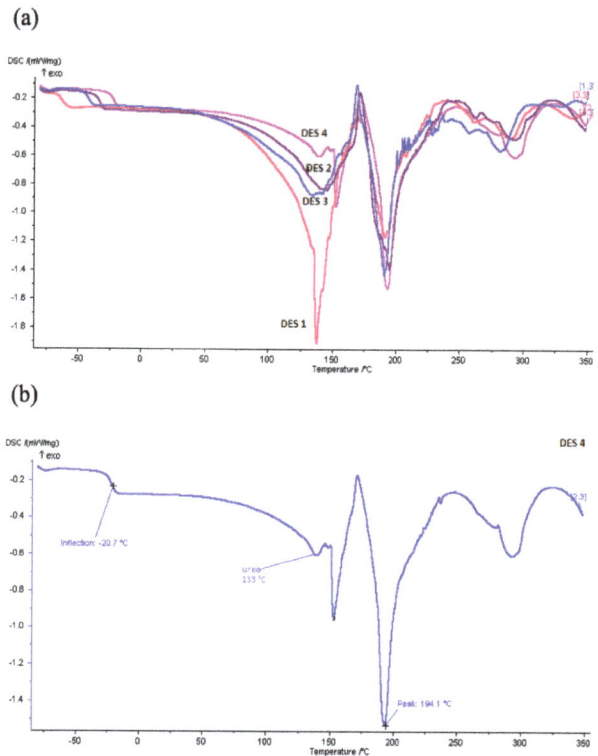

Figure 3. DSC thermograms of the (**a**) prepared DESs and (**b**) single DES-4.

From the thermograms obtained, it was possible to denote how the melting point of the eutectic is different than its individual components, erbium trichloride (776 °C) and urea (133 °C). According to Abbott and collaborators, this change is dependent on the reticular energies of the salt and hydrogen-bond donors (HBD), the way in which the couple interacts, and the entropy changes deriving from the formation of a liquid phase [78,79]. Furthermore, by comparing the four thermograms and observing the intensity of the peaks due to urea as well as the eutectic point, we could deduce that the DES-4 was the best-prepared one. In fact, DES-4 showed a less intense peak relative to urea than that observed in the other thermograms (Figure 3b). This indicated a lower amount of urea in the free state, and consequently, its better coordination with erbium. In this context, the lack of DES formation when using anhydrous erbium trichloride underlines how crystalline water represents a fundamental feature for the formation of this category of DES. Therefore, water molecules are real ligands coordinated at the center of the erbium atom or associated with free chloride ions [80].

From the thermogram in Figure 3b, it is also possible to establish how DES-4 has a high glass-transition temperature ($-20.7\ ^\circ$C) compared to the other experiments, thus confirming the stability of the solvent obtained. These observations led us to hypothesize a structure of the DES formed: erbium is hexacoordinate [81], that is, it forms natural binding interactions with the three chloride ions and coordinative interactions with three urea molecules, assuming a hypothetical structure that is represented in Figure 4.

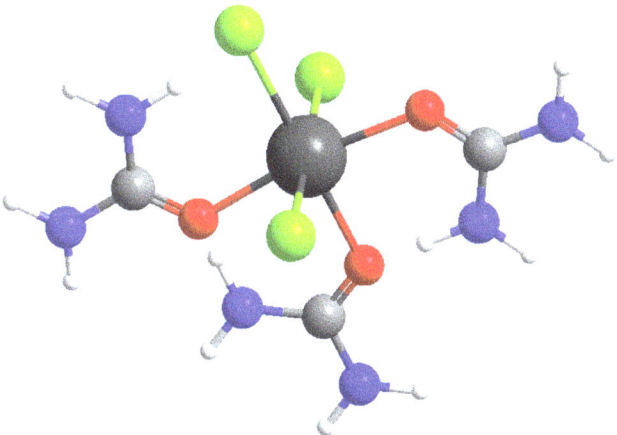

Figure 4. Hypothesis of the hexacoordinated structure of DES-4.

FTIR characterization of the synthesized DES was conducted in order to study the functional groups of the components present in the DES and analyze the possible changes in the structure. The FTIR spectral data of pure urea were in accordance with the literature [82] showing the vibrational bands of the NH_2 group at 3500–3200 cm^{-1}, while the carbonyl stretching vibration is usually observed in the range of 1680–1650 cm^{-1}. The vibrational bands at 1623 and 1461 cm^{-1} correspond to NH and C–N bending, respectively, and C–N stretching vibration is usually found in the range of 1300–1000 cm^{-1}.

The FTIR spectrum of $ErCl_3 \cdot 6H_2O$ (Figure 5a) exhibited an absorption band around 3500–3100 cm^{-1} associated with the stretching vibrations of hydrogen atoms bound to oxygen in water molecules and around 1600 cm^{-1} (H–O–H bending vibration) indicating the presence of water molecules. The stretching vibrations of Er–Cl bonds appear in the lower wavenumber region, below 700 cm^{-1}.

The coordination between erbium chloride hexahydrate and urea in DES-4 can indeed influence the stretching vibrations observed in the FTIR spectrum and are presented in Figure 5b.

In general, when erbium coordinates with urea or any ligand, it can lead to shifts or changes in the vibrational frequencies compared to the free ligand or metal salt. Coordination between erbium trichloride with urea resulted in shifts or splitting of the urea-related bands, such as the carbonyl (C=O) and amino (N–H) stretching vibrations, depending on the coordination mode. The FTIR spectrum for the DES-4 displays all bands corresponding to the functional groups of both constituents; however, a new peak below 2900 cm^{-1} appeared, which might suggest that new bonds are formed by coordination between urea and erbium trichloride. The characteristic vibrational modes in urea, including the carbonyl (C=O) stretching around 1650–1600 cm^{-1} and the N–H stretching around 3400–3200 cm^{-1}, are influenced by the coordination with erbium ions and the presence of water molecules. The carbonyl vibration in urea showed a shift from 1678 cm^{-1} and overlaps with the bands in the FTIR of erbium trichloride, indicating that the coordination bonds were mainly formed between the oxygen of the carbonyl in urea and the metal ion.

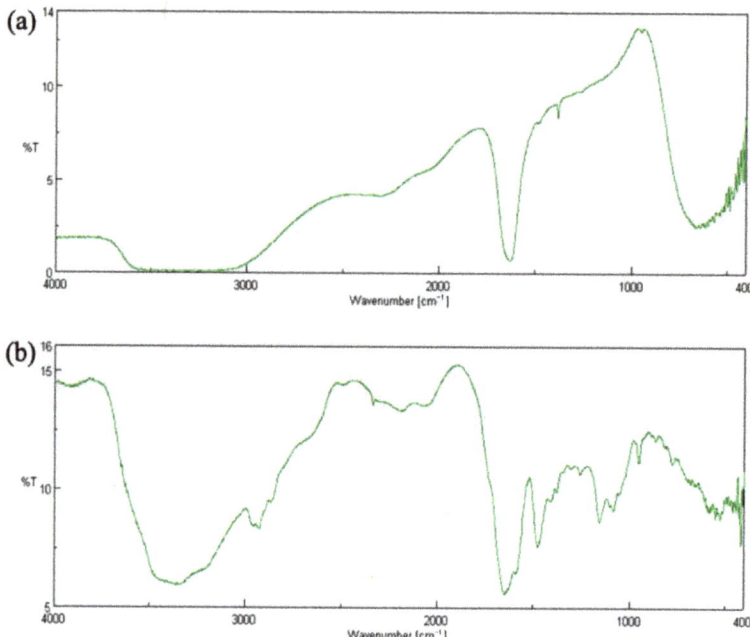

Figure 5. FT-IR spectra of (**a**) ErCl$_3$·6H$_2$O and (**b**) DES-4.

3.3. Synthesis of Cellulose Carbamate

The role of DES-4 as a solvent, reagent, and catalyst was assessed in the preparation of cellulose carbamate. The presence of a large excess of urea in this DES-4 indicates that consumption of urea in the cellulose carbamate preparation does not disrupt the DES solvent system. Various experiments were conducted under different conditions to identify the most suitable procedure for obtaining carbamate, as schematically summarized in Table 2.

Table 2. Optimization data for the synthesis of CC.

Sample	Substrate	CC Preparation Method	Reaction Time (min)	N (%) [a]	Degree of Substitution [b]
1	Cellulose + H_2O	Vacuum evaporation	60	1.59	0.19
2	Cellulose + H_2O	US	30	1.42	0.17
3	Dried cellulose	US	30	0.86	0.10
4 [c]	Cellulose + urea	Vacuum evaporation	60	0.27	0.05

[a] value obtained by elemental analysis; [b] calculated from elemental analysis results; [c] experiment carried out in water and in the absence of $ErCl_3 \cdot 6H_2O$.

In the first attempt, a suspension of microcrystalline cellulose (MCC) in water was added to DES-4. The mixture was stirred using a rotary evaporator under reduced pressure at 50 °C for 1 h, resulting in a pink homogeneous mixture. The obtained mixture was further dried in an oven at 100 °C for 24 h and then in contact with an oil bath at 150 °C for 1 h. The product, dry and adhering to the flask walls, was recovered by washing with a 12 M HCl solution, followed by filtration and additional washing with distilled water. The excess reagents and DES components were removed, and the filtered product was dried using a vacuum pump. A white material was obtained (Table 2, sample 1).

The FTIR spectra of dried cellulose and cellulose carbamate provide valuable insights into the chemical changes associated with the carbamation process. Dried cellulose typically exhibits characteristic peaks in the FTIR spectrum corresponding to the hydroxyl groups at around 3600–3200 cm^{-1} and the cellulose backbone vibrations [83,84]. Furthermore, C–O stretching around 1000–1300 cm^{-1} is predominant.

Upon carbamation, a distinctive peak emerges in the carbonyl stretching region at 1716.34 cm^{-1}, signaling the introduction of carbamate groups (Figure 6). This peak is attributed to the C=O stretching vibration of the carbonyl group in the carbamate functionality. The presence of this carbonyl peak in cellulose carbamate confirms the successful functionalization of cellulose with carbamate groups. The intensity and position of the carbonyl peak can provide information about the extent of functionalization and the nature of the carbamate groups introduced. This observation aligns with published literature data [77]. The absorption of cellulose hydroxyl groups at 3429.78 cm^{-1} decreases during carbamate formation, consistent with the formation of carbamates, especially with the primary OH groups [85].

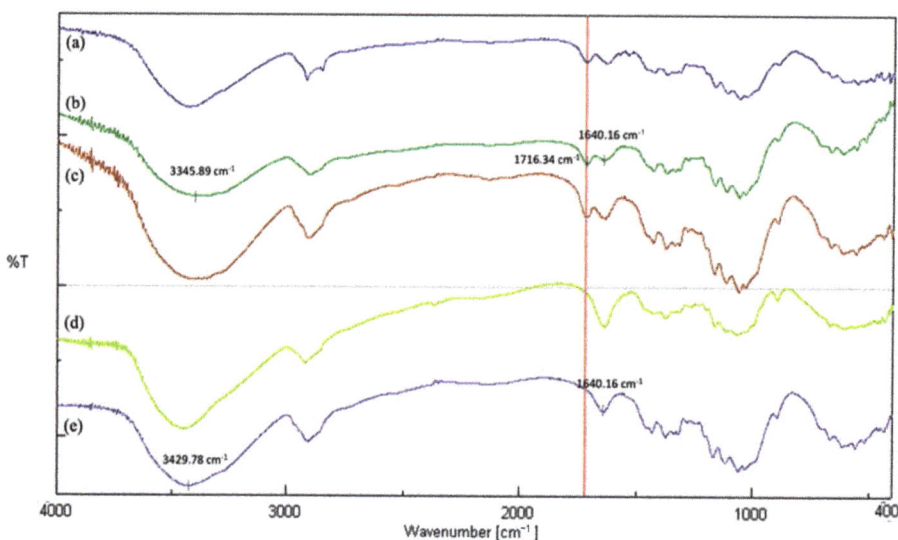

Figure 6. FT-IR spectra of (**a**) sample 2 (N% = 1.42, DS = 0.17), (**b**) sample 1 (N% = 1.59, DS = 0.19), (**c**) sample 3 (N% = 0.86, DS = 0.10), (**d**) sample 4 (N% = 0.27, DS = 0.05), and (**e**) starting dried cellulose.

In the second experiment, the synthesis of CC was attempted using ultrasonication conditions to enhance mixture homogeneity (Table 2, sample 2). A suspension of MCC in water was added to DES-4, and the reaction was assisted by ultrasound for 30 min, followed by stirring under reduced pressure using a rotary evaporator at 50 °C for 1 h. Subsequent purification treatments afforded a solid material that was analyzed by FTIR. The FT-IR spectrum of the modified cellulose revealed, in this case as well, the presence of the characteristic absorption peak due to the stretching vibration of the urethane carbonyl.

To propose an even more eco-sustainable and convenient synthetic procedure, the third experiment was carried out in the absence of water using ultrasonication (Table 2, sample 3). Although water is often suggested as an environmentally friendly additive to regulate DES viscosity and polarity, it can dramatically alter the DES structure due to the rupture of hydrogen bonds between the initial constituents [86]. Cellulose was directly suspended in the previously prepared DES-4, and the mixture was subjected to ultrasonication for 1 h. The final purification treatments afforded a white powder.

To verify the need for the catalyst, erbium trichloride, in the synthesis of cellulose carbamate, a control experiment was conducted in its absence, using only water as the reaction medium (Table 2, sample 4). Cellulose was suspended in water, and urea was subsequently added. The reaction was assisted by ultrasound for 1 h. The product obtained from subsequent purification treatments appeared as a copper-colored material. The FT-IR spectrum of the sample did not reveal the presence of the typical peak due to the urethane carbonyl, indicating that urea did not functionalize the primary hydroxyl of cellulose.

The carbonyl absorption in the FTIR spectra correlates with the nitrogen content, with a higher percentage of nitrogen resulting in a greater absorption peak. To verify this statement, the nitrogen content value (N%) of CC samples was determined using elemental analysis and the N% for each sample has been reported in detail in Table 2. These values were used to calculate the DS of hydroxyl groups on cellulose by carbamate groups and to confirm that the first experiment (sample 1) furnished the best experimental conditions for the preparation of CC with the highest nitrogen content (1.59%) that corresponded to a DS of 0.19.

4. Discussion

Cellulose carbamate, synthesized using DES-4, was characterized by an FT-IR spectrometer. Analyzing the FT-IR spectra related to the different syntheses attempted, it can be concluded that well-defined spectra with a more evident peak due to the stretching of the carbonyl (C=O) of urethane were obtained from the experiments conducted in the presence of water. From these observations, we can hypothesize the formation of cellulose carbamate in a DES solvent according to the scheme described in Scheme 2.

From the reaction scheme, it can be observed how in the presence of water, the chloride ion is replaced in its bond with the erbium and the compound changes from hexacoordinate to an octacoordinated complex [81,82]. Erbium acting as a Lewis acid catalyst makes the amide carbon of urea more electrophilic. The formation of isocyanic acid is promoted by the development of an acid–base reaction that takes place between the amino group of urea and the water molecule coordinated with the erbium. The resulting ammonia causes the loss of a proton from the isocyanic cation with the consequent formation of ammonium chloride. Water favors the displacement of isocyanic acid from its coordination bond with erbium. Isocyanic acid at this point, being strongly reactive, in the free state allows the functionalization of cellulose. The presence of water, according to this hypothetical reaction scheme, is essential to allow the displacement first of the chloride ion and then of the isocyanic acid. Since the coordination bond between erbium and urea is shorter than the bonds with chlorine and the coordination bonds with isocyanic acid, it is evident that the action of water is exerted first in the replacement of chlorine atoms and subsequently in the replacement of the isocyanate formed, while it has no interference with coordination with urea being the strongest bond.

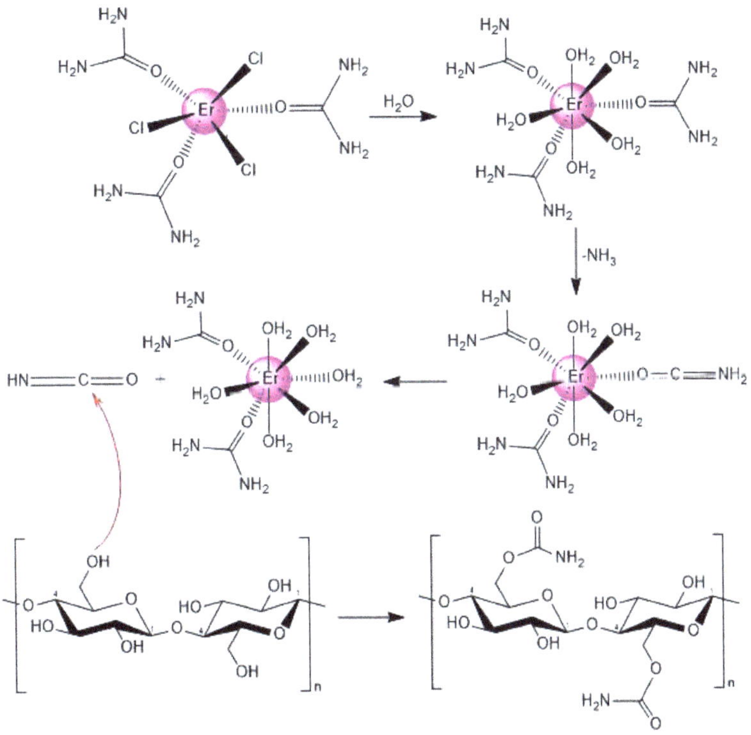

Scheme 2. Hypothetical synthesis of cellulose carbamate using DES-4. In the scheme, for simplicity, the charges have been omitted.

5. Conclusions

In conclusion, a novel type IV DES solvent was effectively prepared and used, exploiting its role as a solvent/catalyst and, at the same time, as an actual reactive component in the preparation of cellulose carbamate. Considering the comparatively lower toxicity of $ErCl_3·6H_2O$ compared to common table salt, we advocate for the adoption of this innovative RDES composed of urea and erbium trichloride as an inexpensive, efficient, and environmentally compatible, non-toxic system for cellulose carbamate production. Notably, samples with a nitrogen content as high as 1.59% were achieved through this method. The potential for scalable implementation of this green synthesis approach in the future holds promise to produce environmentally friendly products, improving cellulose solubility and streamlining its derivatization process.

Author Contributions: Conceptualization, M.L.D.G. and A.D.N.; methodology, V.A., L.M., F.P.N. and S.T.; formal analysis, V.A., D.P., S.T. and P.C.; investigation, M.L.D.G., A.D.N. and L.M.; data curation, L.M., P.C., S.T., D.P. and V.A.; writing—original draft preparation, M.L.D.G., A.D.N. and D.P.; writing—review and editing, M.L.D.G. and A.D.N.; supervision, M.L.D.G. and A.D.N.; funding acquisition, M.L.D.G. and A.D.N. All authors have read and agreed to the published version of the manuscript.

Funding: This research received no external funding.

Institutional Review Board Statement: Not applicable.

Data Availability Statement: Data are contained within the article.

Conflicts of Interest: The authors declare no conflicts of interest.

References

1. Singh, P.; Duarte, H.; Alves, L.; Antunes, F.; Le Moigne, N.; Dormanns, J.; Duchemin, B.; Staiger, M.P.; Medronho, B. From cellulose dissolution and regeneration to added value applications-synergism between molecular understanding and material development. In *Cellulose-Fundamental Aspects and Current Trends*; Intech: Berlin, Germany, 2015; pp. 1–44.
2. Aziz, T.; Farid, A.; Haq, F.; Kiran, M.; Ullah, A.; Zhang, K.; Li, C.; Ghazanfar, S.; Sun, H.; Ullah, R.; et al. A Review on the Modification of Cellulose and Its Applications. *Polymers* 2022, *14*, 3206. [CrossRef]
3. De Nino, A.; Tallarida, M.A.; Algieri, V.; Olivito, F.; Costanzo, P.; De Filpo, G.; Maiuolo, L. Sulfonated Cellulose-Based Magnetic Composite as Useful Media for Water Remediation from Amine Pollutants. *Appl. Sci.* 2020, *10*, 8155. [CrossRef]
4. Sun, Y.; Wang, J.; Li, D.; Cheng, F. The Recent Progress of the Cellulose-Based Antibacterial Hydrogel. *Gels* 2024, *10*, 109. [CrossRef]
5. Tabaght, F.E.; Azzaoui, K.; Idrissi, A.E.; Jodeh, S.; Khalaf, B.; Rhazi, L.; Bellaouchi, R.; Asehraou, A.; Hammouti, B.; Sabbahi, R. Synthesis, characterization, and biodegradation studies of new cellulose-based polymers. *Sci. Rep.* 2023, *13*, 1673. [CrossRef]
6. Kostag, M.; Jedvert, K.; Achtel, C.; Heinze, T.; El Seoud, O.A. Recent advaces in solvents for the dissolution, shaping and derivatization of cellulose: Quaternary ammonium electrolytes and their solutions in water and molecular solvents. *Molecules* 2018, *23*, 511. [CrossRef]
7. Coffey, D.G.; Bell, D.A.; Henderson, A. Cellulose and cellulose derivatives. In *Food Polysaccharides and Their Applications*, 2nd ed.; Stati Uniti, CRC Press Taylor & Francis Group: Boca Raton, FL, USA, 2006.
8. Jedverta, K.; Heinze, T. Cellulose modification and shaping—A review. *J. Polym. Engin.* 2017, *37*, 845–860. [CrossRef]
9. Lindman, B.; Karlström, G.; Stigsson, L. On the mechanism of dissolution of cellulose. *J. Mol. Liq.* 2010, *156*, 76–81. [CrossRef]
10. Michud, A.; Tanttu, M.; Asaadi, S.; Ma, Y.; Netti, E.; Kaariainen, P.; Persson, A.; Berntsson, A.; Hummel, M.; Sixta, H. Ioncell-F: Ionic liquid-based cellulosic textile fibers as an alternative to viscose and Lyocell. *Text. Res. J.* 2016, *86*, 543–552. [CrossRef]
11. Pinkert, A.; Marsh, K.N.; Pang, S.; Staiger, M.P. Ionic Liquids and their interaction with cellulose. *Chem. Rev.* 2009, *109*, 6712–6728. [CrossRef]
12. Striegela, A. Theory and applications of DMAc/LiCl in the analysis of polysaccharides. *Carbohydr. Polym.* 1997, *34*, 267–274. [CrossRef]
13. Isogai, A.; Ishizu, A.; Nakano, J. Dissolution mechanism of cellulose in SO_2-amine-dimethylsulfoxide. *J. Appl. Polym. Sci.* 1987, *33*, 1283–1290. [CrossRef]
14. Fink, H.-P.; Weigel, P.; Purz, H.; Ganster, J. Structure formation of regenerated cellulose materials from NMMO-solutions. *Progr. Polym. Sci.* 2001, *26*, 1473–1524. [CrossRef]
15. Swatloski, R.P.; Spear, S.K.; Holbrey, J.D.; Rogers, R.D. Dissolution of Cellulose with Ionic Liquids. *J. Am. Chem. Soc.* 2002, *124*, 4974–4975. [CrossRef]
16. Heinze, T.; Koschella, A. Solvents applied in the field of cellulose chemistry-A mini review. *Polim.-Cienc. E Tecnol.* 2005, *15*, 84–90. [CrossRef]

17. Fischer, S.; Voigt, W.; Fischer, K. The behaviour of cellulose in hydrated melts of the composition LiX-H$_2$O (X=I-, NO$_3$-, CH$_3$COO-, ClO$_4$-). *Cellulose* **1999**, *6*, 213–219. [CrossRef]
18. Varshney, V.K.; Naithani, S. Chemical functionalization of cellulose derived from nonconventional sources. In *Cellulose Fibers: Bio- and Nano-Polymer Composites*; Springer: Berlin/Heidelberg, Germany, 2011; pp. 43–60.
19. Ali, H. *Cellulose Carbamate: Production and Applications*; VTT Technical Research Centre of Finland: Espoo, Finland, 2019; p. 34.
20. Zhang, Y.; Yin, C.; Zhang, Y.; Wu, H. Synthesis and Characterization of Cellulose Carbamate from Wood Pulp, Assisted by Supercritical Carbon Dioxide. *BioResources* **2013**, *8*, 1398–1408. [CrossRef]
21. Flores, A.; Canamares, D.; Ticona, L.A.; Pablos, J.L.; Pena, J.; Hernaiz, M.J. Highly efficient functionalization of hydroxypropyl cellulose with glucuronic acid through click chemistry under microwave irradiation as potential biomaterial with therapeutic properties. *Catal. Today* **2024**, *429*, 114495. [CrossRef]
22. Xu, M.; Li, T.; Zhang, S.; Li, W.; He, J.; Yin, C. Preparation and characterization of cellulose carbamate membrane with high strength and transparency. *J. Appl. Polym. Sci.* **2021**, *38*, 50068. [CrossRef]
23. Hu, J.; Li, R.; Zhu, S.; Zhang, G.; Zhu, P. Facile preparation and performance study of antibacterial regenerated cellulose carbamate fiber based on N-halamine. *Cellulose* **2021**, *28*, 4991–5003. [CrossRef]
24. Bui, C.V.; Rosenau, T.; Hettegger, H. Synthesis of polyanionic cellulose carbamates by homogeneous aminolysis in an ionic liquid/DMF medium. *Molecules* **2022**, *27*, 1384. [CrossRef]
25. Iller, E.; Stupinska, H.; Starostka, P. Properties of cellulose derivatives produced from radiation-Modified cellulose pulps. *Radiat. Phys. Chem.* **2007**, *76*, 1189–1194. [CrossRef]
26. Klemm, D.; Philipp, B.; Heinze, T.; Heinze, U.; Wagenknecht, W. *Comprehensive Cellulose Chemistry: Fundamentals and Analytical Methods*, 1st ed.; Wiley-VCH Verlag GmbH & Co. KGaA: Weinheim, Germany, 2004; Volume 2, pp. 1–308.
27. Guo, Y.; Zhou, J.; Song, Y.; Zhang, L. An efficient and environmentally friendly method for the synthesis of cellulose carbamate by microwave heating. *Macromol. Rapid Commun.* **2009**, *30*, 1504–1508. [CrossRef]
28. Fu, F.; Zhou, J.; Zhou, X.; Zhang, L.; Li, D.; Kondo, T. Green Method for Production of Cellulose Multifilament from Cellulose Carbamate on a Pilot Scale. *ACS Sustain. Chem. Engin.* **2014**, *2*, 2363–2370.
29. Guo, Y.; Zhou, J.; Wang, Y.; Zhang, L.; Lin, X. An efficient transformation of cellulose into cellulose carbamates assisted by microwave irradiation. *Cellulose* **2010**, *17*, 1115–1125. [CrossRef]
30. Wang, X.; Wu, H.; Yin, C. Preparation and characterization of cellulose carbamate regeneration membrane by supercritical carbon dioxide. *Adv. Mat. Res.* **2013**, *821*, 1031–1034. [CrossRef]
31. Fu, F.; Xu, M.; Wang, H.; Wang, Y.; Ge, H.; Zhou, J. Improved synthesis of cellulose carbamates with minimum urea based on an easy scale-up method. *ACS Sustain. Chem. Engin.* **2015**, *3*, 1510–1517. [CrossRef]
32. Procopio, D.; Siciliano, C.; Trombino, S.; Dumitrescu, D.E.; Suciu, F.; Di Gioia, M.L. Green solvents for the formation of amide linkages. *Org. Biomol. Chem.* **2022**, *20*, 1137–1149. [CrossRef]
33. Kaur, G.; Kumar, H.; Singla, M. Diverse applications of ionic liquids: A comprehensive review. *J. Mol. Liq.* **2022**, *351*, 118556–118574. [CrossRef]
34. De Nino, A.; Maiuolo, L.; Merino, P.; Nardi, M.; Procopio, A.; Roca-Lpez, D.; Russo, B.; Algieri, V. Efficient Organocatalyst Supported on a Simple Ionic Liquid as a Recoverable System for the Asymmetric Diels-Alder Reaction in the Presence of Water. *ChemCatChem* **2015**, *7*, 830–835. [CrossRef]
35. Di Gioia, M.L.; Costanzo, P.; De Nino, A.; Maiuolo, L.; Nardi, M.; Olivito, F.; Procopio, A. Simple and efficient Fmoc removal in ionic liquid. *RSC Adv.* **2017**, *7*, 36482–36491. [CrossRef]
36. Di Gioia, M.L.; Gagliardi, A.; Leggio, A.; Leotta, V.; Romio, E.; Liguori, A. N-Urethane protection of amines and amino acids in an ionic liquid. *RSC Adv.* **2015**, *5*, 63407–63420. [CrossRef]
37. Di Gioia, M.L.; Barattucci, A.; Bonaccorsi, P.; Leggio, A.; Minuti, L.; Romio, E.; Temperini, A.; Siciliano, C. Deprotection/reprotection of the amino group in α-amino acids and peptides. A one-pot procedure in [Bmim][BF$_4$] ionic liquid. *RSC Adv.* **2014**, *4*, 2678–2686. [CrossRef]
38. Gericke, M.; Fardim, M.; Heinze, T. Ionic Liquids-Promising but Challenging Solvents for Homogeneous Derivatization of Cellulose. *Molecules* **2012**, *17*, 7458–7502. [CrossRef]
39. Bubalo, M.C.; Radošević, K.; Redovniković, I.R.; Slivac, I.; Srček, V.G. Toxicity mechanisms of ionic liquids. *Arh. Hig. Rada Toksikol.* **2017**, *68*, 171–179. [CrossRef]
40. Bubalo, M.C.; Radošević, K.; Redovniković, I.R.; Halambek, J.; Srček, V.G. A brief overview of the potential environmental hazards of ionic liquids. *Ecotoxicol. Environ. Safety* **2014**, *99*, 1–12. [CrossRef]
41. De Nino, A.; Merino, P.; Algieri, V.; Nardi, M.; Di Gioia, M.L.; Russo, B.; Tallarida, M.A.; Maiuolo, L. Synthesis of 1,5-functionalized 1,2,3-triazoles using ionic liquid/iron (III) chloride as an efficient and reusable homogeneous catalyst. *Catalysts* **2018**, *8*, 364. [CrossRef]
42. Smith, E.L.; Abbott, A.P.; Ryder, K.S. Deep Eutectic Solvents (DESs) and their applications. *Chem. Rev.* **2014**, *114*, 11060–11082. [CrossRef]
43. Paiva, A.; Craveiro, R.; Aroso, I.; Martins, M.; Reis, R.L.; Duarte, A.R.C. Natural Deep Eutectic Solvents-solvents for the 21st century. *ACS Sustain. Chem. Engin.* **2014**, *2*, 1063–1071. [CrossRef]
44. Perna, F.P.; Vitale, P.; Capriati, V. Deep eutectic solvents and their applications as green solvents. *Curr. Opin. Green Sustain. Chem.* **2020**, *21*, 27–33. [CrossRef]

45. Alonso, D.A.; Baeza, A.; Chinchilla, R.; Guillena, G.; Pastor, I.M.; Ramon, D.J. Deep Eutectic Solvents: The organic reaction medium of the century. *Eur. J. Org. Chem.* **2016**, *4*, 612–632. [CrossRef]
46. Boldrini, C.L.; Manfredi, N.; Perna, F.M.; Capriati, V.; Abbotto, A. Designing Eco-Sustainable Dye-Sensitized Solar Cells by the Use of a Menthol-Based Hydrophobic Eutectic Solvent as an Effective Electrolyte Medium. *Chem.-Eur. J.* **2018**, *24*, 17656–17659. [CrossRef]
47. Procopio, D.; Siciliano, C.; Di Gioia, M.L. Reactive Deep Eutectic Solvents for EDC-mediated Amide Synthesis. *Org. Biomol. Chem.* **2024**, *22*, 1400–1408. [CrossRef]
48. Procopio, D.; Marset, X.; Guillena, G.; Di Gioia, M.L.; Ramon, D.J. Visible-Light-Mediated Amide Synthesis in Deep Eutectic Solvents. *Adv. Synth. Catal.* **2023**, *365*, 1–8. [CrossRef]
49. Taklimi, S.M.; Divsalar, A.; Ghalandari, B.; Ding, X.; Di Gioia, M.L.; Omar, K.A.; Saboury, A.A. Effects of deep eutectic solvents on the activity and stability of enzymes. *J. Mol. Liq.* **2023**, *377*, 121562. [CrossRef]
50. Di Gioia, M.L.; Duarte, A.R.C.; Gawande, M.B. Editorial: Advances in the development and application of deep eutectic solvents. *Front. Chem.* **2023**, *11*, 125871. [CrossRef]
51. Procopio, D.; Siciliano, C.; De Rose, R.; Trombino, S.; Cassano, R.; Di Gioia, M.L. A Bronsted Acidic Deep Eutectic Solvent for N-Boc Deprotection. *Catalysts* **2022**, *12*, 1480. [CrossRef]
52. Gallardo, N.; Saavedra, B.; Guillena, G.; Ramón, D.J. Indium-mediated allylation of carbonyl compounds in deep eutectic solvents. *Appl. Organomet. Chem.* **2021**, *35*, e6418. [CrossRef]
53. Di Gioia, M.L.; Cassano, R.; Costanzo, P.; Cano, N.H.; Maiuolo, L.; Nardi, M.; Nicoletta, F.P.; Oliverio, M.; Procopio, A. Green synthesis of privileged benzimidazole scaffolds using active deep eutectic solvent. *Molecules* **2019**, *24*, 2885. [CrossRef]
54. Sirvio, J.A.; Ukkola, J.; Liimatainen, H. Direct sulfation of cellulose fibers using a reactive deep eutectic solvent to produce highly charged cellulose nanofibers. *Cellulose* **2019**, *26*, 2303–2316. [CrossRef]
55. Willberg-Keyriläinen, P.; Hiltunen, J.; Ropponen, J. Production of cellulose carbamate using urea-based deep eutectic solvents. *Cellulose* **2018**, *25*, 195–204. [CrossRef]
56. Abbott, A.P.; Bell, T.J.; Handaa, S.; Stoddartb, B. Cationic functionalisation of cellulose using a choline based ionic liquid analogue. *Green Chem.* **2006**, *8*, 784–786. [CrossRef]
57. Teng, Y.; Yu, G.; Fu, Y.; Yin, C. The preparation and study of regenerated cellulose fibers by cellulose carbamate pathway. *Int. J. Biol. Macromol.* **2017**, *7*, 383–392. [CrossRef]
58. Sirviö, J.A.; Heiskanen, J.P. Synthesis of alkaline-soluble cellulose methyl carbamate using reactive deep eutectic solvent. *ChemSusChem* **2017**, *10*, 455–460. [CrossRef] [PubMed]
59. Fu, Z.; Wu, F.; Chen, L.; Xu, B.; Feng, C.; Bai, Y.; Liao, H.; Sun, S.; Giesy, J.P.; Guo, W. Copper and zinc, but not other priority toxic metals, pose risks to native aquatic species in a large urban lake in Eastern China. *Environ. Pollut.* **2016**, *219*, 1069–1076. [CrossRef]
60. Erfanifar, E.; Jahanjo, V.; Kasalkhe, N.; Erfanifar, E. Acute toxicity test of Zinc Chloride ($ZnCl_2$) in sobaity seabream (*Sparidebtex hasta*). *Environ. Sci.* **2016**, *1*, 47–51.
61. Walsh, C.T.; Sandstead, H.H.; Prasad, A.S.; Newberne, P.M.; Fraker, P.J. Zinc: Health effects and research priorities for the 1990s. *Environ. Health Perspect.* **1994**, *102*, 5–46.
62. Hjortsø, E.; Ovist, J.; Bud, M.I.; Thomsen, J.L.; Andersen, J.B.; Wiberg-Jørgensen, F.; Jensen, N.K.; Jones, R.; Reid, L.M.; Zapol, W.M. ARDS after accidental inhalation of zinc chloride smoke. *Intensive Care Med.* **1988**, *14*, 17–24. [CrossRef]
63. Amara, S.; Slama, I.B.; Mrad, I.; Rihane, N.; Khemissi, W.; El Mir, L.; Rhouma, K.B.; Abdelmelek, H.; Sakly, M. Effects of zinc oxide nanoparticles and/or zinc chloride on biochemical parameters and mineral levels in rat liver and kidney. *Hum. Exp. Toxicol.* **2014**, *33*, 1150–1157. [CrossRef]
64. García-Gómez, C.; Babin, M.; Obrador, A.; Álvarez, J.M.; Fernández, M.D. Integrating ecotoxicity and chemical approaches to compare the effects of ZnO nanoparticles, ZnO bulk, and $ZnCl_2$ on plants and microorganisms in a natural soil. *Environ. Sci. Pollut. Res. Int.* **2015**, *22*, 16803–16813. [CrossRef]
65. Hirano, S.; Suzuki, K.T. Exposure, metabolism, and toxicity of rare earths and related compounds. *Environ. Health Perspect.* **1996**, *104*, 85–95.
66. Procopio, A.; Dalpozzo, R.; De Nino, A.; Nardi, M.; Oliverio, M.; Russo, B. Er(OTf)$_3$ as a Valuable Catalyst in a Short Synthesis of 2′,3′-Dideoxy Pyranosyl Nucleosides via Ferrier Rearrangement. *Synthesis* **2006**, *15*, 2608–2612. [CrossRef]
67. Maiuolo, L.; Russo, B.; Algieri, V.; Nardi, M.; Di Gioia, M.L.; Tallarida, M.A.; De Nino, A. Regioselective synthesis of 1,5-disubstituted 1,2,3-triazoles by 1,3-dipolar cycloaddition: Role of Er(OTf)$_3$, ionic liquid and water. *Tetrahedron Lett.* **2019**, *60*, 672–674. [CrossRef]
68. Oliverio, M.; Nardi, M.; Costanzo, P.; Di Gioia, M.L.; Procopio, A. Erbium salts as non-toxic catalysts compatible with alternative reaction media. *Sustainability* **2018**, *10*, 721. [CrossRef]
69. Nardi, M.; Di Gioia, M.L.; Costanzo, P.; De Nino, A.; Maiuolo, L.; Oliverio, M.; Olivito, F.; Procopio, A. Selective acetylation of small biomolecules and their derivatives catalyzed by Er(OTf)$_3$. *Catalysts* **2017**, *7*, 269. [CrossRef]
70. Kyung, T.R.; Kwon, H.K.; Jung, S.P. Toxicological Evaluations of Rare Earths and Their Health Impacts to Workers: A Literature Review. *Saf. Health Work* **2013**, *4*, 12–26. [CrossRef]
71. Xiang, X.; Guo, L.; Wu, X.; Ma, X.; Xia, Y. Urea formation from carbon dioxide and ammonia at atmospheric pressure. *Environ. Chem. Lett.* **2012**, *10*, 295–300. [CrossRef]

72. Liesiene, J.; Kazlauske, J. Functionalization of cellulose: Synthesis of water-soluble cationic cellulose derivatives. *Cell. Chem. Technol.* **2013**, *47*, 515.
73. Zhang, M.; Tian, R.; Han, H.; Wu, K.; Wang, B.; Liu, Y.; Zhu, Y.; Lu, H.; Liang, B. Preparation strategy and stability of deep eutectic solvents: A case study based on choline chloride-carboxylic acid. *J. Clean. Prod.* **2022**, *345*, 131028. [CrossRef]
74. Dai, Y.; van Spronsen, J.; Witkamp, G.J.; Verpoorte, R.; Choi, Y.H. Natural deep eutectic solvents as new potential media for green technology. *Anal. Chim. Acta.* **2013**, *766*, 61–68. [CrossRef] [PubMed]
75. Gomez, F.J.; Espino, M.; Fernández, M.A.; Silva, M.F. A greener approach to prepare natural deep eutectic solvents. *ChemistrySelect* **2018**, *3*, 6122–6125. [CrossRef]
76. Hsieha, Y.-H.; Lia, Y.; Pana, Z.; Chenc, Z.; Luc, J.; Yuanc, J.; Zhua, Z.; Zhang, J. Ultrasonication-assisted synthesis of alcohol-based deep eutectic solvents for the extraction of active compounds from ginger. *Ultrason. Sonochem.* **2020**, *63*, 104915. [CrossRef] [PubMed]
77. Mason, T.J. Ultrasound in synthetic organic chemistry. *Chem. Soc. Rev.* **1997**, *26*, 443–451. [CrossRef]
78. Qin, H.; Hu, X.; Wang, J.; Cheng, H.; Chen, L.; Qi, Z. Overview of acidic deep eutectic solvents on synthesis, properties and applications. *Green Energy Environ.* **2020**, *5*, 8–21. [CrossRef]
79. Schick, C. Differential scanning calorimetry (DSC) of semicrystalline polymers. *Anal. Bioanal. Chem.* **2009**, *395*, 1589–1611. [CrossRef] [PubMed]
80. Nada, A.-A.M.; Kamel, S.; El-Sakhawy, M. Thermal behavior and infrared spectroscopy of cellulose carbamates. *Polym. Degrad. Stab.* **2000**, *70*, 347–355. [CrossRef]
81. Cossy, C.; Helm, L.; Merbach, A.E. Oxygen-17 Nuclear Magnetic Resonance Kinetic Study of Water Exchange on the Lanthanide (III). *Aqua Ions Inorg. Chem.* **1988**, *27*, 1973–1979. [CrossRef]
82. Cotton, S.A. Establishing coordination numbers for the lanthanides in simple complexes. *Comptes Rendus Chim.* **2005**, *8*, 129–145. [CrossRef]
83. Mani, M.; Susai, R. Investigation of inhibitive action of urea-Zn^{2+} system in the corrosion control of carbon steel in sea water. *I. J. Eng. Sci. Technol.* **2011**, *3*, 8048–8060.
84. Gupta, V.K.; Carrott, P.J.M.; Singh, R.; Chaudhary, M.; Kushwaha, S. Cellulose: A review as natural, modified, and activated carbon adsorbent. *Bioresour. Technol.* **2016**, *1*, 1–43.
85. Yin, C.; Shen, X. Synthesis of cellulose carbamate by supercritical CO_2-assisted impregnation: Structure and rheological properties. *Eur. Polym. J.* **2007**, *43*, 2111–2116. [CrossRef]
86. Ma, C.; Laaksonen, A.; Liu, C.; Lu, X.; Ji, X. The peculiar effect of water on ionic liquids and deep eutectic solvents. *Chem. Soc. Rev.* **2018**, *47*, 8685–8720. [CrossRef] [PubMed]

Disclaimer/Publisher's Note: The statements, opinions and data contained in all publications are solely those of the individual author(s) and contributor(s) and not of MDPI and/or the editor(s). MDPI and/or the editor(s) disclaim responsibility for any injury to people or property resulting from any ideas, methods, instructions or products referred to in the content.

Article

Crosslinked Polyesters as Fully Biobased Coatings with Cutin Monomer from Tomato Peel Wastes

Eleonora Ruffini [1], Andrea Bianchi Oltolini [1], Mirko Magni [2], Giangiacomo Beretta [2], Marco Cavallaro [1], Raffaella Suriano [1,*] and Stefano Turri [1]

[1] Department of Chemistry, Materials and Chemical Engineering "Giulio Natta", Politecnico di Milano, Piazza Leonardo da Vinci 32, 20133 Milano, Italy; eleonora.ruffini@polimi.it (E.R.); marco.cavallaro@polimi.it (M.C.); stefano.turri@polimi.it (S.T.)

[2] Department of Environmental Science and Policy, Università degli Studi di Milano, Via Celoria 2, 20133 Milano, Italy; mirko.magni@unimi.it (M.M.); giangiacomo.beretta@unimi.it (G.B.)

* Correspondence: raffaella.suriano@polimi.it; Tel.: +39-02-2399-3249

Abstract: Cutin, one of the main structural components of tomato peels, is a waxy biopolymer rich in hydroxylated fatty acids. In this study, 10,16-dihydroxyhexadecanoic acid (10,16-diHHDA) was extracted and isolated from tomato peels and exploited to develop fully crosslinked polyesters as potential candidates for replacing fossil-based metal protective coatings. A preliminary screening was conducted to select the base formulation, and then a design of experiments (DoE) was used as a methodology to identify the optimal composition to develop a suitable coating material. Different formulations containing 10,16-diHHDA and other biorefinery monomers, including 2,5-furandicarboxylic acid, were considered. To this end, all polyesters were characterized through differential scanning calorimetry (DSC) and gel content measurements to determine their T_g value and crosslinking efficiency. Compositions exhibiting the best trade-off between T_g value, chemical resistance, and sufficiently high 10,16-diHHDA content between 39 and 48 wt.% were used to prepare model coatings that were characterized for assessing their wettability, scratch hardness, chemical resistance, and adhesion to metal substrates. These polyester coatings showed a T_g in the range of 45–55 °C, a hydrophobic behavior with a water contact angle of around 100°, a good solvent resistance (>100 MEK double rubs), and an adhesion strength to steel higher than 2 MPa. The results obtained confirmed the potential of cutin-based resins as coatings for metal protection, meeting the requirements for ensuring physicochemical properties of the final product, as well as for optimizing the valorization of such an abundant agri-food waste as tomato peels.

Keywords: cutin; agro-waste; polyester resins; coatings; 2,5-furandicarboxylic acid

Citation: Ruffini, E.; Bianchi Oltolini, A.; Magni, M.; Beretta, G.; Cavallaro, M.; Suriano, R.; Turri, S. Crosslinked Polyesters as Fully Biobased Coatings with Cutin Monomer from Tomato Peel Wastes. *Polymers* **2024**, *16*, 682. https://doi.org/10.3390/polym16050682

Academic Editors: Cristina Cazan and Mihai Alin Pop

Received: 5 January 2024
Revised: 26 February 2024
Accepted: 29 February 2024
Published: 2 March 2024

Copyright: © 2024 by the authors. Licensee MDPI, Basel, Switzerland. This article is an open access article distributed under the terms and conditions of the Creative Commons Attribution (CC BY) license (https://creativecommons.org/licenses/by/4.0/).

1. Introduction

Cutin is a polyfunctional biopolyester constituted of C16 and C18 fatty acids, with dihydroxylated C16 fatty acids being the most abundant ones (more than 60 wt.%) [1–4]. Cutin is one of the main constituents (between 40 and 80 wt.%) of the plant cuticle, the external layer covering and protecting the aerial parts of plants [5–7]. Therefore, it is extensively available and easily recoverable from different agricultural sources, among which tomato peel is the best option [8–10].

Tomato is the second vegetable source produced and consumed worldwide, next to potatoes. Global tomato production is estimated at around 160 Mt/y, of which up to 40% are processed (i.e., 40 Mt/y), with California, China, and some UE countries (Italy, Spain, and Portugal) being major players [11]. A quantity of around 5–30% of tomato pomace is normally lost as food waste and is used as animal feed or disposed of in landfills [12]. Instead, it may become an important biosource of sustainable chemicals and monomers. About 27% of that tomato pomace is represented by skin, of which cutin is a major component (40–80% by weight), leading to an estimated global production potential of 0.2–2.5 Mt/y of cutin [13,14].

Based on these data, cutin is gaining interest as a viable alternative to petroleum-based monomers and polymers in some target applications. Indeed, thanks to its remarkable properties, cutin has been considered as a promising candidate for the development of plant cuticle-like biobased materials to be employed in the food packaging sector, in line with circular bioeconomy principles and guidelines [15–17].

From this point of view, biodegradability is one of the main features that make cutin appealing, since it can be decomposed in soil in a reasonably short time (i.e., three to eight months) at a similar rate compared to bacterial polyhydroxyalkanoates and cellulose, exhibiting higher degradation efficiency than polylactic acid [18–21]. As regards mechanical properties, cutin polymers show higher elongation at break compared to other commercial plastics and bioplastics, despite being less rigid [22]. Indeed, tomato cutin exhibits a mechanical behavior similar to that of some elastomers, with an elongation at break of 27% and Young's modulus of 45 MPa at 23 °C and 40% relative humidity, making it more ductile and less rigid than other polymers such as PLA, P3HB, and cellulose [23]. Furthermore, cutin's non-toxicity and water resistance make it a particularly suitable material for metal packaging containers like cans, where food contact approval, chemical inertness, and mechanical robustness are essential requirements that must be ensured [24,25]. Innovative solutions for converting tomato skin into biodegradable plastic (BIOPROTO EU-funded project) [26] and green lacquer for food packaging (Agrimax EU-funded project [27] and TomaPaint S.r.l. [28]) have been proposed since 2014. However, to the best of the authors' knowledge, high T_g polymer coatings from tomato peel waste were not a concern in previous studies [4,29–31]. A large fraction of industrial coatings is used on metals, and some of the larger end uses are automotive, appliance, container, and coil coatings. Among the different products available for metal protective coatings, epoxy resins based on bisphenol A (BPA) and epichlorohydrin are the most widely employed and have the highest market share of more than 90% [32]. Nevertheless, BPA is a harmful compound to both the environment and human beings. Indeed, it has been demonstrated that BPA can migrate to the human body, causing reproductive anomalies, cardiovascular disease, diabetes, and cancer [33–36].

The global BPA market is around 5.6 Mt/y [37], of which about 20% is used for epoxy resin production (i.e., 1 Mt/y), and cutin can be generated at a rate of the same order of magnitude [13,38]. In this scenario, cutin-based materials could perfectly meet the demand for replacing at least part of BPA-based resins, and their main component, 10,16-dihydroxyhexadecanoic acid (10,16-diHHDA), can be proposed as a building block for the development of fully biobased, sustainable coatings for metal protection.

The general objective of this work is therefore the development of high T_g and high cutin monomer content fully biobased coatings for general metal protection applications. A T_g higher than (at least) 45–50 °C is needed in order to obtain a coating surface with a sufficiently high scratch resistance. On the other hand, a cutin content preferentially higher than 25–30% by weight is advisable to maximize the exploitation of the particular renewable resource. The cutin-derived, dihydroxy hexadecanoic acid monomer was extracted from tomato waste peels and then formulated with other biorefinery monomers (i.e., glycerol, citric acid, succinic acid, and 2,5-furandicarboxylic acid) to develop crosslinked polyesters. Model coatings were prepared and characterized to assess their wettability, scratch hardness, chemical resistance, and adhesion to metal substrates, resulting in good physicochemical properties. Therefore, this work can pave the way for the use of cutin-based crosslinked polyesters as a green alternative to BPA-based resins for industrial coating applications.

2. Materials and Methods

2.1. Materials

Tomato pomace was kindly supplied by Tomato Farm S.r.l. (Tomato Farm S.r.l., Bettole di Pozzollo (AL), Italy). Each seasonal batch of biowaste (ca. 50 kg for each) was collected at two different times, during the first week of September 2022 and 2023. During the sampling, the canning plant was processing the following varieties of tomatoes: grape tomatoes, San Marzano tomatoes, which are a variety of plum tomatoes, vine tomatoes, and cherry tomatoes. All these tomato varieties were grown in fields located in Italy, mostly in northern Italy and a few in the south. All reagents and solvents were purchased from Merck (Merck Life Science S.p.A., Milan, Italy) and used without any further purification if not otherwise specified.

2.2. Synthesis Procedures

2.2.1. Cutin Depolymerization and 10,16-Dihydroxyhexadecanoic Acid Isolation

In our previous work, we described a procedure for recovering 10,16-diHHDA from tomato-peel waste [39]. In this work, our protocol was further optimized in terms of both reaction time and yield of the reaction. Briefly, a known amount (around 30 g) of dried tomato peels is firstly degreased by Soxhlet employing n-hexane. The vegetable matrix is then depolymerized in alkaline conditions (ca. 1 M NaOH in methanol) under reflux for 3 h, filtered to remove solid residue, and left to rest overnight. Eventually, the crude solution is acidified by HCl down to ca. pH 3 to precipitate the desired 10,16-diHHDA compound that is recovered in the dichloromethane phase after extraction in a separatory funnel. In this work, the previously reported protocol was accurately followed, except for immediately processing the crude mixture resulting from the cutin depolymerization reaction. By immediately processing the crude mixture resulting from the cutin depolymerization reaction, a percentage increase of 50% in the yield of 10,16-diHHDA was achieved, when compared to the results obtained using the procedure developed in our previous work, without losing in terms of purity of the compound, as detailed explained in Section 3.1.

2.2.2. Synthesis of Bis(2,3-Dihydroxypropyl) Furan-2,5-Dicarboxylate Prepolymer

A mixture of glycerol and 2,5-furandicarboxylic acid dimethyl ester (FDME) in a molar ratio of 2:1 was put in a three-neck round-bottom flask. The reaction was carried out at 200 °C for 4 h under an inert atmosphere and mechanical stirring, in the presence of a drop of Tin(II) 2-ethylhexanoate as a catalyst. A comparable reaction between 2,5-furan dicarboxylic acid diethyl ester and ethanol was carried out by Zhao et al., employing both conventional heating and microwave irradiation [40]. The time and temperature parameters applied in the current study for the reaction between glycerol and FDME were derived from the aforementioned research by Zhao et al. [40] Specifically, this synthesis was conducted under conventional heating conditions, adapting the time and temperature parameters to the reactants employed and the equipment.

2.2.3. Polyester Resin Formulation and Preparation

The different crosslinked polyester compositions were prepared using the minimum volume of ethanol as a solvent and Ti(IV) isopropoxide (0.3 wt.%) as a catalyst. The mixture was stirred to achieve a homogeneous solution, then poured into a Petri dish and allowed to air-dry to evaporate part of the solvent. The resulting material was transferred into an oven and cured at 150 °C for 2 h at atmospheric pressure, and then kept under vacuum to complete the removal of volatile by-products. The polycondensation temperature employed in this study was derived from a previous paper on cutin-like co-polyester films [41]. Initially, a polycondensation time of 24 h was utilized in this study, as outlined in the abovementioned research. Subsequently, the duration was shortened, as this modification did not negatively affect the T_g and gel content values.

2.3. Design of Experiments (DoE)

To systematically explore the monomer composition effects on the coating's T_g, a design of experiments (DoE) approach was adopted. Specifically, the Box–Behnken design (BBD) and the response surface methodology (RSM) were employed. The experiments were established based on a BBD with three factors (i.e., (1) 10,16-diHHDA number of moles,

(2) glycerol number of moles, and (3) OH/COOH molar ratio) and three levels (10,16-diHHDA content of 0.5–1.5 mol; glycerol content of 1–4 mol; OH/COOH molar ratio of 0.75–1.25) coded as −1, 0, and +1, as reported in Table 1.

Table 1. Three factors and the corresponding three levels were selected for the Box–Behnken design.

Levels	Factors		
	10,16-diHHDA [mol]	Glycerol [mol]	OH/COOH Molar Ratio
−1	0.5	1	0.75
0	1	2.5	1
+1	1.5	4	1.25

Box–Behnken design is a rotatable quadratic design with no embedded factorial or fractional factorial points, where variable combinations represent points lying in the middle of the edges and the center of the variable space. The number of experiments (N) required for the development of a BBD is defined by the following equation:

$$N = k^2 + k + c_p \qquad (1)$$

where k is the number of factors; and c_p is the number of replicates of the central point [42].

In this case, it was chosen to consider three levels and one replicate of the central point, leading to 13 runs (Table 2). Furthermore, each experiment was performed in duplicate, resulting in a total number of 26 runs.

Table 2. Uncoded and coded Box–Behnken design for the three-factor system used in this study.

Std. Order	10,16-diHHDA [mol]	Glycerol [mol]	OH/COOH Molar Ratio
1	0.5 (−1)	1 (−1)	1 (0)
2	1.5 (+1)	1 (−1)	1 (0)
3	0.5 (−1)	4 (+1)	1 (0)
4	1.5 (+1)	4 (+1)	1 (0)
5	0.5 (−1)	2.5 (0)	0.75 (−1)
6	1.5 (+1)	2.5 (0)	0.75 (−1)
7	0.5 (−1)	2.5 (0)	1.25 (+1)
8	1.5 (+1)	2.5 (0)	1.25 (+1)
9	1 (0)	1 (−1)	0.75 (−1)
10	1 (0)	4 (+1)	0.75 (−1)
11	1 (0)	1 (−1)	1.25 (+1)
12	1 (0)	4 (+1)	1.25 (+1)
13	1 (0)	2.5 (0)	1 (0)

The experimental design enables the estimation of the system response at any experimental point within the investigation range [43]. The predicted response can be calculated using the response function, a regression equation in the following form:

$$y = \beta_0 + k\sum_{i=1}^{k}\beta_i x_i + \sum_{i=1}^{k}\beta_{ii} x_i^2 + \sum_{j<i=2}^{k}\beta_{ij} x_i x_j \qquad (2)$$

where y is the response; x_i and x_j are variables (i and j range from 1 to k); β_0 is the model intercept of coefficient; β_i, β_{ii}, and β_{ij} are the interaction coefficients of linear, quadratic, and second-order terms, respectively; and k is the number of independent parameters (in this study, $k = 3$) [44].

Minitab® software (version 21.4.2.0, Minitab, Philadelphia, PA, USA) was used to explore the possibility of obtaining a material exhibiting a T_g value of at least 45–50 °C and with a sufficiently high cutin monomer content, starting from 10,16-diHHDA, glycerol, and succinic acid as reactants. This involved assessing the upper and lower limits of T_g as a function of 10,16-diHHDA content while minimizing the operational time of the analysis. To this end, T_g was set as the response of the experimental design, while gel content was considered as an internal validation parameter to ensure efficient crosslinking—and therefore good physicochemical properties—for the final products. Specifically, a minimum threshold of 98% gel content was set for polyesters to be considered suitable as metal protective coatings.

2.4. Coating Preparation Procedure

The final coatings were prepared by solubilizing all the monomers in ethanol at 70 °C, obtaining a homogeneous solution, which was then deposited on an A1008 steel substrate (Q PANEL, code S, 76 mm × 152 mm × 0.81 mm, ground finish, roughness = 0.51–1.14 μm, Q-Lab Corporation, Bolton, UK) using a K202 Control Coater (RK Print Coat Instruments Ltd., Royston, UK). Following the deposition, the coatings were cured at 150 °C for 2 h at atmospheric pressure and then kept under vacuum to complete the removal of volatile by-products.

2.5. Characterization Techniques

2.5.1. Hydrogen Nuclear Magnetic Resonance (^{1}H-NMR)

^{1}H-NMR spectra were collected using a Bruker AV 400 MHz instrument (Bruker Corporation, Billerica, MA, USA. The samples were prepared by dissolving 1 mg of the sample in 1 mL of dimethyl sulfoxide-d6 (DMSO-d_6).

2.5.2. Differential Scanning Calorimetry (DSC)

DSC curves were collected using a Mettler-Toledo DSC 823e instrument (Mettler-Toledo, Columbus, OH, USA). The measurements were performed on 5–20 mg samples under a nitrogen flux. The thermal history included the following: (i) a first heating run from −50 °C to 150 °C (20 °C/min); (ii) a cooling run from 150 °C to −50 °C (20 °C/min); and (iii) a second heating run from −50 °C to 200 °C (20 °C/min). The glass transition temperature (T_g) was determined as the inflection point of the second heating run.

2.5.3. Gel Content Determination

Gel content measurements were performed by immersing each sample in 30 mL of ethanol and by maintaining it under magnetic stirring for 24 h at ambient temperature. Then, each sample was dried under vacuum for 24 h at 60 °C and then weighed. The gel fraction (%$_{gel}$) was calculated according to the following equation:

$$\%_{gel} = \frac{m_f}{m_i} \times 100 \tag{3}$$

where m_f is the mass of the sample after vacuum drying; and m_i is the initial mass of the sample.

2.5.4. Fourier-Transform Infrared (FTIR) Spectroscopy

FTIR spectra were collected using a Nicolet Nexus 760 FTIR spectrometer (Thermo Fisher Scientific, Waltham, MA, USA). The samples were prepared by dissolving the product in acetone and depositing a drop of the obtained solution on a KBr pellet. The measurements were performed at room temperature, in air, in transmission mode (64 scans at 4 cm^{-1} resolution), and in a range of 4000–1000 cm^{-1}.

2.5.5. Hydroxyl Number Determination

The hydroxyl number determination was performed by chemical titration following a standard procedure reported in the literature [45].

2.5.6. Coating Characterization Tests

Thermogravimetric (TGA) analysis was performed employing a Q500 (TA Instruments, New Castel, DE, USA) instrument by heating from room temperature to 500 °C with a heating rate of 10 °C/min^{-1} in a nitrogen atmosphere.

Coating thickness was measured using a digital external micrometer, 0 ÷ 30 mm, MICROMASTER IP54 (TESA, Renens, Switzerland). The surface wettability of the coatings was determined at room temperature using an OCA 15Plus (DataPhysics Instruments GmbH, Filderstadt, Germany) instrument equipped with a CCD camera and a 500 µL Hamilton syringe, by measuring the static optical contact angle against ultrapure water. The scratch hardness of the coatings was assessed using dry samples through the Wolff–Wilborn method, using a set of 14 pencils (grades 9H to 9B), according to ASTM D3363-05 [46]. The chemical resistance of the coatings was evaluated through the solvent rub test, using methyl ethyl ketone (MEK) as the solvent, according to ASTM D4752 [47]. The adhesive strength of the coatings was estimated at room temperature using an ARW-T05 tester by measuring the pulling force required to detach a 20 mm diameter aluminum dolly adhered to the coating through an epoxy adhesive (EPX/DP460, cured at 25 °C for 24 h).

3. Results and Discussion

3.1. Cutin Depolymerization and 10,16-diHHDA Recovery

In order to assess the performance of cutin depolymerization and 10,16-diHHDA extraction, both the apparent yield (η_g) and the recovery yield (η_r) of the process were calculated. The former is defined as the ratio between recovered monomer mass and unprocessed tomato peel mass, while the latter is defined as the ratio between extracted monomer mass and the maximum amount of recoverable cutin from unprocessed tomato peel. The apparent and recovery yields were calculated according to the following equations:

$$\eta_g = \frac{m_{cm}}{m_{bm}} \times 100 \quad (4)$$

$$\eta_r = \frac{m_{cm}}{(m_{bm} - m_{sr})} \times 100 \quad (5)$$

where η_g is the global yield; η_r is the relative yield; m_{cm} is the mass of recovered monomer obtained after the extraction; m_{bm} is the mass of unprocessed tomato peel weighted before depolymerization; and m_{sr} is the mass of solid residue obtained after depolymerization.

Starting from 30 g of unprocessed tomato peel and performing extraction 24 h after depolymerization [39], 8 ± 1 g of unreacted solid residue was obtained, resulting in average apparent and recovery yields of 10,16-diHHDA up to 30 and 40 wt.%, respectively. The variance in the quantity of the recovered final product is attributed to differences in biomass provenience and growth conditions, as well as to unquantifiable material losses along the single processing steps. A remarkable result was obtained by performing acidification and extraction immediately after depolymerization. In this case, it was possible to increase global and relative yields up to 45 and 60 wt.%, respectively.

At room temperature and atmospheric pressure, the isolated fatty acid appeared as a yellow-orange, waxy solid material. The ^{1}H-NMR analysis demonstrated 96% purity of the extracted 10,16-diHHDA.

3.2. Cutin-Based Polyester Resins Formulation

A preliminary screening was performed to select the best candidates in terms of solvent and monomer mixture composition, to obtain a material exhibiting a T_g value of at least 45–50 °C and with a sufficiently high cutin monomer content.

To this end, solubility tests were performed, and all components showed better solubility in ethanol than in other common solvents tested, leading to the choice of using the former as a diluent in the following experiments. Then, citric acid and succinic acid were investigated as co-sources of –COOH groups as they are both Krebs cycle intermediates, making them easily available bio-based renewable raw materials suitable for food-contact applications. Eventually, the molar ratio between 10,16-diHHDA and glycerol was considered as a variable to study the effect of this parameter on the properties of the final product. Polyester resins were prepared starting from 0.5 g of 10,16-diHHDA, setting the molar ratio between –OH and –COOH groups equal to 1, and using ethanol as a solvent. T_g value and crosslinking efficiency were assessed using DSC and gel content measurements, respectively. The results obtained are reported in Table 3.

Table 3. Glass transition temperature (T_g) and gel content values for coatings were obtained by mixing different percentages of 10,16-dihydroxyhexadecanoic acid (10,16-diHHDA), citric acid (CA), succinic acid (SA), and glycerol (Gly) in the preliminary screening. The molar ratio between –OH and –COOH groups was kept equal to 1 for all the mixtures considered.

Coating Components	Weight Fractions [%]	Molar Fractions [%]	T_g [°C]	Gel Content [%]
10,16-diHHDA	75	67	−11	92
CA	25	33		
10,16-diHHDA	83	67	−7	94
SA	17	33		
10,16-diHHDA	38	19	33	98
SA	43	52		
Gly	18	29		
10,16-diHHDA	33	15	35	99
SA	47	54		
Gly	21	31		
10,16-diHHDA	25	11	43	98
SA	51	56		
Gly	24	33		

Succinic acid-based resins showed a slightly higher T_g and gel content than the citric acid-based ones. Looking at the ternary mixtures, it was concluded that by slightly increasing the glycerol content, the T_g of the material definitely increased as well. Nevertheless, the role played by the single components (as well as by the interactions among them) in determining the properties of the final product still needs to be extensively investigated.

To this end, a design of experiments (DoE) methodology was used to find the optimal composition for obtaining the most suitable material to be used as a coating in food applications. 10,16-diHHDA, succinic acid, and glycerol were selected as monomers (Figure 1).

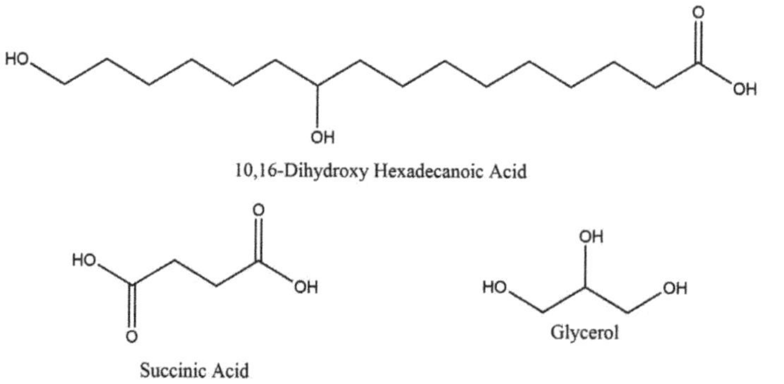

Figure 1. Chemical structures of the selected biobased monomers: 10,16-dihydroxyhexadecanoic acid, succinic acid, and glycerol.

3.3. Box–Behnken Design

To analyze the response surface design of the experiments, Minitab® software (version 21.4.2.0) was employed. Specifically, collected data corresponding to the combinations identified in Table 2 were imported into the tool. This allowed the validation of the assumed significance of the main factors and formulation components through statistical evidence. To enhance accuracy, all experiments were performed in duplicate, following the principles of repetition and randomization. The functions "Create Response Surface Design" and "Analyze Response Surface Design" within the software were utilized.

The response function generated as the output of the analysis is given by the following regression equation:

$$T_g = \begin{aligned} &-65.0486 - 4.16667x_1 + 2.72222x_2 + 169.0x_3 - 4.25x_1^2 - 0.361111x_2^2 \\ &-81.0x_3^2 + 0.166667x_1x_2 - 7.00x_1x_3 + 5.00x_2x_3 \end{aligned} \quad (6)$$

where T_g is the glass transition temperature; and x_1, x_2, and x_3 are 10,16-diHHDA content, glycerol content, and OH/COOH molar ratio, respectively.

To evaluate the goodness-of-fit and the significance of the model, the determination coefficient (R^2), the correlation coefficient (R), and the adjusted determination coefficient (R^2_{adj}) were calculated, and the Analysis of Variance (ANOVA) was performed (Table 4).

Table 4. Analysis of Variance (ANOVA) and Model Summary of the quadratic regression model for the prediction of glass transition temperature (T_g).

Source	DF	Adj SS	Adj MS	F-Value	p-Value
Model	9	3135.62	348.40	112.61	0.000
$R = 0.992$		$R^2 = 0.985$		$R^2_{adj} = 0.975$	

The determination coefficient ($R^2 = 0.985$) indicates that 98.5% of the variation in the response is explained by the model. The adjusted determination coefficient ($R^2_{adj} = 0.975$) is very high and close to the determination coefficient, confirming that the model was highly significant. Furthermore, the correlation coefficient ($R = 0.992$) is very high, indicating a good correlation between the actual and predicted responses. The statistical significance of the model is also confirmed by the very high F-value (F-value = 112.61) and the very low p-value (p-value << 0.05). The F-value and p-value are used to assess the null hypothesis for the regression, i.e., that the model does not explain any of the variation in the response. The higher than 1 the F-value and the lower than 0.05 the p-value, the stronger the evidence against the null hypothesis. Furthermore, the adequacy of the model can be validated in a visual way using a parity plot, displaying how accurate the estimated responses are against the experimentally observed ones (Figure 2). Each point has a pair of Cartesian coordinates (x, y) such that its actual and predicted T_g values represent its abscissa and ordinate, respectively. A good correlation between collected and estimated data is evidenced by the fact that the linear fit for the data points, namely the red line ($y = 0.984x + 0.322$, with $R^2 = 0.984$) plotted in the graph, is very close to a 45-degree line ($y = x$), representing the ideal case where predicted values match actual ones.

The regression equation was used to calculate the T_g values predicted by the response function and then compare them to the experimental results for all the samples. Furthermore, once the model was validated, it was used as a predictive tool to identify the cutin monomer content for a composition leading to the target T_g value of 50 °C (Simulation 1), as well as to estimate the glass transition temperature for the composition containing the highest 10,16-diHHDA weight fraction (Simulation 2). In both cases, the accuracy of the model was verified by preparing duplicates of the corresponding polyesters and characterizing them using DSC and gel content measurements. Run experiments and their corresponding collected data are reported in Table 5.

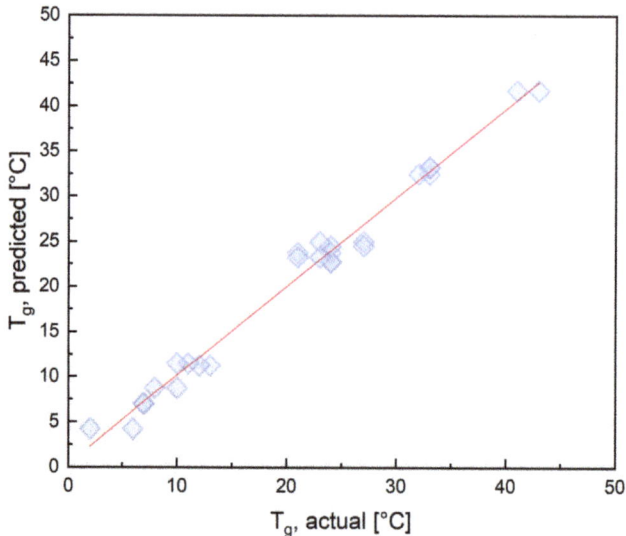

Figure 2. Parity plot for T_g with the linear fit for the data set.

Table 5. Box–Behnken design and simulation experiments with the actual and predicted responses ("T_g, actual" and "T_g, predicted", respectively) and 10,16-dihydroxyhexadecanoic acid weight fraction, sorted by glass transition temperature from the highest to the lowest.

Run No.	10,16-diHHDA [mol]	Glycerol [mol]	OH/COOH Molar Ratio	T_g, Actual [°C]	T_g, Predicted [°C]	10,16-diHHDA Weight Fraction [%]
Simulation 1 [a]	0.5	6	1	48	50.13	8
16	0.5	4	1	47	41.75	12
3	0.5	4	1	43	41.75	12
7	0.5	2.5	1.25	41	32.50	19
12	1	4	1.25	33	33.25	23
25	1	4	1.25	33	33.25	23
20	0.5	2.5	1.25	33	32.50	19
13	1	2.5	1	32	25.00	32
5 [b]	0.5	2.5	0.75	27	24.50	14
4	1.5	4	1	27	25.00	27
14	0.5	1	1	24	24.50	33
17	1.5	4	1	24	22.75	27
18 [b]	0.5	2.5	0.75	24	24.50	14
26	1	2.5	1	24	25.00	32
10 [b]	1	4	0.75	23	23.25	17
1	0.5	1	1	23	23.75	33
23 [b]	1	4	0.75	21	23.25	17
11	1	1	1.25	21	11.25	52
24	1	1	1.25	13	11.25	52
8	1.5	2.5	1.25	12	11.50	40
21	1.5	2.5	1.25	11	11.50	40
9 [b]	1	1	0.75	10	8.75	41
22 [b]	1	1	0.75	10	8.75	41
19 [b]	1.5	2.5	0.75	8	7.00	31
6 [b]	1.5	2.5	0.75	7	7.00	31
15	1.5	1	1	7	4.25	55
2	1.5	1	1	6	4.25	55
Simulation 2	1.5	1	1.25	2	−0.44	60

[a] This point falls outside the investigation range defined in this Box–Behnken design. [b] This composition showed a gel content lower than 98%.

As evident from Table 5, the Box–Behnken design demonstrated that there is no possibility of achieving the target T$_g$ value of 50 °C within the investigation range defined in this study in terms of selected monomers (factors) and molar ratios between them (levels). As for the compositions identified by the BBD, the combination of 0.5 mol of 10,16-diHHDA, 4 mol of glycerol, and a stoichiometric OH/COOH molar ratio led to the highest glass transition temperature (i.e., 41–43 °C), but a too low cutin monomer content. The collected data outline a clear trend such that T$_g$ is negatively affected by increasing 10,16-diHHDA and decreasing glycerol contents, as well as by OH/COOH molar ratios deviating from the ideal stoichiometric value (OH/COOH = 1). In the case of unitary value, the expectation is to achieve a closer-to-perfect network structure, free from defects and unreacted functional groups that could create defect points. To explore the effects of an off-stoichiometric value on the network structure, the OH/COOH molar ratio was selected as a factor, with its three levels (i.e., 0.75, 1, and 1.25), to assess its impact on glass transition temperature. The highlighted trend can be graphically visualized using the main effects plot, which displays the main effects of the analyzed factors on T$_g$ (Figure 3).

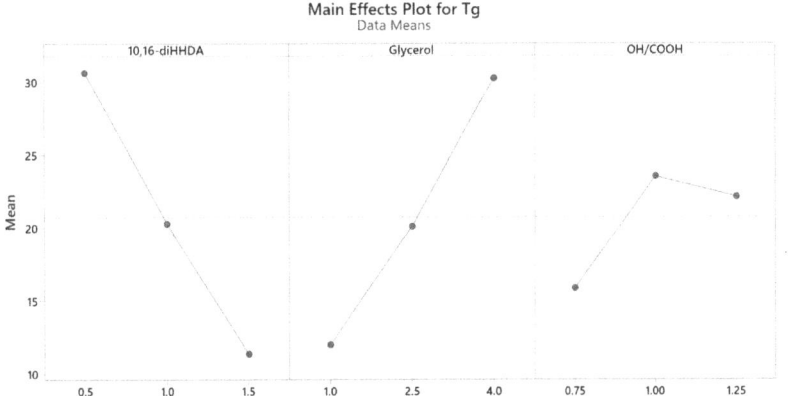

Figure 3. Main effects plot for T$_g$, showing the mean for the T$_g$ values computed for each level of the three factors, namely 10,16-diHHDA, Glycerol, and OH/COOH molar ratio.

Different levels of factors affect the response differently. The steeper the slope of the line, the greater the magnitude of the main effect. 10,16-diHHDA and glycerol monomers show constant influence on glass transition temperature in both the passages between levels −1 and 0 and between levels 0 and +1. Furthermore, their respective middle levels and means of T$_g$ are almost coincident. A rather different behavior is underlined for the OH/COOH molar ratio, showing how glass transition temperature is way more dramatically affected by an excess of –COOH groups (OH/COOH = 0.75) than by an excess of –OH groups (OH/COOH = 1.25) in the system, since the slope of the line passing through levels −1 and 0 is steeper than the one passing through levels 0 and +1. In the case of an OH/COOH = 1.25, there were probably defects in the polymer networks induced by the excess of unreacted –OH groups that caused a slight decrease in the conversion degree of the crosslinking reaction—and therefore in the T$_g$ value. Nevertheless, all of the compositions with an over-stoichiometric OH/COOH molar ratio showed a gel content greater than or equal to the threshold value of 98%, indicating a sufficient extent of crosslinking. Conversely, in the case of OH/COOH = 0.75, the excess of –COOH groups led to a surplus of mono- and bi-functional monomers and a lack of trifunctional monomers with –OH groups (i.e., glycerol). As a result, the system did not fulfill the following two conditions that generally maximize the extent of crosslinking: (1) a stoichiometric balance between reacting groups (in this case, –OH and –COOH), and (2) a mean functionality of the mixture higher than 2, favored by increasing polyfunctional reacting species and reducing monofunctional ones. Indeed, as reported in Table 5, all of the compositions with an under-stoichiometric OH/COOH molar ratio showed a gel content lower than the threshold value of 98%, which is undesirable for high-performance protective coatings. For the above-discussed remarks, polyesters obtained from the formulations with OH/COOH = 0.75 were considered unsuitable for the target application.

As for the model, the collected data demonstrate that the estimation is reliable not only within the study range defined in this Box–Behnken design but also outside. Indeed, the accuracy of the prediction was verified for both Simulation 1 (entry 1 in Table 5) and Simulation 2 (last entry in Table 5)—the latter being a combination of levels encompassed in the BBD; the former presenting a glycerol content of 6 mol, which is higher than 4 mol, corresponding to the glycerol level coded as "+1". In both cases, the experimentally observed T_g values were fairly consistent with the ones calculated using the regression equation generated as the output of the analysis. Therefore, the model is a valid and useful predictive tool to estimate glass transition temperature for any combination of 10,16-diHHDA, glycerol, and succinic acid monomers that is represented by a point falling within the investigation range and/or reasonably close to its boundaries.

Considering this statistical analysis, the variation in the initial formulation can represent a good strategy to modify the T_g of the final coating, adapting it to the desired requirements. As already stated in Section 3.2, a minimum T_g value of at least 45–50 °C was fixed, because this enabled a good trade-off between sufficiently high scratch hardness, commonly associated with high T_g values, and good post-cure formability of the already coated metal strips without cracks and defects, usually achieved with lower T_g. In any case, the highest value of the T_g obtained with this statistical analysis is around 40 °C, which is still lower than the target value of 50 °C. On the other hand, any attempt to further increase the T_g value—even by considering different levels rather than the ones defined in this Box–Behnken design (e.g., a glycerol content of 6 mol, as in Simulation 1)—would result in a material containing a too low 10,16-diHHDA weight fraction, preventing a proper valorization of such an abundant agri-food waste as tomato peels.

To increase the T_g and optimize the 10,16 diHHDA content in the final coating, it was decided to introduce a stiffer heterocyclic biobased prepolymer in place of the aliphatic, highly mobile glycerol, i.e., the bis(2,3-dihydroxypropyl) furan-2,5-dicarboxylate monomer (Section 3.4). Thanks to the presence of a furan ring, this compound can form strong intermolecular interactions due to its ring conformational constraints, leading to a reduction in the mobility of the polymer chains in crosslinked coatings. This can cause a decrease in the free volume in the crosslinked polymeric network, increasing its glass transition temperature. Furthermore, this prepolymer is a derivative of 2,5-furandicarboxylic acid (FDCA), one of the "hottest" biobased monomers in modern green chemistry, available from various integrated biorefinery routes, and a candidate to replace fossil-based terephthalic acid in many polymer-related applications [48].

3.4. Synthesis of the Bis(2,3-Dihydroxypropyl) Furan-2,5-Dicarboxylate Prepolymer

To further increase the T_g of the final product, the base formulation of the cutin-based resins was therefore modified by replacing glycerol with an FDCA prepolymer constituted by the bis(2,3-dihydroxypropyl) furan-2,5-dicarboxylate. A simplified reaction scheme for the synthesized prepolymer is shown in Figure 4.

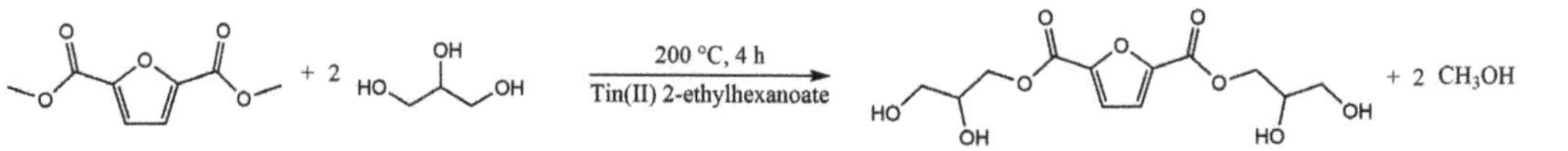

Figure 4. Simplified reaction scheme of the transesterification between 2,5-furandicarboxylic acid dimethyl ester and glycerol, yielding bis(2,3-dihydroxypropyl) furan-2,5-dicarboxylate as the main product and methanol as polycondensation by-product.

The reaction was run in a stoichiometric excess of glycerol (1 mole of FDME to 2 moles of glycerol). The progress of the reaction was monitored by periodically sampling the crude mixture and measuring FT-IR spectra (Figure 5).

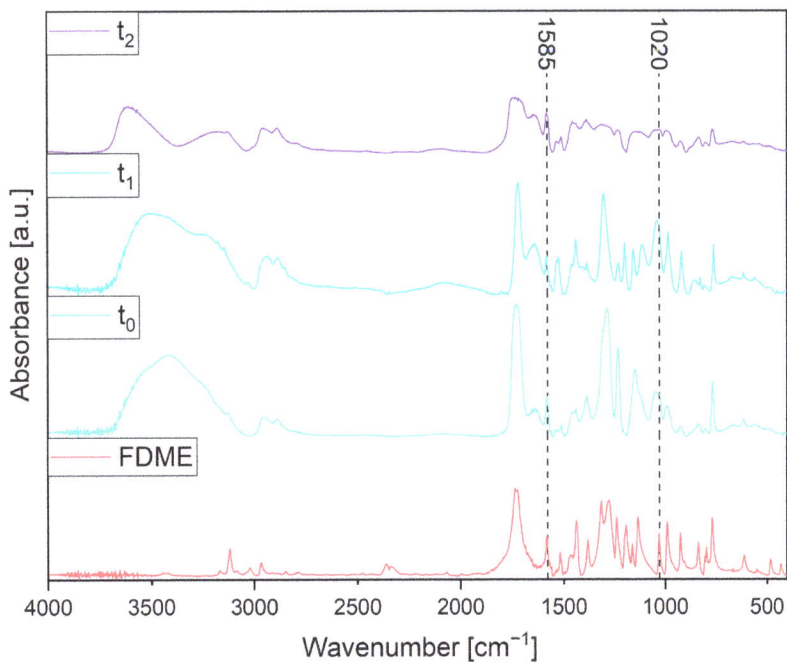

Figure 5. FT-IR spectra: 2,5-furandicarboxylic acid dimethyl ester (FDME); the system at the beginning of the reaction (t_0); the system after two hours since the beginning of the reaction (t_1); and the system after four hours (t_2).

The signal at 1585 cm^{-1}, attributed to the stretching vibration of the C=C bonds in the furan ring, was used as an invariant reference peak to normalize all the spectra. The broad absorption band in the region 3700–3100 cm^{-1} is ascribed to the stretching vibration of terminal O–H groups in the glycerol and the final FDCA prepolymer. The peaks detected at 3131 cm^{-1} are associated with the stretching vibration of the =C–H bonds in the furan ring. The absorption band in the region 3000–2840 cm^{-1} is related to the stretching vibration of –CH$_2$ and C–H in the backbone. The signal at 1735 cm^{-1} is attributed to the stretching vibration of the C=O bond in ester groups. The broad absorption band in the region 1310–1200 cm^{-1} is ascribed to the stretching vibration of the C–O–C bond in ester groups. The peak detected at 1020 cm^{-1} is associated with the stretching vibration of the C–O–C bond in the furan ring.

The progress of the reaction can be assessed by looking at the broad adsorption band in the region 3700–3200 cm^{-1}. As the reactants are converted into the products, methanol is released and evaporated because of the operating conditions (the reaction was run at 200 °C, and methanol boils at 65 °C). The decrease in the total number of –OH groups within the system caused by the polycondensation led to a decrease in the intensity of the adsorption band related to the stretching vibration of terminal O–H groups. Another interesting marker of the condensation reaction is the splitting of the unique, unstructured very broad signal into two well-distinct bands at around 3700–3400 and 3400–3100 cm^{-1}, starting from the unique broader band.

^{1}H-NMR was also performed to check the prepolymer structure (Figure 6).

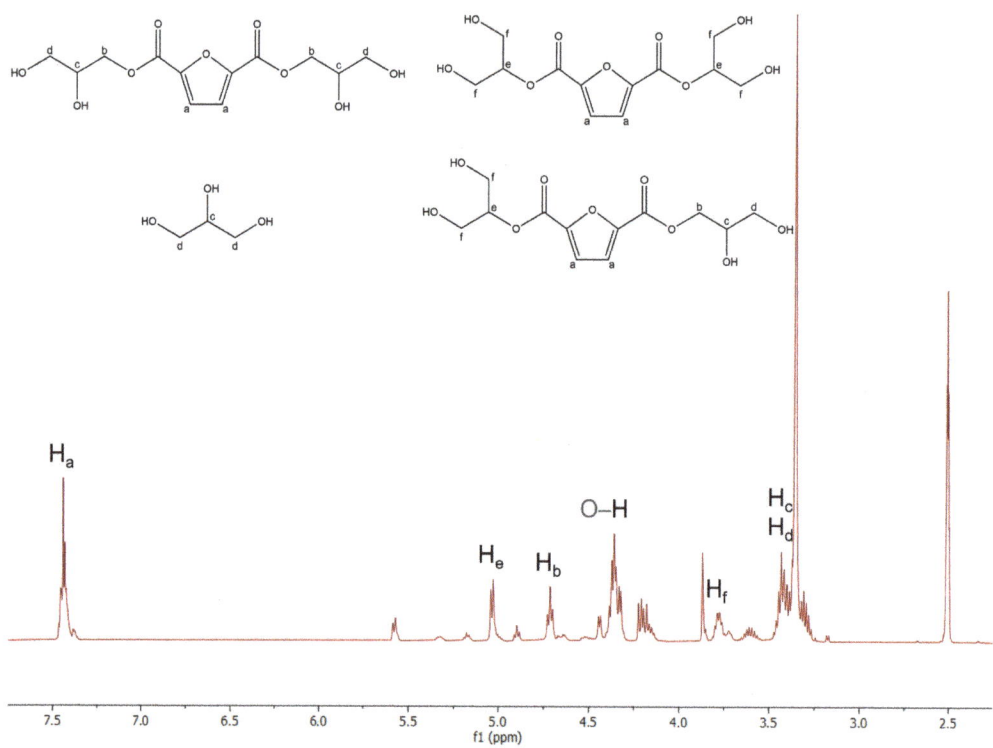

Figure 6. ^{1}H-NMR spectrum in DMSO-d$_6$ of the synthesized FDCA prepolymer.

The peak detected at 7.44 ppm is attributed to the two ideally homotopic H$_a$ atoms of the furan ring. Due to the different isomers available for the condensation reaction of FDCA, the signal is not a singlet but a more complex multiplet. The doublet centered at 5.03 ppm and the triplet centered at 4.72 ppm are ascribed to H$_e$ and H$_b$ atoms, respectively. The complex signal in the region 4.41–4.29 ppm is associated with the H atoms of the terminal hydroxyl groups in glycerol and the co-products. The multiplet in the region of 3.80–3.75 ppm is attributed to H$_f$ atoms. The complex signal centered at 3.43 ppm is ascribed to H$_c$ and H$_d$ atoms in glycerol and its co-products. Except for the characteristic residual not-deuterated solvent signal, centered at 2.50 ppm, namely dimethyl sulfoxide-d$_6$ (DMSO-d$_6$), the unassigned signals are likely attributed to some oligomers formed during the reaction.

The equivalent hydroxyl number of the synthesized prepolymer was experimentally determined in view of its use to replace glycerol and supply each system with the same number of –OH groups provided in the glycerol-based polyester resins. To this end, the following equation was used:

$$m_{prepolymer,i} = \frac{n_{glycerol,i} \times 3}{OH_{exp}} \quad (7)$$

where $m_{prepolymer,i}$ is the mass content of the FDCA prepolymer needed in the ith formulation; $n_{glycerol,i}$ is the molar content of glycerol in the ith formulation; and OH_{exp} is the experimentally determined equivalent hydroxyl number of the synthesized prepolymer.

To assess the possible beneficial effects of the higher rigidity of the central furan ring on glass transition temperature, two combinations of levels encompassed in the study range defined by this BBD were selected and modified. Specifically, glycerol was replaced with the FDCA prepolymer in the formulations exhibiting the highest and lowest 10,16-diHHDA content, namely, compositions coded as "Simulation 2" (last entry in Table 5) and "3" (entry 3 in Table 2), respectively. As demonstrated by the results reported in Table 6, the introduction of the prepolymer into the reaction system led to a remarkable increment of around 40 °C in the glass transition temperature of the final crosslinked coatings when compared to previously obtained materials. Based on this evidence, two additional formulations showing a T_g around 10–15 °C and a gel content not lower than 98% were modified as well, expecting the same increment of around 40 °C in the glass transition temperature of the final crosslinked coatings as in the previous cases, as a consequence of replacing glycerol with the FDCA prepolymer. The compositions coded as "11" (entry 11 in Table 2) and "8" (entry 8 in Table 5) were selected to prepare model polyester coatings. These concerning data are reported in Table 6.

Table 6. Glass transition temperature (T_g) and gel content for coatings were obtained by mixing different weight percentages of 10,16-dihydroxyhexadecanoic acid (10,16-diHHDA), the prepolymer synthesized from 2,5-furandicarboxylic acid dimethyl ester (FDME) and glycerol (FDCA prepolymer), and succinic acid (SA).

Resin Components	Weight Fractions [%]	–OH Fraction from 10,16-diHHDA [%]	–COOH Fraction from 10,16-diHHDA [%]	T_g [°C]	Gel Content [%]
10,16-diHHDA FDCA prepolymer SA	48 31 21	50	31	41	100
10,16-diHHDA FDCA prepolymer SA	39 38 23	40	25	54	100
10,16-diHHDA FDCA prepolymer SA	28 45 27	29	18	53	100
10,16-diHHDA FDCA prepolymer SA	7 56 37	8	4	83	100

The composition constituted by 39 wt.% of 10,16-diHHDA, 38 wt.% of FDCA prepolymer, and 23 wt.% of succinic acid (entry 2 in Table 6, from here on referred to as "C39F38") was selected as the most promising candidate for coating applications. Indeed, it exhibited the best trade-off between glass transition temperature and 10,16-diHHDA content, meeting the requirements for both ensuring good physicochemical properties of the final material and quantitative valorization of tomato-peel waste. To assess the potential effects of a 10,16-diHHDA-enriched formulation on the chemical, mechanical, optical, and physical properties of the coating, the composition constituted by 48 wt.% of 10,16-diHHDA, 31 wt.% of FDCA prepolymer, and 21 wt.% of succinic acid (entry 1 in Table 6, from here on referred to as "C48F31") was investigated as well, resulting in a T_g lower than 50 °C, set as a threshold. DSC thermograms and thermogravimetric analyses for C39F38 and C48F31 coatings are shown in Figure A1. The two cutin- and FDCA-based crosslinked polyesters showed less than 2% weight loss at 200 °C during TGA in an N_2 atmosphere, indicating good thermal stability even at high temperatures. Both coatings were also characterized by assessing their wettability, scratch hardness, chemical resistance, and adhesion to the substrate (Table 7).

The chemical resistance of the cutin-based polyester coatings was evaluated by a solvent rub test using methyl ethyl ketone (MEK) as the solvent. Both materials successfully passed more than 100 double rubs without failure or breakthrough of the surface, demonstrating high chemical inertness. An adhesive strength higher than 2 MPa for the prepared films on steel substrates was estimated by performing pull-off adhesive tests, confirming their potential applicability as the inner lining of food and beverage cans. As a result, even a percentage of around 50 wt.% of 10,16-diHHDA does not negatively impact the functionality of the material in terms of the properties here investigated by technological tests.

Table 7. Key technological features of selected coatings.

Coating Components	Weight Fractions [%]	Thickness [μm]	Water Contact Angle [°]	Pencil Hardness	MEK Test [Double Rubs]	Pull-Off Strength [MPa]
10,16-diHHDA	39					
FDCA prepolymer	38	42	101 ± 1	HB	>100	>2.15
SA	23					
10,16-diHHDA	48					
FDCA prepolymer	31	44	100 ± 1	HB	>100	>2.20
SA	21					

4. Conclusions

In this paper, we showed that the monomer derived by the depolymerization of tomato's cutin present in the peels of this fruit (i.e., a biowaste from the canning industry) can be converted into fully biobased crosslinked networks by a high-temperature polyesterification reaction with other biorefinery co-monomers bearing hydroxyl and carboxylic groups such as succinic acid, glycerol, and 2,5-furandicarboxylic acid. To investigate the effect of the co-monomer contents on the T_g of cutin-derived coatings, a Box–Behnken design and response surface methodology were utilized for coatings comprising 10,16-diHHDA, glycerol, and succinic acid. Among the compositions identified by the Bex–Behnken design, a combination of 0.5 mol of 10,16-diHHDA, 4 mol of glycerol, and a stoichiometric OH/COOH molar ratio led to a higher glass transition temperature (i.e., 41–43 °C) but a cutin monomer content lower than desired. To obtain a coating with a T_g higher than 45–50 °C and a cutin monomer content exceeding 25–30 wt.% to exploit the agro-waste renewable sources, a furan dicarboxylic acid-based prepolymer was developed and used as a substitute for glycerol. Through careful optimization of the monomer mix composition, this study showed how coating films characterized by remarkable properties determined from technological tests, like chemical resistance and good adhesion to metals, as well as relatively high levels of scratch hardness and T_g values of around 50 °C, can be obtained. The employed formulations completely avoid the use of fossil-based and toxic external curing agents like polyisocyanates and melamines. The "green" profile and the technological properties of cutin-based polymer films make them potential candidates as internal can coatings for the food industry, a market that is nowadays still dominated by BPA-based epoxy coatings. While the results obtained in this study from technological characterizations are promising, it is important to note that further research is essential to comprehensively assess the potential of cutin-based metal protective coatings. This includes an in-depth examination of their anticorrosion properties, toxicological profiles, durability, and post-cure metal formability through rigorous testing.

Author Contributions: Conceptualization, M.M., R.S. and S.T.; data curation, E.R., A.B.O., M.C., R.S. and S.T.; formal analysis, E.R., M.C. and R.S.; funding acquisition, M.M., G.B., R.S. and S.T.; investigation, E.R. and A.B.O.; methodology, M.C., R.S. and S.T.; project administration, G.B. and R.S.; resources, G.B., R.S. and S.T.; supervision, M.C., R.S. and S.T.; validation, E.R., M.C. and R.S.; visualization, E.R., M.C. and R.S.; writing—original draft preparation, E.R., M.C., R.S. and S.T.; writing—review and editing, E.R., M.M., G.B., M.C., R.S. and S.T. All authors have read and agreed to the published version of the manuscript.

Funding: This research was funded by Fondazione Cariplo in the framework of the 2021 Call "Circular Economy for a Sustainable Future, for the project entitled "Cutin from Tomato-Peel Waste: Green Source for Plurality of Engineered Polymer Products (CutToPro)", grant number: 2021-0651. M.M. acknowledges the Italian Ministry of Education, University, and Research (MIUR) and the European Social Fund (ESF) for supporting his research through the PON 2014–2020 program, action IV.6 'Research contracts on green issues'.

Institutional Review Board Statement: Not applicable.

Data Availability Statement: Publicly available datasets analyzed in this study can be found here: https://polimi365-my.sharepoint.com/:f:/g/personal/10542832_polimi_it/EqB6cZ0lcV1FvQQbzn6_IqsBTKhEJ-OClJ43JBOML0xveQ?e=jfw4j1, accessed on 24 February 2024.

Acknowledgments: The authors thank Tomato Farm S.r.l. for the availability to provide tomato peels, Beatrice Tagliabue for her support in preparing samples for DSC analyses and gel content measurements, and Rita Nasti for all the technical work and support for the cutin depolymerization and 10,16-dihydroxyhexadecanoic acid recovery.

Conflicts of Interest: The authors declare no conflict of interest. The funders had no role in the design of the study; in the collection, analyses, or interpretation of data; in the writing of the manuscript; or in the decision to publish the results.

Appendix A

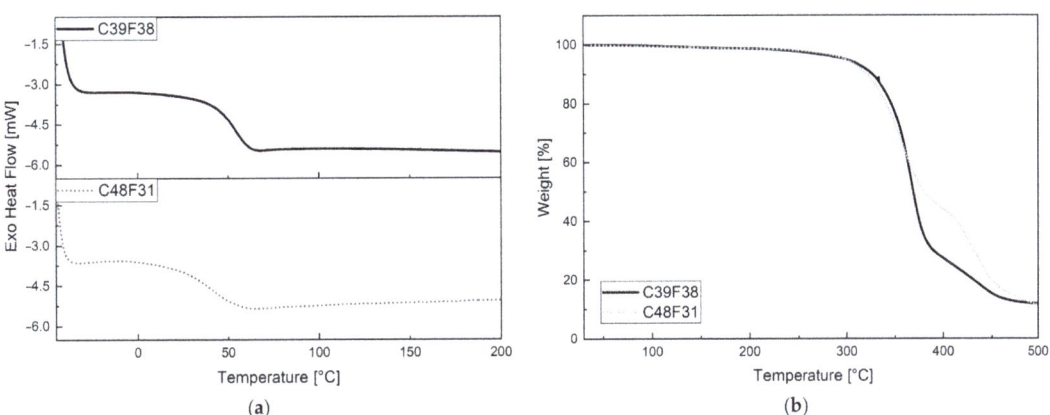

Figure A1. DSC thermograms (**a**) and thermogravimetric analyses (**b**) for the two selected coatings, named as "C39F38" (constituted by 39 wt.% of 10,16-diHHDA, 38 wt.% of FDCA prepolymer, and 23 wt.% of succinic acid) and "C48F31" (constituted by 48 wt.% of 10,16-diHHDA, 31 wt.% of FDCA prepolymer, and 21 wt.% of succinic acid), in a nitrogen atmosphere.

References

1. Domínguez, E.; Heredia-Guerrero, J.A.; Heredia, A. The Biophysical Design of Plant Cuticles: An Overview. *New Phytol.* **2011**, *189*, 938–949. [CrossRef]
2. Heredia, A. Biophysical and Biochemical Characteristics of Cutin, a Plant Barrier Biopolymer. *Biochim. Biophys. Acta BBA-Gen. Subj.* **2003**, *1620*, 1–7. [CrossRef] [PubMed]
3. Chatterjee, S.; Matas, A.J.; Isaacson, T.; Kehlet, C.; Rose, J.K.C.; Stark, R.E. Solid-State ^{13}C NMR Delineates the Architectural Design of Biopolymers in Native and Genetically Altered Tomato Fruit Cuticles. *Biomacromolecules* **2016**, *17*, 215–224. [CrossRef] [PubMed]
4. Cifarelli, A.; Cigognini, I.M.; Bolzoni, L.; Montanari, A. Physical–Chemical Characteristics of Cutin Separated from Tomato Waste for the Preparation of Bio-Lacquers. *Adv. Sci. Eng.* **2019**, *11*, 1–33. [CrossRef]
5. Domínguez, E.; Heredia-Guerrero, J.A.; Heredia, A. Plant Cutin Genesis: Unanswered Questions. *Trends Plant Sci.* **2015**, *20*, 551–558. [CrossRef]
6. España, L.; Heredia-Guerrero, J.A.; Segado, P.; Benítez, J.J.; Heredia, A.; Domínguez, E. Biomechanical Properties of the Tomato (*Solanum lycopersicum*) Fruit Cuticle during Development Are Modulated by Changes in the Relative Amounts of Its Components. *New Phytol.* **2014**, *202*, 790–802. [CrossRef]

7. Heredia-Guerrero, J.A.; Benítez, J.J.; Domínguez, E.; Bayer, I.S.; Cingolani, R.; Athanassiou, A.; Heredia, A. Infrared and Raman Spectroscopic Features of Plant Cuticles: A Review. *Front. Plant Sci.* **2014**, *5*, 305. [CrossRef]
8. Luque, P.; Bruque, S.; Heredia, A. Water Permeability of Isolated Cuticular Membranes: A Structural Analysis. *Arch. Biochem. Biophys.* **1995**, *317*, 417–422. [CrossRef]
9. Pollard, M.; Beisson, F.; Li, Y.; Ohlrogge, J.B. Building Lipid Barriers: Biosynthesis of Cutin and Suberin. *Trends Plant Sci.* **2008**, *13*, 236–246. [CrossRef]
10. Rock, C.; Yang, W.; Goodrich-Schneider, R.; Feng, H. Conventional and Alternative Methods for Tomato Peeling. *Food Eng. Rev.* **2012**, *4*, 1–15. [CrossRef]
11. World Processing Tomato Council. Available online: https://www.wptc.to/production/ (accessed on 24 January 2024).
12. Cifarelli, A.; Cigognini, I.; Bolzoni, L.; Montanari, A. Cutin Isolated from Tomato Processing By-Products: Extraction Methods and Characterization. In Proceedings of the CYPRUS 2016 4th International Conference on Sustainable Solid Waste Management, Limassol, Cyprus, 23–25 June 2016; pp. 1–20.
13. Taofiq, O.; González-Paramás, A.; Barreiro, M.; Ferreira, I. Hydroxycinnamic Acids and Their Derivatives: Cosmeceutical Significance, Challenges and Future Perspectives, a Review. *Molecules* **2017**, *22*, 281. [CrossRef]
14. Saini, R.K.; Moon, S.H.; Keum, Y.-S. An Updated Review on Use of Tomato Pomace and Crustacean Processing Waste to Recover Commercially Vital Carotenoids. *Food Res. Int.* **2018**, *108*, 516–529. [CrossRef]
15. Heredia-Guerrero, J.A.; Benítez, J.J.; Cataldi, P.; Paul, U.C.; Contardi, M.; Cingolani, R.; Bayer, I.S.; Heredia, A.; Athanassiou, A. All-Natural Sustainable Packaging Materials Inspired by Plant Cuticles. *Adv. Sustain. Syst.* **2017**, *1*, 1600024. [CrossRef]
16. Zhang, B.; Uyama, H. Biomimic Plant Cuticle from Hyperbranched Poly(Ricinoleic Acid) and Cellulose Film. *ACS Sustain. Chem. Eng.* **2016**, *4*, 363–369. [CrossRef]
17. D'Amato, D.; Droste, N.; Allen, B.; Kettunen, M.; Lähtinen, K.; Korhonen, J.; Leskinen, P.; Matthies, B.D.; Toppinen, A. Green, Circular, Bio Economy: A Comparative Analysis of Sustainability Avenues. *J. Clean. Prod.* **2017**, *168*, 716–734. [CrossRef]
18. De Vries, H.; Bredemeijer, G.; Heinen, W. The Decay of Cutin And Cuticular Components By Soil Microorganisms In Their Natural Environment. *Acta Bot. Neerlandica* **1967**, *16*, 102–110. [CrossRef]
19. Katayama, A.; Kuwatsuka, S. Effect of Pesticides on Cellulose Degradation in Soil under Upland and Flooded Conditions. *Soil Sci. Plant Nutr.* **1991**, *37*, 1–6. [CrossRef]
20. Lu, H.; Madbouly, S.A.; Schrader, J.A.; Srinivasan, G.; McCabe, K.G.; Grewell, D.; Kessler, M.R.; Graves, W.R. Biodegradation Behavior of Poly(Lactic Acid) (PLA)/Distiller's Dried Grains with Solubles (DDGS) Composites. *ACS Sustain. Chem. Eng.* **2014**, *2*, 2699–2706. [CrossRef]
21. Yew, S.-P.; Tang, H.-Y.; Sudesh, K. Photocatalytic Activity and Biodegradation of Polyhydroxybutyrate Films Containing Titanium Dioxide. *Polym. Degrad. Stab.* **2006**, *91*, 1800–1807. [CrossRef]
22. Mark, J.E. (Ed.) *Physical Properties of Polymers Handbook*; Springer: New York, NY, USA, 2007.
23. Guzmán-Puyol, S.; Heredia, A.; Heredia-Guerrero, J.A.; Benítez, J.J. Cutin-Inspired Polymers and Plant Cuticle-like Composites as Sustainable Food Packaging Materials. In *Sustainable Food Packaging Technology*; Athanassiou, A., Ed.; Wiley: Hoboken, NJ, USA, 2021; pp. 161–198.
24. Benítez, J.J.; Castillo, P.M.; del Río, J.C.; León-Camacho, M.; Domínguez, E.; Heredia, A.; Guzmán-Puyol, S.; Athanassiou, A.; Heredia-Guerrero, J.A. Valorization of Tomato Processing By-Products: Fatty Acid Extraction and Production of Bio-Based Materials. *Materials* **2018**, *11*, 2211. [CrossRef]
25. Kershaw, P.J. *Exploring the Potential for Adopting Alternative Materials to Reduce Marine Plastic Litter*; UN: New York, NY, USA, 2018. [CrossRef]
26. Bioplastic Production from Tomato Peel Residues I BIOPROTO Project I Fact Sheet I FP7. Available online: https://cordis.europa.eu/project/id/625297/it (accessed on 24 January 2024).
27. Agrimax EU-Funded Project. Available online: https://agrimax-project.eu/ (accessed on 24 January 2024).
28. TomaPaint—From Tomato Waste to Natural Paint. Available online: https://www.tomapaint.com/ (accessed on 24 January 2024).
29. Tedeschi, G.; Benitez, J.J.; Ceseracciu, L.; Dastmalchi, K.; Itin, B.; Stark, R.E.; Heredia, A.; Athanassiou, A.; Heredia-Guerrero, J.A. Sustainable Fabrication of Plant Cuticle-Like Packaging Films from Tomato Pomace Agro-Waste, Beeswax, and Alginate. *ACS Sustain. Chem. Eng.* **2018**, *6*, 14955–14966. [CrossRef]
30. Heredia-Guerrero, J.A.; Heredia, A.; García-Segura, R.; Benítez, J.J. Synthesis and Characterization of a Plant Cutin Mimetic Polymer. *Polymer* **2009**, *50*, 5633–5637. [CrossRef]
31. Benítez, J.J.; Heredia-Guerrero, J.A.; Guzmán-Puyol, S.; Barthel, M.J.; Domínguez, E.; Heredia, A. Polyhydroxyester Films Obtained by Non-Catalyzed Melt-Polycondensation of Natural Occurring Fatty Polyhydroxyacids. *Front. Mater.* **2015**, *2*, 59. [CrossRef]
32. Arnich, N.; Canivenc-Lavier, M.-C.; Kolf-Clauw, M.; Coffigny, H.; Cravedi, J.-P.; Grob, K.; Macherey, A.-C.; Masset, D.; Maximilien, R.; Narbonne, J.-F.; et al. Conclusions of the French Food Safety Agency on the Toxicity of Bisphenol A. *Int. J. Hyg. Environ. Health* **2011**, *214*, 271–275. [CrossRef]
33. Carwile, J.L. Canned Soup Consumption and Urinary Bisphenol A: A Randomized Crossover Trial. *JAMA* **2011**, *306*, 2218. [CrossRef]

34. LaKind, J.S.; Naiman, D.Q. Daily Intake of Bisphenol A and Potential Sources of Exposure: 2005–2006 National Health and Nutrition Examination Survey. *J. Expo. Sci. Environ. Epidemiol.* **2011**, *21*, 272–279. [CrossRef]
35. Rudel, R.A.; Gray, J.M.; Engel, C.L.; Rawsthorne, T.W.; Dodson, R.E.; Ackerman, J.M.; Rizzo, J.; Nudelman, J.L.; Brody, J.G. Food Packaging and Bisphenol A and Bis(2-Ethyhexyl) Phthalate Exposure: Findings from a Dietary Intervention. *Environ. Health Perspect.* **2011**, *119*, 914–920. [CrossRef]
36. Vandenberg, L.N.; Hauser, R.; Marcus, M.; Olea, N.; Welshons, W.V. Human Exposure to Bisphenol A (BPA). *Reprod. Toxicol.* **2007**, *24*, 139–177. [CrossRef] [PubMed]
37. Bisphenol A (BPA) Market Size, Growth & Forecast, 2032. Available online: https://www.chemanalyst.com/industry-report/bisphenol-a-market-57 (accessed on 24 January 2024).
38. Epoxy Resin Market Size & Share Analysis—Growth Trends & Forecasts (2024–2029). Available online: https://www.mordorintelligence.com/industry-reports/global-epoxy-resin-market-industry (accessed on 24 January 2024).
39. Righetti, G.I.C.; Nasti, R.; Beretta, G.; Levi, M.; Turri, S.; Suriano, R. Unveiling the Hidden Properties of Tomato Peels: Cutin Ester Derivatives as Bio-Based Plasticizers for Polylactic Acid. *Polymers* **2023**, *15*, 1848. [CrossRef]
40. Zhao, D.; Delbecq, F.; Len, C. One-Pot FDCA Diester Synthesis from Mucic Acid and Their Solvent-Free Regioselective Polytransesterification for Production of Glycerol-Based Furanic Polyesters. *Molecules* **2019**, *24*, 1030. [CrossRef] [PubMed]
41. Marc, M.; Risani, R.; Desnoes, E.; Falourd, X.; Pontoire, B.; Rodrigues, R.; Escórcio, R.; Batista, A.P.; Valentin, R.; Gontard, N.; et al. Bioinspired Co-Polyesters of Hydroxy-Fatty Acids Extracted from Tomato Peel Agro-Wastes and Glycerol with Tunable Mechanical, Thermal and Barrier Properties. *Ind. Crops Prod.* **2021**, *170*, 113718. [CrossRef]
42. Prakash Maran, J.; Sivakumar, V.; Sridhar, R.; Prince Immanuel, V. Development of Model for Mechanical Properties of Tapioca Starch Based Edible Films. *Ind. Crops Prod.* **2013**, *42*, 159–168. [CrossRef]
43. Hamed, E.; Sakr, A. Application of Multiple Response Optimization Technique to Extended Release Formulations Design. *J. Control. Release* **2001**, *73*, 329–338. [CrossRef] [PubMed]
44. Prakash Maran, J.; Manikandan, S. Response Surface Modeling and Optimization of Process Parameters for Aqueous Extraction of Pigments from Prickly Pear (*Opuntia ficus-indica*) Fruit. *Dyes Pigm.* **2012**, *95*, 465–472. [CrossRef]
45. E222-00; Standard Test Methods for Hydroxyl Groups Using Acetic Anhydride Acetylation. ASTM International: West Conshohocken, PA, USA, 2000.
46. D3363-05; Standard Test Method for Film Hardness by Pencil Test. ASTM International: West Conshohocken, PA, USA, 2005.
47. D4752-20; Standard Practice for Measuring MEK Resistance of Ethyl Silicate (Inorganic) Zinc-Rich Primers by Solvent Rub. ASTM International: West Conshohocken, PA, USA, 2020.
48. Davidson, M.G.; Elgie, S.; Parsons, S.; Young, T.J. Production of HMF, FDCA and Their Derived Products: A Review of Life Cycle Assessment (LCA) and Techno-Economic Analysis (TEA) Studies. *Green Chem.* **2021**, *23*, 3154–3171. [CrossRef]

Disclaimer/Publisher's Note: The statements, opinions and data contained in all publications are solely those of the individual author(s) and contributor(s) and not of MDPI and/or the editor(s). MDPI and/or the editor(s) disclaim responsibility for any injury to people or property resulting from any ideas, methods, instructions or products referred to in the content.

Article

Stabilization of Black Locust Flower Extract via Encapsulation Using Alginate and Alginate–Chitosan Microparticles

Ivana A. Boškov [1], Ivan M. Savić [1], Nađa Đ. Grozdanić Stanisavljević [2], Tatjana D. Kundaković-Vasović [3], Jelena S. Radović Selgrad [3] and Ivana M. Savić Gajić [1,*]

[1] Faculty of Technology in Leskovac, University of Nis, Bulevar oslobodjenja 124, 16000 Leskovac, Serbia; savicivan@tf.ni.ac.rs (I.M.S.)
[2] Institute for Oncology and Radiology of Serbia, Pasterova 14, 11000 Belgrade, Serbia
[3] Department of Pharmacognosy, Faculty of Pharmacy, University of Belgrade, Vojvode Stepe 450, 11221 Belgrade, Serbia; tatjana.kundakovic@pharmacy.bg.ac.rs (T.D.K.-V.); jelena.radovic@pharmacy.bg.ac.rs (J.S.R.S.)
* Correspondence: savicivana@tf.ni.ac.rs

Citation: Boškov, I.A.; Savić, I.M.; Grozdanić Stanisavljević, N.Đ.; Kundaković-Vasović, T.D.; Radović Selgrad, J.S.; Savić Gajić, I.M. Stabilization of Black Locust Flower Extract via Encapsulation Using Alginate and Alginate–Chitosan Microparticles. *Polymers* **2024**, *16*, 688. https://doi.org/10.3390/polym16050688

Academic Editors: Cristina Cazan and Mihai Alin Pop

Received: 7 February 2024
Revised: 28 February 2024
Accepted: 29 February 2024
Published: 2 March 2024

Copyright: © 2024 by the authors. Licensee MDPI, Basel, Switzerland. This article is an open access article distributed under the terms and conditions of the Creative Commons Attribution (CC BY) license (https://creativecommons.org/licenses/by/4.0/).

Abstract: Black locust flower extract contains various polyphenols and their glucosides contribute to the potential health benefits. After intake of these bioactive compounds and passage through the gastrointestinal tract, their degradation can occur and lead to a loss of biological activity. To overcome this problem, the bioactive compounds should be protected from environmental conditions. This study aimed to encapsulate the black flower extract in the microparticles based on biodegradable polysaccharides, alginate, and chitosan. In the extract, the total antioxidant content was found to be 3.18 ± 0.01 g gallic acid equivalent per 100 g of dry weight. Also, the presence of lipids (16), phenolics (27), organic acids (4), L-aspartic acid derivative, questinol, gibberellic acid, sterol, and saponins (2) was confirmed using the UHPLC–ESI-MS analysis. In vitro assays showed that the extract has weak anti-α-glucosidase activity and moderate antioxidant and cytotoxic activity against the HeLa cell line. The extrusion method with secondary air flow enabled the preparation of microparticles (about 270 µm) encapsulated with extract. An encapsulation efficiency of over 92% was achieved in the alginate and alginate–chitosan microparticles. The swelling study confirmed a lower permeability of alginate–chitosan microparticles compared with alginate microparticles. For both types of microparticles, the release profile of antioxidants in the simulated gastrointestinal fluids at 37 °C followed the Korsmeyer–Peppas model. A lower diffusion coefficient than 0.5 indicated the simple Fick diffusion of antioxidants. The alginate–chitosan microparticles enabled a more sustained release of antioxidants from extract compared to the alginate microparticles. The obtained results indicated an improvement in the antioxidant activity of bioactive compounds from the extract and their protection from degradation in the simulated gastric conditions via encapsulation in the polymer matrixes. Alginate–chitosan showed slightly slower cumulative antioxidant release from microparticles and better antioxidant activity of the extract compared to the alginate system. According to these results, alginate–chitosan microparticles are more suitable for further application in the encapsulation of black locust flower extract. Also, the proposed polymer matrix as a drug delivery system is safe for human use due to its biodegradability and non-toxicity.

Keywords: black locust flower; UHPLC–ESI-MS analysis; release profile; antioxidant activity; cytotoxic activity; anti-α-glucosidase activity

1. Introduction

Free radicals that represent highly reactive molecules naturally occur in the human body during various metabolic processes. They serve as a defense mechanism against harmful substances. However, when their production becomes excessive and surpasses the body's antioxidant capacity, it leads to a state known as oxidative stress [1]. Oxidative stress has a negative impact on all biological systems and is responsible for the pathogenesis of various diseases: atherosclerosis [2], diabetes and related complications [3], cancer [4], etc. In recent years, there has been considerable interest in natural antioxidants derived from plant sources, including phenolic acids, flavonoids, and other bioactive molecules [5]. These compounds are valued for their multifaceted effects on health and functionality. In particular, they exhibit antioxidant, antimicrobial, and other bioactive properties, which makes them very attractive compounds with broad applications in the pharmaceutical, cosmetic, and food industries. Unfortunately, these compounds are subject to degradation under the influence of external factors, which limits their application [6]. In addition, simultaneous gastrointestinal digestion leads to a decrease in antioxidant levels and antioxidant activity [7]. Antioxidant compounds often exhibit astringent effects and have a bitter taste that limits their use in oral formulations [8]. An effective approach to preserve the health-beneficial attributes of antioxidants involves their encapsulation within polymer matrices [9]. This strategy not only enhances their stability and bioavailability but also mitigates unpleasant tastes and elevates the overall quality of the final product. Among the existing encapsulation methods, microencapsulation is considered one of the most efficient techniques to improve the delivery of phytochemicals [10]. It ensures their controlled or sustained release while improving their physicochemical properties. Various biopolymer materials are used for the encapsulation of bioactive compounds, among which polysaccharides have attracted the attention of many researchers [11]. Due to its natural origin, availability, non-toxicity, biocompatibility, and thermal and chemical stability, sodium alginate is a suitable polysaccharide for the encapsulation of bioactive compounds, including plant extracts [12].

Alginates are anionic polysaccharides extracted from different types of algae. It is composed of 1–4 linked α-L-guluronic (G) and β-D-mannuronic acids (M), alternately arranged in homopolymeric (poly-M and poly-G) and mixed blocks (MG). The G/M ratio determines the permeability and swelling properties of alginate gel [13]. The alginate solution can gel in the presence of divalent cations, such as Ca^{2+}, Zn^{2+}, or Sr^{2+}. The hydrogel is formed by selectively binding two adjacent guluronates in poly-G or MG blocks with multivalent cations (e.g., Ca^{2+} ions), forming a compact structure known as an "egg-box" [14]. Alginate microparticles are characterized by high porosity, which enables easy and fast diffusion of water or other fluids. This phenomenon leads to a decrease in the yield of encapsulated compounds [15]. Due to this reason, the alginate structure is modified using cationic polyelectrolytes (chitosan). Chitosan, a copolymer of β-(1-4)-2-acetamido-D-glucose and β-(1-4)-2-amino-D-glucose, is a partially deacetylated product of chitin. It is a mucopolysaccharide isolated from the waste shells of crabs and shrimps. Through electrostatic interactions, chitosan forms a strong complex membrane between its amino residues and carboxyl residues of alginate, with which it coats the surface of alginate microparticles [12]. The prepared microparticles exhibit controlled permeability and mucoadhesive properties. They are stable at pH > 3, making them acceptable for application in the gastrointestinal tract [16]. Alginate–chitosan microparticles can be produced using two methods: either by combining a sodium alginate solution with a $CaCl_2$ solution containing chitosan or by incubating calcium alginate microparticles in a chitosan solution [17].

The black locust (*Robinia pseudoacacia* L.) is one of the most widespread, woody, deciduous species around the world [18]. The raw parts of black locusts represent a promising source of antioxidants. Only the flower is used in medicine because the other organs of this plant contain high concentrations of robinin, which is toxic to humans [19]. Robinin is a thermally sensitive compound, so the extract can be safe for use after the preparation at higher temperatures. In traditional folk medicine, the black locust flower is employed to address various health concerns [20]. It is used to treat conditions, such as hemoptysis, metrorrhagia (abnormal uterine bleeding), and other gynecological diseases. Additionally, it is believed to alleviate symptoms associated with colds, fever, migraines, skin diseases, and even bleeding in the colon. Pharmacological studies confirm the antimicrobial [21], antioxidant [20], and anticancer activity of black locust flower extract [22]. These properties of the extract can be attributed to rutin, hyperoside, epigallocatechin, ferulic acid, quercetin, and others [23,24]. Also, the presence of acacetin, secundiflorol, mucronulatol, isomucronulatol, and isovestitol flavonoids in all parts of the black locust was confirmed [25].

Due to the presence of different functional groups in the structure (carboxylic, keto, aldehydic, alcoholic, etc.), the bioactive compounds found in black locust flower extract are susceptible to degradation within the gastrointestinal tract following oral administration. Such degradation can lead to a decrease in their pharmacological activity or to the formation of various degradation products that could pose risks to human health. In this study, the encapsulation of the ethanolic extract of black locust flower in alginate and alginate–chitosan microparticles was carried out to retain the antioxidant activity of the extract and prevent the degradation of antioxidants in gastrointestinal fluids.

2. Materials and Methods

2.1. Chemicals and Reagents

In this study, absolute ethanol (Sani-Hem doo, Novi Bečej, Serbia), Folin–Ciocalteu's reagent (Carlo Erba Reagents, Val de Reuil, France), calcium chloride dihydrate, methanol, acetic acid (Zorka Pharm, Šabac, Serbia), gallic acid (purity of 97%) (Merck, Darmstadt, Germany), α-glucosidase, 4-nitrophenyl α-D-glucopyranoside (PNP-G), acarbose, 2,2-diphenyl-1-picrylhydrazyl (DPPH), butylhydroxytoluene (BHT), 3-(4,5-dimethylthiazol-2-yl)-2,5-dyphenyl tetrazolium bromide (MTT), phosphate-buffered saline (PBS), dimethyl sulfoxide (DMSO), sodium dodecyl sulfate (SDS), RPMI 1640 nutrient medium, simulated gastric fluid (pH 1.2, SGF), and simulated intestinal fluid (pH 7.4, SIF) without enzyme (Sigma-Aldrich, St. Louis, MO, USA), alginic acid sodium salt (very low viscosity), water (LC-MS grade), acetonitrile (LC-MS grade), formic acid (Thermo Scientific™, Waltham, MA, USA), chitosan (molecular weight from 100,000 to 300,000) (Acros Organics, Thermo Fisher Scientific, Geel, Belgium) were used. Other used chemicals were pro analysis quality.

2.2. Plant Material

A dried black locust flower (*Robinia pseudoacacia* L., Fabaceae) was obtained from the Institute for Medicinal Plants Research "Dr. Josif Pančić" (Belgrade, Serbia). Before extraction, the plant material was ground in an electric mill and sieved on a sieve shaker. A fraction of 0.5 mm was used for further analysis. The moisture content of 10.6% (m/m) was determined after drying the plant material at 105 °C in a laboratory oven until a constant mass was achieved.

2.3. Preparation and Characterization of Black Locust Flower Extract

2.3.1. Extraction Procedure

Black locust flower extract was prepared using ultrasound-assisted extraction in an ultrasonic bath (Sonic, Niš, Serbia). The total power of the device was 3×50 W, while the frequency was 40 kHz. The extraction was carried out using 60% (v/v) ethanol at the extraction temperature of 60 °C and the liquid-to-solid ratio of 10 mL/g for 30 min. The liquid extract was separated from the solid matrix via vacuum filtration.

2.3.2. Total Antioxidant Content

The total antioxidant content (TAC) in the obtained extract was determined according to the spectrophotometric method using Folin–Ciocalteu's reagent [24]. Its content was expressed as a gram equivalent of gallic acid per 100 g of dry weight (g GAE/100 g d.w.).

2.3.3. Antioxidant Activity According to DPPH Assay

The antioxidant activity of the extract was determined using the DPPH assay. The inhibition of DPPH radicals (I_{DPPH}) was calculated according to Equation (1):

$$I_{DPPH}\ (\%) = \frac{A_c - (A_s - A_b)}{A_c} \times 100 \tag{1}$$

where A_c, A_s, and A_b are the absorbances of the control solution, sample solution, and blank solution at 517 nm, respectively. The synthetic antioxidant BHT was used as a positive control solution. The half maximum inhibitory concentration (IC_{50}) was obtained via interpolation from the functionality between the inhibition of DPPH radicals and the sample concentration [24].

2.3.4. Anti α Glucosidase Activity

The anti-α-glucosidase activity of black locust flower extract was determined using the 400 mU/mL of α-glucosidase enzyme solution in a 0.1 M phosphate buffer (pH 6.8). Initially, the stock of the analyzed extract was dissolved in DMSO and then diluted using 0.1 M phosphate buffer (pH 6.8), so the final concentrations were 333.33, 166.67, 83.33, 41.67, 20.83, 10.42, and 5.21 µg/mL. In the 96-well plates, 50 µL of the extract dilutions were preincubated with 50 µL of enzyme solution for each well at 37 °C for 15 min. Afterwards, 50 µL of the substrate solution, 4-nitrophenyl α-D-glucopyranoside (1.5 mg/mL PNP-G in phosphate buffer), was added and absorbance A_1 was measured at 405 nm. This solution was incubated at 37 °C for another 15 min whereupon second absorbance A_2 was measured at 405 nm. Acarbose was used as a positive control. The percentage of enzyme inhibition (I_e) was calculated according to the following formula (Equation (2)):

$$I_e\ (\%) = \frac{A_{2s} - A_{1s}}{A_{2b} - A_{1b}} \times 100 \quad (2)$$

where A_{1b}, A_{2b} and A_{1s}, A_{2s} are the absorbances of the blanks (phosphate buffer, DMSO, enzyme solution, and substrate PNP-G) and sample, respectively. The IC_{30} value (estimated concentration of compounds that caused 30% inhibition of α-glucosidase activity) was determined using linear regression analysis [26].

2.3.5. Cytotoxic Activity According to MTT Assay

MTT assay was used to estimate the viability, proliferation, and cytotoxicity of the black locust flower extract. The stock solution was prepared by dissolving the extract in DMSO to the concentration of 400 µg/mL. A solution series (12.5–200 µg/mL) was prepared via dilution of the stock solution in the RPMI 1640 nutrient medium supplemented with 10% fetal bovine serum, 3 mM L-glutamine, 1% penicillin-streptomycin, and 25 mM HEPES buffer and adjusted to pH 7.2 using bicarbonate solution in 96-well microtiter plates (Nunc, Nalgene, Denmark). The HeLa cells (human cervical adenocarcinoma), LS-174 cells (cells of human colon carcinoma), A549 (non-small cell lung carcinoma), and normal MRC-5 cells (human lung fibroblasts) from American-Type Culture Collection (ATCC, Manassas, VA, USA) were seeded in a nutrient medium at 37 °C in a humidified atmosphere (95% air, 5% CO_2) and incubated for 72 h. The density of HeLa, LS-174, MRC-5, and A549 cells was 2000, 7000, 5000, and 5000 cells/well, respectively. Yellow MTT reagent (20 µL) at the concentration of 5 mg/mL PBS (phosphate buffered saline) was added into each well and incubated for 4 h. After that, the colourimetric reaction occurred and a purple formazan product was formed due to the reduction of MTT dye. Formazan was extracted by adding 100 µL of 10% (m/v) SDS into the wells. On the next day, the absorbance of the sample was measured at 570 nm on a Multiskan EX reader (Thermo Labsystems, Beverly, MA, USA). A blank solution contained a nutrient medium. The antiproliferative effect of the extract was monitored in relation to the control culture of cells.

The cell's survival was calculated according to Equation (3) [27]:

$$S(\%) = \frac{A_t - A_b}{A_c - A_b} \times 100 \quad (3)$$

where A_t, A_b, and A_c are the absorbances of treated cells with the extract, blank solution, and control solution at 570 nm, respectively. The IC_{50} value represented the sample concentration that inhibited 50% of the cell's proliferation relative to the untreated control.

2.3.6. Chromatographic Analysis

Validated ultra-high-performance liquid chromatography coupled with mass detection (UHPLC–MS) was used to identify bioactive compounds of black locust flower according to the previously described method with slight modification [28]. The Dionex Ultimate 3000 UHPLC+ system was equipped with a quaternary pump and diode-array detector (DAD). The LCQ Fleet Ion Trap spectrometer (Thermo Fisher Scientific, San Jose, CA, USA) had a heated electrospray ionization probe installed (HESI-II, ThermoFisher Scientific, Bremen, Germany). Xcalibur (version 2.2 SP1.48) and LCQ Fleet (version 2.1) software were used for data acquisition, collection, and analysis. The separation of bioactive compounds was achieved using a Hypersil gold C18 (50 mm × 2.1 mm, 1.9 μm) column at 40 °C. A mobile phase consisted of phase A (water + 0.1% (v/v) formic acid) and phase B (acetonitrile + 0.1% (v/v) formic acid). Gradient elution was used as follows: 0–10 min (5–95% B), 10–12 min (95% B), 12–12.1 min (95–5% B), and 12.1–15 min (5% B). The flow rate of the mobile phase was 0.3 mL/min, while the injection volume was 3 μL. The wavelengths of 280, 320, 340, and 360 nm were chosen for spectrum scanning of bioactive compounds. The negative ion mode mass spectra were obtained via full range acquisition from m/z 110 to 2000 using HESI with the following parameters: source voltage, 5 kV; capillary voltage, −40 V; tube lens voltage, −80 V; capillary temperature, 275 °C; and sheath and auxiliary gas flow (N_2), 42 and 11 (arbitrary units). A data-dependent scan based on collision-induced dissociation (CID) was used to fragment ions. The normalized collision energy of CID was 35 eV. The bioactive compounds were identified via comparison of the mass of molecular and fragment ions with the available literature data.

2.4. Encapsulation of Black Locust Flower Extract in the Microparticles

The alginate microparticles were encapsulated with black locust flower extract using the coaxial air flow extrusion method. An aqueous solution of alginate (1.5%, m/v) was prepared by mixing alginate overnight to completely dissolve it. Black locust flower extract was added to the alginate solution. The obtained homogeneous solution was transferred to a plastic syringe of 100 mL and added dropwise through a metal needle with a straight-cut tip and a diameter of 26 G (0.45 × 12 mm). The volumetric flow rate of the prepared solution was 33.3 mL/h, while a coaxial air flow pressure was 0.8 bar. The formed drops broke off from the tip of the needle in the form of a stream of tiny droplets under the influence of gravity and coaxial airflow. The spherical microparticles were dropped and solidified in the crosslinking solution of calcium chloride (2%, w/v) to a total volume of 200 mL. The volumetric ratio of the alginate solution to the calcium chloride solution was 1:2. According to the same procedure, alginate–chitosan microparticles encapsulated with black locust flower extract were prepared. The chitosan solution (0.5%, m/v) was dissolved in 0.5% (v/v) acetic acid (acidity regulator).

2.5. Characterization of Microparticles

The prepared alginate and alginate–chitosan microparticles encapsulated with the extract were dried before further analysis. The drying process was carried out in a laboratory oven heated to 50 °C for 24 h.

2.5.1. Determination of the Shape and Size of Microparticles

The shape and size of the microparticles were determined using an optical microscope (Leica DM 750, Leica Microsystems, Wetzlar, Germany) equipped with a digital camera. The sphericity factor (SF) used to evaluate the roundness of the formed microparticles was calculated according to Equation (4) [29]:

$$SF = \frac{D_{max} - D_{per}}{D_{max} + D_{per}} \quad (4)$$

where D_{max} is a maximum diameter (mm) passing through the central part of the microparticle; D_{per} is a diameter directed at D_{max} that passes through the central part of the microparticle (mm).

Microparticles are ideal spheres if SF is around zero. A greater degree of distortion of the shape of the microparticle is achieved when SF has higher values.

2.5.2. Determination of Encapsulation Efficiency

Fresh microparticles were vortexed with sodium citrate in a mass ratio of 1:5 for 15 min to destroy their structure. The TAC in the citrate solutions was determined according to the previously described procedure in Section 2.3.2. The encapsulation efficiency of antioxidants was calculated as the ratio of TAC in the citrate solution of destroyed microparticles to the initial extract of black locust flower used for encapsulation.

2.5.3. Swelling Study

The swelling ability of encapsulated alginate and alginate–chitosan microparticles was evaluated gravimetrically under the conditions of SGF and SIF. The swelling of microparticles was analyzed at 37 °C as follows: 0.05 g of dry microparticles were weighed and immersed in 20 mL of the SIF for 2 h, after which the microparticle samples were transferred to 20 mL of the SGF for the next 22 h. Measurements of the mass of the swollen microparticles were performed on an analytical balance. The swelling degree (SD) of microparticles was calculated according to Equation (5) [30]:

$$SD(\%) = \frac{m_t - m_i}{m_i} \times 100 \tag{5}$$

where m_t is the mass of swollen microparticles at time t and m_i is the mass of dry microparticles (xerogel).

2.6. In Vitro Release of Antioxidants from Microparticles

Antioxidant release from dry encapsulated microparticles was monitored under SGF and SIF conditions at 37 °C. About 0.6 g of microparticles were immersed and stirred in 30 mL of SGF at 100 rpm and 37 °C. At certain time intervals, 2 mL of aliquots were taken from the medium. Instead of the aliquot, the equivalent volume of fresh SGF was added to the analyzed solution. The release of antioxidants in SGF was monitored for up to 2 h. After that, the microparticles were filtered, dried, and then combined with 30 mL of SIF preheated to 37 °C with constant stirring at the same temperature. The sampling procedure was repeated according to the previously described method until the microparticles were completely disintegrated. In the collected aliquots (2 mL), the TAC was determined. The obtained results were expressed as the cumulative antioxidant release from microparticles through time. The various kinetic models, including zero order, first order, Higuchi, Hixson–Crowell, Korsmeyer–Peppas, and Baker–Lonsdale models were applied to describe the antioxidant release process from the polymer matrix. Excel add-in DD Solver was used to model the obtained data.

2.7. Determination of Antioxidant Activity of the Extract after Its Release from Microparticles in Gastrointestinal Fluids

The antioxidant activity of the extract was also determined after the release of antioxidants from alginate and alginate–chitosan microparticles in SGF and SIF at 37 °C. About 0.1 g of dry microparticles encapsulated with the extract were immersed in 10 mL of SGF or SIF. After 2 h of antioxidants release, 100 µL of the sample was taken and the antioxidant activity was determined.

2.8. Statistical Analysis

All data were carried out in triplicate and expressed to be mean ± standard deviation (SD). Analysis of variance (ANOVA) according to multiple range tests was used to determine statistical differences among samples. IBM SPSS Statistics software (version 27, Chicago, IL, USA) was applied to evaluate the statistical differences among samples. The statistically significant differences were those whose $p < 0.05$.

3. Results and Discussion

3.1. Characterization of Black Locust Flower Extract

3.1.1. Total Antioxidant Content

Before the encapsulation of the black locust flower extract, the TAC of 3.18 ± 0.01 g GAE/100 g d.w. was determined according to the spectrophotometric method. Hallmann [31] reported a tenfold lower TAC of about 3.91 mg/g d.w. for the 80% (v/v) methanolic extract obtained via maceration with stirring. Unlike this extract, the lyophilized 70% (v/v) ethanolic extract had almost tenfold higher TAC of 311.93 ± 0.01 mg GAE/g d.w. according to Jurca et al. [32]. These data indicate that the TAC depends on the extraction technique and solvent used. Bratu et al. [33] also determined a TAC of 0.72 ± 0.02 mg GAE/mL for the 50% (v/v) ethanolic extract obtained via maceration at room temperature for 72 h. This data is not comparable since the unit for the TAC is not expressed as in this study.

3.1.2. Anti-α-Glucosidase Activity

The black locust flower extract showed weak anti-α-glucosidase activity. Its IC_{30} value of 291.24 ± 21.23 μg/mL was less than the standard substance, acarbose (IC_{50} of 119.72 ± 4.64 μg/mL). Unlike the weak anti-α-glucosidase activity of the black locust flower extract, the prepared extracts in other similar studies did not show significant activity. The 40% (v/v) ethanolic extract of flowers from Ganghwa island (Republic of Korea) expressed weaker α-glucosidase inhibitory activity compared to our extract, with an IC_{50} value of 2.39 mg/mL [34]. According to Sarikurkcu et al. [35], among the black locust flower extracts from Turkey, ethyl acetate extract showed the highest anti-α-glucosidase activity (109.01 mg acarbose equivalent/g extract) compared to acetone, methanolic, and aqueous extracts (51.69, 36.76, and 44.68 mg acarbose equivalent/g extract, respectively). These results were correlated with a significant amount of total polyphenols found in the extract: 46.9 mg gallic acid equivalent/g extract, 51.99 mg rutin equivalent/g extract were flavonoids, and even 3.96 mg catechin equivalent/g extract were condensed tannin [35]. So far, it has been shown that polyphenols participate in glucose absorption through α-amylase and α-glucosidase inhibition, which are key enzymes in the digestion of carbohydrates [36]. More precisely, due to their ability to precipitate proteins, hydrolysable and condensed tannins are good enzyme inhibitors. Tannins, especially condensed tannins, are effective inhibitors of α-glucosidase, and their effectiveness can be compared with synthetic inhibitors (acarbose), while the inhibition of α-amylase is mediated by hydrolyzing tannins [37]. On the other hand, the black locust leaf from Japan showed better anti-α-glucosidase activity compared to the flower, where at 50 μg/mL the inhibitory activity was only 15% and at 200 μg/mL activity reached 93%, which was far more than the standard, acarbose (IC_{50} of 13 mg/mL) [38].

3.1.3. Cytotoxic Activity

The cytotoxic activity of the prepared extract was estimated based on an MTT assay. In Table 1, the IC_{50} values of the extract against tumor and healthy cell lines expressed as micrograms per milliliter are depicted. As can be noticed, the extract was only efficient against the HeLa cell line, while in all other cases these values were higher than 200 μg/mL. The compounds were classified based on the IC_{50} values. According to the National Cancer Institute (NCI) of the United States, the extract had moderate cytotoxic activity since the IC_{50} value ranged between 21 and 200 μg/mL [39]. This extract was presented as a tumor-selective agent because it had a weak cytotoxic activity against the MRC-5 cell line ($IC_{50} > 200$ μg/mL). Uzelac et al. [40] also analyzed the cytotoxic activity of the extract of black locust flower from the territory of Istria (Croatia) during vegetation in

2021. They concluded that the 70% (v/v) methanolic and 80% (v/v) ethanolic extracts did not show cytotoxic activity against Vero African green monkey kidney cells because the IC$_{50}$ values (in that paper, IC$_{50}$ is presented as the LC$_{50}$ value) were higher than 1 mg/mL. Bratu et al. [33] also confirmed the cytotoxic activity of 50% (v/v) ethanolic extract of black locust flower prepared via maceration against HeLa cells. That extract did not express the cytotoxic effect against the palatal mesenchymal stem cells. Cvetković et al. [41] analyzed the cytotoxic activity of methanolic extract of black locust flower obtained using Soxhlet extraction. They concluded that the extract had no significant cytotoxic activity on MRC-5 and MDA-MB-231 (human breast cancer) cell lines. Also, the results of the analysis indicated a great anti-invasive potential in MDA-MB-231 cells.

Table 1. IC$_{50}$ values (n = 3) of black locust flower extract according to the MTT assay.

Cell Lines	IC$_{50}$ (µg/mL)
HeLa	161.99 ± 6.88
MRC-5	>200
LS-174	>200
A549	>200

3.1.4. UHPLC–MS Analysis

The qualitative content of black locust flower extract was determined using the UHPLC–MS method. A base peak chromatogram of the extracts recorded in the negative mode is presented in Figure 1.

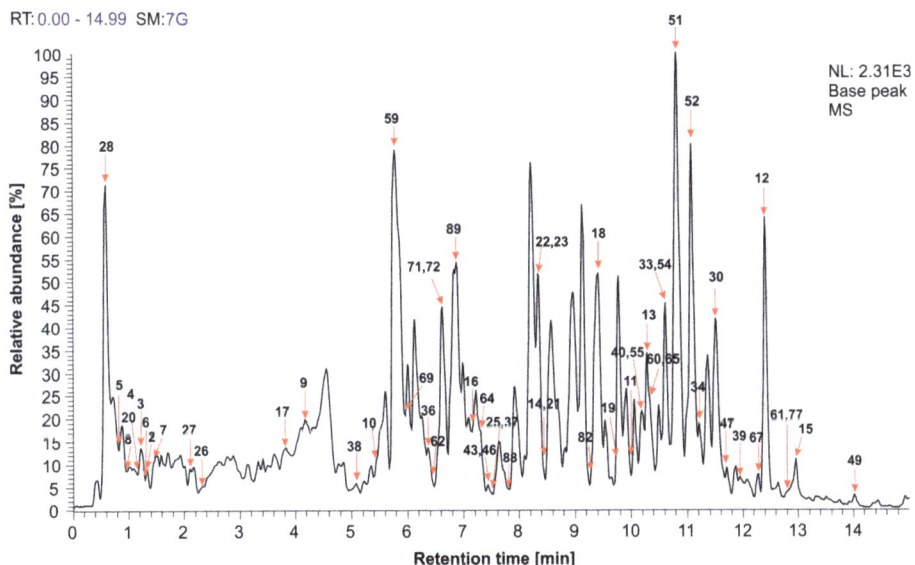

Figure 1. The UHPLC–MS chromatogram of the black locust flower extract (x-axis = retention time in minutes; y-axis = relative abundance of negative ions in percentages).

The molecular or adduct ions and their fragment ions for 89 quantified compounds, of which 59 are identified compounds, are depicted in Table 2. Among the identified compounds are lipids (16), phenolics (27), organic acids (4), L-aspartic acid derivative, questinol, gibberellic acid, sterol, and saponins (2).

Table 2. The identified compounds in the black locust flower extract using the UHPLC–MS method.

No.	R_t, min	Molecular or Adduct Ion, m/z	Fragment Ions, m/z	Compound	Class of Compounds	References
1	1.41	116.96	89 (100%)	unknown		
2	1.35	136.67	91 (100%)	protocatechuic aldehyde	phenolic aldehyde	[42]
3	1.27	136.97	109, 91 (100%)	2,4-dihydroxybenzaldehyde	phenolic aldehyde	[42]
4	0.95	161.07	143, 115, 99 (100%), 89, 57	3- or 4-hydroxy-2-oxoglutaric acid	keto organic acid	[43]
5	0.81	191.13	173, 145, 129, 111 (100%), 101	citric acid	organic acid	[43]
6	1.29	218.13 [a]	200, 130, 99, 88 (100%)	D-(+)-pantothenic acid	organic acid (vitamin)	
7	1.44	224.96 [M−H + HCOOH + HOH]−	179 (100%), 161, 143, 119, 89	3- or 4-hydroxy-2-oxoglutaric acid	phenolic aldehyde	
8	0.98	239.04 [a] [M−H + HCOOH]−	193 (100%), 165, 124, 114	ferulic acid	hydroxycinnamic acid	
9	4.13	250.08	204, 146, 132 (100%), 115, 91, 88	L-aspartic acid derivative	α-Amino acid	
10	5.44	252.12 [a]	234, 137, 136 (100%), 92	4-amino benzoic acid derivative	organic acid	
11	10.01	265.28 [a]	97 (100%)	oxidized fatty acid	fatty acid	
12	12.4	271.3	253, 225 (100%)	pinobanksin	flavanonol	[44]
13	10.29	285.39	267, 223 (100%)	hexadecanedioic acid	fatty acid	[45]
14	8.41	287.47	269 (100%), 223, 211, 169, 155, 139	fustin	flavanonol	[46]
15	12.94	299.36	253 (100%)	questinol	quinone	[47]
16	7.17	301.35	273, 255, 239, 193, 179 (100%), 151, 107	quercetin	flavonol	[48]
17	4.07	305.01	261 (100%), 224, 201, 181, 128	gallocatechin	flavanol	[49]
18	9.43	311.45	293 (100%), 275, 253, 235, 223, 201, 183, 171	hydroxydioxoheptadecenoic acid	fatty acid	[50]
19	9.75	313.39	295 (100%), 277, 213, 201, 195, 183, 179, 171, 129	dihydroxy-octadecenoic acid	fatty acid	[51]
20	1.18	321.06	277 (100%), 259, 215, 128	12-hydroxy-6, 8, 10, 13-octatetraenedioic acid	fatty acid	[52]
21	8.44	327.47	309 (100%), 291, 263, 251, 225, 209, 197, 183	9,12,13-trihydroxyoctadecadienoic acid	fatty acid	[50]
22	8.33	327.5	309, 291 (100%), 239, 197, 171	hirsutenone	diarylheptanoid	[53]
23	8.3	329.43	311, 293, 229, 211, 171 (100%)	11,12,13-trihydroxyoctadecenoic acid	fatty acid	[50]
24	12	343.41	325, 283 (100%), 254, 225, 211	unknown		
25	7.52	345.58 [a]	327, 317, 301, 291, 285, 271, 245, 239, 229, 215 (100%), 195, 181	gibberellic acid	plant hormone	
26	2.27	355.04	337, 216, 209, 191 (100%), 173, 129	coumaroylglucaric acid isomer	phenolic acid	[54]
27	2.07	355.27	337, 322, 309, 209, 191 (100%), 173, 147	coumaroylglucaric acid isomer	phenolic acid	[54]
28	0.61	387.2 [M−H + HCOOH]−	341 (100%), 251, 195, 179	caffeic acid hexoside	hydroxycinnamic acid	[55]
29	4.73	395.14	377, 349 (100%), 179, 143	unknown		
30	11.47	409.4	361, 251, 171, 153 (100%)	lyso-phosphatidic acid (16:0)	phospholipid	[56]
31	12.86	411.32	368, 281, 147, 129 (100%)	unknown		
32	6.95	423.57	405, 279 (100%), 249, 235, 205, 169, 139, 122	unknown		
33	10.57	431.42	171, 153 (100%)	N-acylglycerophosphatidylethanolamine (18:3)	phospholipid	[56]
34	11.19	433.45	329, 313, 279, 171, 153 (100%)	N-acylglycerophosphatidylethanolamine (18:2)	phospholipid	[56]
35	12.56	437.37	420, 313, 285, 279, 263, 251, 171, 153 (100%)			
36	6.35	447.55 [a]	357, 327, 285 (100%), 255, 241, 165	kaempferol-3-O-glucoside	flavonol	
37	7.55	449.58 [a]	431, 413, 403, 353, 327, 301, 287 (100%), 269, 251, 239	isookanin-7-O-glucoside	flavanone	
38	5.1	459.23	441, 399, 381, 295, 287, 242, 173, 157 (100%)	sterol-hexose conjugates (ST21:3,O;Hex)/sterol (ST 27:4;O6)	sterol	[50]
39	11.93	463.21	445, 426, 417 (100%), 399, 356, 345, 301, 255, 161	quercetin-3-O-glucoside	flavonol	[48]
40	10.23	476.41	402, 384, 277, 233, 171, 153 (100%)	N-acylglycerophosphatidylethanolamine (18:2)	phospholipid	[56]
41	12.71	480.46	434, 412, 390, 350, 279, 200 (100%)	unknown		
42	7.4	483.57	465, 439, 421, 391 (100%), 229, 172, 153	unknown		
43	7.43	491.36 [M−H + HCOOH]−	445, 343, 303, 283 (100%)	glycitein-o-hexoside	isoflavone	[57]
44	13.52	499.42	313, 261, 255 (100%), 243, 187	unknown		
45	5.65	501.18	483, 403, 250, 206 (100%), 164, 147	unknown		
46	7.37	503.36	459 (100%), 441, 365, 345, 327, 281, 187	liquiritigenin derivative	flavanone	[57]
47	11.72	505.43	487, 467, 361, 267, 255 (100%), 249, 231, 205, 189	monoacylglyceryl glucuronides (16:0)	glycerolipid	[50]
48	6.56	511.47	493, 452, 431 (100%), 285	unknown		
49	14.05	547.51 [a]	287 (100%), 277, 269	fustin derivative	flavanonol	
50	11.81	555.13	509, 486, 475 (100%), 280	unknown		

Table 2. Cont.

No.	R_t, min	Molecular or Adduct Ion, m/z	Fragment Ions, m/z	Compound	Class of Compounds	References
51	10.82	555.52 [a]	538, 495, 476, 390, 285 (100%), 269, 223, 195	cyanidin derivative	anthocyanin	
52	11.16	557.66 [a]	287 (100%), 269, 239, 221	fustin derivative	flavanonol	
53	10.76	564.37	520, 504 (100%), 279, 251	unknown		
54	10.7	571.46	409, 391, 315, 283, 255 (100%), 241	palmitoyl-glycerophosphoinositol	phospholipid	[58]
55	10.13	577.43	532, 514, 475, 317, 299 (100%), 225, 207, 165	sulfoquinovosyl monoacylglycerols (18:3)	polar lipid	[50]
56	10.07	577.55	521, 469, 371, 299 (100%), 277, 225, 207, 189, 165	unknown		
57	12.37	579.34	445, 410, 392 (100%), 323, 256, 187	unknown		
58	11.9	583.46	537, 442, 299 (100%), 287, 225, 207, 183, 165	unknown		
59	6.22	593.44	549, 339, 327, 285 (100%), 255, 239, 227, 211, 186	kaempferol rutinoside	flavonol	[54]
60	10.32	595.5	507, 415, 341, 315, 279 (100%), 261, 241, 223	lyso-phosphatidylinositols (18:2)	phospholipid	[56]
61	12.82	607.32	575 (100%), 563, 531, 487, 475, 329, 311, 295, 277	diosmetin-7-O-glucuronide-3'-O-pentoside	flavone	[59]
62	6.44	623.06	577 (100%), 507, 350, 300	isorhamnetin-O-rutinoside	flavonol	[48]
63	11.22	625.28	579 (100%), 557, 341, 306, 287	unknown		
64	7.34	637.43 [M−H + HCOOH]−	591 (100%), 335, 283, 268	acacetin-rhamnoglucoside isomer	flavone	[59]
65	10.38	647.28	601 (100%), 485, 323	galacturonoglucan	glucan	[60]
66	11	649.33	621, 603 (100%), 423	unknown		[56]
67	12.24	653.25 [M−H + HCOOH]−	607 (100%), 311	diosmetin-7-O-glucuronide-3'-O-pentoside	flavone	[59]
68	13.14	721.56	683, 678, 595, 465 (100%), 416, 409, 391, 329, 255	unknown		
69	5.94	739.41	593 (100%), 431, 369, 285, 246	kaempferol-3-O-robinoside-7-O-rhamnoside (robinin)	flavonol	[61]
70	9.51	763.82	632, 613, 571, 551, 525, 498, 455 (100%), 437, 407, 358, 249	unknown		
71	6.69	767.46	749, 707, 657, 483, 325, 283 (100%), 268	kaempferol-3-O-(4-coumaroyl)-(feruloyl)-glucoside (isomer)	flavonol	[62]
72	6.66	767.52 [a]	749, 592, 483, 283 (100%), 268, 240	acacetin derivative	flavone	
73	6.98	769.25	723 (100%), 415, 283	unknown		
74	13.6	774.26	728 (100%), 579	unknown		
75	9.61	793.78	750, 613, 603, 454 (100%), 436	unknown		
76	6.79	799.24	753 (100%), 579	unknown		
77	12.79	827.75	810, 720, 626, 558, 539, 287 (100%), 269	fustin derivative	flavanonol	[46]
78	8.8	881.82	838 (100%), 778, 723, 381	unknown		
79	12.21	883.55	721, 391, 335, 329 (100%), 291	unknown		
80	9.65	921.73	876, 822 (100%), 741, 652, 585, 564, 457, 401	unknown		
81	5.37	947.08	991 (100%), 787	unknown		
82	8.91	957.36	822, 797, 708, 498 (100%), 453	triterpenoid saponin	saponin	
83	9.29	970.98	924 (100%), 827, 719	unidentified saponin	saponin	
84	9.25	971.3	840 (100%), 714	unknown		
85	8.87	985.44	819 (100%), 595, 447	unknown		
86	13.04	997.28	966, 746, 718 (100%)	unknown		
87	8.15	1001.05	992, 982, 934, 363 (100%)	unknown		
88	7.78	1003.37 [M−H + HCOOH]−	959, 896 (100%), 640	triterpenoid saponin	saponin	
89	6.85	1155.44	577 (100%)	sulfoquinovosyl monoacylglycerols (18:3) dimer	polar lipid	

Numbers in parentheses (C:N) indicate the number of carbon atoms (C) and double bonds (N) in the fatty acid side chains. R_t—retention time. [a] https://massbank.eu/MassBank/ (available 1 February 2024).

Lipids. Polar lipids (2), phospholipids (6), glycerolipids (1), and fatty acids (7) were detected in the extract. Of the polar lipids, monoacylglycerol in conjugation with sulfoquinovosyl moiety had an [M−H]− ion at m/z 577 (**55**); i.e., its dimer had an [M−H]− ion at m/z 1155 (**89**). Compounds **30**, **33**, **34**, **40**, **54**, and **60** were identified as phospholipids with an [M−H]− ion at m/z 409.4, 431.42, 433.45, 476.41, 571.46, and 595.5, respectively. Among the glycerolipids, monoacylglyceryl glucuronides (C:N 16:0) (**47**) were detected at t_R 11.72 min with an [M−H]− ion at m/z 505.43. Fatty acids represent the most abundant compounds among lipids. Hexadecanedioic acid, which represents a long-chain fatty acid, was assigned as compound **13** with an [M−H]− ion at m/z 285.39 and t_R 10.29 min. The peak at t_R 10.01 min originated from oxidized fatty acid (**11**), which had an [M−H]− ion at m/z 265.28. Compounds with [M−H]− ions at 311.45 (**18**) and 321.06 (**20**) were identified as hydroxy fatty acids. Dihydroxy (**19**) and trihydroxy (**23**) octadecenoic acids had [M−H]−

ions at m/z 313.39 and m/z 329.43, respectively. Another trihydroxy fatty acid (21) with an [M−H]⁻ ion at m/z 327.47 occurred at t_R 8.44 min. Tian et al. [63] also confirmed the presence of various fatty acids in the extract of black locust flower from China.

Phenolics. In the extract, the main polyphenolic classes identified were flavonoids (flavanonols (5), flavonols (7), flavones (3), isoflavones (1), flavanones (1), flavanols (1), anthocyanins (1)), hydroxycinnamic acids (2), phenolic acids (2), phenolic aldehydes (3), and diarylheptanoids (1). Fustin (14) with an [M−H]⁻ ion at m/z 287.47 and its three various derivatives (49, 52, 77) were from the flavanonol class. A strong antioxidant pinobanksin (12), which had the characteristic fragment ions at m/z 253 and 225, also belongs to this class of compounds. Quercetin, kaempferol, and isorhamnetin as flavonols were noticed in the glycosidic form. Compounds 16 ([M−H]⁻ ion at m/z 301.35) and 39 ([M−H]⁻ ion at m/z 463.21) were assigned as quercetin and its glycosidic derivative, respectively. Kaempferol-3-O-glucoside (36), kaempferol rutinoside (59), kaempferol-3-O-robinoside-7-O-rhamnoside (robinin) (69), and kaempferol-3-O-(4-coumaroyl)-(feruloyl)-glucoside (isomer) (71) had molecular ions at m/z 447.55, 593.44, 739.41, and 767.46, respectively. The peak at t_R 6.44 min with an [M−H]⁻ ion at m/z 623.06 originated from isorhamnetin-O-rutinoside (62). The three various flavones (64, 67, and 72) were identified in the extract, with two being acacetin derivatives (64 and 72). Compound 64 quantified at t_R 7.34 min gave an adduct ion with formic acid [M−H + HCOOH]⁻ at m/z 637.43. The effect of collision energy caused the loss of a rhamnoglucoside moiety from this molecule resulting in the appearance of a fragment ion at m/z 283. Consequently, compound 64 was tentatively assigned to be an acacetin-rhamnoglucoside isomer. The molecular ion at m/z 767.52 was noted as an acacetin derivative (72). According to the literature data, the peak at t_R 12.24 min that resulted from diosmetin-7-O-glucuronide-3'-O-pentoside (67) had an [M−H + HCOOH]⁻ ion at m/z 653.25 and fragment ions at m/z 607 and 311. Compound 43 ([M−H + HCOOH]⁻ ion at m/z 491.36) was characterized as isoflavone glycitein-O-hexoside. Compound 46 exhibited a molecular ion at m/z 503.36. This compound corresponds to a liquiritigenin derivative. Flavanol gallocatechin (17, [M−H]⁻ at m/z 305.01) produced fragment ions at m/z 261, corresponding to the loss of CO_2 (−44 Da). Among flavonoids, cyanidin derivative (51) was also identified based on an [M−H]⁻ ion at m/z 555.52 and classified as anthocyanin. Ferulic acid (8), which belongs to hydroxycinnamic acids, had a peak at t_R 0.98 min. Its mass spectrum had a molecular ion at m/z 239.04. The glycosidic derivative of caffeic acid with an [M−H]⁻ ion at m/z 387.2 was also found in the chromatogram at t_R 0.61 min. Two different peaks with almost the same molecular ions at about m/z 355 and fragment ions at m/z 191 (100%) were noted as coumaroylglucaric acid isomers (26 and 27). Compounds 2, 3, and 7 were identified as phenolic aldehydes, wherein compounds 2 and 3 had approximately the same [M−H]⁻ ions at about m/z 137 and a main fragment ion at m/z 91. Compound 7 was the third identified phenolic aldehyde with an adduct ion [M−H + HCOOH + HOH]⁻ at m/z 224.96 and fragmentation ions at m/z 179, 161, 143, 119, and 89. This compound was characterized as 3- or 4-hydroxy-2-oxoglutaric acid. A relatively small class of secondary metabolites (diarylheptanoid) belonging to the phenolics group was also identified. Compound 22 (hirsutenone) exhibited an [M−H]⁻ ion at m/z 327.5 and fragmentation ions at m/z 309, 291, 239, 197, and 171. Due to their potential therapeutic and organoleptic features, diarylhepatanoids can be considered nutraceuticals.

The identified phenolic compounds in the black locust flower extract are in accordance with previously reported data [40]. The phenolic content of the extract was studied for the black locust flower from different areas of cultivation. Uzelac et al. [40] and Tian et al. [63] analyzed the phenolic content of 70% (v/v) ethanolic extracts of black locust flowers from the territory of the Istra region (Croatia) and China, respectively. Via mutual comparison of their results, as well as the results obtained in this study, it can be concluded that the climate conditions of plant cultivation and extraction conditions are of crucial significance to the phenolic content. It is known that these phenolic compounds generally have a beneficial effect on human health.

Organic acids. Compound **4** ([M−H]⁻ ion at m/z 161.07) was assigned as hydroxy oxoglutaric acid (keto acid). The molecular ions at m/z 191.13, 218.13, and 252.12 (**5**, **6**, and **10**) belong to the organic acids which are used as vitamins.

Amino acids. L aspartic acid derivative (compound **9**) had an [M−H]⁻ ion at m/z 250.08.

Quinones. Questinol (compound **15**) with an [M−H]⁻ ion at m/z 299.36 was noticed at t_R 12.94 min.

Hormones. Compound **23** was identified as gibberellic acid with an [M−H]⁻ ion at m/z 345.58.

Sterols. Hexose conjugate of sterol (compound **38**) with an [M−H]⁻ ion at m/z 459.23 has occurred at t_R 5.1 min.

Glucans. The peak of galacturonoglucan (compound **65**) with an [M−H]⁻ ion at m/z 647.28 was found at t_R 10.38 min.

Saponins. The peaks at t_R 8.91 min and t_R 7.78 min with an [M−H]⁻ ion at m/z 957.36 (compound **82**) and an [M−H + HCOOH]⁻ ion at m/z 1003.37 (compound **88**) were due to the presence of triterpene saponins. Compound **83** (t_R = 9.29) had an [M−H]⁻ ion at m/z 970.98 and fragment ions at m/z 924, 827, and 719. According to the literature data, the detected compound is most likely saponin [64].

3.1.5. Antioxidant Activity

The antioxidant activity of the extract was estimated based on its IC$_{50}$ value which was found to be 120.9 ± 0.08 μg/mL. The synthetic antioxidant BHT had the IC$_{50}$ value of 35.31 ± 0.12 μg/mL. Antioxidants can be characterized based on their IC$_{50}$ value according to the following classification: "very strong" if the IC$_{50}$ value is less than 50 μg/mL, "strong" if the IC$_{50}$ value is between 50 μg/mL and 100 μg/mL, "moderate" if the IC$_{50}$ value is between 100 μg/mL and 150 μg/mL, and "weak" if the IC$_{50}$ is 150–200 μg/mL [65]. Based on this classification, the black locust flower extract can be considered a moderate antioxidant, while BHT belongs to the group of very strong antioxidants. Although BHT has better antioxidant activity, the extract is a source of natural antioxidants that, due to its origin, may be more suitable for human health. In the literature, the antioxidant activity of black locust flower is also described regardless of the used solvent and extraction technique. The comparison of obtained data is very hard because of the different units used. According to the DPPH assay, the lyophilized 70% (v/v) ethanolic extract of black locust flower had an antioxidant capacity of 3.58 ± 0.11% [32]. Bratu et al. [33] determined the antioxidant activity of ethanolic extract to be 0.141 ± 0.02 μM Trolox equivalent/mL using DPPH assay.

3.2. Characterization of Microparticles

3.2.1. Shape and Size of Microparticles

The size, texture, and shape of encapsulated microparticles, as well as proof of chitosan binding, were estimated based on microscopic analysis. The extract was homogenously distributed within the alginate part of the microparticles. The average particle size of alginate and alginate–chitosan microparticles was 228.0 ± 8.5 and 273.0 ± 10.0 μm, respectively (Figure 2). The increase in microparticle size can be justified by forming a polyelectrolytic membrane as a result of electrostatic interactions between alginate and chitosan. Yousefi et al. [66] also noticed an increase in the size of the alginate–chitosan microparticle encapsulated with *Viola odorata* Linn. extract.

The calculated *SF* values for alginate and alginate–chitosan microparticles were 0.141 ± 0.2 and 0.165 ± 0.3, respectively. These values indicate the absence of an ideal spherical morphology in the majority of prepared microparticles; i.e., the prepared microparticles belong to an elongated shape (*SF* > 0.07). This is in correlation with the other available literature data [67]. The absence of sphericity in the microparticles can lead to a reduction in their mechanical and chemical resistance [68]. It is known that the size and sphericity of alginate–chitosan microparticles, produced via the extrusion dripping method, depend on the process variables (needle size, encapsulation flow surface tension of crosslinking solution, viscosity, and mixing velocity) [68,69]. Having this in mind, the lack of total sphericity for prepared microparticles can be attributed to any mentioned factors. However, since the microparticles are designed for oral administration, the size and shape are not as crucial as those intended for intravenous or intraperitoneal application.

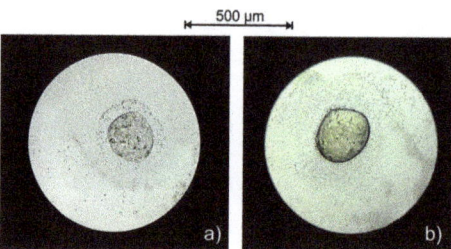

Figure 2. The optical microscopic views of alginate microparticles (**a**) and alginate–chitosan microparticles (**b**) with black locust flower extract at the magnification of 10×.

3.2.2. Encapsulation Efficiency

The encapsulation efficiency of black locust flower extract in the alginate and alginate–chitosan microparticles was 92.56 ± 3.21% and 92.05 ± 2.88%, respectively. A significant change in encapsulation efficiency has not occurred due to the presence of the chitosan membrane. The remaining amount of extract was not encapsulated because of its loss and the rapid degradation of unstable compounds during the encapsulation process. Villate et al. [67] also obtained a similar encapsulation efficiency.

3.2.3. Swelling Study

One of the most important features of hydrophilic microparticles, such as alginate and alginate–chitosan microparticles, is swelling when coming into contact with water or other fluids at physiological pH values. Kanokpanont et al. [17] reported the swelling ability to be even over 2000% or 20 g of water per 1 g of alginate–chitosan xerogel. The *SD* of microparticles depends on the pH value of the solution [70]. This factor has a significant influence on the release mechanism of encapsulated compounds from the extract. The swelling ability of encapsulated alginate and alginate–chitosan microparticles was monitored in conditions of different pH values at 37 °C for 24 h. Firstly, the microparticles were stored in SGF for 2 h; after that, they were filtered, transferred, and stored in SIF for the next 22 h. The dependences of the *SD* of encapsulated microparticles on time for pH 1.2 and pH 7.4 are depicted in Figure 3.

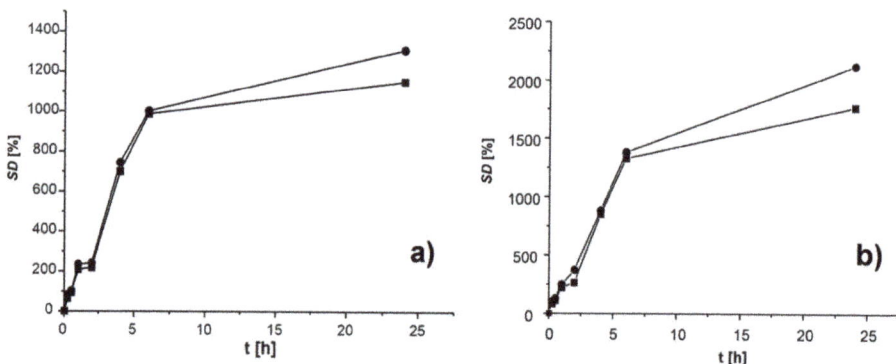

Figure 3. The dependence of swelling degree (*SD*) of empty (■) and encapsulated (•) microparticles of) alginate-chitosan (**a**) and alginate (**b**) on time in SGF and SIF at 37 °C.

At pH 1.2, the *SD* of alginate microparticles was intensively grown in the first 2 h reaching a maximal value of 372.2 g of water per g of xerogel (372.2%) (Figure 3). This is probably the result of the hydration of hydrophilic groups in alginate molecules. After transferring the microparticles in SIF (pH 7.4), the *SD* continuously increased up to 2051.6 g/g of xerogel. The *SD* of alginate–chitosan microparticles grew to 241.4 g of water per g of xerogel (241.4%) in the first 2 h (Figure 3). This increase probably occurred due to the hydration of the hydrophilic group in alginate and chitosan molecules as well as the protonation of amino groups of chitosan molecules at lower pH values [71]. In that case, repulsive forces are created, which affect the partial separation of the two polymers and the creation of pores that allow easier penetration of water. The *SD* reached 1309.1 g of water per g of xerogel (1309.1%) after transferring microparticles in SIF (pH 7.4) for 24 h. The high *SD* value indicated the possible separation of polymer chains due to the ionization of carboxylic groups without breaking the chemical bonds in the polymeric net. A similar change in *SD* has been noticed for both types of microparticles in SGF and SIF. The *SD* values for alginate microparticles were significantly higher than those for alginate-chitosan microparticles. The chitosan membrane reduces the permeability of alginate microparticles due to the formation of a polyelectrolytic complex between the amino groups of chitosan and the hydrophilic groups of alginate. The formed complex does not allow easy penetration of fluid inside the alginate–chitosan microparticles. Pravilović et al. [72] also showed a lower *SD* of alginate–chitosan microparticles encapsulated with thyme extracts compared to the alginate microparticles. The results of swelling studies indicated that *SD* increased as the pH value of the medium increased. This behavior of microparticles is desirable since they should be resistant to the conditions of the gastric environment (pH 1.2) but also enable the release of the extract compounds in the conditions of the intestinal environment (pH 7.4).

3.3. In Vitro Studies of Antioxidants Release in Simulated Gastrointestinal Fluids

A cumulative antioxidant release (expressed in percentage) from alginate and alginate–chitosan microparticles under the conditions of SGF and SIF is depicted in Figure 4. Initially, the rapid release of 7.01% and 6.16% of antioxidants from alginate and alginate–chitosan microparticles in SGF occurred within the first 6 min, respectively. This release profile is due to the difference in antioxidant concentrations between the interior and exterior mediums of the microparticles. The sustained release of antioxidants from alginate (55.96%) and alginate–chitosan (51.95%) microparticles was noticed up to 2 h in SGF. The slower release of antioxidants occurred after transferring microparticles in the conditions of SIF. After 6 h in SIF, exactly 60.11% and 56.56% of antioxidants were released from alginate and alginate–chitosan microparticles, respectively. Villate et al. [67] reported an analogous in vitro release profile of cannabinoids from alginate–chitosan microparticles using a similar model. In their study, the cumulative release of cannabinoids in SGF during the first 2 h was significantly lower (about 20% of the total encapsulated compound). They obtained almost the same percentage of cumulative release as in this study after 6 h. The low pH value of SGF impacted the formation of alginic acid in alginate which interfered with antioxidant release from microparticles; i.e., it led to the low cumulative release [73]. The acidic degradation of antioxidants in the gastric environment is one of the factors that negatively impacts their oral bioavailability [74].

In this study, the antioxidants release profiles in both types of microparticles were quite similar. The percentage of released antioxidants from alginate microparticles was higher than from alginate–chitosan microparticles, which was caused by the presence of a chitosan membrane [75]. The membrane reduced the diffusion of antioxidants from the inner parts of microparticles to the exterior environment and extended the drug release profile [76]. The zero order, first order, Higuchi, Hixson–Crowell, Korsmeyer–Peppas, and Baker–Lonsdale models were applied to fit the obtained data for antioxidant release from the polymeric matrix. The fit goodness of the experimental data was estimated based on the statistical parameters, such as the coefficient of determination (R^2), adjusted coefficient of determination (R_{adj}^2), root mean square error (*RMSE*), and Akaike information criterion (*AIC*). It is recommended that the model has a higher coefficient of determinations and lower *RMSE* and *AIC*. The Korsmeyer–Peppas model was proved to be the best model for describing both release profiles because the highest R^2 value (around 0.84) and lowest *AIC* value (around 44) were found in this model (Table 3). The R^2 value of 0.84 showed that 84% of the variance in antioxidant release could be explained using this model. The Korsmeyer–Peppas model is suitable for use in drug release from hydrogels or other systems which change shape and volume during this process. In this model, the diffusion coefficients of 0.305 and 0.322 were obtained for antioxidant release from alginate and alginate–chitosan microparticles, respectively. The values of *n* less than 0.5 indicate the simple Fick diffusion. In the case of values greater than 0.5, a non-Fickian release occurs indicating diffusion in the hydrated matrix and relaxation of the polymer. Since the diffusion exponents were lower than 0.5, the release of antioxidants was subjected to the simple Fickian diffusion. This fact implied the controlled release of antioxidants via a diffusion process [77]. Fickian diffusion is defined by a high rate of solvent diffusion into the matrix and a low rate of polymeric relaxation.

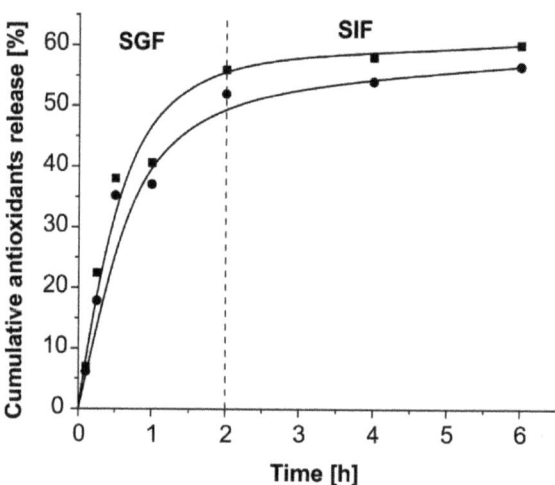

Figure 4. Cumulative antioxidants release (%) from alginate (■) and alginate-chitosan (•) microparticles in simulated gastric fluid (SGF) and simulated intestinal fluid (SIF) at 37 °C.

Table 3. Kinetic models and statistics parameters of fitting the data of antioxidant release from alginate and alginate–chitosan microparticles at 37 °C.

Kinetic Model	Equation	Alginate Microparticle		Alginate–Chitosan Microparticle	
		Parameter	Goodness of Fit	Parameter	Goodness of Fit
Zero order	$F = k_0 t$	$k_0 = 13.446$	$R^2 = -0.4241$ $R_{adj}^2 = -0.4241$ $RMSE = 23.7987$ $AIC = 58.9172$	$k_0 = 12.547$	$R^2 = -0.2469$ $R_{adj}^2 = -0.2469$ $RMSE = 21.4682$ $AIC = 57.4743$
First order	$F = 100 \times \left(1 - e^{-k_1 t}\right)$	$k_1 = 0.299$	$R^2 = 0.2264$ $R_{adj}^2 = 0.2264$ $RMSE = 17.5400$ $AIC = 54.6451$	$k_1 = 0.247$	$R^2 = 0.2719$ $R_{adj}^2 = 0.2719$ $RMSE = 16.4048$ $AIC = 53.7084$
Higuchi model	$F = k_H \sqrt{t}$	$k_H = 30.572$	$R^2 = 0.6575$ $R_{adj}^2 = 0.6575$ $RMSE = 11.6704$ $AIC = 48.9411$	$k_H = 28.372$	$R^2 = 0.7112$ $R_{adj}^2 = 0.7112$ $RMSE = 10.3328$ $AIC = 47.2368$

Table 3. Cont.

Kinetic Model	Equation	Alginate Microparticle		Alginate–Chitosan Microparticle	
		Parameter	Goodness of Fit	Parameter	Goodness of Fit
Korsmeyer–Peppas model	$F = k_{KP} t^n$	$k_{KP} = 38.296$ $n = 0.305$	$R^2 = 0.8675$ $R_{adj}^2 = 0.8410$ $RMSE = 7.9514$ $AIC = 44.2930$	$k_{KP} = 34.892$ $n = 0.322$	$R^2 = 0.8701$ $R_{adj}^2 = 0.8442$ $RMSE = 7.5899$ $AIC = 43.6414$
Hixson–Crowell model	$F = 100 \times \left(1 - (1 - k_{HC} t)^3\right)$	$k_{HC} = 0.076$	$R^2 = 0.0204$ $R_{adj}^2 = 0.0204$ $RMSE = 19.7379$ $AIC = 56.2979$	$k_{HC} = 0.065$	$R^2 = 0.1062$ $R_{adj}^2 = 0.1062$ $RMSE = 18.1759$ $AIC = 55.1437$
Baker–Lonsdale model	$\frac{3}{2} \times \left(1 - \left(1 - \frac{F}{100}\right)^{\frac{2}{3}}\right) - \frac{F}{100} = k_{BL} t$	$k_{BL} = 0.024$	$R^2 = 0.7877$ $R_{adj}^2 = 0.7877$ $RMSE = 9.1895$ $AIC = 45.5952$	$k_{BL} = 0.019$	$R^2 = 0.8105$ $R_{adj}^2 = 0.8105$ $RMSE = 8.3700$ $AIC = 44.2875$

R^2—coefficient of determination; R_{adj}^2—adjusted coefficient of determination; $RMSE$—root mean square error; AIC—Akaike information criterion; F—cumulative antioxidants release; k_0, k_1, k_H, k_{KP}, k_{HC}, k_{BL}—release constant of zero order, first order, Higuchi, Korsmeyer–Peppas, Hixson–Crowell, Baker–Lonsdale, respectively.

3.4. Antioxidant Activity of the Extract after Its Release from Microparticles in Gastrointestinal Fluids

The antioxidant activity of black locust flower extract encapsulated in alginate and alginate–chitosan microparticles was determined after 2 h in SGF or SIF. The calculated IC_{50} values of the samples are depicted in Table 4. With both types of microparticles, a higher antioxidant activity of the extract was noticed in the conditions of SIF. The extract encapsulated in alginate–chitosan microparticles expressed a better antioxidant activity compared to the extract encapsulated in alginate microparticles in the conditions of SIF. In this case, the chitosan membrane also had an important role in preventing the loss of antioxidant activity of the extract in the conditions of SGF. The obtained data are expected considering the alginate microparticles are degraded at the higher pH values, while they are stable in an acidic environment. The chitosan membrane enhances the microparticle stability, causing the reduction in antioxidants released in SGF [17]. Comparing the IC_{50} values of the extract before (120.9 ± 0.08 μg/mL) and after encapsulation, it can be noticed that the antioxidant activity significantly increased. The reason for such behavior is probably due to the synergistic effect of the extract, alginate [78], and chitosan [79], which is also known to possess antioxidant activity.

Table 4. IC_{50} values (μg/mL) of encapsulated extract of black locust flowers in alginate–chitosan and alginate microparticles in SGF and SIF.

Microparticle	Simulated Gastric Fluid	Simulated Intestinal Fluid
Alginate	109.4 ± 0.06	89.6 ± 0.07
Alginate–chitosan	115.7 ± 0.03	68.3 ± 0.05

According to the carried-out analyses, the alginate–chitosan microparticles were shown as a more suitable drug delivery system compared to the pure alginate microparticles due to the slower antioxidants release in simulated gastrointestinal fluids and better antioxidant activity of the sample. The main contribution of this study is the preparation of microparticles encapsulated with black locust flower extract at the laboratory level by protecting its antioxidant activity. To provide the scale enlargement of microparticle production, the procedure should be adjusted to the appropriate equipment at the industrial level.

4. Conclusions

The ethanolic extract of black locust flowers represents a source of antioxidant compounds with a TAC of 3.18 ± 0.01 g GAE/100 g d.w. The UHPLC–ESI–MS method also indicated the presence of 27 phenolic compounds. According to MTT and DPPH assays, the extract had moderate antioxidant and cytotoxic activity against the HeLa cell line. The extract can be considered a tumor-selective agent due to having no cytotoxic effect against the normal MRC-5 cell line. To protect and stabilize the extract from the conditions of the gastrointestinal tract in vitro, the encapsulation via extrusion method was used for the preparation of microparticles based on alginate and alginate–chitosan. In alginate–chitosan microparticles, the chitosan membrane reduced the permeability of alginate microparticles and directly impacted the protection of antioxidants and their low diffusion from the inner part of microparticles to the exterior medium. Also, the antioxidant activity of the extract was improved after its encapsulation in the microparticles. This is most likely caused by the synergistic effects of chitosan and alginate, which have expressed antioxidant activity. In summary, the prepared microparticles had standard quality characteristics in gastrointestinal environments. The procedure for microparticle preparation did not require the use of expensive equipment and toxic solvents and did not produce any hazardous waste harmful to human health or the environment. All research was carried out at the laboratory level. For further commercial application of these microparticles, it is necessary to study new industrial equipment and procedures for their preparation.

Author Contributions: Conceptualization, I.M.S.G. and I.M.S.; methodology, I.A.B., I.M.S.G. and I.M.S.; software, I.M.S.; validation, I.M.S.G. and I.M.S.; formal analysis, I.A.B., J.S.R.S., I.M.S.G. and N.Đ.G.S.; investigation, I.M.S.G., I.M.S., N.Đ.G.S. and T.D.K.-V.; resources, I.M.S.G.; data curation, I.M.S.; writing—original draft preparation, I.M.S.G. and I.M.S.; writing—review and editing, I.M.S., J.S.R.S. and T.D.K.-V.; visualization, I.M.S.G.; supervision, I.M.S.G.; project administration, I.M.S.G.; funding acquisition, I.M.S.G. All authors have read and agreed to the published version of the manuscript.

Funding: This research was funded by the Ministry of Science, Technological Development and Innovation of the Republic of Serbia (grant No. 451-03-65/2024-03/ 200133).

Institutional Review Board Statement: Not applicable.

Data Availability Statement: Data are contained within the article.

Conflicts of Interest: The authors declare no conflicts of interest.

References

1. Kruk, J.; Aboul-Enein, H.Y.; Kładna, A.; Bowser, J.E. Oxidative stress in biological systems and its relation with pathophysiological functions: The effect of physical activity on cellular redox homeostasis. *Free Radic. Res.* **2019**, *53*, 497–521. [CrossRef]
2. Khosravi, M.; Poursaleh, A.; Ghasempour, G.; Farhad, S.; Najafi, M. The effects of oxidative stress on the development of atherosclerosis. *Biol. Chem.* **2019**, *400*, 711–732. [CrossRef] [PubMed]
3. Zhang, P.; Li, T.; Wu, X.; Nice, E.C.; Huang, C.; Zhang, Y. Oxidative stress and diabetes: Antioxidative strategies. *Front. Med.* **2020**, *14*, 583–600. [CrossRef] [PubMed]
4. Hayes, J.D.; Dinkova-Kostova, A.T.; Tew, K.D. Oxidative stress in cancer. *Cancer Cell* **2020**, *38*, 167–197. [CrossRef] [PubMed]
5. Amarowicz, R.; Pegg, R.B. Natural antioxidants of plant origin. *Adv. Food Nutr. Res.* **2019**, *90*, 1–81. [PubMed]
6. Enaru, B.; Drețcanu, G.; Pop, T.D.; Stănilă, A.; Diaconeasa, Z. Anthocyanins: Factors affecting their stability and degradation. *Antioxidants* **2021**, *10*, 1967. [CrossRef]
7. Spínola, V.; Llorent-Martínez, E.J.; Castilho, P.C. Antioxidant polyphenols of Madeira sorrel (*Rumex maderensis*): How do they survive to in vitro simulated gastrointestinal digestion? *Food Chem.* **2018**, *259*, 105–112. [CrossRef] [PubMed]
8. Soares, S.; Silva, M.S.; Garcia-Estevez, I.; Grosmann, P.; Brás, N.; Brandão, E.; Mateus, N.; de Freitas, V.; Behrens, M.; Meyerhof, W. Human bitter taste receptors are activated by different classes of polyphenols. *J. Agric. Food Chem.* **2018**, *66*, 8814–8823. [CrossRef] [PubMed]
9. Casadey, R.; Broglia, M.; Barbero, C.; Criado, S.; Rivarola, C. Controlled release systems of natural phenolic antioxidants encapsulated inside biocompatible hydrogels. *React. Funct. Polym.* **2020**, *156*, 104729. [CrossRef]
10. Paulo, F.; Paula, V.; Estevinho, L.M.; Santos, L. Propolis microencapsulation by double emulsion solvent evaporation approach: Comparison of different polymeric matrices and extract to polymer ratio. *Food Bioprod. Process.* **2021**, *127*, 408–425. [CrossRef]

11. Romero-González, J.; Ah-Hen, K.S.; Lemus-Mondaca, R.; Muñoz-Fariña, O. Total phenolics, anthocyanin profile and antioxidant activity of maqui, *Aristotelia chilensis* (Mol.) Stuntz, berries extract in freeze-dried polysaccharides microcapsules. *Food Chem.* **2020**, *313*, 126115. [CrossRef]
12. Toprakçı, İ.; Şahin, S. Encapsulation of olive leaf antioxidants in microbeads. Application of alginate and chitosan as wall materials. *Sustain. Chem. Pharm.* **2022**, *27*, 100707. [CrossRef]
13. Lozano-Vazquez, G.; Lobato-Calleros, C.; Escalona-Buendia, H.; Chavez, G.; Alvarez-Ramirez, J.; Vernon-Carter, E.J. Effect of the weight ratio of alginate-modified tapioca starch on the physicochemical properties and release kinetics of chlorogenic acid containing beads. *Food Hydrocoll.* **2015**, *48*, 301–311. [CrossRef]
14. Iskandar, L.; Rojo, L.; Di Silvio, L.; Deb, S. The effect of chelation of sodium alginate with osteogenic ions, calcium, zinc, and strontium. *J. Biomater. Appl.* **2019**, *34*, 573–584. [CrossRef] [PubMed]
15. Pereira, A.D.S.; Diniz, M.M.; De Jong, G.; Gama Filho, H.S.; dos Anjos, M.J.; Finotelli, P.V.; Fontes-Sant'Ana, G.C.; Amaral, P.F.F. Chitosan-alginate beads as encapsulating agents for *Yarrowia lipolytica* lipase: Morphological, physico-chemical and kinetic characteristics. *Int. J. Biol. Macromol.* **2019**, *139*, 621–630. [CrossRef] [PubMed]
16. Cook, M.T.; Tzortzis, G.; Charalampopoulos, D.; Khutoryanskiy, V.V. Production and evaluation of dry alginate-chitosan microcapsules as an enteric delivery vehicle for probiotic bacteria. *Biomacromolecules* **2011**, *12*, 2834–2840. [CrossRef]
17. Kanokpanont, S.; Yamdech, R.; Aramwit, P. Stability enhancement of mulberry-extracted anthocyanin using alginate/chitosan microencapsulation for food supplement application. *Artif. Cells Nanomed. Biotechnol.* **2018**, *46*, 773–782. [CrossRef] [PubMed]
18. Martin, G.D. Addressing geographical bias: A review of *Robinia pseudoacacia* (black locust) in the Southern Hemisphere. *S. Afr. J. Bot.* **2019**, *125*, 481–492. [CrossRef]
19. Veitch, N.C.; Elliott, P.C.; Kite, G.C.; Lewis, G.P. Flavonoid glycosides of the black locust tree, *Robinia pseudoacacia* (Leguminosae). *Phytochem.* **2010**, *71*, 479–486. [CrossRef] [PubMed]
20. Ji, H.F.; Du, A.L.; Zhang, L.W.; Xu, C.Y.; Yang, M.D.; Li, F.F. Effects of drying methods on antioxidant properties in *Robinia pseudoacacia* L. flowers. *J. Med. Plant Res.* **2012**, *6*, 3233–3239.
21. Bhalla, P.; Bajpai, V.K. Chemical composition and antibacterial action of *Robinia pseudoacacia* L. flower essential oil on membrane permeability of foodborne pathogens. *J. Essent. Oil-Bear. Plants* **2017**, *20*, 632–645. [CrossRef]
22. Stefova, M.; Kulevanova, S.; Stafilov, T. Assay of flavonols and quantification of quercetin in medicinal plants by HPLC with UV-diode array detection. *J. Liq. Chromatogr. Relat.* **2001**, *24*, 2283–2292. [CrossRef]
23. Călina, D.; Olah, N.K.; Pătru, E.; Docea, A.; Popescu, H.; Bubulica, M.V. Chromatographic analysis of the flavonoids from *Robinia pseudoacacia* species. *Curr. Health Sci. J.* **2013**, *39*, 232–236.
24. Savic Gajic, I.; Savic, I.; Boskov, I.; Žerajić, S.; Markovic, I.; Gajic, D. Optimization of ultrasound-assisted extraction of phenolic compounds from black locust (*Robiniae pseudoacaciae*) flowers and comparison with conventional methods. *Antioxidants* **2019**, *8*, 248. [CrossRef]
25. Tian, F.; McLaughlin, J.L. Bioactive flavonoids from the black locust tree, *Robinia pseudoacacia*. *Pharm. Biol.* **2000**, *38*, 229–234. [CrossRef] [PubMed]
26. Grozdanić, N.; Djuričić, I.; Kosanić, M.; Zdunić, G.; Šavikin, K.; Etahiri, S.; Assobhei, O.; Benba, J.; Petović, S.; Matić, I.Z.; et al. *Fucus spiralis* extract and fractions: Anticancer and pharmacological potentials. *JBUON* **2020**, *25*, 1219–1229. [PubMed]
27. Kolundžić, M.; Grozdanić, N.Đ.; Dodevska, M.; Milenković, M.; Sisto, F.; Miani, A.; Farronato, G.; Kundaković, T. Antibacterial and cytotoxic activities of wild mushroom *Fomes fomentarius* (L.) Fr., Polyporaceae. *Ind. Crops Prod.* **2016**, *79*, 110–115. [CrossRef]
28. Gašić, U.; Kečkeš, S.; Dabić, D.; Trifković, J.; Milojković-Opsenica, D.; Natić, M.; Tešić, Ž. Phenolic profile and antioxidant activity of Serbian polyfloral honeys. *Food Chem.* **2014**, *145*, 599–607. [CrossRef] [PubMed]
29. Chan, E.S.; Lee, B.B.; Ravindra, P.; Poncelet, D. Prediction models for shape and size of ca-alginate macrobeads produced through extrusion–dripping method. *J. Colloid Interface Sci.* **2009**, *338*, 63–72. [CrossRef] [PubMed]
30. Frent, O.D.; Duda-Seiman, D.M.; Vicas, L.G.; Duteanu, N.; Nemes, N.S.; Pascu, B.; Teusdea, A.; Pallag, A.; Micle, O.; Marian, E. Study of the influence of the excipients used for the synthesis of microspheres loaded with quercetin: Their characterization and antimicrobial activity. *Coatings* **2023**, *13*, 1376. [CrossRef]
31. Hallmann, E. Quantitative and qualitative identification of bioactive compounds in edible flowers of black and bristly locust and their antioxidant activity. *Biomolecules* **2020**, *10*, 1603. [CrossRef] [PubMed]
32. Jurca, T.; Pallag, A.; Vicaș, L.; Marian, E.; Mureșan, M.; Ujhelyi, Z.; Fehér, P.; Bacskay, I. Formulation and antioxidant investigation of creams containing *Robiniae pseudacaciae flos* L. ethanolic extract. *Farmacia* **2021**, *69*, 697–704. [CrossRef]
33. Bratu, M.M.; Birghila, S.; Stancu, L.M.; Mfflai, C.C.; Emoke, P.; Popescu, A.; Radu, M.D.; Zglimbea, L. Evaluation of the antioxidant, cytotoxic and antitumoral activities of a polyphenolic extract of *Robinia pseudoacacia* L. flowers. *J. Sci. Arts* **2021**, *21*, 547–556. [CrossRef]
34. Han, M.G.; Park, Y.J.; In, M.J.; Kim, D.C. Biological activities of ethanolic extract from *Robinia pseudoacacia* L. flower. *J. Appl. Biol. Chem.* **2022**, *65*, 107–111. [CrossRef]
35. Sarikurkcu, C.; Kocak, M.S.; Tepe, B.; Uren, M.C. An alternative antioxidative and enzyme inhibitory agent from Turkey: *Robinia pseudoacacia* L. *Ind. Crops Prod.* **2015**, *78*, 110–115. [CrossRef]
36. Ćorković, I.; Gašo-Sokač, D.; Pichler, A.; Šimunović, J.; Kopjar, M. Dietary polyphenols as natural inhibitors of α-amylase and α-glucosidase. *Life* **2022**, *12*, 1692. [CrossRef] [PubMed]
37. Sieniawska, E. Activities of tannins-From in vitro studies to clinical trials. *Nat. Prod. Commun.* **2015**, *10*, 1877–1884. [CrossRef]

38. Niger, T.; Ohtani, K.; Ahamed, B.F. Inhibitory effects of Japanese plant leaf extracts on α-glucosidase activity. *J. Mol. Stud. Med.* **2018**, *3*, 161–168.
39. Sajjadi, S.E.; Ghanadian, M.; Haghighi, M.; Mouhebat, L. Cytotoxic effect of *Cousinia verbascifolia* Bunge against OVCAR-3 and HT-29 cancer cells. *J. Herbmed Pharmacol.* **2015**, *4*, 15–19.
40. Uzelac, M.; Sladonja, B.; Šola, I.; Dudaš, S.; Bilić, J.; Famuyide, I.M.; McGaw, L.; Eloff, J.; Mikulic-Petkovsek, M.; Poljuha, D. Invasive alien species as a potential source of phytopharmaceuticals: Phenolic composition and antimicrobial and cytotoxic activity of *Robinia pseudoacacia* L. leaf and flower extracts. *Plants* **2023**, *12*, 2715. [CrossRef]
41. Cvetković, D.M.; Jovankić, J.V.; Milutinović, M.G.; Nikodijević, D.D.; Grbović, F.J.; Ćirić, A.R.; Topuzović, M.D.; Marković, S.D. The anti-invasive activity of *Robinia pseudoacacia* L. and *Amorpha fruticosa* L. on breast cancer MDA-MB-231 cell line. *Biologia* **2019**, *74*, 915–928. [CrossRef]
42. Sanz, M.; de Simón, B.F.; Esteruelas, E.; Muñoz, Á.M.; Cadahía, E.; Hernández, M.T.; Estrella, I.; Martinez, J. Polyphenols in red wine aged in acacia (*Robinia pseudoacacia*) and oak (*Quercus petraea*) wood barrels. *Anal. Chim. Acta* **2012**, *732*, 83–90. [CrossRef]
43. Fraternale, D.; Ricci, D.; Verardo, G.; Gorassini, A.; Stocchi, V.; Sestili, P. Activity of *Vitis vinifera* tendrils extract against phytopathogenic fungi. *Nat. Prod. Commun.* **2015**, *10*, 1037–1042.
44. Gardana, C.; Scaglianti, M.; Pietta, P.; Simonetti, P. Analysis of the polyphenolic fraction of propolis from different sources by liquid chromatography–tandem mass spectrometry. *J. Pharm. Biomed. Anal.* **2007**, *45*, 390–399. [CrossRef] [PubMed]
45. Zengin, G.; Mahomoodally, F.; Picot-Allain, C.; Diuzheva, A.; Jekő, J.; Cziáky, Z.; Cvetanović, A.; Aktumsek, A.; Zeković, Z.; Rengasamy, K.R. Metabolomic profile of *Salvia viridis* L. root extracts using HPLC–MS/MS technique and their pharmacological properties: A comparative study. *Ind. Crops Prod.* **2019**, *131*, 266–280. [CrossRef]
46. Destandau, E.; Charpentier, J.P.; Bostyn, S.; Zubrzycki, S.; Serrano, V.; Seigneuret, J.M.; Breton, C. Gram-scale purification of dihydrorobinetin from *Robinia pseudoacacia* L. wood by centrifugal partition chromatography. *Separations* **2016**, *3*, 23. [CrossRef]
47. Shaker, K.H.; Zohair, M.M.; Hassan, A.Z.; Sweelam, H.T.M.; Ashour, W.E. LC–MS/MS and GC–MS based phytochemical perspectives and antimicrobial effects of endophytic fungus *Chaetomium ovatoascomatis* isolated from *Euphorbia milii*. *Arch. Microbiol.* **2022**, *204*, 661. [CrossRef]
48. Vuković, N.L.; Vukić, M.D.; Đelić, G.T.; Kacaniova, M.M.; Cvijović, M. The investigation of bioactive secondary metabolites of the methanol extract of *Eryngium amethystinum*. *Kragujevac J. Sci.* **2018**, *40*, 113–129. [CrossRef]
49. Zhong, B.; Robinson, N.A.; Warner, R.D.; Barrow, C.J.; Dunshea, F.R.; Suleria, H.A. LC-ESI-QTOF-MS/MS characterization of seaweed phenolics and their antioxidant potential. *Mar. Drugs* **2020**, *18*, 331. [CrossRef] [PubMed]
50. Fahmy, N.M.; El-Din, M.I.G.; Salem, M.M.; Rashedy, S.H.; Lee, G.S.; Jang, Y.S.; Kim, K.; Kim, C.; El-Shazly, M.; Fayez, S. Enhanced expression of p53 and Suppression of PI3K/Akt/mTOR by three red sea algal extracts: Insights on their composition by LC-MS-based metabolic profiling and molecular networking. *Mar. Drugs* **2023**, *21*, 404. [CrossRef]
51. Kang, J.; Price, W.E.; Ashton, J.; Tapsell, L.C.; Johnson, S. Identification and characterization of phenolic compounds in hydromethanolic extracts of sorghum wholegrains by LC-ESI-MS^n. *Food Chem.* **2016**, *211*, 215–226. [CrossRef] [PubMed]
52. Mbakidi-Ngouaby, H.; Pinault, E.; Gloaguen, V.; Costa, G.; Sol, V.; Millot, M.; Mambu, L. Profiling and seasonal variation of chemical constituents from *Pseudotsuga menziesii* wood. *Ind. Crops Prod.* **2018**, *117*, 34–39. [CrossRef]
53. Ghareeb, M.; Saad, A.; Ahmed, W.; Refahy, L.; Nasr, S. HPLC-DAD-ESI-MS/MS characterization of bioactive secondary metabolites from *Strelitzia nicolai* leaf extracts and their antioxidant and anticancer activities in vitro. *Pharmacogn. Res.* **2018**, *10*, 368–378. [CrossRef]
54. Chandradevan, M.; Simoh, S.; Mediani, A.; Ismail, N.H.; Ismail, I.S.; Abas, F. UHPLC-ESI-Orbitrap-MS analysis of biologically active extracts from *Gynura procumbens* (Lour.) Merr. and *Cleome gynandra* L. leaves. *eCAM* **2020**, *2020*, 3238561. [CrossRef] [PubMed]
55. Grati, W.; Samet, S.; Bouzayani, B.; Ayachi, A.; Treilhou, M.; Téné, N.; Mezghani-Jarraya, R. HESI-MS/MS analysis of phenolic compounds from *Calendula aegyptiaca* fruits extracts and evaluation of their antioxidant activities. *Molecules* **2022**, *27*, 2314. [CrossRef] [PubMed]
56. Cerulli, A.; Napolitano, A.; Hošek, J.; Masullo, M.; Pizza, C.; Piacente, S. Antioxidant and in vitro preliminary anti-inflammatory activity of *Castanea sativa* (Italian Cultivar "Marrone di Roccadaspide" PGI) burs, leaves, and chestnuts extracts and their metabolite profiles by LC-ESI/LTQOrbitrap/MS/MS. *Antioxidants* **2021**, *10*, 278. [CrossRef]
57. Spínola, V.; Llorent-Martínez, E.J.; Gouveia-Figueira, S.; Castilho, P.C. *Ulex europaeus*: From noxious weed to source of valuable isoflavones and flavanones. *Ind. Crops Prod.* **2016**, *90*, 9–27. [CrossRef]
58. Żuchowski, J.; Skalski, B.; Juszczak, M.; Woźniak, K.; Stochmal, A.; Olas, B. LC/MS analysis of saponin fraction from the leaves of *Elaeagnus rhamnoides* (L.) A. Nelson and its biological properties in different in vitro models. *Molecules* **2020**, *25*, 3004. [CrossRef]
59. Al-Yousef, H.M.; Hassan, W.H.; Abdelaziz, S.; Amina, M.; Adel, R.; El-Sayed, M.A. UPLC-ESI-MS/MS profile and antioxidant, cytotoxic, antidiabetic, and antiobesity activities of the aqueous extracts of three different *Hibiscus* species. *J. Chem.* **2020**, *2020*, 6749176. [CrossRef]
60. Huang, Y.X.; Liang, J.; Chai, J.H.; Kuang, H.X.; Xia, Y.G. Structure of a highly branched galacturonoglucan from fruits of *Schisandra chinensis* (Turcz.) Baill. *Carbohydr. Polym.* **2023**, *313*, 120844. [CrossRef]
61. March, R.E.; Miao, X.S.; Metcalfe, C.D. A fragmentation study of a flavone triglycoside, kaempferol-3-O-robinoside-7-O-rhamnoside. *Rapid Commun. Mass Spectrom.* **2004**, *18*, 931–934. [CrossRef]

62. Matkovits, A.; Nagy, K.; Fodor, M.; Jókai, Z. Analysis of polyphenolic components of Hungarian acacia (*Robinia pseudoacacia*) honey; method development, statistical evaluation. *J. Food Compos. Anal.* **2023**, *120*, 105336. [CrossRef]
63. Tian, J.; Gong, Y.; Li, J. Nutritional attributes and phenolic composition of flower and bud of *Sophora japonica* L. and *Robinia pseudoacacia* L. *Molecules* **2022**, *27*, 8932. [CrossRef]
64. Liu, R.; Cai, Z.; Xu, B. Characterization and quantification of flavonoids and saponins in adzuki bean (*Vigna angularis* L.) by HPLC–DAD–ESI–MS^n analysis. *Chem. Cent. J.* **2017**, *11*, 1–17. [CrossRef]
65. Ervina, M.; Nawu, Y.E.; Esar, S.Y. Comparison of in vitro antioxidant activity of infusion, extract and fractions of Indonesian Cinnamon (*Cinnamomum burmannii*) bark. *Int. Food Res. J.* **2016**, *23*, 1346–1350.
66. Yousefi, M.; Khanniri, E.; Shadnoush, M.; Khorshidian, N.; Mortazavian, A.M. Development, characterization and in vitro antioxidant activity of chitosan-coated alginate microcapsules entrapping *Viola odorata* Linn. extract. *Int. J. Biol. Macromol.* **2020**, *163*, 44–54. [CrossRef] [PubMed]
67. Villate, A.; San Nicolas, M.; Olivares, M.; Aizpurua-Olaizola, O.; Usobiaga, A. Chitosan-coated alginate microcapsules of a full-spectrum *Cannabis* extract: Characterization, long-term stability and in vitro bioaccessibility. *Pharmaceutics* **2023**, *15*, 859. [CrossRef] [PubMed]
68. Lee, B.B.; Ravindra, P.; Chan, E.S. Size and shape of calcium alginate beads produced by extrusion dripping. *Chem. Eng. Technol.* **2013**, *36*, 1627–1642. [CrossRef]
69. Lim, G.P.; Lee, B.B.; Ahmad, M.S.; Singh, H.; Ravindra, P. Influence of process variables and formulation composition on sphericity and diameter of Ca-alginate-chitosan liquid core capsule prepared by extrusion dripping method. *Part. Sci. Technol.* **2016**, *34*, 681–690. [CrossRef]
70. Ling, K.; Wu, H.; Neish, A.S.; Champion, J.A. Alginate/chitosan microparticles for gastric passage and intestinal release of therapeutic protein nanoparticles. *J. Control Release* **2019**, *295*, 174–186. [CrossRef] [PubMed]
71. Unagolla, J.M.; Jayasuriya, A.C. Drug transport mechanisms and in vitro release kinetics of vancomycin encapsulated chitosan-alginate polyelectrolyte microparticles as a controlled drug delivery system. *Eur. J. Pharm. Sci.* **2018**, *114*, 199–209. [CrossRef] [PubMed]
72. Pravilović, R.N.; Balanč, B.D.; Djordjević, V.B.; Bošković-Vragolović, N.M.; Bugarski, B.M.; Pjanović, R.V. Diffusion of polyphenols from alginate, alginate/chitosan, and alginate/inulin particles. *J. Food Process Eng.* **2019**, *42*, e13043. [CrossRef]
73. Mulia, K.; Singarimbun, A.C.; Krisanti, E.A. Optimization of chitosan–alginate microparticles for delivery of mangostins to the colon area using Box–Behnken experimental design. *Int. J. Mol. Sci.* **2020**, *21*, 873. [CrossRef]
74. Gonçalves, G.A.; Corrêa, R.C.; Barros, L.; Dias, M.I.; Calhelha, R.C.; Correa, V.G.; Bracht, A.; Peralta, R.M.; Ferreira, I.C. Effects of in vitro gastrointestinal digestion and colonic fermentation on a rosemary (*Rosmarinus officinalis* L.) extract rich in rosmarinic acid. *Food Chem.* **2019**, *271*, 393–400. [CrossRef]
75. Vinceković, M.; Jurić, S.; Vlahoviček-Kahlina, K.; Martinko, K.; Šegota, S.; Marijan, M.; Krčelić, A.; Svečnjak, L.; Majdak, M.; Nemet, I.; et al. Novel zinc/silver ions-loaded alginate/chitosan microparticles antifungal activity against *Botrytis cinerea*. *Polymers* **2023**, *15*, 4359. [CrossRef] [PubMed]
76. El Maghraby, G.M.; Arafa, M.F. Alginate-chitosan combinations in controlled drug delivery. In *Natural Polysaccharides in Drug Delivery and Biomedical Applications*; Hasnain, M.S., Nayak, A.K., Eds.; Academic Press: Cambridge, MA, USA, 2019; pp. 339–361.
77. Fu, Y.; Kao, W.J. Drug release kinetics and transport mechanisms of non-degradable and degradable polymeric delivery systems. *Expert Opin. Drug Deliv.* **2010**, *7*, 429–444. [CrossRef]
78. Kelishomi, Z.H.; Goliaei, B.; Mahdavi, H.; Nikoofar, A.; Rahimi, M.; Moosavi-Movahedi, A.A.; Mamashli, F.; Bigdeli, B. Antioxidant activity of low molecular weight alginate produced by thermal treatment. *Food Chem.* **2016**, *196*, 897–902. [CrossRef]
79. Yen, M.T.; Yang, J.H.; Mau, J.L. Antioxidant properties of chitosan from crab shells. *Carbohydr. Polym.* **2008**, *74*, 840–844. [CrossRef]

Disclaimer/Publisher's Note: The statements, opinions and data contained in all publications are solely those of the individual author(s) and contributor(s) and not of MDPI and/or the editor(s). MDPI and/or the editor(s) disclaim responsibility for any injury to people or property resulting from any ideas, methods, instructions or products referred to in the content.

Article

Stability in Aqueous Solution of a New Spray-Dried Hydrocolloid of High Andean Algae *Nostoc sphaericum*

David Choque-Quispe [1,2,3,4,*], Carlos A. Ligarda-Samanez [2,3,4,5], Yudith Choque-Quispe [1,3,4,6], Sandro Froehner [7], Aydeé M. Solano-Reynoso [4,8], Elibet Moscoso-Moscoso [5], Yakov Felipe Carhuarupay-Molleda [8] and Ronald Peréz-Salcedo [2]

1. Water and Food Treatment Materials Research Laboratory, Universidad Nacional José María Arguedas, Andahuaylas 03701, Peru; ychoque@unajma.edu.pe
2. Department of Agroindustrial Engineering, Universidad Nacional José María Arguedas, Andahuaylas 03701, Peru; caligarda@unajma.edu.pe (C.A.L.-S.); rperez@unajma.edu.pe (R.P.-S.)
3. Research Group in the Development of Advanced Materials for Water and Food Treatment, Universidad Nacional José María Arguedas, Andahuaylas 03701, Peru
4. Nutraceuticals and Biopolymers Research Group, Universidad Nacional José María Arguedas, Andahuaylas 03701, Peru; amsolano@unajma.edu.pe
5. Food Nanotechnology Research Laboratory, Universidad Nacional José María Arguedas, Andahuaylas 03701, Peru; eibetmm22@gmail.com
6. Department of Environmental Engineering, Universidad Nacional José María Arguedas, Andahuaylas 03701, Peru
7. Department of Environmental Engineering, Federal University of Parana, Curitiba 80010, Brazil; froehner@ufpr.br
8. Department of Basic Sciences, Universidad Nacional José María Arguedas, Andahuaylas 03701, Peru; ycarhuarupay@unajma.edu.pe
* Correspondence: dchoque@unajma.edu.pe

Citation: Choque-Quispe, D.; Ligarda-Samanez, C.A.; Choque-Quispe, Y.; Froehner, S.; Solano-Reynoso, A.M.; Moscoso-Moscoso, E.; Carhuarupay-Molleda, Y.F.; Peréz-Salcedo, R. Stability in Aqueous Solution of a New Spray-Dried Hydrocolloid of High Andean Algae *Nostoc sphaericum*. *Polymers* 2024, *16*, 537. https://doi.org/10.3390/polym16040537

Academic Editors: Cristina Cazan and Mihai Alin Pop

Received: 4 January 2024
Revised: 5 February 2024
Accepted: 9 February 2024
Published: 16 February 2024

Copyright: © 2024 by the authors. Licensee MDPI, Basel, Switzerland. This article is an open access article distributed under the terms and conditions of the Creative Commons Attribution (CC BY) license (https://creativecommons.org/licenses/by/4.0/).

Abstract: There is a growing emphasis on seeking stabilizing agents with minimal transformation, prioritizing environmentally friendly alternatives, and actively contributing to the principles of the circular economy. This research aimed to assess the stability of a novel spray-dried hydrocolloid from high Andean algae when introduced into an aqueous solution. *Nostoc sphaericum* freshwater algae were subject to atomization, resulting in the production of spray-dried hydrocolloid (SDH). Subsequently, suspension solutions of SDH were meticulously prepared at varying pH levels and gelling temperatures. These solutions were then stored for 20 days to facilitate a comprehensive evaluation of their stability in suspension. The assessment involved a multifaceted approach, encompassing rheological analysis, scrutiny of turbidity, sedimentation assessment, ζ-potential, and measurement of particle size. The findings from these observations revealed that SDH exhibits a dilatant behavior when in solution, signifying an increase in with higher shear rate. Furthermore, it demonstrates commendable stability when stored under ambient conditions. SDH is emerging as a potential alternative stabilizer for use in aqueous solutions due to its easy extraction and application.

Keywords: stability; high Andean algae; hydrocolloid; SDH solution

1. Introduction

Suspension stabilizers find widespread applications in various industrial processes, ranging from the formulation of consumable foods, and pharmaceuticals, to water treatment. Maintaining the suspension of solids in an aqueous medium not only enhances the visual quality of foods and drugs [1,2] but, in water treatment, it also improves the efficiency of removing both suspended and dissolved materials. Achieving these objectives relies on the utilization of stabilizing agents, which can be of chemical origin or derived from biological origin [3,4]. Nevertheless, the use or production of these stabilizers may generate waste, posing potential environmental challenges. Consequently, there is a growing interest in developing stabilizers from biological origins that are environmentally friendly and contribute to the principles of the circular economy.

A diverse array of stabilizing agents, including guar gum, xanthan gum, sodium alginate, pectins, carrageenans, gelatins, and locust bean gum, are extensively employed due to their exceptional functionality. These ingredients have significantly transformed the landscapes of the food and pharmaceutical and water treatment industries, with their availability in the market reflecting their synthetic, semi-synthetic, or natural origins [3,5–7]. With a heightened consumer awareness of environmental conservation and sustainability, there is a discernible preference for natural hydrocolloids, with algae emerging as prominent contenders in this category [8,9].

On the other hand, the extraction processes of commercial gums or hydrocolloids in many cases require the use of organic solvents and strong acidic or basic media [10,11], generating waste that negatively impacts the environment. Given this, the proposal of extraction with environmentally friendly methods is recurrent, so an alternative is extraction by atomization, and nostoc, being an algae with high moisture content (around 98%), is ideal to be subjected to this process.

While the majority of marine algae have undergone extensive study and exploitation in the food industry, the scientific knowledge and utilization of freshwater algae, particularly their extracted derivatives like the hydrocolloids of *Nostoc sphaericum*, remain limited. *Nostoc sphaericum*, an algae thriving as a renewable natural resource in the lagoons and wetlands of the Peruvian Andes, represents an area of untapped potential [12–14]; it has a protein content of 24.01%, fat 1.88%, ash 6.19%, fiber 8.84%, carbohydrates 57.32%, humidity 10.57%, and pH 6.91 in dehydrated samples on a wet basis [14].

Introducing novel stabilizers with broad applications poses a formidable challenge for the current industry, and the exploration of high Andean algae, such as *Nostoc spaericum*, holds promise in meeting these expectations. However, understanding the key parameters of the stabilizers or hydrocolloids, including ζ potential, particle size, and rheological behavior, is imperative [5,15–17]. Additionally, comprehending their behavior during storage is essential. The stability of stabilizing agents in an aqueous medium over time is critical during application. This aspect becomes particularly significant in preventing the undesired sedimentation of suspended solids, a concern in various liquid products such as nectars, juices, and dairy-derived concentrates, and especially in the context of suspended drugs. Conversely, in applications where agglomeration is desired, such as certain processes [18–22], sedimentation becomes a favorable outcome.

The stability of hydrocolloids in aqueous solution hinges on various factors, including hydrocolloid concentration, pH, temperature, and mixing speed. This stability can be characterized by understanding key parameters such as ζ potential, particle size, molecular weight, shear stress, strain rate, and activation energy [18–22]. While these control parameters are well-established for commercial hydrocolloids, determining them for a new hydrocolloid is essential to define its potential.

In the high-altitude lagoons of Andahuaylas, a province of the Peruvian Andes, algae of the genus *Nostoc* flourishes at elevations above 4000 m. Remarkably, these algae serve as a folkloric food source for the residents of the local communities.

Despite its content of hydrocolloids with favorable techno-functional properties [14,23], this algae currently lacks commercial significance, potentially serving as a valuable alternative to commonly used industry hydrocolloids. The research is directed towards evaluating the aqueous stability of a newly developed spray-dried hydrocolloid derived from high Andean algae, specifically *Nostoc sphaericum*.

2. Materials and Methods

2.1. Raw Material

The samples of atomized hydrocolloid (SDH) of *Nostoc sphaericum* were supplied by the Laboratory of Research in Advanced Materials for Water Treatment of the National University José María Arguedas, Peru. They were extracted by atomization according to the methodology proposed by Choque et al. [14], which consists of liquefying the algae with distilled water in a 1/1 ratio, then being sieved at 45 microns, then atomized at 100 °C inlet temperature, air speed of 600 L/s, and suction speed of 38 m^3/h in a mini spray dryer model B-290, Buchi brand (Flawil, Switzerland).

2.2. Preparation of Hydrocolloid Suspensions

Suspensions were formulated in accordance with Table 1. The process involved adjusting the pH of distilled water with 0.1 M citric acid. Subsequently, 100 mL of the solution was taken, and 1 mg of SDH along with potassium sorbate 1 mg was added (added to prevent microbial growth). The mixture was stirred at 1500 rpm for 1 min for homogenization and then left in agitation at 60 rpm for 24 h. To achieve uniform temperature, the suspensions were heated at 60 and 80 °C under continuous agitation at 60 rpm until a constant temperature was reached (refer to Figure 1). Following this, the suspensions were cooled down to room temperature for subsequent evaluations.

Table 1. Experimental design matrix.

Treatment	Factor 1: pH	Factor 2: Concentration (ppm)	Factor 3: Temperature (°C)
T1	6.5	100	60
T2	6.5	100	80
T3	4.5	100	60
T4	4.5	100	80
T5	6.5	70	60
T6	6.5	70	80
T7	4.5	70	60
T8	4.5	70	80
T9	6.5	100	40
T10	6.5	70	40
T11	4.5	100	40
T12	4.5	70	40

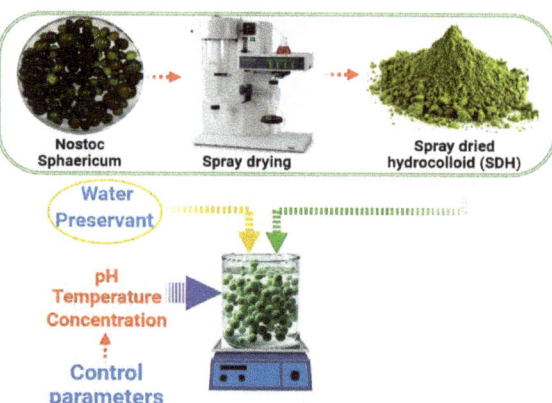

Figure 1. SDH solution preparation and analysis flowchart.

2.3. SDH Characterization

The determination of the point of zero charge (PZC) involved preparing solutions with varying pH values (ranging from 2 to 12). In each case, 0.05 g of SDH was added to 50 mL of each solution and stirred at 150 rpm for 24 h at room temperature. The resulting solution's pH was then measured, and the PZC was determined by identifying the intersection point of the initial pH and final pH curve [24].

In a separate analysis, SDH was subject to examination using a Thermo Fisher (Waltham, MA, USA) FTIR spectrometer in ATR mode. This analysis focused on identifying

Although the diffractogram of SDH reveals an overall amorphous structure (Figure 2b), two discernible peaks, at 9.17° and 19.52° of 2θ, respectively, indicate the presence of crystalline zones. These zones are associated with materials featuring non-covalent bonds and gelling qualities, such as CMC, citrus pectin, guar gum, and alginates among others [36,38,40–43].

The XRD pattern of SDH reports a degree of crystallinity at 71.98% and amorphous content at 28.02%, signifying a substantial level of crystallinity. This heightened crystallinity is likely attributed to the protein content within SDH [44], rendering it a stable material during storage and thereby enhancing its shelf life significantly [36,39].

3.3. FTIR Analysis

Fourier transform infrared analysis (FTIR) is instrumental in unraveling the intricate interactions involving stretching, bending, and torsion of chemical bonds within materials [14,45]. A pronounced peak of high intensity centered around 3400 cm^{-1}. (Figure 2c) signifies the stretching of hydroxyl groups present in water, amides, carbohydrates, and carboxylic acids. Simultaneously, intense peaks at approximately 2900 cm^{-1} are indicative of vibrations from the methyl groups of carbohydrates, a characteristic feature of biological origin material [46].

The peak around 1600 cm^{-1} is attributed to the -OH stretching of adsorbed water molecules, suggesting a highly hygroscopic material as a result of its hydrocolloid content [7]. Furthermore, the peak at 1410 cm^{-1} is associated with the stretching of the C-O, C-H, and -OH single bonds, predominantly from carbohydrates [47].

Conversely, the peak observed at approximately 1050 cm^{-1} signifies the stretching of the C-O, C-O-C, and C-OH bonds within the polymeric chains of carbohydrates and proteins [2,47]. Peaks registering below 1000 cm^{-1} constitute the distinctive fingerprint of SDH, attributed to the stretching of the C-H and C-O bonds inherent to carbohydrates, a characteristic feature of hydrocolloids [47,48].

3.4. SEM Analysis

Particle morphology serves as a macroscopic representation of molecular arrangements within a material [49]. In the SEM image (Figure 2d), spherical shapes predominate, a characteristic tendency of materials containing carbohydrate-protein-lipid complexes. This configuration arises from interactions among nonpolar groups within the particle, leading to them agglomerating in spherical form [47]. The splashed or dented shapes in the photomicrograph are attributed to the rapid vaporization of water during the atomization process, causing contractions on the particle surface [43]. Such morphological features are traits of spray-dried materials [44,50].

3.5. Rheological Analysis

A noticeable increase in shear stress corresponding to shear rate was observed for all the formulations (Figure 3), displaying a consistent trend, especially among formulations with a pH of 4.5. Although the temperature allows the shear stress to be slightly increased, the opposite effect was observed at pH 6.5.

The concave curve fitted to the Power Law model yielded R^2 values exceeding 0.9898, and the Herschel–Bulkley model resulted in R^2 values surpassing 0.9929. In both models, behavior index values (n) exceed 1.0 (Table 3), which suggests that the solution behaves as a dilatant fluid. This behavior aligns with shear thickening dispersions [23,51,52], a characteristic observed when subject to a wide range of cutting rates [53–55]. Notably, this behavior intensified with significantly increasing temperature (Figure 4a), likely attributed to the thickening capacity of the hydrogels and gums [53,56,57], although no significant differences of n with SDH concentration and pH of the solution have been evidenced (p-value < 0.05).

$$CI^* = \frac{a^* \cdot 1000}{L^* \cdot b^*},\tag{9}$$

For the ζ potential measurements, a 2 mL aliquot was extracted from each treatment and transferred to a polystyrene cell. Then, the samples were subject to analysis using dynamic light scattering equipment (DLS, Zetasizer ZSU3100, Malvern Instruments, Worcestershire, UK). The instrument operated at 632.8 nm, a scattering angle of 90°, and an electric field strength of 5 V/cm. To ensure accuracy and reproducibility, readings were performed in triplicate.

2.7. Statistical Analysis

The data on the stability properties of the SDH solutions were collected in Excel sheets and were evaluated by measuring the main effects and interactions of the input variables using the Statistica V12 software. The PCA analysis was carried out by standardizing the data of the response variables to integer values through the Origin Pro 2022b Software.

3. Results and Discussion

3.1. Zero Charge Point (ZCP)

The zero charge point (ZCP), also known as the isoelectric point, represents the pH value at which a substance in a solution carries no electrical charge. This parameter significantly influences the stability of substances in aqueous media [33,34]. Values below pH_{ZCP} indicate a higher availability of functional groups with a positive charge in the substance [33,35]. In the case of SDH, the reported pH_{ZCP} was approximately 8.1 (Figure 2a). This finding suggests that in the aqueous medium, SDH has the capacity to adsorb negatively charged molecules by physisorption or chemisorption at a pH lower than the pH_{ZCP}.

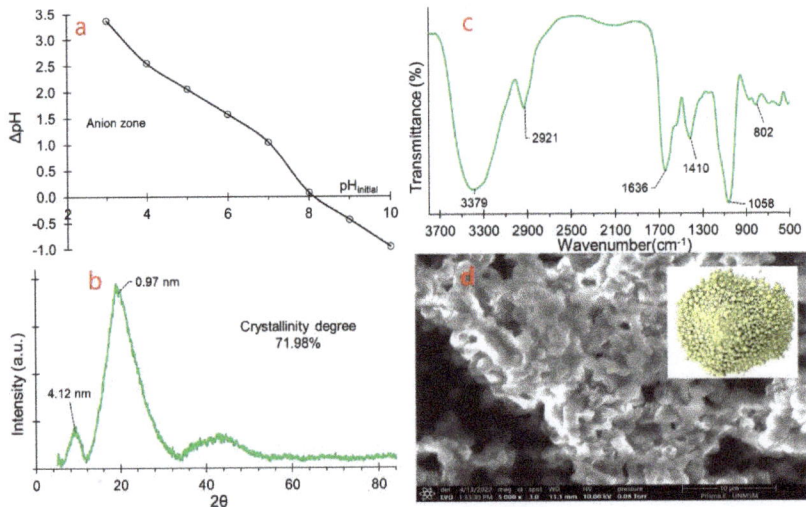

Figure 2. Characteristics of SDH (**a**) Zero charge point, (**b**) XRD diffractogram, (**c**) FTIR spectrogram, (**d**) SEM image.

3.2. Diffractometric Analysis and Degree of Crystallinity

Powdered materials are often favored for their enhanced stability and extended shelf life, especially when processing high crystallinity; this is in contrast to amorphous materials, which tend to retain more water [36,37], which could lead to microstructural collapse and microbiological instability [38,39].

$$MAPE = \frac{1}{N}\sum_{i=1}^{N}\frac{|x_i - \hat{x}_i|}{x_i} \qquad (7)$$

In the given equations, x_i represents the observed value; $\hat{x}_i$ denotes the predicted value; n signifies the number of observations, and N represents the total number of experimental observations.

Similarly, an assessment of the dispersion of residuals was conducted, employing criteria such as Random (R), Slightly Random (SR), and Tendentious (T). Models demonstrating the best fit exhibited a random distribution of residuals.

These analyses were conducted at a significance level of 5%, utilizing Excel sheets, the Solver utility, and Statistica V12 software (Statsoft, Tulsa, OK, USA).

2.5. Determination of Temperature Dependence

The influence of temperature on the rheological behavior was examined by assessing the activation energy (E_a), providing insights into the behavior of colloidal solutions, interpenetrating networks, and nanofluid flow. This parameter is linked to the energy necessary for the interchain displacement of polymers, with higher E_a values indicating elevated crosslinking [29,30]. The calculation of E_a was carried out using the Arrhenius equation (Equation (8)) based on the consistency index values.

$$k = k_0 e^{-\frac{E_a}{R}\cdot\frac{1}{T}}, \qquad (8)$$

where k is the consistency index of the fitted model; k_0 is the pre-exponential factor; R is the universal gas constant (8.314 kJ/kmol.K), and T is absolute temperature, K.

2.6. SDH Suspension Stability Evaluation

The stability of the suspensions was evaluated for 20 days, employing indicators such as turbidity, sedimentation, color, and ζ potential.

For each SDH suspension formulation, 20 mL was dispensed into flat-bottomed tubes with a diameter of 1.6 cm. Subsequently, 10 mL of suspension was extracted from the top of each treatment at rest, and turbidity was measured at intervals of 4 days for 20 days. The measurements were conducted by recording the transmittance at 560 nm using a UV-Vis spectrophotometer, specifically the Thermo Fisher Genesys 150 UV model (Waltham, MA, USA). Distilled water with potassium sorbate at the study pH served as a control. It is noteworthy that samples were discarded after each measurement.

To assess sedimentation, 8 mL of the tube's remaining volume was extracted and discarded, leaving 2 mL. The residual content was vigorously vortexed at 3000 rpm for 2 min to achieve sediment homogenization. Subsequently, the homogenized sample was subject to spectrophotometric analysis, measuring transmittance at 560 nm. This process was repeated over a span of days, with readings taken at 4-day intervals.

The color stability of the suspensions was evaluated in the CIE L* a* b* color space, employing specific criteria. Luminosity (L*) was gauged on a scale from 0 = black and 100 = white, while chroma values a* and b* were utilized to determine color characteristics (+a = red, −a = green, +b = yellow and −b = blue) [31]. For this analysis, treatment samples were examined using a Konica Minolta colorimeter, model CR-5 (Japan), and readings were recorded in the reflectance module. Additionally, the color index (CI*) was calculated using Equation (9), providing a singular numerical representation of the color index as follows [32]:

- If CI* −40 to −20, colors range from blue-violet to deep green.
- If CI* −20 to −2, the colors range from deep green to yellowish green.
- If CI* −2 to +2, represents greenish yellow.
- If CI* +2 to +20, colors range from pale yellow to deep orange.
- If CI* +20 to +40, colors range from deep orange to deep red.

influential functional groups within the range of 4000 to 400 cm^{-1}, with a resolution of 4 cm^{-1}.

Additionally, X-ray diffraction analysis was conducted using a Bruker diffractometer, model D8-Focus (Karlsruhe, Germany), (Cu Kα1 = 1.5406 Å°) at 40 kV and 40 mA, with a PSD Lynxeye detector. The degree of crystallinity was determined by calculating the ratio between the area corresponding to the crystalline phase and the total area under the XRD curve using Equation (1) [25].

$$CD\ (\%) = \frac{S_{cr.p}}{S_t} \times 100, \qquad (1)$$

Crystal size is not interchangeable with particle size, as crystals are contained within particles. Therefore, the average crystal size (D) was calculated from the diffractogram using Scherrer's formula (Equation (2)) [26].

$$D = \frac{k.\lambda}{\beta.\cos\theta}, \qquad (2)$$

where k represents Scherrer's constant (0.9), λ is the wavelength of the X-ray source (0.15406 nm), β denotes the peak width of the diffraction peak profile at half maximum height, a result of the small size of the crystallites (in radians), and θ signifies the position of the peak (in radians). The data were analyzed using Origin Pro 2023 software.

The morphology of the SDH was examined using a scanning electron microscope (SEM), particularly the Prism E model by Thermo Fisher (Waltham, MA, USA), operating at an acceleration voltage of 25 kV and a magnification of 1000×.

2.4. Analysis of Rheological Behavior

The experimental samples underwent continuous testing using an Anton Paar rotational rheometer, specifically the MCR702e model (Graz, Austria). The rheometer featured a concentric cylinder arrangement, and the tests were conducted at controlled shear rates ranging from 1 to 300 s^{-1} and at temperatures of 40, 60, and 80 °C. The acquired data were analyzed using shear stress models designed for non-Newtonian fluids, specifically the Power Law, Herschel–Bulkley, and Casson models, the details of which are presented in Table 2.

Table 2. Rheological models for non-Newtonian fluids.

Model	Equation	Parameters	
Power law	$\tau = k\gamma^n$	k, n	(3)
Herschel–Bulkley	$\tau = \tau_y + k_H\gamma^n$	τ_y, k_H, n	(4)
Casson	$\tau^{1/2} = \tau_y^{1/2} + (k\gamma)^{1/2}$	τ_y, k	(5)

Donde: τ, yield stress (Pa); γ, shear rate, (s^{-1}); k, consistency index (Pa.s^n); n, behavioral index; τ_y, elastic limit or yield point (Pa); η_B, plastic viscosity (Pa.s); k_H, consistency index (Pa.s^n).

The rheological models underwent adjustment via non-linear regression, employing the least squares difference as the convergence criterion and evaluated using the Quasi-Newton (QN), Simplex/Quasi-Newton (SQN), and Rosenbrock/Quasi-Newton (RQN) methods [27,28]. To assess the model's quality, key metrics including the adjusted correlation coefficient (R^2), the residual mean square of the error (MSE) calculated using Equation (6), and the mean absolute percentage of the error ($MAPE$) determined through Equation (7) were considered.

$$MSE = \frac{\sum_i (x_i - \hat{x}_i)^2}{n}, \qquad (6)$$

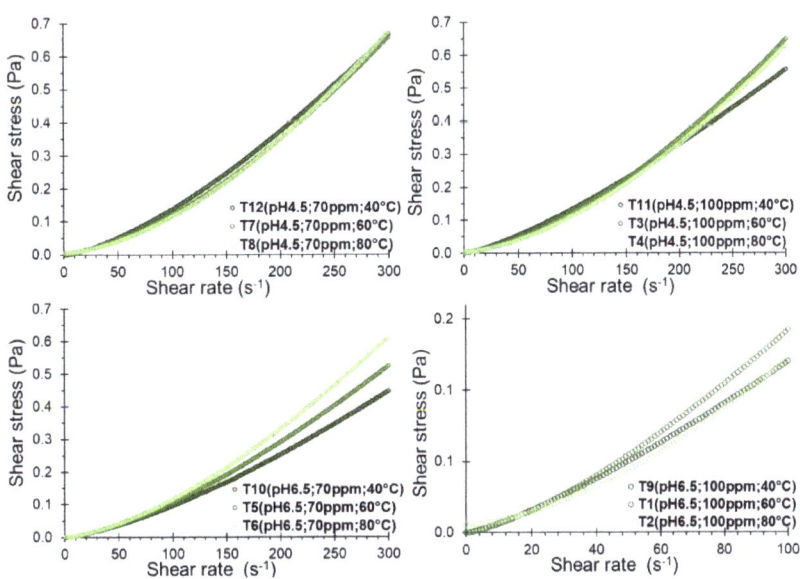

Figure 3. Rheological behavior of SDH in solution.

Table 3. Parameters and statistical values of rheological models for SDH solutions.

Model	T1	T2	T3	T4	T5	T6	T7	T8	T9	T10	T11	T12
Power Law												
$k(Pa.s^n \times 10^{-4})$	2.5042	1.7931	1.1154	0.7935	1.3742	1.1315	0.9657	0.6444	3.6937	1.5130	3.4568	1.8615
n	1.3772	1.4125	1.5196	1.5741	1.4457	1.5068	1.5473	1.6229	1.2557	1.4010	1.2951	1.4350
R^2	0.9942	0.9898	0.9996	0.9999	0.9995	0.9997	0.9996	0.9999	0.9883	0.9998	0.9950	0.9986
MSE	0.0000	0.0000	0.0000	0.0000	0.0000	0.0000	0.0000	0.0000	0.0000	0.0000	0.0001	0.0001
MAPE	0.0892	0.1149	0.0703	0.0568	0.0495	0.0599	0.1522	0.0659	0.3849	0.0587	0.1107	0.0747
Residuals	R	R	R	R	R	R	R	R	R	R	R	R
EM	QN	QN	QN	QN	QN	QN	QN	QN	QN	QN	QN	QN
Herschel-Bulkley												
$\tau_y (Pa \times 10^{-4})$	53.4769	56.4827	44.5463	31.8798	0.0000	18.7968	55.7519	53.5837	0.0000	0.0000	0.0000	0.1239
$k_H (Pa.s^n \times 10^{-4})$	1.2746	0.7209	0.9518	0.7023	1.3742	1.0556	0.7876	0.5272	3.6938	1.5131	3.4568	1.8579
n	1.5185	1.6041	1.5466	1.5949	1.4457	1.5187	1.5821	1.6573	1.2557	1.4010	1.2951	1.4354
R^2	0.9961	0.9929	0.9997	0.9999	0.9995	0.9997	0.9997	0.9999	0.9968	0.9997	0.9958	0.9988
MSE	0.0000	0.0000	0.0000	0.0000	0.0000	0.0000	0.0000	0.0000	0.0000	0.0000	0.0001	0.0001
MAPE	0.0579	0.0990	0.0320	0.0216	0.0495	0.0452	0.4167	0.0179	0.3850	0.0587	0.1107	0.0745
Residuals	R	R	R	R	R	R	R	R	R	R	T	R
EM	QN	QN	QN	QN	QN	QN	QN	QN	QN	QN	QN	QN
Casson												
$\tau_y (Pa \times 10^{-4})$	0.0109	0.0423	0.2735	0.4306	2.0557	0.3171	0.1104	0.0620	0.0158	0.3009	0.1378	0.0247
$k(Pa.s^n \times 10^{-4})$	11.834	10.060	17.718	17.189	14.458	17.119	18.020	18.452	10.530	13.066	16.298	19.236
R^2	0.9456	0.9426	0.9353	0.9270	0.9433	0.9385	0.9329	0.9211	0.9630	0.9545	0.9656	0.9507
MSE	0.0001	0.0001	0.0024	0.0026	0.0014	0.0021	0.0026	0.0032	0.0001	0.0008	0.0010	0.0020
MAPE	0.2130	0.2187	0.4176	0.5376	0.4910	0.4587	1.1148	0.5534	0.6942	0.4808	0.3843	0.4534
Residuals	T	T	T	T	T	T	T	T	R	T	T	T
EM	SQN	SQN	QN	SQN	SQN	SQN	QN	SQN	SQN	SQN	SQN	SQN

Where EM, estimation method; R, random; T, trending; QN, Quasi-Newton method; SQN, Simplex and Quasi-Newton method; MSE, mean square of the error; MAPE, mean absolute percentage of the error.

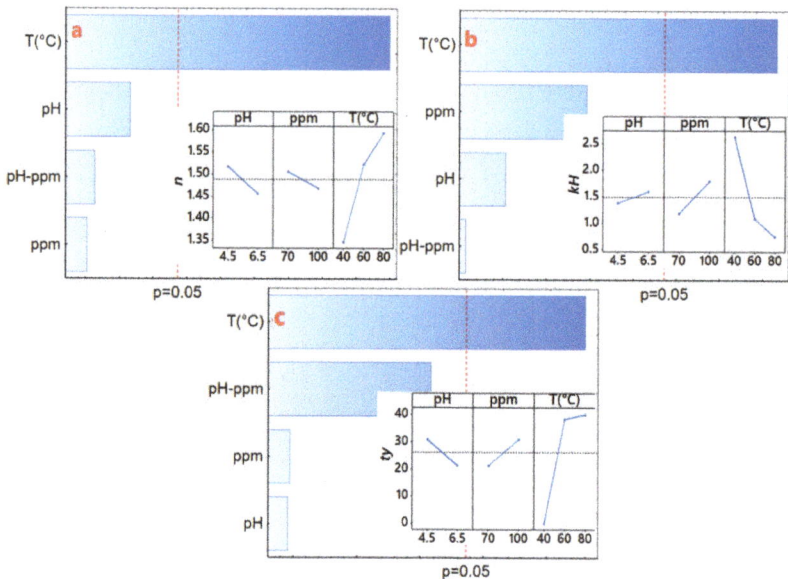

Figure 4. Effect diagram, (**a**) consistency index (k_H), (**b**) behavior index (n), (**c**) elastic limit (τ_y).

The consistency index, denoted as k in the Power Law model or k_H in the Herschel–Bulkley model, serves as an indirect measure of viscosity, indicating that as the fluid density increases, it becomes thicker or more viscous. Notably, a significant decrease in the consistency index was observed with increasing temperature (Figure 4b). It exhibited an increase with the concentration of SDH and pH of the solution (p-value < 0.05), implying that higher concentrations of SDH lead to thicker solutions, exemplified by higher values at concentrations of 100 ppm of SDH as seen in T1 and T9 (Table 3).

Regarding the elastic limit or yield point (τ_y), no consistent trend was observed. However, in some instances, it increased with the temperature and the solute concentration (Figure 4c), indicating a greater shear stress requirement to initiate the deformation of the SDH solution. Higher values of τ_y have been observed for the Herschel–Bulkley model compared to the Casson model (Table 3). Though these values depend only on the fit of the model, both present the same trends.

Concerning viscosity behavior, an increase was noted in correlation with the shear rate for all SDH formulations. However, no clear trend was evident with temperature (Figure 5). This indicates that as the shear rate increases, the fluid undergoes rapid deformation, exhibiting rheopectic behavior. This manifestation is indicative of crystallization induced by continuous shear, a characteristic feature of dilatant fluids [57,58]. As gelatinization progresses, SDH granules swell and become more deformable, contributing to this observed rheopectic behavior.

To study the changes of the hydrocolloid in solution, it is essential to heat the mixture under controlled conditions while continuously recording the viscosity changes over time [59]. This can be measured through the rapid viscoanalyzer (RVA); this method is characterized by a faster mixing action, which allows understanding of the rheopectic behavior through pasting curves, providing information on the initial gelatinization temperature and range, maximum or peak viscosity, swelling capacity (during heating), retrogradation and syneresis (during cooling), final viscosity or in a state of equilibrium, and capacity to form gels (in resistance to heat and mechanical agitation). These parameters allow us to know the useful life or the behavior of hydrocolloids in solution during storage [60,61].

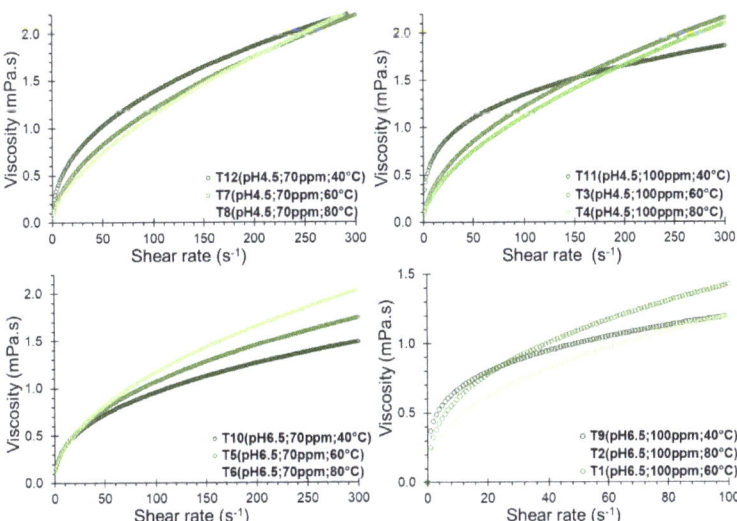

Figure 5. SDH viscosity in solution as a function of shear rate.

The rheopectic behavior of SDH solutions becomes apparent with heating, as the apparent viscosity shows a tendency to increase over time. Notably, samples heated within the initial 4 min exhibit peak viscosity values, reaching levels between 8.5 to 10 cP in the temperature range of 60 to 90 °C (Figure 6). However, for durations exceeding 20 min, higher viscosity values are consistently reported across all aqueous formulations, particularly around 30 °C. This phenomenon can be attributed to the elevated content of total solids dissolved in the dispersant solution produces, leading to an increase in viscosity. This increase restricts intermolecular movement driven by hydrodynamic forces due to the complex and branched structure of SDH. Consequently, the solution establishes intermolecular bonds, contributing to the observed rise in viscosity [62].

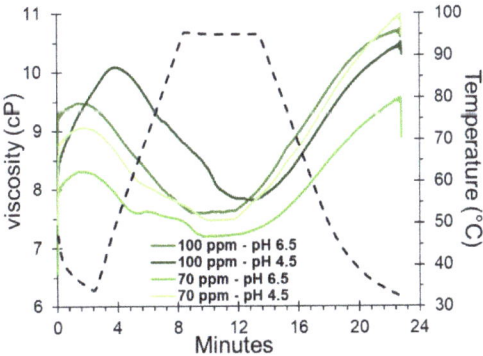

Figure 6. Rheopectic behavior of SDH in solution as a function of temperature.

3.6. Activation Energy of Solutions with SDH

The activation energy serves as a metric for quantifying the energy required to overcome resistance in the flow of a viscous fluid [30,63]. It can be calculated using the consistency index, an indirect measure of viscosity. Our study reveals a significant increase in E_a in the presence of the SDH, which is more pronounced at higher pH levels. Specifically, at 100 ppm of SDH, the E_a value was determined to be 37.74 kJ/mol (Table 4).

Table 4. Activation energy in SDH solutions.

Treatment			E_a (kJ/mol)	R^2
pH	ppm	T (°C)		
4.5	70	40		
4.5	70	60	29.127	0.9708
4.5	70	80		
4.5	100	40		
4.5	100	60	37.043	0.9079
4.5	100	80		
6.5	70	40		
6.5	70	60	8.190	0.9144
6.5	70	80		
6.5	100	40		
6.5	100	60	37.740	0.981
6.5	100	80		

This rise in E_a can be attributed to the increased viscosity of the solution. Consequently, a higher shear stress is required to induce deformation in the parallel plates of the water-SDH system. This observation underscores the interdependence of temperature and viscosity, a behavior commonly noted in various hydrocolloid materials and gums [64–66]. Elevated E_a values signify a greater demand for external energy to facilitate molecular movement. This correlation aligns with heightened viscosity attributed to the presence of macromolecules with gel-forming capabilities, resulting in more stable during thermal processing and exhibiting non-Newtonian behavior [62,67–69]

3.7. SDH Suspension Stability

The stability of SDH was evaluated through the variation in turbidity and sedimentation over time. Undesirable occurrences such as phase separation, precipitation, and agglomeration of hydrocolloids in solution are due to the thermodynamic incompatibility between phases, and these are particularly problematic in industries like food where appearance is crucial [33,70].

These phenomena are influenced by various factors intrinsic to the hydrocolloid, including concentration, particle size, electric charge, solubility, molecular weight, and ionic strength. Simultaneously, the solvent medium is sensitive to pH and temperature conditions. Achieving stability necessitates a delicate balance of attractive and repulsive interactions between the hydrocolloid and the solvent medium [71,72].

In Figure 7a, it is evident that there is high turbidity in the SDH solution on the initial day, signifying the extensive dispersion of SDH particles. Subsequently, from day 1 to day 8, there was a noticeable decline in transmittance, indicative of agglomeration, coalescence, and sedimentation of the hydrated SDH particles. From day 8 onwards, the transmittance increases, which translates into lower turbidity of the SDH solution. This suggests that the sedimented agglomerates are resuspended due to the hydration of the hydrocolloid molecules, producing a decrease in the sedimented particles, giving, as a result, high transmittance values for the sediments (Figure 7c). Notably, high transmittance values are reported for the sediments (Figure 7c). Furthermore, the observation from day 16 onwards indicates the stability of the SDH, suggesting that the hydrocolloid remains stable over extended periods. This stability is attributed to the hydration of SDH particles, establishing a delicate equilibrium between the attractive and repulsive forces governing the interactions between SDH particles and water.

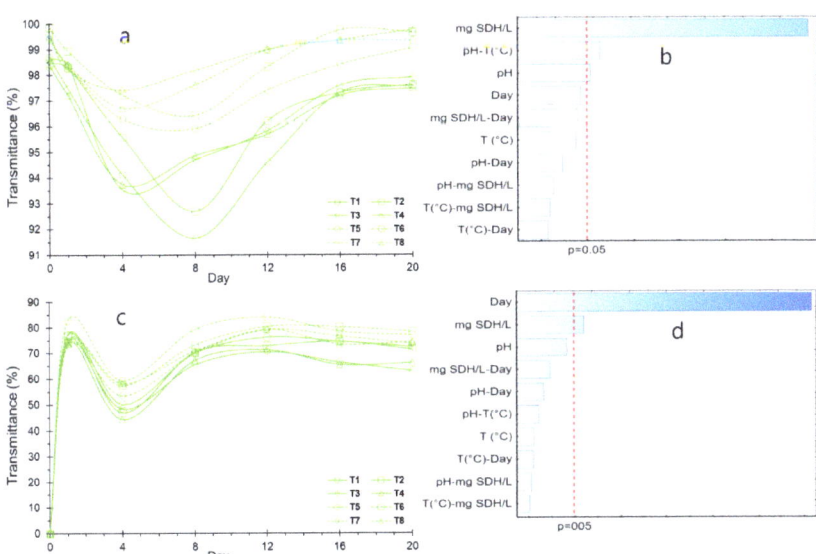

Figure 7. Suspension stability, (**a**) Turbidity variation; (**b**) Effects for turbidity; (**c**) Sedimentation variation; (**d**) Effects for sedimentation.

The impact of the application dosage of SDH is notably significant in influencing the turbidity and sedimentation of the solution, as illustrated in Figure 7b,d. Higher concentrations exhibit reduced stability, attributed to the increased availability of SDH particles for agglomeration. It is worth noting that this susceptibility to agglomeration may also be influenced by the presence of additional constituents [71]. Conversely, pH emerges as a critical determinant of stability, with optimal stability observed in an acid medium, specifically at pH 4.5, Notably, significant differences among treatments are evident during the assessment of stability over various days, as detailed in Table 5.

Table 5. Significant difference between treatments for turbidity and sedimentation.

Treatment	Turbidity							p-Value *	Sedimentation							p-Value *
	Day								Day							
	0	1	4	8	12	16	20		0	1	4	8	12	16	20	
T1	a	a	c	e	d	d	c	<0.05	--	b	a	c	c	d	a	<0.05
T2	b	ab	b	e	c	d	c	<0.05	--	b	b	c	d	d	b	<0.05
T3	c	ab	c	d	c	cd	c	<0.05	--	c	c	e	e	e	c	<0.05
T4	d	a	c	d	c	d	c	<0.05	--	d	d	d	e	e	d	<0.05
T5	e	ab	ab	cd	b	bc	b	<0.05	--	a	e	a	a	b	e	<0.05
T6	f	b	a	bc	a	a	a	<0.05	--	c,d	f	b	b	a	f	<0.05
T7	g	ab	a	a	a	ab	ab	<0.05	--	e	g	c	b	c	g	<0.05
T8	h	ab	ab	ab	a	ab	a	<0.05	--	c,d	h	c	b	d	h	<0.05

* Different letters indicate a significant difference, evaluated through the Tukey test at 5%, n = 5.

3.8. ζ Potential and Particle Size of SDH Suspension

The stability of the SDH hydrocolloid particles in an aqueous medium was assessed through the ζ potential, a measure of the electrical charge indicating either repulsion or attraction [73]. Negative values ranging from −10 to −40 mV signify robust stability by fostering the repulsion of negatively charged particles. However, it is crucial to acknowledge that this stability is subject to influence from factors such as pH, concentration, and ionic strength [20,74,75].

The SDH in solution displayed zeta potential values ranging between −17 to −31 mV, as depicted in Figure 8a. Initial values fell within the range of −23 to −30 mV, experiencing a notable increase by day 8. This upward shift suggests a decline in stability, as evidenced by the current rise in sedimentation (Figure 7a). After 20 days, zeta potential values were noted within the range of −24 to −28 mV, indicative of sustained stability for SDH throughout the storage period. Notably, there were significant differences between treatments, as indicated by the p-value (<0.05) in Table 6. These ζ potential align with observations in other hydrocolloids such as pectins, carrageenans, xanthan gum, and alginates, among others [14,20,76].

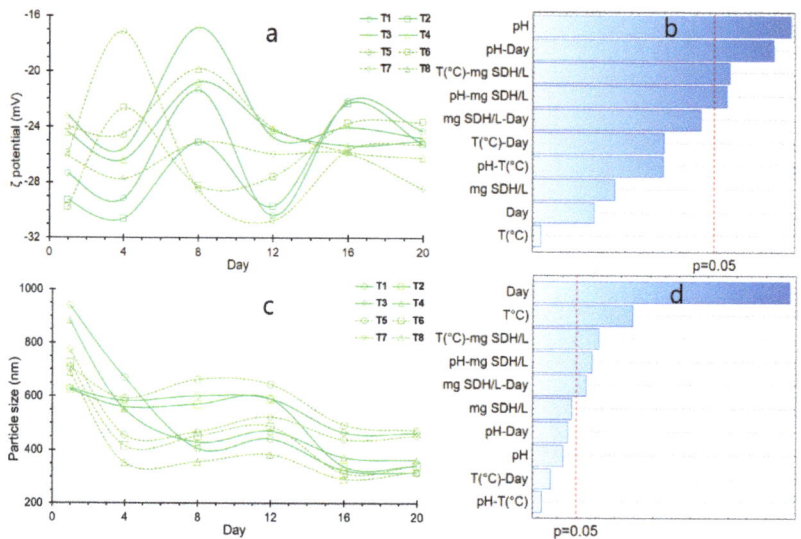

Figure 8. (**a**) variation of ζ potential; (**b**) Effects for ζ potential; (**c**) Variation of particle size; (**d**) Effects for particle size.

Table 6. Significant difference between treatments for ζ potential and particle size.

Treatment	ζ Potential (mV)							Size Particle (nm)						
	Day						p-Value *	Day						p-Value *
	1	4	8	12	16	20		1	4	8	12	16	20	
T1	a	a,b	c	a	b	b	<0.05	a	a	c,d	c,d	c	b	<0.05
T2	a	a	b	a,b	b	b	<0.05	d	b	b	a,b	c	b	<0.05
T3	a	a,b,c	d	c	a,b	b	<0.05	d	a,b	a,b	a,b	a	a	<0.05
T4	a	a,b,c	c	c	a,b	b	<0.05	a,b	b,c	c,d	c,d	b,c	b	<0.05
T5	a	d	a	a	a	b	<0.05	c,d	a,b	a	a	a	a	<0.05
T6	a	c	a	a,b,c	a,b	b	<0.05	c,d	d,e	c	c	c	b	<0.05
T7	a	a,b,c	b	b,c	a	a	<0.05	b,c	c,d	c	b,c	a,b	a	<0.05
T8	a	b,c	c,d	c	a	a,b	<0.05	c,d	e	d	d	c	b	<0.05

* Different letters indicate a significant difference, evaluated through the Tukey test at 5%, n = 3.

Its commendable stability is attributed to the abundance of negatively charged functional groups, such as carboxyl, carbonyl, and hydroxyl in SDH, as elucidated by the FTIR analysis (Figure 2c). The increase in pH enhances the prevalence of negative charges, augmenting the stability of SDH in solution, as evidenced by a significant effect (Figure 8b). Intriguingly, this contrasts with the behavior observed in gelation temperature and the impact of SDH addition, a departure from the typical response seen in analogous materials [74,77,78].

On the other hand, the particle size closely correlates with ζ potential, with nanometric sizes facilitating optimal dispersion of hydrocolloids and thereby promoting stability [21,22]. It has been observed that the particle size of SDH in solution decreases considerably with the progression of storage time (Figure 8c), initially ranging between 650 to 900 nm. By days 16 and 20, it fluctuates between 300 to 500 nm, displaying a tendency to maintain uniformity, albeit with slight differences observed between treatments (p-value < 0.05) (Table 6). This behavior signifies a high degree of dispersion for SDH, contrasting with relatively low values of ζ potential. Although no significant effect has been observed with pH, storage time and gelation temperature do have a considerable influence (Figure 8d).

3.9. PCA for the Treatments and Properties of the SDH Solution

PCA allows the exploration of data grouping based on principal components (PC) that explain the most significant variability in the dataset. Two primary components were identified, with the first PC1 (44.99%) grouping T1 and T2 on the left side of Figure 9. These treatments are notably influenced by the consistency index (k), indicating a tendency toward lower viscosity. On the right side, T7 and T8 are clustered, characterized by a pH of 4.5 and 0.07 g of SDH/L. These treatments exhibit higher turbidity, zeta potential, and particle size; these properties are desirable in aqueous food systems.

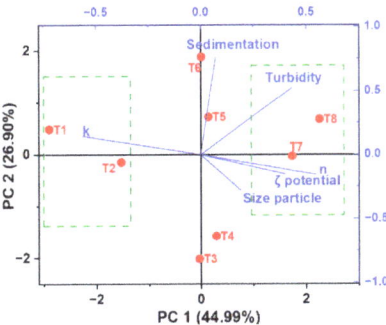

Figure 9. PCA for treatments.

Additionally, PC2 (26.90%) reveals that T6 displays greater transmittance (measured as turbidity), while T4 and T3, situated in the negative part below, do not exhibit influential properties. This comprehensive analysis provides valuable insights into relationships between different treatments and their key properties.

Thus, the SDH hydrocolloid reported good stability in an aqueous medium, so this material could be used as an active ingredient in food products, pharmaceuticals, cosmetics, textiles, and paints due to the high solubility it presents. Likewise, it presents good stability against variations in pH (between 4.5 and 6.5) and temperature (between 60 and 80 °C). However, there are aspects to overcome, such as the residual color of SDH, which would considerably affect food products and drugs.

4. Conclusions

Nostoc sphaericum, a freshwater algae native to the Peruvian Andes, serves as a valuable source for extracting spray-drying hydrocolloid (SDH). This hydrocolloid exhibits favorable characteristics, rendering it highly stable in aqueous medium, with zero charge point at approximately pH 8.1, a crystallinity degree of 71.98%, and an average particle size of 4.12 nm. In the aqueous medium within the pH range of 4.5 to 6.5, SDH demonstrates dilatant behavior conforming to the Power Law ($R^2 > 0.99$). Notably, its viscosity ranges from 8.5 to 10 cP at temperatures spanning 60 to 90 °C, displaying an activation energy fluctuating between 8.19 to 37.74 kJ/mol. Extended storage stability tests conducted up to day 20 reveal consistent turbidity, minimal sedimentation, ζ potential ranging between

−31 to −17 mV, and particle size maintaining a steady range of 300 to 500 nm with low variability over time. Given these attributes, SDH emerges as a potential alternative stabilizer for application in aqueous media due to its exceptional stability. Although there is still the challenge of overcoming the depigmentation of the hydrocolloid, as well as the sensory study applied in food systems.

Author Contributions: Conceptualization, D.C.-Q. and S.F.; methodology, D.C.-Q. and Y.F.C.-M.; software, E.M.-M. and R.P.-S.; validation, D.C.-Q.; C.A.L.-S., S.F. and Y.F.C.-M.; formal analysis, D.C.-Q., C.A.L.-S., Y.C.-Q. and E.M.-M.; investigation, D.C.-Q., Y.C.-Q., A.M.S.-R. and R.P.-S.; resources, A.M.S.-R.; data curation, A.M.S.-R.; writing—original draft preparation, D.C.-Q. and Y.C.-Q.; writing—review and editing, D.C.-Q., C.A.L.-S. and S.F.; visualization, D.C.-Q. and S.F.; supervision, D.C.-Q.; project administration, Y.C.-Q.; funding acquisition, D.C.-Q. All authors have read and agreed to the published version of the manuscript.

Funding: This research received no external funding.

Institutional Review Board Statement: Not applicable.

Data Availability Statement: The data presented in this study are available in this same article.

Acknowledgments: The authors would like to thank the Vice Rectorate of Research of the Universidad Nacional José María Arguedas (UNAJMA), for the financing and use of the Materials for the Treatment of Water and Food Research Laboratory and Food Nanotechnology Research Laboratory, in the same way to Instituto de Investigación—UNAJMA.

Conflicts of Interest: The authors declare that they have no known competing financial interests or personal relationships that could have appeared to influence the work reported in this paper.

References

1. Culetu, A.; Duta, D.E.; Papageorgiou, M.; Varzakas, T. The role of hydrocolloids in gluten-free bread and pasta; rheology, characteristics, staling and glycemic index. *Foods* **2021**, *10*, 3121. [CrossRef] [PubMed]
2. Aftab, K.; Hameed, S.; Umbreen, H.; Ali, S.; Rizwan, M.; Alkahtani, S.; Abdel-Daim, M.M. Physicochemical and Functional Potential of Hydrocolloids Extracted from Some Solanaceae Plants. *J. Chem.* **2020**, *2020*, 3563945. [CrossRef]
3. Costa, J.A.V.; Lucas, B.F.; Alvarenga, A.G.P.; Moreira, J.B.; de Morais, M.G. Microalgae polysaccharides: An overview of production, characterization, and potential applications. *Polysaccharides* **2021**, *2*, 759–772. [CrossRef]
4. Xu, J.; Zhang, Y.; Wang, W.; Li, Y. Advanced properties of gluten-free cookies, cakes, and crackers: A review. *Trends Food Sci. Technol.* **2020**, *103*, 200–213. [CrossRef]
5. Gomes Gradíssimo, D.; Pereira Xavier, L.; Valadares Santos, A. Cyanobacterial polyhydroxyalkanoates: A sustainable alternative in circular economy. *Molecules* **2020**, *25*, 4331. [CrossRef]
6. Jensen, S.; Petersen, B.O.; Omarsdottir, S.; Paulsen, B.S.; Duus, J.Ø.; Olafsdottir, E.S. Structural characterisation of a complex heteroglycan from the cyanobacterium Nostoc commune. *Carbohydr. Polym.* **2013**, *91*, 370–376. [CrossRef]
7. Rodriguez, S.; Torres, F.G.; López, D. Preparation and characterization of polysaccharide films from the cyanobacteria Nostoc commune. *Polym. Renew. Resour.* **2017**, *8*, 133–150. [CrossRef]
8. Fernández, L.A.G.; Ramos, V.C.; Polo, M.S.; Castillo, N.A.M. Fundamentals in applications of algae biomass: A review. *J. Environ. Manag.* **2023**, *338*, 117830. [CrossRef]
9. Pérez-Lloréns, J.L. Microalgae: From staple foodstuff to avant-garde cuisine. *Int. J. Gastron. Food Sci.* **2020**, *21*, 100221. [CrossRef]
10. Khalil, H.P.S.; Lai, T.K.; Tye, Y.Y.; Rizal, S.; Chong, E.W.N.; Yap, S.W.; Hamzah, A.A.; Fazita, M.R.; Paridah, M.T. A review of extractions of seaweed hydrocolloids: Properties and applications. *Express Polym. Lett.* **2018**, *12*, 296–317. [CrossRef]
11. Ritzoulis, C.; Marini, E.; Aslanidou, A.; Georgiadis, N.; Karayannakidis, P.D.; Koukiotis, C.; Filotheou, A.; Lousinian, S.; Tzimpilis, E. Hydrocolloids from quince seed: Extraction, characterization, and study of their emulsifying/stabilizing capacity. *Food Hydrocoll.* **2014**, *42*, 178–186. [CrossRef]
12. Corpus-Gomez, A.; Alcantara-Callata, M.; Celis-Teodoro, H.; Echevarria-Alarcón, B.; Paredes-Julca, J.; Paucar-Menacho, L.M. Cushuro (*Nostoc sphaericum*): Habitat, physicochemical characteristics, nutritional composition, forms of consumption and medicinal properties. *Agroind. Sci.* **2021**, *11*, 231–238. [CrossRef]
13. Ponce, E. Nostoc: A different food and their presence in the precordillera of Arica. *Idesia* **2014**, *32*, 115–118. [CrossRef]
14. Choque-Quispe, D.; Mojo-Quisani, A.; Ligarda-Samanez, C.A.; Calla-Florez, M.; Ramos-Pacheco, B.S.; Zamalloa-Puma, L.M.; Peralta-Guevara, D.E.; Solano-Reynoso, A.M.; Choque-Quispe, Y.; Zamalloa-Puma, A. Preliminary characterization of a spray-dried hydrocolloid from a high Andean algae (*Nostoc sphaericum*). *Foods* **2022**, *11*, 1640. [CrossRef]
15. Pirsa, S.; Hafezi, K. Hydrocolloids: Structure, preparation method, and application in food industry. *Food Chem.* **2023**, *399*, 133967. [CrossRef]

16. Peppas, N.A.; Bures, P.; Leobandung, W.S.; Ichikawa, H. Hydrogels in pharmaceutical formulations. *Eur. J. Pharm. Biopharm.* **2000**, *50*, 27–46. [CrossRef]
17. Rodriguez, S.; Gonzales, K.N.; Romero, E.G.; Troncoso, O.P.; Torres, F.G. Unusual reversible elastomeric gels from Nostoc commune. *Int. J. Biol. Macromol.* **2017**, *97*, 411–417. [CrossRef] [PubMed]
18. Wang, H.; Nobes, D.S.; Vehring, R. Particle surface roughness improves colloidal stability of pressurized pharmaceutical suspensions. *Pharm. Res.* **2019**, *36*, 43. [CrossRef]
19. Nep, E.I.; Conway, B.R. Evaluation of grewia polysaccharide gum as a suspending agent. *Int. J. Pharm. Pharm. Sci.* **2011**, *3*, 168–173.
20. Koko, M.Y.F.; Hassanin, H.A.M.; Qi, B.; Han, L.; Lu, K.; Rokayya, S.; Harimana, Y.; Zhang, S.; Li, Y. Hydrocolloids as promising additives for food formulation consolidation: A short review. *Food Rev. Int.* **2023**, *39*, 1433–1439. [CrossRef]
21. Zhang, Z. A new method for estimating zeta potential of carboxylic acids' functionalised particles. *Mol. Phys.* **2023**, e2260014. [CrossRef]
22. Averkin, D.V.; Stakheev, A.A.; Vishnevetskii, D.V.; Pakhomov, P.M. Characterization of particles of the dispersed system based on low-concentrated aqueous solutions of L-cysteine and silver acetate. *Proc. J. Phys. Conf. Ser.* **2022**, *2192*, 012030. [CrossRef]
23. Torres-Maza, A.; Yupanqui-Bacilio, C.; Castro, V.; Aguirre, E.; Villanueva, E.; Rodríguez, G. Comparison of the hydrocolloids Nostoc commune and Nostoc sphaericum: Drying, spectroscopy, rheology and application in nectar. *Sci. Agropecu.* **2020**, *11*, 583–589. [CrossRef]
24. Ligarda-Samanez, C.A.; Choque-Quispe, D.; Palomino-Rincón, H.; Ramos-Pacheco, B.S.; Moscoso-Moscoso, E.; Huamán-Carrión, M.L.; Peralta-Guevara, D.E.; Obregón-Yupanqui, M.E.; Aroni-Huamán, J.; Bravo-Franco, E.Y.; et al. Modified Polymeric Biosorbents from Rumex acetosella for the Removal of Heavy Metals in Wastewater. *Polymers* **2022**, *14*, 2191. [CrossRef]
25. Dome, K.; Podgorbunskikh, E.; Bychkov, A.; Lomovsky, O. Changes in the crystallinity degree of starch having different types of crystal structure after mechanical pretreatment. *Polymers* **2020**, *12*, 641. [CrossRef] [PubMed]
26. Monshi, A.; Foroughi, M.R.; Monshi, M.R. Modified Scherrer equation to estimate more accurately nano-crystallite size using XRD. *World J. Nano Sci. Eng.* **2012**, *2*, 154–160. [CrossRef]
27. Emiola, I.; Adem, R. Comparison of Minimization Methods for Rosenbrock Functions. In Proceedings of the 2021 29th Mediterranean Conference on Control and Automation (MED), Puglia, Italy, 22–25 June 2021; pp. 837–842.
28. Novati, P. Some secant approximations for Rosenbrock W-methods. *Appl. Numer. Math.* **2008**, *58*, 195–211. [CrossRef]
29. He, H.; Lee, J.; Jiang, Z.; He, Q.; Dinic, J.; Chen, W.; Narayanan, S.; Lin, X.-M. Kinetics of Shear-Induced Structural Ordering in Dense Colloids. *J. Phys. Chem. B* **2023**, *127*, 7408–7415. [CrossRef]
30. Martínez-Padilla, L.P. Rheology of liquid foods under shear flow conditions: Recently used models. *J. Texture Stud.* **2024**, *55*, e12802. [CrossRef]
31. Choque-Quispe, D.; Froehner, S.; Ligarda-Samanez, C.A.; Ramos-Pacheco, B.S.; Palomino-Rincón, H.; Choque-Quispe, Y.; Solano-Reynoso, A.M.; Taipe-Pardo, F.; Zamalloa-Puma, L.M.; Calla-Florez, M. Preparation and chemical and physical characteristics of an edible film based on native potato starch and nopal mucilage. *Polymers* **2021**, *13*, 3719. [CrossRef]
32. Hadimani, L.; Mittal, N. Development of a computer vision system to estimate the colour indices of Kinnow mandarins. *J. Food Sci. Technol.* **2019**, *56*, 2305–2311. [CrossRef] [PubMed]
33. Mohamed, S.K.; Alazhary, A.M.; Al-Zaqri, N.; Alsalme, A.; Alharthi, F.A.; Hamdy, M.S. Cost-effective adsorbent from arabinogalactan and pectin of cactus pear peels: Kinetics and thermodynamics studies. *Int. J. Biol. Macromol.* **2020**, *150*, 941–947. [CrossRef] [PubMed]
34. Salminen, H.; Weiss, J. Effect of pectin type on association and pH stability of whey protein—Pectin complexes. *Food Biophys.* **2014**, *9*, 29–38. [CrossRef]
35. Ahmad, R.; Mirza, A. Synthesis of Guar gum/bentonite a novel bionanocomposite: Isotherms, kinetics and thermodynamic studies for the removal of Pb (II) and crystal violet dye. *J. Mol. Liq.* **2018**, *249*, 805–814. [CrossRef]
36. Arumugham, T.; Krishnamoorthy, R.; AlYammahi, J.; Hasan, S.W.; Banat, F. Spray dried date fruit extract with a maltodextrin/gum arabic binary blend carrier agent system: Process optimization and product quality. *Int. J. Biol. Macromol.* **2023**, *238*, 124340. [CrossRef] [PubMed]
37. Wang, W.; Zhou, W. Characterisation of spray dried soy sauce powders made by adding crystalline carbohydrates to drying carrier. *Food Chem.* **2015**, *168*, 417–422. [CrossRef]
38. Nazir, S.; Wani, I.A. Fractionation and characterization of mucilage from Basil (*Ocimum basilicum* L.) seed. *J. Appl. Res. Med. Aromat. Plants* **2022**, *31*, 100429. [CrossRef]
39. Karrar, E.; Mahdi, A.A.; Sheth, S.; Ahmed, I.A.M.; Manzoor, M.F.; Wei, W.; Wang, X. Effect of maltodextrin combination with gum arabic and whey protein isolate on the microencapsulation of gurum seed oil using a spray-drying method. *Int. J. Biol. Macromol.* **2021**, *171*, 208–216. [CrossRef]
40. Chuenkaek, T.; Nakajima, K.; Kobayashi, T. Water-soluble pectin films prepared with extracts from Citrus maxima wastes under acidic conditions and their moisturizing characteristics. *Biomass Convers. Biorefinery* **2023**, 1–14. [CrossRef]
41. Supreetha, R.; Bindya, S.; Deepika, P.; Vinusha, H.M.; Hema, B.P. Characterization and biological activities of synthesized citrus pectin-MgO nanocomposite. *Results Chem.* **2021**, *3*, 100156. [CrossRef]
42. Ubeyitogullari, A.; Ciftci, O.N. Fabrication of bioaerogels from camelina seed mucilage for food applications. *Food Hydrocoll.* **2020**, *102*, 105597. [CrossRef]

43. Alpizar-Reyes, E.; Carrillo-Navas, H.; Gallardo-Rivera, R.; Varela-Guerrero, V.; Alvarez-Ramirez, J.; Pérez-Alonso, C. Functional properties and physicochemical characteristics of tamarind (*Tamarindus indica* L.) seed mucilage powder as a novel hydrocolloid. *J. Food Eng.* **2017**, *209*, 68–75. [CrossRef]
44. González-Martínez, D.A.; Carrillo-Navas, H.; Barrera-Díaz, C.E.; Martínez-Vargas, S.L.; Alvarez-Ramírez, J.; Pérez-Alonso, C. Characterization of a novel complex coacervate based on whey protein isolate-tamarind seed mucilage. *Food Hydrocoll.* **2017**, *72*, 115–126. [CrossRef]
45. Zhao, Y.; Li, B.; Li, C.; Xu, Y.; Luo, Y.; Liang, D.; Huang, C. Comprehensive review of polysaccharide-based materials in edible packaging: A sustainable approach. *Foods* **2021**, *10*, 1845. [CrossRef] [PubMed]
46. Chakravartula, S.S.N.; Soccio, M.; Lotti, N.; Balestra, F.; Dalla Rosa, M.; Siracusa, V. Characterization of composite edible films based on pectin/alginate/whey protein concentrate. *Materials* **2019**, *12*, 2454. [CrossRef] [PubMed]
47. Hamrun, N.; Talib, B.; Ruslin, M.; Pangeran, H.; Hatta, M.; Marlina, E.; Yusuf, A.S.H.; Saito, T.; Ou, K.-L. A promising potential of brown algae Sargassum polycystum as irreversible hydrocolloid impression material. *Mar. Drugs* **2022**, *20*, 55. [CrossRef] [PubMed]
48. Sebeia, N.; Jabli, M.; Ghith, A.; Elghoul, Y.; Alminderej, F.M. Production of cellulose from Aegagropila Linnaei macro-algae: Chemical modification, characterization and application for the bio-sorptionof cationic and anionic dyes from water. *Int. J. Biol. Macromol.* **2019**, *135*, 152–162. [CrossRef] [PubMed]
49. Guo, X.; Duan, H.; Wang, C.; Huang, X. Characteristics of two calcium pectinates prepared from citrus pectin using either calcium chloride or calcium hydroxide. *J. Agric. Food Chem.* **2014**, *62*, 6354–6361. [CrossRef]
50. Timilsena, Y.P.; Wang, B.; Adhikari, R.; Adhikari, B. Preparation and characterization of chia seed protein isolate–chia seed gum complex coacervates. *Food Hydrocoll.* **2016**, *52*, 554–563. [CrossRef]
51. Chuquilín-Goicochea, R.; Ccente-Lulo, F.; Arteaga-Llacza, P.; Huayta, F.; Tinoco, H.A.O. Chemical and rheological characterization of Senna birostris hydrocolloid. *Manglar* **2021**, *18*, 11–115. [CrossRef]
52. Temsiripong, T.; Pongsawatmanit, R.; Ikeda, S.; Nishinari, K. Influence of xyloglucan on gelatinization and retrogradation of tapioca starch. *Food Hydrocoll.* **2005**, *19*, 1054–1063. [CrossRef]
53. Wang, L.; Wu, Q.; Zhao, J.; Lan, X.; Yao, K.; Jia, D. Physicochemical and rheological properties of crude polysaccharides extracted from Tremella fuciformis with different methods. *CyTA-J. Food* **2021**, *19*, 247–256. [CrossRef]
54. Okechukwu, P.E.; Rao, M.A. Role of granule size and size distribution in the viscosity of cowpea starch dispersions heated in excess water. *J. Texture Stud.* **1996**, *27*, 159–173. [CrossRef]
55. Hernández-Morales, M.d.l.Á.; Maldonado-Astudillo, Y.I.; Jiménez-Hernández, J.; Salazar, R.; Ramírez-Sucre, M.O.; Ibarz, A.; Utrilla-Coello, R.G.; Ortuño-Pineda, C. Physicochemical and rheological properties of gum seed and pulp from *Hymenaea courbaril* L. *CyTA-J. Food* **2018**, *16*, 986–994. [CrossRef]
56. Yekta, M.; Ansari, S. Jujube mucilage as a potential stabilizer in stirred yogurt: Improvements in the physiochemical, rheological, and sensorial properties. *Food Sci. Nutr.* **2019**, *7*, 3709–3721. [CrossRef] [PubMed]
57. León-Martínez, F.M.; Cano-Barrita, P.F.J.; Lagunez-Rivera, L.; Medina-Torres, L. Study of nopal mucilage and marine brown algae extract as viscosity-enhancing admixtures for cement based materials. *Constr. Build. Mater.* **2014**, *53*, 190–202. [CrossRef]
58. Rivera-Corona, J.L.; Rodríguez-González, F.; Rendón-Villalobos, R.; García-Hernández, E.; Solorza-Feria, J. Thermal, structural and rheological properties of sorghum starch with cactus mucilage addition. *LWT-Food Sci. Technol.* **2014**, *59*, 806–812. [CrossRef]
59. Zhou, M.; Robards, K.; Glennie-Holmes, M.; Helliwell, S. Structure and pasting properties of oat starch. *Cereal Chem.* **1998**, *75*, 273–281. [CrossRef]
60. Mohamed, I.O. Effects of processing and additives on starch physicochemical and digestibility properties. *Carbohydr. Polym. Technol. Appl.* **2021**, *2*, 100039. [CrossRef]
61. Rady, A.M.; Soliman, S.N.; El-Wersh, A. Effect of mechanical treatments on creep behavior of potato tubers. *Eng. Agric. Environ. Food* **2019**, *10*, 282–291. [CrossRef]
62. Alpizar-Reyes, E.; Román-Guerrero, A.; Gallardo-Rivera, R.; Varela-Guerrero, V.; Cruz-Olivares, J.; Pérez-Alonso, C. Rheological properties of tamarind (*Tamarindus indica* L.) seed mucilage obtained by spray-drying as a novel source of hydrocolloid. *Int. J. Biol. Macromol.* **2018**, *107*, 817–824. [CrossRef]
63. Chin, N.L.; Chan, S.M.; Yusof, Y.A.; Chuah, T.G.; Talib, R.A. Modelling of rheological behaviour of pummelo juice concentrates using master-curve. *J. Food Eng.* **2009**, *93*, 134–140. [CrossRef]
64. García-Cruz, E.E.; Rodriguez-Ramirez, J.; Lagunas, L.L.M.; Medina-Torres, L. Rheological and physical properties of spray-dried mucilage obtained from *Hylocereus undatus* cladodes. *Carbohydr. Polym.* **2013**, *91*, 394–402. [CrossRef] [PubMed]
65. Xu, X.; Xu, G.; Liu, T.; Chen, Y.; Gong, H. The comparison of rheological properties of aqueous welan gum and xanthan gum solutions. *Carbohydr. Polym.* **2013**, *92*, 516–522. [CrossRef] [PubMed]
66. Dak, M.; Verma, R.C.; Jaaffrey, S.N.A. Effect of temperature and concentration on rheological properties of "Kesar" mango juice. *J. Food Eng.* **2007**, *80*, 1011–1015. [CrossRef]
67. Quispe-Chambilla, L.; Pumacahua-Ramos, A.; Choque-Quispe, D.; Curro-Pérez, F.; Carrión-Sánchez, H.M.; Peralta-Guevara, D.E.; Masco-Arriola, M.L.; Palomino-Rincón, H.; Ligarda-Samanez, C.A. Rheological and functional properties of dark chocolate with partial substitution of peanuts and sacha inchi. *Foods* **2022**, *11*, 1142. [CrossRef]
68. Kassem, I.A.A.; Ashaolu, T.J.; Kamel, R.; Elkasabgy, N.A.; Afifi, S.M.; Farag, M.A. Mucilage as a functional food hydrocolloid: Ongoing and potential applications in prebiotics and nutraceuticals. *Food Funct.* **2021**, *12*, 4738–4748. [CrossRef] [PubMed]

69. Medina-Torres, L.; García-Cruz, E.E.; Calderas, F.; Laredo, R.F.G.; Sánchez-Olivares, G.; Gallegos-Infante, J.A.; Rocha-Guzmán, N.E.; Rodriguez-Ramirez, J. Microencapsulation by spray drying of gallic acid with nopal mucilage (*Opuntia ficus indica*). *LWT-Food Sci. Technol.* **2013**, *50*, 642–650. [CrossRef]
70. Dickinson, E. Hydrocolloids at interfaces and the influence on the properties of dispersed systems. *Food Hydrocoll.* **2003**, *17*, 25–39. [CrossRef]
71. Zhang, M.; Chen, M.; Yang, C.; Wang, Z. The Growth of Precipitates in A Solution with Cross Diffusion Between Solutes. *Surf. Rev. Lett.* **2022**, *29*, 2250148. [CrossRef]
72. Alexander, K.S.; Biskup, M.; Chayes, L. Colligative properties of solutions: II. Vanishing concentrations. *J. Stat. Phys.* **2005**, *119*, 509–537. [CrossRef]
73. Kosmulski, M.; Mączka, E. Zeta potential and particle size in dispersions of alumina in 50–50 w/w ethylene glycol-water mixture. *Colloids Surf. A Physicochem. Eng. Asp.* **2022**, *654*, 130168. [CrossRef]
74. Ishaq, A.; Nadeem, M.; Ahmad, R.; Ahmed, Z.; Khalid, N. Recent advances in applications of marine hydrocolloids for improving bread quality. *Food Hydrocoll.* **2023**, *148*, 109424. [CrossRef]
75. Chatterjee, A.; Lal, S.; Manivasagam, T.G.; Batabyal, S.K. Surface charge induced bioelectricity generation from freshwater macroalgae *Pithophora*. *Bioresour. Technol. Rep.* **2023**, *22*, 101379. [CrossRef]
76. Niknam, R.; Mousavi, M.; Kiani, H. Intrinsic viscosity, steady and oscillatory shear rheology of a new source of galactomannan isolated from Gleditsia caspica (*Persian honey locust*) seeds in aqueous dispersions. *Eur. Food Res. Technol.* **2021**, *247*, 2579–2590. [CrossRef]
77. Strieder, M.M.; Silva, E.K.; Mekala, S.; Meireles, M.A.A.; Saldaña, M.D.A. Barley-Based Non-dairy Alternative Milk: Stabilization Mechanism, Protein Solubility, Physicochemical Properties, and Kinetic Stability. *Food Bioprocess Technol.* **2023**, *16*, 2231–2246. [CrossRef]
78. Marsiglia-Fuentes, R.; Quintana, S.E.; García Zapateiro, L.A. Novel Hydrocolloids Obtained from Mango (*Mangifera indica*) var. Hilaza: Chemical, Physicochemical, Techno-Functional, and Structural Characteristics. *Gels* **2022**, *8*, 354. [CrossRef]

Disclaimer/Publisher's Note: The statements, opinions and data contained in all publications are solely those of the individual author(s) and contributor(s) and not of MDPI and/or the editor(s). MDPI and/or the editor(s) disclaim responsibility for any injury to people or property resulting from any ideas, methods, instructions or products referred to in the content.

Article

Cholesteric Liquid Crystals with Thermally Stable Reflection Color from Mixtures of Completely Etherified Ethyl Cellulose Derivative and Methacrylic Acid

Kazuma Matsumoto, Naoto Iwata and Seiichi Furumi *

Department of Chemistry, Graduate School of Science, Tokyo University of Science, 1-3 Kagurazaka, Shinjuku, Tokyo 162-8601, Japan; 1322628@ed.tus.ac.jp (K.M.); n-iwata@rs.tus.ac.jp (N.I.)
* Correspondence: furumi@rs.tus.ac.jp; Tel.: +81-3-3260-4271

Abstract: Cellulose derivatives have attracted attention as environmentally friendly materials that can exhibit a cholesteric liquid crystal (CLC) phase with visible light reflection. Previous reports have shown that the chemical structures and the degrees of substitution of cellulose derivatives have significant influence on their reflection properties. Although many studies have been reported on CLC using ethyl cellulose (EC) derivatives in which the hydroxy groups are esterified, there have been no studies on EC derivatives with etherified side chains. In this article, we optimized the Williamson ether synthesis to introduce pentyl ether groups in the EC side chain. The degree of substitution with pentyl ether group (DS_{Pe}), confirmed via ^{1}H-NMR spectroscopic measurements, was controlled using the solvent and the base concentration in this synthesis. All the etherified EC derivatives were soluble in methacrylic acid (MAA), allowing for the preparation of lyotropic CLCs with visible reflection. Although the reflection peak of lyotropic CLCs generally varies with temperature, the reflection peak of lyotropic CLCs of completely etherified EC derivatives with MAA could almost be preserved in the temperature range from 30 to 110 °C even without the aid of any crosslinking. Such thermal stability of the reflection peak of CLCs may be greatly advantageous for fabricating new photonic devices with eco-friendliness.

Keywords: ethyl cellulose; Williamson ether synthesis; cholesteric liquid crystal; Bragg reflection; methacrylic acid

Citation: Matsumoto, K.; Iwata, N.; Furumi, S. Cholesteric Liquid Crystals with Thermally Stable Reflection Color from Mixtures of Completely Etherified Ethyl Cellulose Derivative and Methacrylic Acid. *Polymers* **2024**, *16*, 401. https://doi.org/10.3390/polym16030401

Academic Editors: Cristina Cazan and Mihai Alin Pop

Received: 10 January 2024
Revised: 25 January 2024
Accepted: 29 January 2024
Published: 31 January 2024

Copyright: © 2024 by the authors. Licensee MDPI, Basel, Switzerland. This article is an open access article distributed under the terms and conditions of the Creative Commons Attribution (CC BY) license (https://creativecommons.org/licenses/by/4.0/).

1. Introduction

Given the concerns about the mass consumption of the finite petroleum resources that remain on the Earth, the promotion of research and development on functional materials prepared from biomass resources is of prime importance for the realization of a sustainable society. Cellulose and its derivatives have recently attracted renewed interest as one of the biomass resource options because of their safety and biocompatibility. Cellulose is the naturally occurring homopolymer of β-D-glucopyranose, and it has been widely used as a raw material for paper dating back to the times of ancient civilizations. As one of their unique properties, cellulose derivatives can demonstrate a liquid crystal phase, allowing for dissolving them in solutions [1] or suspensions [2]. The emergence of the liquid crystal phase depends on the solubility of the polymer in the solvent and the polymer concentration. Such a liquid crystal phase is called the lyotropic liquid crystal phase. Due to these interesting properties of cellulose and its derivatives, numerous studies have been conducted to explore their application potential as functional materials [1,2].

Hydroxypropyl cellulose (HPC) and ethyl cellulose (EC), as depicted in Figure 1, are among the representative cellulose derivatives that can be obtained by reacting natural cellulose, in this case, with propylene oxide and ethylene oxide, respectively. Owing to their safety, both HPC and EC are nowadays utilized as not only ink [3,4] and pharmaceutical additives [5,6] but also thickeners and coating agents in interdisciplinary industrial fields [7].

More interestingly, EC has been reported as a biodegradable material, which fits into the concept of the circular economy [8].

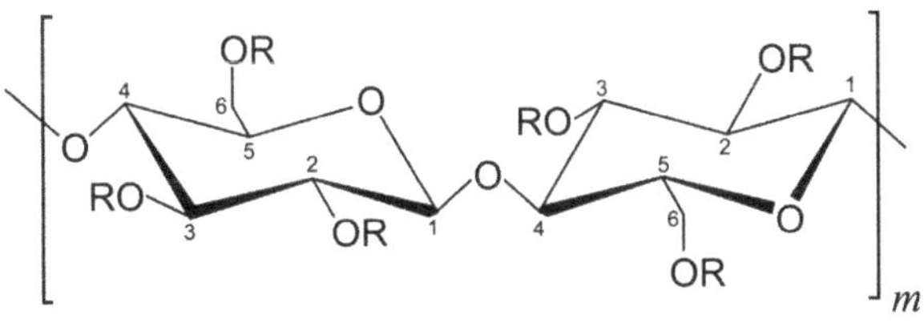

(A) **HPC**: R = –H or –CH$_2$CHOR (with CH$_3$ on the CH)

(B) **EC**: R = –H or –CH$_2$CH$_3$
EC-Pe: R = –H, –CH$_2$CH$_3$ or –(CH$_2$)$_4$CH$_3$

Figure 1. (**A**) Chemical structure of hydroxypropyl cellulose (HPC). (**B**) Chemical structures of ethyl cellulose (EC) and its derivative tethering pentyl ether side chains (EC-Pe) synthesized in this study.

Another interesting property of HPC and EC is their ability to exhibit a cholesteric liquid crystal (CLC) phase with visible light reflection in solutions [1,9–12]. When powdery HPC or EC is dissolved in solvent(s) at high concentrations, the viscous solution shows a lyotropic CLC phase with light reflection characteristics depending on the concentration of cellulose derivative. For instance, the lyotropic CLC phase appears when dissolving pristine HPC in water [1,12–14] or methanol [15]. As another precedent, solutions of pristine EC in organic solvents such as chloroform [11,16] or acrylic acid (AA) [17] also exhibit the lyotropic CLC phase. Such a light reflection phenomenon is regarded as a kind of Bragg reflection [18–20]. The mechanism of the Bragg reflection color change of HPC or EC solutions with their concentration can be explained by the difference in CLC structure. The CLC structure is characterized by periodic helical molecular assemblages. This molecular structure causes periodic modulation of the reflective index of the CLC, thereby leading to the emergence of light reflection. The maximum wavelength (λ) of the Bragg reflection peak is numerically expressed as the following equation:

$$\lambda = n \cdot p \tag{1}$$

where n denotes the average reflective index of CLC and p is the helical pitch length [21]. The reflection peak wavelength of CLC is significantly dependent on the concentrations of HPC [13] or EC [17] solutions, leading to their wide range of potential applications as concentration indicators, reflective color displays [22], full-color recording media [23], tunable lasers [24], and so forth.

In this context, HPC can exhibit a CLC phase depending on the temperature when the hydroxy groups in its side chains are chemically modified [25]. A substance that exhibits a liquid crystal phase within a certain temperature range is called a thermotropic liquid crystal. It has been reported that when the hydroxy groups in the side chains of pristine HPC are esterified [9,25–27] or etherified [25], the reflection peak wavelength of CLC can be controlled by temperature in either case. Moreover, the reflection peak of CLCs from HPC derivatives generally shifts to the longer wavelength side upon undergoing a heating process. Such changes in the reflection peak wavelength can be ascribed to the modulation of p depending on the temperature [19]. Interestingly, the reflection properties of esterified HPC derivatives and etherified HPC derivatives differ. In the case of esterified HPC derivatives, the reflection peak appears at the longer wavelengths at the same temperature when the substitution degree of hydroxyl groups in the side chain of pristine HPC decreases [28]. However, the etherified HPC derivatives show a reflection peak at the shorter wavelength side at the same temperature, accompanied by a decline in the substitution degree [25]. Based on these precedents, it can be understood that the p value of CLC from HPC derivatives greatly depends not only on the concentration or temperature but also their substitution degree.

Previously, Gray and Guo reported that esterified EC derivatives exhibit a lyotropic CLC phase when dissolved in solvents at appropriate concentrations [11,16,18]. Like esterified or etherified HPC derivatives, the reflection peak wavelength of the solutions of esterified EC derivatives can be controlled by temperature. However, the research progress on the CLCs of EC derivatives has lagged behind that of HPC derivatives because the chemical modification of EC is not as easy as that of HPC due to the inferior solubility of pristine EC to organic solvents. In addition, etherification with alkyl halides has low reaction efficiency, whereas esterification with alkanoyl chlorides proceeds with high yields. Therefore, no reports have yet been released on CLCs prepared from etherified EC derivatives. As mentioned above, the reflection properties of CLC change significantly in the case of HPC derivatives depending on the presence or absence of carbonyl groups as well as the substitution degree. Accordingly, this situation motivates us to investigate the optical properties of etherified EC derivatives. This is extremely important to comprehend the CLC behaviors of EC derivatives for the fabrication of new cellulose-based photonic materials. Moreover, it is also essential to elucidate the effect of the degree of etherification of EC derivatives on the reflection properties of CLCs.

In this study, we optimized the etherification of hydroxy groups of pristine EC in order to obtain EC derivatives possessing pentyl ether groups (EC-Pe), as presented in Figure 1B. During the course of our systematic syntheses, we found that the degree of substitution with a pentyl ether group (DS_{Pe}) rises with the increase in the base concentration in N,N-dimethylacetamide (DMAc) as a reaction solvent. The CLC phase appeared when dissolving the etherified EC derivatives in methacrylic acid (MAA) or AA. Such lyotropic CLCs with visible light reflection could be prepared regardless of the DS_{Pe} of EC derivatives only when MAA was used as the solvent. In general, lyotropic CLCs are highly susceptible to slight temperature fluctuation, so that the reflection peak wavelength is readily shifted in wide wavelength ranges by changing the temperature. Therefore, efforts have been made to tune the reflection peak wavelength by altering the temperature; however, these have involved cumbersome handling, which presents a serious problem from the technological viewpoint. Against that background, we serendipitously found a unique reflection characteristic whereby the reflection peak wavelength of CLC from completely etherified EC derivatives dissolved in MAA is almost maintained in a wide temperature range of 30–110 °C even though MAA is not polymerized. Such thermally stable Bragg reflection colors of CLCs are expected to be highly advantageous for the creation of next-generation photonic devices with eco-friendliness from cellulose.

2. Experimental Section
2.1. Materials

Three kinds of pristine EC substances with different molecular weights were purchased from Tokyo Chemical Industry (Tokyo, Japan) and used as the starting materials to synthesize EC derivatives. The viscosities of 5 wt% solutions of pristine ECs in the mixed solvent of toluene and ethanol (volume ratio 8:2), respectively, were 90–110, 45–55, and 9–11 mPa·s according to the datasheet of the manufacturer. Hereafter, each pristine EC is coded as ECx, where x is the sample number when arranged in descending order of viscosity. The number average molecular weight (M_n) and weight average molecular weight (M_w) of each EC were found to be 6.74×10^5 and 4.29×10^6 for EC1, 5.02×10^5 and 2.43×10^6 for EC2, and 2.58×10^5 and 0.762×10^6 for EC3, respectively, as determined using size exclusion chromatography (SEC) equipped with a reflective index detector (HLC-8220GPC, Tosoh, Tokyo, Japan) calibrated using the polystyrene standards. In the SEC measurements, tetrahydrofuran (THF) was employed as the eluent. The molar amount of chemically combined ethylene oxide per anhydroglucose unit, that is, the molecular substitution (*MS*), was found to be 2.50 via the ^{1}H-NMR spectrum measurement of pristine EC in CDCl$_3$. Therefore, the average molecular weight per anhydroglucose monomer unit could be calculated to be 232. EC was dried under vacuum at room temperature for over 24 h before use.

Dehydrated DMAc, dehydrated *N*-methyl-2-pyrrolidone (NMP), *N*,*N*-dimethylformamide (DMF), acetonitrile, and dehydrated dimethyl sulfoxide (DMSO) as solvents for the etherification of EC were obtained from Fujifilm Wako Pure Chemical Co., Inc. (Tokyo, Japan). 1-Bromopentane (PeBr), used for the synthesis of EC-Pe. MAA and AA, used as solvents of lyotropic CLCs, were purchased from Tokyo Chemical Industry (Tokyo, Japan). Sodium hydroxide (NaOH) and potassium iodide (KI), used as the catalysts of etherification, were acquired from Fujifilm Wako Pure Chemical Co., Inc. (Tokyo, Japan). All reagents except EC were used as received.

2.2. Syntheses of the EC Derivative Possessing Pentyl Ether Groups (EC-Pe)

EC-Pe underwent the Williamson ether synthesis in a manner similar to our previous procedures to prepare the etherified HPC derivatives [29,30]. The reaction conditions are listed in Table 1. The typical etherification procedure of EC-Pe is described as follows (Table 1, Sample code: EC1-Pe$_{0.15}$).

Table 1. Synthesis conditions of EC-Pe and DS_{Pe} values.

Sample Code	Solvent	NaOH Concentration (g/mL)	Viscosity of Pristine EC (mPa·s)	DS_{Pe} [a]
EC1-Pe$_{0.15}$	NMP	0.03	90–110	0.15
EC1-Pe$_{0.34}$	DMAc	0.03	90–110	0.34
EC1-Pe$_{0.36}$	DMAc	0.05	90–110	0.36
EC1-Pe$_{0.50}$	DMAc	0.10	90–110	0.50
EC2-Pe$_{0.50}$	DMAc	0.10	45–55	0.50
EC3-Pe$_{0.50}$	DMAc	0.10	9–11	0.50
EC3-Pe$_{0.12}$	DMAc	0.02	9–11	0.12
EC3-Pe$_{0.29}$	DMAc	0.03	9–11	0.29

[a] The maximum DS_{Pe} value of 0.50 represents the completely etherified EC derivative with pentyl ether side chains.

In a 100 mL round-bottomed flask, 3.00 g of EC was completely dissolved in 24.0 mL of dehydrated NMP. After that, we subsequently added 4.01 mL of PeBr to the EC solution (5.00 eq. to hydroxy groups of EC). After stirring for 30 min at 65 °C, 0.72 g of powdered NaOH (0.03 g/mL for reaction solvents) and 0.27 g of KI (5.00 mol% to PeBr) were added. Subsequently, this reaction mixture was refluxed at 65 °C for 48 h. The reaction mixture was then purified via two rounds of centrifugation at 1.0×10^4 rpm for 5 min to remove any sediment such as sodium bromide. Then, the supernatant was dialyzed against an equivolume mixture of methanol and water for 3 h, and the dialysis was prolonged for an additional 48 h in an equivolume mixture of THF and methanol by using a Visking dialysis tube with pore sizes of ~5 nm and a molecular weight cutoff of $1.2–1.4 \times 10^4$. The product was achieved through evaporation at 35 °C in vacuo for approximately 30 min and was finally vacuum dried for a few days to obtain purified EC-Pe. The characterization of EC-Pe was carried out via FT-IR spectroscopy using an attenuated total reflection module (FT-IR4700 and ATR PRO ONE, JASCO, Tokyo, Japan), ^{1}H-NMR spectroscopy (JNM-ECZ400S, JEOL, Tokyo, Japan) for the molecular structure, and SEC analysis for the M_n and M_w values.

2.3. Fabrication Procedure of Lyotropic CLC Cells

The EC derivatives were completely dissolved in MAA or AA using a planetary centrifugal mixer (AR-100, Thinky, Tokyo, Japan). The lyotropic CLC mixture was sandwiched between two glass substrates. The cell gap was adjusted using polytetrafluoroethylene film spacers with a thickness of ~200 µm. The edge of each cell was sealed with epoxy resin to prevent the evaporation of MAA or AA upon heating.

2.4. Optical Measurements of Lyotropic CLC Cells

The transmission spectrum of the lyotropic CLC cell was determined on a compact charge-coupled (CCD) spectrometer (USB2000+, Ocean Optics, Orlando, FL, USA) equipped with an optical fiber. The CLC cell was illuminated with white light from a tungsten halogen light source (Ocean Optics, HL2000). The transmitted light from the CLC cell was focused through two pieces of achromatic doublet lenses, and it was collected into the entrance of an optical fiber connected with the CCD spectrometer in a collinear arrangement with respect to both the white light source and CLC cell. The temperature of the CLC cell was precisely controlled using a hot-stage system for the optical microscope (HS82 and HS1, Mettler Toledo, Columbus, OH, USA). Polarized optical microscope (POM) images were taken with a CCD camera (EO-5012, Edmund, Barrington, NJ, USA) equipped on the microscope (IX71, Olympus, Tokyo, Japan).

2.5. Rheological Measurements of the Lyotropic CLCs of EC Derivatives

Viscosity measurements were conducted using a stress-controlled rheometer (MCR102, Anton Paar, Graz, Austria) equipped with a stainless-steel parallel plate with a diameter of 8 mm. The temperature was tuned by using a forced convection oven (CTD450, Anton Paar). The lyotropic CLCs were sandwiched at a gap of ~1.0 mm. The angular frequency (ω) dependence of the storage modulus (G') and the loss modulus (G'') was taken on the above-mentioned rheometer. The measurements were performed in the ω range between 0.1 and 100 rad/s at 30 °C. The strain amplitude was adjusted in the range between 0.2 and 1.2%, which was small enough to measure the linear viscoelasticity. Prior to these measurements, the lyotropic CLCs were pre-sheared to erase any orientational history of CLC structures. The pre-shear treatment was conducted by shearing the samples at a constant shear rate of 3 s^{-1} for 350 s at 30 °C, then leaving to stand for 1 h at the same temperature after stopping the shearing force.

3. Results and Discussion

3.1. Characterization of EC-Pe

After the Williamson ether synthesis, we measured both the FT-IR and ^{1}H-NMR spectra to confirm that hydroxy groups of pristine EC are substituted with pentyl ether groups. Figure 2 shows comparative FT-IR spectra between pristine EC and EC3-Pe$_{0.5}$ (Figure 2A) and the representative ^{1}H-NMR spectrum of EC3-Pe$_{0.5}$ (Figure 2B).

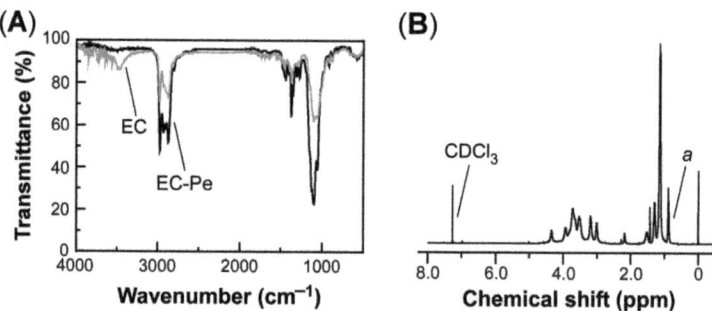

Figure 2. (**A**) FT-IR spectra of pristine EC (gray line) and EC3-Pe$_{0.5}$ (black line). (**B**) ^{1}H-NMR spectrum of EC3-Pe$_{0.5}$ in CDCl$_3$. The peak *a* is assigned to terminal methyl groups in the pentyl ether side chains of EC3-Pe$_{0.5}$.

As shown in Figure 2A, the FT-IR spectrum of pristine EC showed an intense peak from the O-H stretching vibration of the hydroxy groups of EC in a broad wavenumber range of 3000–3600 cm^{-1}. In contrast to pristine EC, the FT-IR spectrum of the completely etherified EC derivative did not show a broad peak in the same wavenumber region. Furthermore, the intense peak from the C-H stretching vibration at 2840–3000 cm^{-1} became stronger, suggesting that etherification had proceeded. These results suggest that the hydroxy groups of pristine EC are completely substituted with pentyl ether groups. Figure 2B represents the ^{1}H-NMR spectrum of EC3-Pe$_{0.5}$ in CDCl$_3$. The proton peak at 0.89 ppm can be assigned as the signal "a" corresponding to the terminal methyl groups in the pentyl ether groups of EC-Pe. The DS_{Pe} value, that is, the degree of substitution with pentyl ether group, can be quantitatively analyzed using the following equation:

$$DS_{Pe} = A(7 + 5MS)/(3W - 11A) \tag{2}$$

where A is the integrated value of the signal peak "a" and W is the sum of the integrated values of all protons in EC-Pe. The mathematical derivation of Equation (2) is available in Appendix A, as described below. As mentioned in Section 2.1, we adopted the MS value of 2.50 from the ^{1}H-NMR spectrum of pristine EC. By applying the experimental results to Equation (2), the DS_{Pe} value of EC-Pe was estimated to be 0.50. Moreover, the values of M_n and M_w of pristine EC and EC-Pe were evaluated based on the SEC measurements using the polystyrene standards. As compiled in Table 2, both M_n and M_w of pristine EC and EC-Pe were in the same order of magnitude as those of pristine EC. This implies that no depolymerization in the main chain of EC might occur in the etherification process.

Table 2. SEC results of pristine EC and EC-Pe.

Sample Code	$M_n \times 10^{-5}$	$M_w \times 10^{-6}$	M_w/M_n
EC1	6.74	4.29	6.36
EC2	5.02	2.43	4.84
EC3	2.58	0.762	2.95
EC1-Pe$_{0.15}$	4.88	2.51	5.15
EC1-Pe$_{0.34}$	4.92	2.74	5.58
EC1-Pe$_{0.36}$	4.54	2.08	4.59
EC1-Pe$_{0.50}$	5.02	2.28	4.53
EC2-Pe$_{0.50}$	4.28	1.57	3.68
EC3-Pe$_{0.50}$	2.61	0.770	2.95
EC3-Pe$_{0.12}$	2.15	0.742	3.45
EC3-Pe$_{0.29}$	2.16	0.666	3.09

Previously, Gray and co-workers revealed that the reflection peak wavelength of CLCs from esterified EC derivatives is greatly affected by the degree of esterification, that is, the number of hydroxy groups of the EC backbone that remains [11,16,18]. These precedents motivated us to investigate the effect of the degree of etherification on optical properties even for the etherified EC derivatives. Therefore, it was necessary to find synthetic conditions that allow us to control the number of residual hydroxyl groups in the side chain of pristine EC.

3.2. Synthesis of Etherified EC Derivatives

A series of etherified EC derivatives were synthesized by changing the reaction conditions such as solvent or NaOH concentration, as listed in Table 1. First, we examined the effect of the solvent on the etherification of EC1 with high M_n and M_w values. Since the Williamson ether synthesis is the S$_N$2 reaction, aprotic and high-polarity solvents are known to promote the reaction. Therefore, we synthesized EC-Pe using five kinds of solvents, that is, acetonitrile, DMSO, DMF, NMP, and DMAc. In all cases, the NaOH concentration in the reaction solvent was fixed at 0.03 g/mL. When acetonitrile, DMSO, and DMF were used as reaction solvents, etherification hardly proceeded because the pristine EC and NaOH were not soluble in these solvents. Alternatively, when NMP and DMAc were adopted in the Williamson ether synthesis, etherified EC derivatives could be synthesized because EC and NaOH were easily dissolved in NMP or DMAc. However, it was found that the DS_{Pe} can be improved from 0.15 to 0.34 by changing the solvent from NMP to DMAc (Table 1, Sample codes: EC1-Pe$_{0.15}$ and EC1-Pe$_{0.34}$). When considering our overall results, we concluded that DMAc is the most appropriate solvent for the etherification of EC.

Subsequently, we examined the effect of NaOH concentration in DMAc on the etherification of pristine EC. For this purpose, the NaOH concentration was increased from 0.03 g/mL to 0.05 g/mL or 0.10 g/mL. As a result, DS_{Pe} became 0.36 or 0.50 when the NaOH concentration in DMAc was 0.05 g/mL or 0.10 g/mL, respectively (Table 1, Sample codes: EC1-Pe$_{0.36}$ and EC1-Pe$_{0.50}$). These results indicated that the hydroxyl groups in the side chains of pristine EC can be completely modified by examining the base concentration. This can be attributed to the increased efficiency of the reaction between the alkoxide ion and the brominated terminal carbon of PeBr at higher base concentrations. Figure S1 in the Supplementary Materials shows a comparison of the FT-IR spectra of pristine EC and etherified EC derivatives. The peak intensity at 3000–3600 cm^{-1}, which was assigned to the O-H stretching vibration of the hydroxy groups in the EC side chain, was greatly affected by the difference in DS_{Pe}. This peak disappeared when the base concentration in the reaction solvent was 0.10 g/mL. Therefore, it turned out that a NaOH concentration of 0.10 g/mL in DMAc enables the synthesis of a completely pentyl-etherified EC derivative with a maximum DS_{Pe} value of 0.50. As explained in Section 2.1, ^{1}H-NMR spectrum measurement of pristine EC revealed that the molar amount of chemically combined ethylene oxide per anhydroglucose unit, that is, the MS value, was 2.50. Considering the chemical structure of pristine EC, the maximum DS_{Pe} value was 0.50 since the monomer unit of cellulose has three hydroxy groups, as depicted in Figure 1B.

Furthermore, we found that the DS_{Pe} value of etherified EC derivatives can be easily controlled from 0.34 to 0.50 by changing the NaOH concentration in DMAc at 0.03–0.10 g/mL. To prove the versatility of this reaction condition, we synthesized another series of EC-Pe from pristine EC with different molecular weights under the same conditions. Consequently, the completely etherified EC derivatives could be prepared regardless of the molecular weight of pristine EC (Table 1, Sample codes: EC2-Pe$_{0.50}$, EC3-Pe$_{0.50}$). Additionally, two kinds of EC-Pe derivatives with DS_{Pe} values of 0.12 and 0.29 were also synthesized from EC3 when the base concentration was adjusted to 0.02 g/mL and 0.03 g/mL, respectively (Table 1, Sample codes: EC3-Pe$_{0.12}$, EC3-Pe$_{0.29}$). These results highlight that the DS_{Pe} of EC derivatives could be easily tuned by changing the NaOH concentration.

3.3. Molecular Weight Dependence of EC-Pe on the Reflection Property of Its Lyotropic CLC

We found that the derivatives from pristine EC with a smaller molecular weight were suitable for preparing lyotropic CLCs with a sharp reflection peak. In this study, three kinds of completely etherified EC derivatives with different molecular weights were synthesized from EC1, EC2, and EC3 (Table 1, Sample codes: EC1-Pe$_{0.50}$, EC2-Pe$_{0.50}$, and EC3-Pe$_{0.50}$). All the EC derivatives were dissolved in MAA at a polymer concentration of 65 wt%, whereupon reflection peaks appeared in the wavelength range between 420 nm and 520 nm at 30 °C for the lyotropic CLCs (Figure 3). Here, we discovered two phenomena related to the molecular weight dependence of EC-Pe on the reflection property of CLC. First, the reflection peak of CLC appeared at the shorter wavelengths with the increase in the molecular weight of EC-Pe when compared at the same temperature. The reflection peak wavelengths of the lyotropic CLCs were 517 nm for EC1-Pe$_{0.50}$, 435 nm for EC2-Pe$_{0.50}$, and 426 nm for EC3-Pe$_{0.50}$. The shorter wavelength shift of the reflection peak with decreasing molecular weight suggests the decrease in the helical pitch length of CLC, corresponding to p in Equation (1).

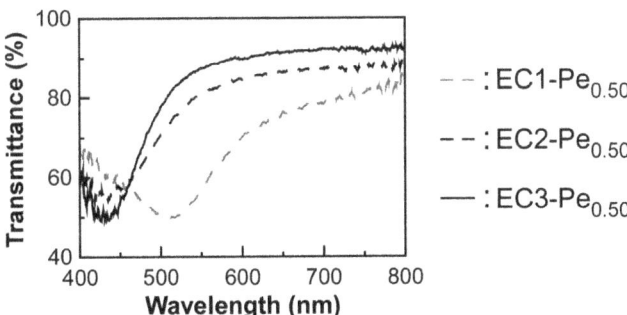

Figure 3. Transmission spectra of the CLC cells of EC1-Pe$_{0.50}$, EC2-Pe$_{0.50}$, and EC3-Pe$_{0.50}$ at 30 °C.

As another phenomenon, the baseline in the transmission spectrum increased from ~80% to ~90%, and a sharp reflection peak emerged, accompanied by the decrease in the molecular weights of EC derivatives. To quantitatively evaluate the spectral sharpness of the reflection peaks, we focused on the half width at half maximum in the reflection peak of each lyotropic CLC. The half width at half maximum of lyotropic CLC was 103 nm for EC1-Pe$_{0.50}$, 61 nm for EC2-Pe$_{0.50}$, and 45 nm for EC3-Pe$_{0.50}$. Therefore, it can be considered that the lyotropic CLC of EC3-Pe$_{0.50}$ is most likely to be self-assembled in the molecular helical manner. Such a difference in the orientation behaviors of lyotropic CLCs can be explained by their viscosity. As shown in Figure S2 of the Supplementary Materials, we explored the changes in the viscosities of EC1-Pe$_{0.50}$, EC2-Pe$_{0.50}$, and EC3-Pe$_{0.50}$ as a function of shearing time. The measurements were conducted at a constant shear rate of 3 s^{-1} for 350 s at 30 °C. From the experimental results, the steady-state viscosity of EC-Pe in MAA was determined as follows: 1.83 kPa·s for EC1-Pe$_{0.50}$, 1.03 kPa·s for EC2-Pe$_{0.50}$, and 0.66 kPa·s for EC3-Pe$_{0.50}$. These fit with our expectations as the viscosities of polymer solutions generally increase with the molecular weight. It can be considered that the lowest viscosity, of EC3-Pe$_{0.50}$, enables the formation of highly oriented CLCs to improve their optical properties because the liquid crystal molecules are more likely to move than in the other cases. It should be noted that the low viscosity of a lyotropic CLC is very advantageous for the emergence of a vivid Bragg reflection color because air bubbles in the lyotropic CLC, which cause light scattering, can be easily removed by degassing. Therefore, we concluded that EC3 is the most suitable pristine EC for the preparation of lyotropic CLCs. From these results, the optical properties of CLCs from EC3-Pe were further investigated, and the results are presented in the following section.

3.4. Reflection Properties of Lyotropic CLC of EC3-Pe in MAA or AA

When lyotropic CLCs were prepared from a series of EC3-Pe with different DS_{Pe}, we found that MAA is far more suitable than AA as a solvent to form lyotropic CLCs with visible Bragg reflection. The preparation conditions and sample definitions of lyotropic CLCs are shown in Table 3. Figure 4 shows the transmission spectra of lyotropic CLCs of a series of EC3-Pe with different DS_{Pe} dissolved in MAA or AA, which were determined at 30 °C. The insets of this figure also present the light reflection images.

Table 3. Sample definition of lyotropic CLCs and their CLC properties relevant for changing the temperature.

Sample	DS_{Pe}	Solvent	Polymer Conc. (wt%) [a]	$\lambda_{30\,°C}$ (nm) [b]	λ_{end} (nm) [c]	$\lambda_{shift}/10\,°C$ (nm) [d]	d_{change} [e]
1	0.12	MAA	54	480	402	25	0.92
2	0.12	MAA	52	533	434	33	0.97
3	0.12	MAA	50	663	543	60	0.90
4	0.29	MAA	60	455	415	7	0.96
5	0.29	MAA	58	520	418	16	0.96

Table 3. *Cont.*

Sample	DS_{Pe}	Solvent	Polymer Conc. (wt%) [a]	$\lambda_{30\,°C}$ (nm) [b]	λ_{end} (nm) [c]	$\lambda_{shift}/10\,°C$ (nm) [d]	d_{change} [e]
6	0.29	MAA	56	705	497	35	0.93
7	0.50	MAA	69	426	436	1	1.00
8	0.50	MAA	67	524	495	4	1.05
9	0.50	MAA	63	639	545	12	0.94
10	0.12	AA	69	440	428	– §	– §
11	0.12	AA	56	526	455	– §	– §
12	0.12	AA	53	771	486	– §	– §
13	0.29	AA	65	429	415	– §	– §
14	0.29	AA	62	500	410	– §	– §
15	0.29	AA	60	656	417	– §	– §
16	0.50	AA	65	– *	– *	– §	– §

[a] Weight concentration. [b] Reflection peak wavelength at 30 °C. [c] Reflection peak wavelength at the end of the measurement. [d] The average wavelength shift of the reflection peak for every 10 °C. [e] The relative change ratio in the diameter of the lyotropic CLC. * The values were not known because the reflection peaks could not be measured. § Not measured.

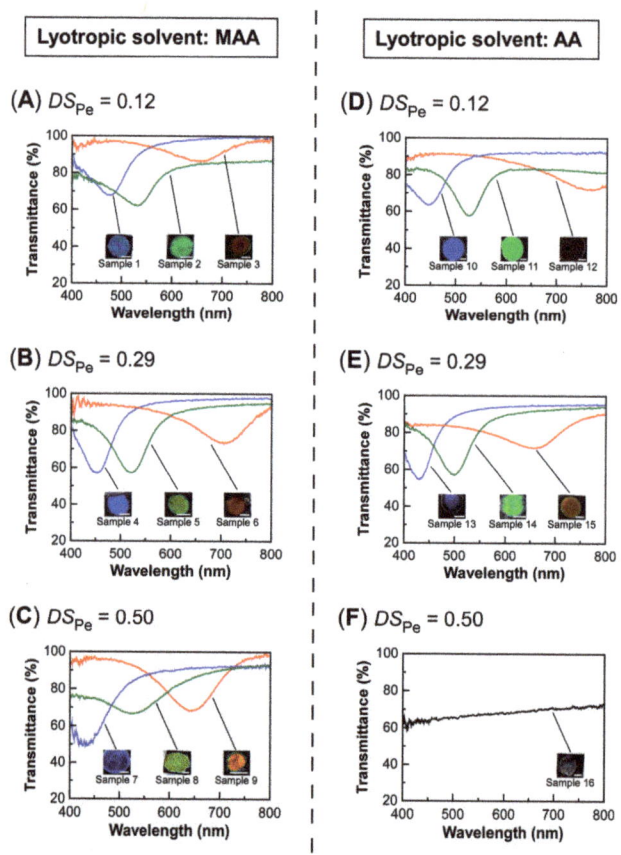

Figure 4. Changes in transmission spectra of CLC cells of lyotropic EC3-Pe in MAA or EC3-Pe in AA mixtures as a function of the concentration of EC3-Pe. The measurement temperature is 30 °C. The insets present the reflection images of CLC cells at the same temperature with white scale bars of 5.0 mm. (**A**) EC3-Pe$_{0.12}$ in MAA, **Samples 1–3**. (**B**) EC3-Pe$_{0.29}$ in MAA, **Samples 4–6**. (**C**) EC3-Pe$_{0.50}$ in MAA, **Samples 7–9**. (**D**) EC3-Pe$_{0.12}$ in AA, **Samples 10–12**. (**E**) EC3-Pe$_{0.29}$ in AA, **Samples 13–15**. (**F**) EC3-Pe$_{0.50}$ in AA, **Sample 16**.

For example, when EC3-Pe$_{0.12}$ was completely dissolved in MAA at the concentration of 50–54 wt%, a lyotropic CLC phase with visible reflection was observed (Figure 4A, Table 3, **Samples 1–3**). It was found that the Bragg reflection peak wavelength can be tuned by changing the concentration of EC3-Pe in the lyotropic CLC mixture. Although the reflection peak wavelength of the lyotropic CLC was 480 nm at 30 °C when the polymer concentration was 54 wt% (Figure 4A, **Sample 1**), the reflection peak wavelength shifted to 533 nm and 663 nm when measured at the same temperature accompanied by the decrease in polymer concentration to 52 wt% or 50 wt%, respectively (Figure 4A, **Samples 2** and **3**). This plausibly happens due to the decrease in the helical twisting power of CLC or the expansion of its helical pitch length. Similarly, the red-shift of reflection peak with the decrease in polymer concentration was also observed for EC3-Pe$_{0.29}$ (Figure 4B, **Samples 4–6**) and EC3-Pe$_{0.50}$ (Figure 4C, **Samples 7–9**). These results fit with our expectations, which were based on the similar red-shift of the reflection peak wavelength observed for lyotropic CLCs from aqueous solutions of pristine HPC [1,12] and for a pristine EC solution in AA [17]. In both HPC and EC systems, the reflection peaks were broadened at lower polymer concentrations. The phenomenon can be explained by the decrease in the helical twisting power.

When AA was used instead of MAA as a solvent of lyotropic CLCs, the reflection peak also shifted to the longer wavelength side as the polymer concentration decreased for EC3-Pe$_{0.12}$ (Figure 4D, **Samples 10–12**) or EC3-Pe$_{0.29}$ (Figure 4E, **Samples 13–15**). However, the reflection peak of EC3-Pe$_{0.50}$ dissolved in AA at a polymer concentration of 65 wt% did not appear in its transmission spectrum in the visible wavelength range of 400–800 nm, which was the detectable range of the measurement instrument (Figure 4F, **Sample 16**). To investigate this phenomenon, we conducted POM observation of **Sample 16**. The POM image at 30 °C showed transmitted light under cross-Nicols, revealing the emergence of optical birefringence through liquid crystallinity (Supplementary Materials, Figure S3A) [17]. When the temperature was increased from 30 °C to 110 °C, blue light of ~400 nm was found to be reflected (Supplementary Materials, Figure S3B). According to the precedent by Nishio and co-workers, the reflection peak wavelength of AA solutions of esterified EC derivatives shifts to the shorter wavelength upon heating [31]. Additionally, the reflection peak of lyotropic CLCs of EC3-Pe$_{0.12}$ and EC3-Pe$_{0.29}$ in AA (**Samples 10–15**) also shifted to the shorter wavelength side upon heating (Table 3). These results contextualized why the reflection peak appeared within the near-infrared wavelength region at 30 °C for **Sample 16**. Therefore, further research was conducted using the lyotropic CLC mixtures of EC3-Pe in MAA (**Samples 1–9**) to investigate the dependence of optical properties on DS_{Pe} value.

3.5. Reflection Peak Wavelength of EC3-Pe in MAA Dependence on DS_{Pe}

Interestingly, we found that simply changing the DS_{Pe} value of EC-Pe gives rise to a significant difference in the shifting wavelength range of the reflection peak of its lyotropic CLC mixture in MAA upon conducting a stepwise heating process. In this study, lyotropic CLCs exhibiting blue, green, and red reflection colors at 30 °C were prepared from a series of EC3-Pe in MAA with different DS_{Pe} values (**Samples 1–9**). The changes in transmission spectra of these lyotropic CLCs upon heating are summarized in Figure S4 in the Supplementary Materials. The sample temperature was increased from 30 °C at temperature intervals of 10 °C. The measurements were conducted in the temperature range at which the reflection peak of CLC appeared and in the wavelength range between 400 nm and 800 nm by considering the detectable wavelength limit of the CCD photodetector. The reflection peak wavelength at 30 °C, corresponding to the starting temperature for the measurements, was defined as $\lambda_{30\,°C}$ (Table 3, 5th column from left side). The reflection peak wavelength at the end of the measurements was denoted as λ_{end} (Table 3, 6th column).

Figure 5 plots the change in the reflection peak wavelength of lyotropic CLCs from EC3-Pe in MAA with a DS_{Pe} value of 0.12 (Figure 5A, **Samples 1–3**) to a DS_{Pe} of 0.29 (Figure 5B, **Samples 4–6**) and of 0.50 (Figure 5C, **Samples 7–9**) upon heating from 30 °C in a stepwise manner. In Figure 5, the plots are color-coded, and each color represents the reflection of the lyotropic CLC at 30 °C. For example, red plots in Figure 5 are those where the reflection color of lyotropic CLC was red at 30 °C.

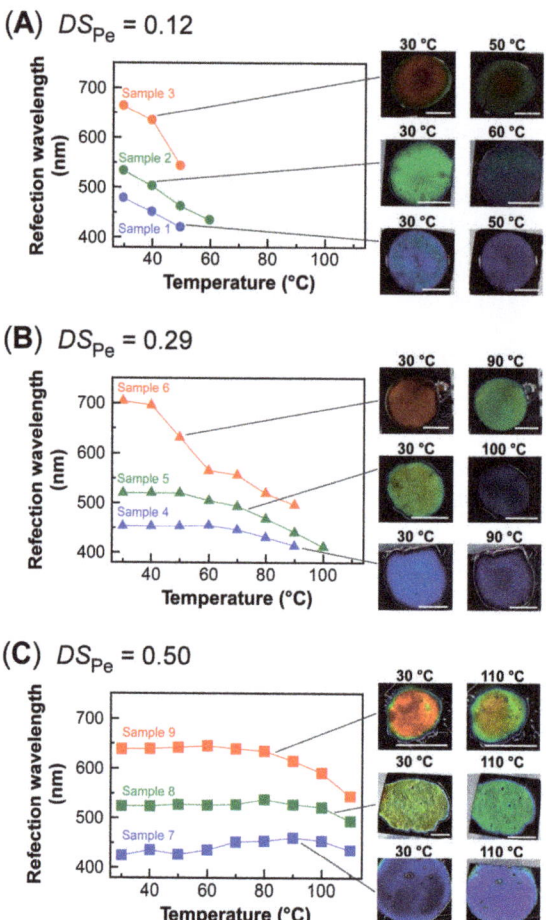

Figure 5. Temperature dependences of the reflection peak wavelength of lyotropic CLC from EC3-Pe with different DS_{Pe}. (**A**) EC3-Pe$_{0.12}$ in MAA, **Samples 1–3**. (**B**) EC3-Pe$_{0.29}$ in MAA, **Samples 4–6**. (**C**) EC3-Pe$_{0.50}$ in MAA, **Samples 7–9**. The insets are the reflection images of each lyotropic CLC at indicated temperatures with white scale bars of 5.0 mm. The plots are color-coded, and each color represents the reflection color of the lyotropic CLC at 30 °C.

In the case of **Samples 1–3**, the reflection peak wavelength of each lyotropic CLC continuously shifted to the shorter wavelength side upon heating (Figure 5A). When heated from 30 °C, the reflection peak wavelengths of **Samples 1–3** were changed from 480 nm to 402 nm (**Sample 1**), from 533 nm to 434 nm (**Sample 2**), and from 663 nm to 543 nm (**Sample 3**). Such changes in the reflection peak wavelength were especially apparent for **Samples 2** and **3** as their Bragg reflection color drastically changed upon heating (Figure 5A, images). These results were in good agreement with many precedents on CLCs from pristine EC or esterified EC derivatives [11,18,31]. Regardless of the difference in $\lambda_{30°C}$, the reflection peaks of each lyotropic CLC disappeared when heated above 60 °C. Because the isotropic phase transition temperature (T_i) of **Sample 3** was determined to be 44–68 °C through POM observation, we considered that such disappearance of the Bragg reflection color may be due to the transition of molecular orientation from liquid crystal phase to isotropic phase under heating. In order to quantitatively discuss the DS_{Pe} dependence of the wavelength shift of reflection peak upon heating, the measurement temperature range must be normalized. Therefore, the following equation was adopted to calculate the average wavelength shift range of reflection peak for every 10 °C, corresponding to the value of $\lambda_{shift}/10$ °C.

$$\lambda_{shift}/10\ °C = \{(\lambda_{30\ °C} - \lambda_{end})/(T_{end} - 30)\} \times 10 \quad (3)$$

In Equation (3), T_{end} is the highest temperature of the measurement. The calculations show that at the DS_{Pe} value of 0.12, the $\lambda_{shift}/10$ °C values for **Sample 1**, **Sample 2**, and **Sample 3** are estimated to be 25 nm, 33 nm, and 60 nm, respectively.

For the lyotropic CLCs from EC3-Pe in MAA with a DS_{Pe} value of 0.29 (**Samples 4–6**), the temperature dependence of the reflection peak wavelength was smaller than those from EC3-Pe in MAA with a DS_{Pe} of 0.12 (Figure 5B). The reflection peak shifted from 455 nm to 415 nm (**Sample 4**), from 520 nm to 418 nm (**Sample 5**), and from 705 nm to 497 nm (**Sample 6**) upon heating from 30 °C to T_{end}. In particular, the change in the reflection peak wavelength was not so apparent for **Samples 4** and **5** in the temperature range of 30 °C to 60 °C (Figure 5B, green and blue triangles). Furthermore, the $\lambda_{shift}/10$ °C of **Samples 4–6** were 7 nm, 16 nm, and 35 nm, respectively. These values were much smaller than those of **Samples 1–3**, as summarized in Table 3. Therefore, it was found that the $\lambda_{shift}/10$ °C of lyotropic CLCs can be reduced by utilizing EC3-Pe with a relatively higher DS_{Pe}. Notably, the measured temperature range was expanded to 30–90 °C. The mechanism for the expansion of temperature range for visible light reflection was the increase in T_i of these lyotropic CLCs. For example, the T_i of **Sample 6** was 90–98 °C, which was ~40 °C higher than that of **Sample 3**. Thus, the disappearance of the reflection peak upon heating can be ascribed to the transition from the CLC phase to isotropic phase.

The temperature dependence of the reflection peak wavelength of lyotropic CLC was smallest when using EC3-Pe in MAA with a DS_{Pe} vale of 0.50, that is, the completely etherified EC derivative (Figure 5C). The reflection peak wavelength only changed from 426 nm to 436 nm (**Sample 7**), from 524 nm to 495 nm (**Sample 8**), and from 639 nm to 545 nm (**Sample 9**) upon heating from 30 °C to T_{end}. In addition, the values of $\lambda_{shift}/10$ °C were 1 nm (**Sample 7**), 4 nm (**Sample 8**), and 12 nm (**Sample 9**). In these cases, the reflection peak disappeared on heating to 110 °C, irrespective of the reflection color at the starting temperature of the measurement. Considering that the T_i of **Sample 9** emerges above 155 °C, it was considered that the disappearance of reflection peak is due to light scattering caused by bubbles in the liquid crystal cell generated by heating above 110 °C, probably because of the boiling point of the MAA (160 °C) or the expulsion of MAA encapsulated in the EC-Pe polymer chain, which will be explained in the following section. As mentioned above, the $\lambda_{shift}/10$ °C values of **Samples 7–9** were much smaller than those of **Samples 1–6**. The Supporting Video file shows the difference in Bragg reflection color change of **Samples 2** and **8** upon heating in the temperature range of 30–110 °C at 20 °C/min (Supplementary Materials, Video S1). After the reflection color of **Sample 2** was green at 30 °C, it changed to blue at ~50 °C and finally disappeared at ~85 °C. The disappearance of Bragg reflection color is ascribed to the reflection of ultraviolet light. In contrast, we found a unique reflection behavior whereby the green reflection color of **Sample 8** was hardly changed upon heating. Indeed, the reflection peak of **Sample 8** was almost maintained at ~520 nm even when heating (Figure 5C, green plots). Such a peculiar thermal stability of the Bragg reflection color of **Sample 8** expands the potential for CLCs from EC derivatives to be applied to new photonic devices with eco-friendliness.

During systematic studies, we found a tendency for the T_i points of lyotropic CLCs to rise when utilizing EC3-Pe with higher DS_{Pe}. For instance, the T_i points of lyotropic CLCs showing a red reflection color at 30 °C were found to be 44–68 °C (**Sample 3**), 90–98 °C (**Sample 6**), and above 155 °C (**Sample 9**). Considering that each lyotropic CLC consisted of EC-Pe with the DS_{Pe} values of 0.12, 0.29, and 0.50, as given in Table 3, it was suggested that the T_i point of CLC is greatly dependent on the DS_{Pe} of EC3-Pe. This stems from the difference in MAA concentration of each lyotropic CLC, because the lyotropic CLC phase is more likely to be retained with the reduction in the number of solvent molecules that can flow upon heating.

The viscoelastic behavior of the lyotropic CLC of EC3-Pe dissolved in MAA was investigated to unravel the mechanism for the difference in the amount of reflection peak wavelength shift upon heating depending on the DS_{Pe} of EC3-Pe. Figure S5 shows the angular frequency (ω) dependence of the storage modulus (G') and the loss modulus (G'') of **Samples 1**, **4**, and **7** (Supplementary Materials, Figure S5, red plots for **Sample 1**, blue plots for **Sample 4**, and black plots for **Sample 7**). To obtain reproducible data, pre-treatment was performed before each measurement based on previous studies [30,32]. As shown in Figure S5A, G' was greater than G'' in the entire ω range of 0.1–100 rad/s, implying the gel-like behavior of lyotropic CLCs. Although this behavior is not consistent with our previous rheological studies on the CLCs from HPC derivatives [30,32,33], these results are reasonable if we consider that the hydrogen bonding between MAA prohibits flow behavior. Additionally, the inflection points of G'' for the EC3-Pe solution in MAA were found at 6.3 rad/s for **Sample 1**, 2.5 rad/s for **Sample 4**, and 1.0 rad/s for **Sample 7**. These inflection points were more apparent when the loss tangent of G''/G' was plotted against ω (Supplementary Materials, Figure S5B). As shown in Figure S5, the inflection point appeared shifted toward the lower ω side as DS_{Pe} increased. Based on the precedent from the literature [33], it is plausible that the inflection points appeared in the ω dependence of G'' are inseparably related to the tilting motion of the helical axis of the CLC, which is likely to be hindered with the increase in DS_{Pe} value as they shift to the lower ω side. These results lead us to consider that the reflection peak wavelength of the lyotropic CLC of EC3-Pe with a higher DS_{Pe} is maintained upon heating because of the restriction of molecular motion.

However, even when using EC3-Pe in MAA with the DS_{Pe} value of 0.50, the reflection peak wavelength of **Sample 9** shifted to 94 nm when heating from 30 °C to the T_{end} point (Figure 5C, red plots). This could be due to the expulsion of solvent molecules to the outside of CLC layers when heating. To prove this hypothesis, the changes in diameter of the lyotropic CLC sandwiched between glass plates were evaluated upon heating. The relative change ratio in the diameter (d_{change}) of the lyotropic CLC was calculated as follows:

$$d_{change} = d_{110\,°C}/d_{30\,°C} \tag{4}$$

where $d_{30\,°C}$ and $d_{110\,°C}$ are the diameter of the CLC at 30 °C and 110 °C, respectively. The d_{change} values of **Samples 7–9** were estimated to be 1.00, 1.05, and 0.94, respectively. It should be noted that only in **Sample 9** is d_{change} smaller than 1.0. This is because the solvent molecules inserted into the polymer networks of EC3-Pe are released from the CLC helical pitches due to heating. As discussed earlier in Section 3.4, the helical twisting power of **Sample 9** was weaker than those of **Samples 7** and **8**. In addition, the motility of liquid crystal molecules is enhanced by heating, which prohibits their orientation. Therefore, it can be considered that solvent molecules inserted into the polymer chain are more likely to be ejected in **Sample 9** rather than those in **Samples 7** and **8**. The solvent molecules inserted between the CLC layers are ejected, which shortens the helical pitch length of CLC, resulting in a shorter wavelength shift of the reflection peak wavelength.

4. Conclusions

In this study, we successfully prepared completely etherified EC derivatives through the Williamson ether synthesis. The EC derivatives with pentyl ether groups showed a lyotropic CLC phase with visible Bragg reflection when we dissolved them in MAA. The sharp reflection peak appeared only when utilizing pristine EC with a relatively low molecular weight because the lower viscosity of CLC enabled its better orientation. Moreover, the reflection peak wavelength of CLC could be maintained upon heating by utilizing the completely etherified EC derivatives. This can be ascribed to the restriction of the molecular movement of the liquid crystal structure, which was greatly affected by the differences in the substitution degrees of EC derivatives. These experimental results support us not only to understand the fundamental physical properties of CLCs from EC derivatives but also to create environmentally friendly CLC photonic devices using cellulose derivatives, which may help us to achieve a sustainable society.

Supplementary Materials: The following supporting information can be downloaded at: https://www.mdpi.com/article/10.3390/polym16030401/s1, Figure S1: FT-IR spectra of pristine EC (black line), EC1-Pe$_{0.36}$ (gray line), and EC1-Pe$_{0.50}$ (blue line); Figure S2: Changes in the viscosity of EC1-Pe$_{0.50}$ (red points), EC2-Pe$_{0.50}$ (black points), and EC3-Pe$_{0.50}$ (blue points) as a function of shearing time; Figure S3: (**A**) Polarized optical microscopic image at 30 °C of **Sample 16** under cross-Nicols. (**B**) Transmission spectra of **Sample 16** measured at 100 °C (green line), 105 °C (red line), and 110 °C (blue line); Figure S4: Changes in transmission spectra of these lyotropic CLCs upon conducting the heating process; Figure S5: (**A**) Angular frequency (ω) dependence of the storage modulus (G'; closed squares) and loss modulus (G''; open triangles) of **Sample 1** (red plots), **Sample 4** (blue plots), and **Sample 7** (black plots) measured at 30 °C. (**B**) Angular frequency (ω) dependence of the loss tangent (G''/G') of **Sample 1** (red plots), **Sample 4** (blue plots), and **Sample 7** (black plots) measured at 30 °C; Video S1: The difference in Bragg reflection color change of **Samples 2** and **8** upon heating in the temperature range of 30–110 °C at 20 °C/min.

Author Contributions: K.M. conducted most of the experiments and wrote the original draft of the manuscript. N.I. analyzed the results and revised the manuscript. S.F. supervised this project and prepared the final version of the manuscript. All authors have read and agreed to the published version of the manuscript.

Funding: This work was financially supported in part by the JSPS Grant-in-Aid for Scientific Research (B) (Grant No. 21H02261), the Sugar Industry Public Association, and the Descente and Ishimoto Memorial Foundation for the Promotion of Sports Science.

Institutional Review Board Statement: Not applicable.

Data Availability Statement: Data are contained within the article.

Acknowledgments: All authors deeply thank Takeo Sasaki and Koha V. Le (Tokyo University of Science) for their support with SEC measurements.

Conflicts of Interest: The authors declare no conflicts of interest.

Appendix A. Mathematical Derivation of Equation (2)

As mentioned in Section 3.1, Equation (2) enables us to quantitatively evaluate the DS_{Pe} value, that is, the average number of pentyl ether groups in a monomer unit of EC-Pe. Equation (2) was obtained according to our previously reported procedure [34].

When measuring ^{1}H-NMR spectrum of EC-Pe, a peak of terminal methyl protons in the pentyl ether groups was observed at ~0.89 ppm. In the preceding section, this ^{1}H-NMR spectrum was shown in Figure 2B. Successively, we defined the integrated values as A and the sum of integrated values of the peaks derived from all protons of EC-Pe as W. Therefore, the W value satisfied Equation (A1) as follows:

$$W = 7 + 5MS + 11DS_{Pe} \tag{A1}$$

Because the DS_{Pe} value was calculated from the ratio of $A/3$ and W, the relation was expressed as Equation (A2).

$$DS_{Pe}/(7 + 5MS + 11DS_{Pe}) = A/3W \tag{A2}$$

Consequently, Equation (A2) was rearranged so as to be Equation (2) relating to the DS_{Pe} value, as addressed in Section 3.1.

References

1. Charlet, G.; Gray, D.G. Chiroptical Filters from Aqueous (Hydroxypropyl) Cellulose Liquid Crystals. *J. Appl. Polym. Sci.* **1989**, *37*, 2517–2527. [CrossRef]
2. Revol, J.F.; Bradford, H.; Giasson, J.; Marchessault, R.H.; Gray, D.G. Helicoidal Self-Ordering of Cellulose Microfibrils in Aqueous Suspension. *Int. J. Biol. Macromol.* **1992**, *14*, 170–172. [CrossRef] [PubMed]
3. Ebers, L.S.; Laborie, M.P. Direct Ink Writing of Fully Bio-Based Liquid Crystalline Lignin/Hydroxypropyl Cellulose Aqueous Inks: Optimization of Formulations and Printing Parameters. *ACS Appl. Bio Mater.* **2020**, *3*, 6897–6907. [CrossRef]
4. Adams, D.; Ounaies, Z.; Basak, A. Printability Assessment of Ethyl Cellulose Biopolymer Using Direct Ink Writing. *JOM* **2021**, *73*, 3761–3770. [CrossRef]
5. Murtaza, G. Ethylcellulose Microparticles: A Review. *Acta Pol. Pharm.-Drug Res.* **2012**, *69*, 11–22.
6. Arca, H.C.; Mosquera-Giraldo, L.I.; Bi, V.; Xu, D.; Taylor, L.S.; Edgar, K.J. Pharmaceutical Applications of Cellulose Ethers and Cellulose Ether Esters. *Biomacromolecules* **2018**, *19*, 2351–2376. [CrossRef]
7. Tsuji, R.; Tanaka, K.; Oishi, K.; Shioki, T.; Satone, H.; Ito, S. Role and Function of Polymer Binder Thickeners in Carbon Pastes for Multiporous-Layered-Electrode Perovskite Solar Cells. *Chem. Mater.* **2023**, *35*, 8574–8589. [CrossRef]
8. Zhu, J.; Dong, X.T.; Wang, X.L.; Wang, Y.Z. Preparation and Properties of a Novel Biodegradable Ethyl Cellulose Grafting Copolymer with Poly(*p*-dioxanone) Side-Chains. *Carbohydr. Polym.* **2010**, *80*, 350–359. [CrossRef]
9. Tseng, S.L.; Laivins, G.V.; Gray, D.G. Propanoate Ester of (2-Hydroxypropyl)cellulose: A Thermotropic Cholesteric Polymer that Reflects Visible Light at Ambient Temperatures. *Macromolecules* **1982**, *15*, 1262–1264. [CrossRef]
10. Zugenmaier, P.; Haurand, P. Structural and Rheological Investigations on the Lyotropic, Liquid-crystalline System: *o*-Ethylcellulose-Acetic Acid-Dichloroacetic Acid. *Carbohydr. Res.* **1987**, *160*, 369–380. [CrossRef]
11. Guo, J.X.; Gray, D.G. Chiroptical Behavior of (Acetyl)(ethyl)cellulose Liquid Crystalline Solutions in Chloroform. *Macromolecules* **1989**, *22*, 2086–2090. [CrossRef]
12. Almeida, A.P.C.; Canejo, J.P.; Fernandes, S.N.; Echeverria, C.; Almeida, P.L.; Godinho, M.H. Cellulose-Based Biomimetics and their Applications. *Adv. Mater.* **2018**, *30*, 1703655. [CrossRef] [PubMed]
13. Werbowyj, R.S.; Gray, D.G. Liquid Crystalline Structure in Aqueous Hydroxypropyl Cellulose Solutions. *Mol. Cryst. Liq. Cryst.* **1976**, *34*, 97–103. [CrossRef]
14. Werbowyj, R.S.; Gray, D.G. Ordered Phase Formation in Concentrated Hydroxypropylcellulose Solutions. *Macromolecules* **1980**, *13*, 69–73. [CrossRef]
15. Werbowyj, R.S.; Gray, D.G. Optical Properties of (Hydroxypropyl)cellulose Liquid Crystals. Cholesteric Pitch and Polymer Concentration. *Macromolecules* **1984**, *17*, 1512–1520. [CrossRef]
16. Guo, J.X.; Gray, D.G. Preparation and Liquid-Crystalline Properties of (Acetyl)(ethyl)cellulose. *Macromolecules* **1989**, *22*, 2082–2086. [CrossRef]
17. Nishio, Y.; Fujiki, Y. Liquid-Crystalline Characteristics of Cellulose Derivatives: Binary and Ternary Mixtures of Ethyl Cellulose, Hydroxypropyl Cellulose, and Acrylic Acid. *J. Macromol. Sci. Part B* **1991**, *30*, 357–384. [CrossRef]
18. Guo, J.-X.; Gray, D.G. Effect of Degree of Acetylation and Solvent on the Chiroptical Properties of Lyotropic (Acetyl)(ethyl)cellulose Solutions. *J. Polym. Sci. Part B Polym. Phys.* **1994**, *32*, 2529–2537. [CrossRef]
19. Seddon, J.M. Structural Studies of Liquid Crystals by X-ray Diffraction. *Handb. Liq. Cryst. Set.* **1998**, *1*, 635–679.
20. Shimamoto, S.; Uraki, Y.; Sano, Y. Optical Properties and Photopolymerization of Liquid Crystalline (Acetyl)(ethyl)cellulose/Acrylic Acid System. *Cellulose* **2000**, *7*, 347–358. [CrossRef]
21. De Vries, H. Rotatory Power and Other Optical Properties of Certain Liquid Crystals. *Acta Crystallogr.* **1951**, *4*, 219–226. [CrossRef]
22. Zhang, Z.; Chen, Z.; Wang, Y.; Zhao, Y.; Shang, L. Cholesteric Cellulose Liquid Crystals with Multifunctional Structural Colors. *Adv. Funct. Mater.* **2022**, *32*, 2107242. [CrossRef]
23. Moriyama, M.; Song, S.; Matsuda, H.; Tamaoki, N. Effects of Doped Dialkylazobenzenes on Helical Pitch of Cholesteric Liquid Crystal with Medium Molecular Weight: Utilisation for Full-Colour Image Recording. *J. Mater. Chem.* **2001**, *11*, 1003–1010. [CrossRef]
24. Manabe, T.; Sonoyama, K.; Takanishi, Y.; Ishikawa, K.; Takezoe, H. Toward Practical Application of Cholesteric Liquid Crystals to Tunable Lasers. *J. Mater. Chem.* **2008**, *18*, 3040–3043. [CrossRef]
25. Yamagishi, T.A.; Guittard, F.; Godinho, M.H.; Martins, A.F.; Cambon, A.; Sixou, P. Comparison of Thermal and Cholesteric Mesophase Properties among the Three Kind of Hydroxypropylcellulose (HPC) Derivatives. *Polym. Bull.* **1994**, *32*, 47–54. [CrossRef]
26. Tseng, S.L.; Valente, A.; Gray, D.G. Cholesteric Liquid Crystalline Phases Based on (Acetoxypropyl)cellulose. *Macromolecules* **1981**, *14*, 715–719. [CrossRef]
27. Kosho, H.; Hiramatsu, S.; Nishi, T.; Tanaka, Y.; Kawauchi, S.; Watanabe, J. Thermotropic Cholesteric Liquid Crystals in Ester Derivatives of Hydroxypropylcellulose. *High Perform. Polym.* **1999**, *11*, 41–48. [CrossRef]
28. Hou, H.; Reuning, A.; Wendorff, J.H.; Greiner, A. Tuning of the Pitch Height of Thermotropic Cellulose Esters. *Macromol. Chem. Phys.* **2000**, *201*, 2050–2054. [CrossRef]
29. Baba, Y.; Saito, S.; Iwata, N.; Furumi, S. Synthesis and Optical Properties of Completely Etherified Hydroxypropyl Cellulose Derivatives. *J. Photopolym. Sci. Technol.* **2021**, *34*, 549–554. [CrossRef]
30. Matsumoto, K.; Ogiwara, Y.; Iwata, N.; Furumi, S. Rheological Properties of Cholesteric Liquid Crystal with Visible Reflection from an Etherified Hydroxypropyl Cellulose Derivative. *Polymers* **2022**, *14*, 2059. [CrossRef]

31. Nishio, Y.; Nada, T.; Hirata, T.; Fujita, S.; Sugimura, K.; Kamitakahara, H. Handedness Inversion in Chiral Nematic (Ethyl)cellulose Solutions: Effects of Substituents and Temperature. *Macromolecules* **2021**, *54*, 6014–6027. [CrossRef]
32. Ogiwara, Y.; Iwata, N.; Furumi, S. Viscoelastic Properties of Cholesteric Liquid Crystals from Hydroxypropyl Cellulose Derivatives. *J. Photopolym. Sci. Technol.* **2021**, *34*, 537–542. [CrossRef]
33. Ogiwara, Y.; Iwata, N.; Furumi, S. Dominant Factors Affecting Rheological Properties of Cellulose Derivatives Forming Thermotropic Cholesteric Liquid Crystals with Visible Reflection. *Int. J. Mol. Sci.* **2023**, *24*, 4269. [CrossRef] [PubMed]
34. Ishizaki, T.; Uenuma, S.; Furumi, S. Thermotropic Properties of Cholesteric Liquid Crystal from Hydroxypropyl Cellulose Mixed Esters. *Kobunshi Ronbunshu* **2015**, *72*, 737–745. [CrossRef]

Disclaimer/Publisher's Note: The statements, opinions and data contained in all publications are solely those of the individual author(s) and contributor(s) and not of MDPI and/or the editor(s). MDPI and/or the editor(s) disclaim responsibility for any injury to people or property resulting from any ideas, methods, instructions or products referred to in the content.

Article

Fluid Mechanics of Droplet Spreading of Chitosan/PVA-Based Spray Coating Solution on Banana Peels with Different Wettability

Endarto Yudo Wardhono [1,*], Nufus Kanani [1], Mekro Permana Pinem [2], Dwinanto Sukamto [2], Yenny Meliana [3], Khashayar Saleh [4] and Erwann Guénin [4]

1 Chemical Engineering, University of Sultan Ageng Tirtayasa, Cilegon 42435, Indonesia; nufus.kanani@untirta.ac.id
2 Mechanical Engineering, University of Sultan Ageng Tirtayasa, Cilegon 42435, Indonesia; mekro_pinem@untirta.ac.id (M.P.P.); dwinanto@untirta.ac.id (D.S.)
3 Research Center for Chemistry, National Research and Innovation Agency, BRIN, Kawasan Puspiptek, Serpong, South Tangerang 15314, Banten, Indonesia; yenn005@brin.go.id
4 Université de Technologie de Compiègne, ESCOM, TIMR (Integrated Transformations of Renewable Matter), Centre de Recherche Royallieu, CS 60 319, 60 203 Compiègne CEDEX, France; khashayar.saleh@utc.fr (K.S.); erwann.guenin@utc.fr (E.G.)
* Correspondence: endarto.wardhono@untirta.ac.id

Citation: Wardhono, E.Y.; Kanani, N.; Pinem, M.P.; Sukamto, D.; Meliana, Y.; Saleh, K.; Guénin, E. Fluid Mechanics of Droplet Spreading of Chitosan/PVA-Based Spray Coating Solution on Banana Peels with Different Wettability. *Polymers* 2023, *15*, 4277. https://doi.org/10.3390/polym15214277

Academic Editor: Cristina Cazan

Received: 2 October 2023
Revised: 25 October 2023
Accepted: 27 October 2023
Published: 31 October 2023

Copyright: © 2023 by the authors. Licensee MDPI, Basel, Switzerland. This article is an open access article distributed under the terms and conditions of the Creative Commons Attribution (CC BY) license (https://creativecommons.org/licenses/by/4.0/).

Abstract: The spreading behavior of a coating solution is an important factor in determining the effectiveness of spraying applications. It determines how evenly the droplets spread on the substrate surface and how quickly they form a uniform film. Fluid mechanics principles govern it, including surface tension, viscosity, and the interaction between the liquid and the solid surface. In our previous work, chitosan (CS) film properties were successfully modified by blending with polyvinyl alcohol (PVA). It was shown that the mechanical strength of the composite film was significantly improved compared to the virgin CS. Here we propose to study the spreading behavior of CS/PVA solution on fresh bananas. The events upon droplet impact were captured using a high-speed camera, allowing the identification of outcomes as a function of velocity at different surface wettabilities (wetting and non-wetting) on the banana peels. The mathematical model to predict the maximum spreading factor, β_{max}, was governed by scaling law analysis using fitting experimental data to identify patterns, trends, and relationships between β_{max} and the independent variables, Weber (We) numbers, and Reynolds (Re) numbers. The results indicate that liquid viscosity and surface properties affect the droplet's impact and spreading behavior. The Ohnesorge (Oh) numbers significantly influenced the spreading dynamics, while the banana's surface wettability minimally influenced spreading. The prediction model reasonably agrees with all the data in the literature since the $R^2 = 0.958$ is a powerful goodness-of-fit indicator for predicting the spreading factor. It scaled with $\beta_{max} = a + 0.04(We.Re)^{1/3}$, where the "a" constants depend on Oh numbers.

Keywords: CS/PVA solution; banana peels; spreading behavior; β_{max}; scaling law analysis

1. Introduction

In recent years, developing bio films and coatings that protect fresh foods while maintaining their quality has been a crucial area of research and innovation. In this regard, bio-based polymers are an excellent solution to the challenges posed by synthetic polymers [1], and they can be derived from renewable resources such as plant-based feed stocks, agricultural waste, or algae [2]. Chitosan (CS) is a versatile biopolymer [3,4] that can be potentially applied as a preservative coating because it has an excellent film-forming ability [5]. Nevertheless, poor mechanical and gas barrier properties restrict its potential for widespread use. Blending CS with biodegradable synthetic polymers is one method to modify its characteristics and enhance flexibility [6]. To fully understand and harness its potential as a coating material, delving into the fluid mechanics and rheological properties of the chitosan-based solution is important.

Fluid mechanics plays a vital role in food engineering, especially when understanding or manipulating the characteristics of liquid droplets in processes involving spraying for coating applications [6]. Spray technology is widely used for various purposes, such as coating foods with flavorings, colorings, preservatives, or protective films. It allows for a controlled distribution of substances onto the food surface and provides a desired functional layer [7]. A spray is a dynamic collection of liquid droplets dispersed in a gas medium, usually created by fragmenting bulk liquid into smaller droplets [8]. Different devices can produce the spray, such as pressure nozzles, ultrasonic atomizers, or air across a liquid's surface. The resulting droplet size and spray pattern can impact process efficiency, which is controlled by the microscopic properties of a single droplet [9]. Spreading, rebounding, splashing, and penetration are physical phenomena that might occur as a droplet of liquid impinges on a solid surface, depending on the fluid properties, impact conditions, wettability, and roughness of a surface [10]. The Weber (We) number We (inertia or surface tension forces), Ohnesorge (Oh) number (viscous forces and surface tension forces), and Reynolds (Re) number Re (inertia or viscous forces) are dimensionless parameters used to quantify different aspects of these phenomena [11].

The efficiency of the coating, particularly in terms of film thickness and barrier qualities, is greatly influenced by the spreading behavior of the liquid coating, which determines how evenly it spreads on the substrate surface and how quickly it forms a uniform film. In fresh food products, cuticle and epicuticular waxes act as a substantial barrier to wetting on the solid surface, causing droplets to bead up, bounce, or partially splash rather than spreading out and wetting the surface [12]. These may reduce the efficacy of the coating and render spray applications ineffective. Thus, controlling the surface characteristics to get uniform and continuous coating is important since it affects the surface's ability to repel or absorb liquid. Several forces come into play, in which the competition between spreading and viscous forces is crucial in determining droplet dynamics [13]. Spreading is driven by the force that arises from the droplet's attempt to minimize its surface energy and causes the droplet to flatten and increase its contact area with the solid surface. On the other hand, viscous forces oppose the spreading process, which tends to maintain the droplet's structure and resist deformation [14]. If the spreading force is strong enough, the droplet will spread out, creating a thin film on the solid surface. The larger viscous forces might prevent the droplet from spreading completely. In this case, the droplet may keep a more spherical form, with a restricted contact area with the surface. The balance between the spreading and viscous forces determines the droplet's final form and behavior. By controlling these forces, the processes involving droplet deposition, wetting, and coating may be manipulated [15]. However, it is important to note that other factors, such as surface roughness, gravity, and external flows, can also influence droplet dynamics.

The idea behind this work is to investigate the spreading behavior of a CS/PVA solution on the surface coating of an organic substrate (lady finger bananas) with different wettability in order to compare the adherence of the formulation to the fruit if the surface is adequately washed. Bananas are a highly perishable fruit, and improper handling after harvest can result in rapid deterioration, loss of quality, and reduced shelf life. Postharvest management techniques like coating can enhance shelf life by creating a protective barrier around the product [16]. Spray coating of fresh food requires a combination of technical knowledge, careful planning, and adherence to the products that ensures they are coated effectively while maintaining their quality and safety. Upon impact, cuticle and epicuticular waxes act as a substantial barrier to wetting on fruit peels, causing droplets to bead up, bounce, or partially splash rather than spreading out and wetting the surface [12].

In our previous work, we studied CS/polyvinyl alcohol (PVA) composite films fabricated using the solution casting technique with enhanced properties [17]. Here, to study the fluid mechanics of liquid droplet impacts on two banana surfaces, the evolution of the spreading droplets was observed as a function of the impact velocity and the wettability of the banana peels. The surface's wettability was controlled by washing or not washing the fruit with tap water. The velocity of the droplets was estimated by measuring the distance between the needle tip and the peels. The viscosity and surface tension effects were determined by comparing the liquid droplet properties of the CS/PVA coating solution and water as a reference. A mathematical model is proposed by fitting experimental data regarding the spreading factor, which involves finding a statistical relationship that describes the data accurately.

2. Materials and Methods

2.1. Plant Substrates

Lady finger bananas were collected from a local market (Cilegon, Indonesia) and kept in a refrigerator at 15 °C before use. Prior to testing, the fruits were left at room temperature and washed or not washed with running tap water. The samples were peeled and divided into two treatments: unwashed (non-wetting) and washed (wetting surface). The peels were subsequently trimmed into a 4 cm × 4 cm rectangle and adhered to a nine cm diameter Petri dish.

2.2. Liquid Droplets

The coating solution for liquid droplets consisted of CS/PVA blends and deionized water (referred to as water). The methodology for preparing the CS/PVA solution has been described in our prior research [17], in which CS solution (1% w/v in 0.1 M of acetic acid) was mixed with PVA at the optimum ratio of CS/PVA (75/25) together with glycerol 10% w/w and succinic acid 5% w/w in a dry-basis CS/PVA blend.

2.3. Impact Measurement

The experimental setup of the impact measurement is described in detail in our prior work [18], and the outline sketch of the experiment is shown in a diagram in Figure 1.

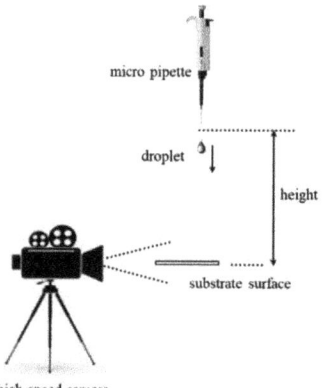

Figure 1. Experimental setup of impact measurement.

Experiments were conducted at various vertical velocities. A droplet was generated by pushing liquid through a micropipette, and it was released from a predetermined vertical height, in which the velocity of the droplet (v_0) was determined by measuring the distance (h) between the tip and the substrate surface, $v_0 = \sqrt{2gh}$. The droplet-impacting process was recorded using a high-speed camera. The acquisition rate was set to 2000 frames per second (fps), and the shutter speed was adjusted to 1/2000 s.

2.4. Impact Measurement

Liquid density at room temperature was measured using the pycnometric method using American Society for Testing and Materials (ASTM D854) [19]. Surface tension of the samples was evaluated at room temperature using a Kruss tensiometer and a Wilhelmy plate (KRUSS GmbH, Hamburg, Germany). The viscosity measurements of the liquid solution were conducted at 25 °C with a Physica MCR 301 rheometer (Anton Par GmbH, Graz, Austria), using concentric cylinder measuring system according to DIN 53019. The surface structure of the banana peel was observed using a VHX-5000 digital microscope (Olympus, Tokyo, Japan). The wettability of the organic substrate was evaluated via the contact angle measurement using a Drop shape Analyzer (DSA 100; KRUSS GmbH, Hamburg, Germany). ImageJ software NIH or equivalent was used to measure the droplet data, including initial diameter (D_0), droplet spreading diameter $D(t)$, maximum diameter (D_{max}), and droplet height (h_D).

3. Results and Discussion

3.1. Substrate Properties

The surface properties of the substrates were characterized using image processing techniques to observe the roughness and irregularities on a surface, while the wettability was determined by measuring mean contact angles and calculating Gibbs Adsorption Energy. A visual comparison of the two banana peels is illustrated in Figure 2.

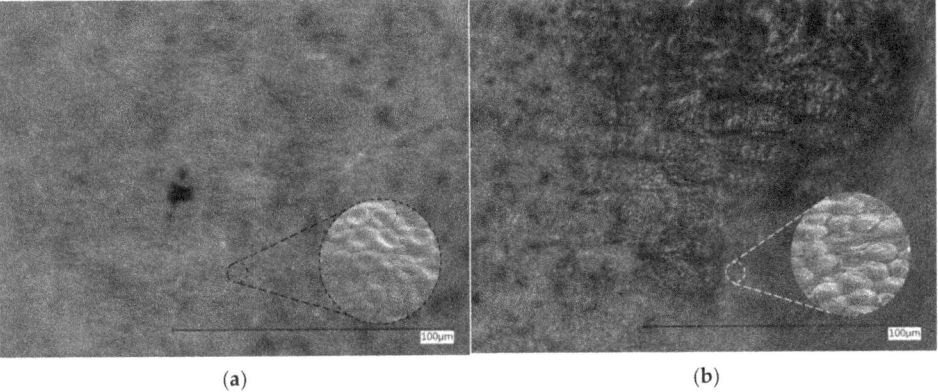

Figure 2. Surface morphology (2000× magnification) of the banana peels: (**a**) unwashed (non-wetting); (**b**) washed (wetting) surfaces.

The images reveal that the unwashed texture of the banana's peel is smoother than the washed one. It exhibits a thin layer covering the surface, and some fractures and wrinkled structures were observed in the fruit that had been washed. The sessile drop method was used to determine the mean contact angle (CA), and the measurements were carried out three times for each sample to ensure reproducibility. The Gibbs adsorption energy (ΔG_{ads}) was calculated using the contact angle data and the Young–Laplace equation [20]. The results are presented in Table 1.

Table 1. Wettability determination on the banana surface for the two liquids.

Liquid	Banana Peels			
	Washed Surface		Unwashed Surface	
	CA (°)	ΔG (kJ/mole)	CA (°)	ΔG (kJ/mole)
Water	81.4 ± 1.8	−0.21	108.1 ± 2.1	0.44
CS/PVA	63.8 ± 2.5	−0.64	98.4 ± 2.4	0.23

Liquid water and CS/PVA demonstrate wetting behavior on the washed banana peel. CS/PVA exhibits better wettability with CA~63.8° than water CA~81.4°. Furthermore, both liquids display non-wetting characteristics on unwashed peel, where water has a higher contact angle with CA~108.1° than CS/PVA CA~98.4°. The same results are shown with ΔG_{ads} calculation, representing a thermodynamic quantity of molecule adsorption on the solid surface. Water has higher energies on both surfaces, -0.21 and 0.44 kJ/mole, compared to CS/PVA, which is -0.64 and 0.23 kJ/mole, respectively, for the washed and unwashed surfaces. A higher ΔG_{ads} might decrease the contact angle, suggesting that the liquid is better at wetting the surface due to stronger interactions with the surface molecules. On the other hand, for some systems, a higher ΔG_{ads} might increase the contact angle, indicating reduced wettability due to a higher degree of surface coverage by the adsorbed molecules onto a solid [21].

3.2. Droplet Properties and Impact Conditions

Physicochemical properties (density, ρ; viscosity, μ; and surface tension, σ) and the impact conditions (diameter, D_0 and height, h) are listed in Table 2. Using water as the reference, the density of both liquids is quite similar, and the surface tension of CS/PVA is lower than water. The apparent viscosity of CS/PVA solution is approximately 12 times higher than water, which exhibits a shear-thinning (pseudo-plastic) flow behavior.

Table 2. Liquid properties at room temperature and droplet impact conditions.

Liquids	ρ g/cm^3	μ cps	σ mN/m	D_0 mm	h cm	$Oh = \frac{\mu}{\sqrt{\rho v D_0}}$	$W_e = \frac{\rho D_0 v_0^2}{\sigma}$
Water	0.998	1.00	72.00	3.00 ± 0.05	5–45	0.002	40–400
CS/PVA	1.125	12.25	51.62	2.85 ± 0.05	5–45	0.030	60–600

It is characteristic of a non-Newtonian fluid for the viscosity to decrease with increased shear rate. At the same time, the water has a constant viscosity independent of the applied shear rate characteristic of a Newtonian fluid (see Figure 3a,b).

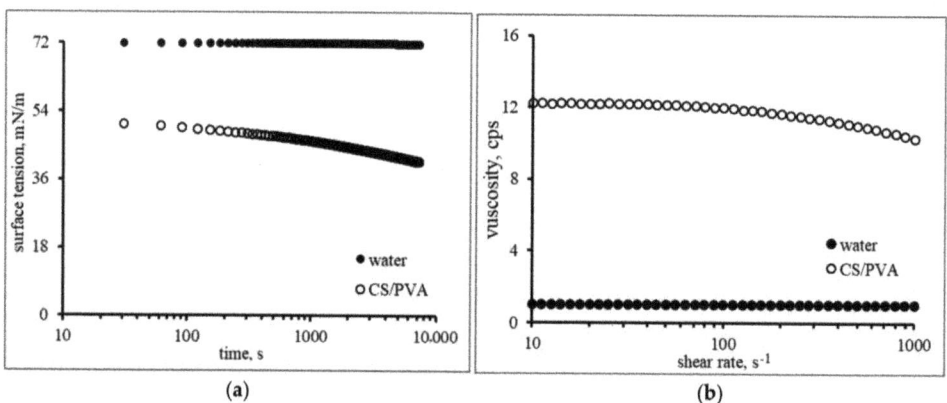

Figure 3. Liquid properties: (a) surface tension; (b) and viscosity at room temperature.

For the impact parameters, the mean initial droplet diameter is $D_0 = 3.00$ mm for liquid water and $D_0 = 2.85$ mm for CS/PVA solution. Experiments were conducted at various heights ranging from 5 to 45 cm, with an impact velocity of $1 < v_0 < 3$ m/s. The impact dynamics characteristics acquired are $Oh = 0.002$ ($40 < We < 400$) for water and $Oh = 0.030$ ($60 < We < 600$) for CS/PVA. Both liquid droplets show a low Oh number ($Oh < 1$), indicating that viscous forces dominate, and surface tension effects are less significant [22].

The maximum We number for all tests is 600, which means that the impact is sufficiently low not to induce splashing [23].

3.3. Spreading Behavior

3.3.1. Spreading on the Wetting Surface

The time series of droplet impacts and subsequent spreading stages on the wetting surface for both liquids with Oh = 0.002 and Oh = 0.030 are shown in Figure 4a.

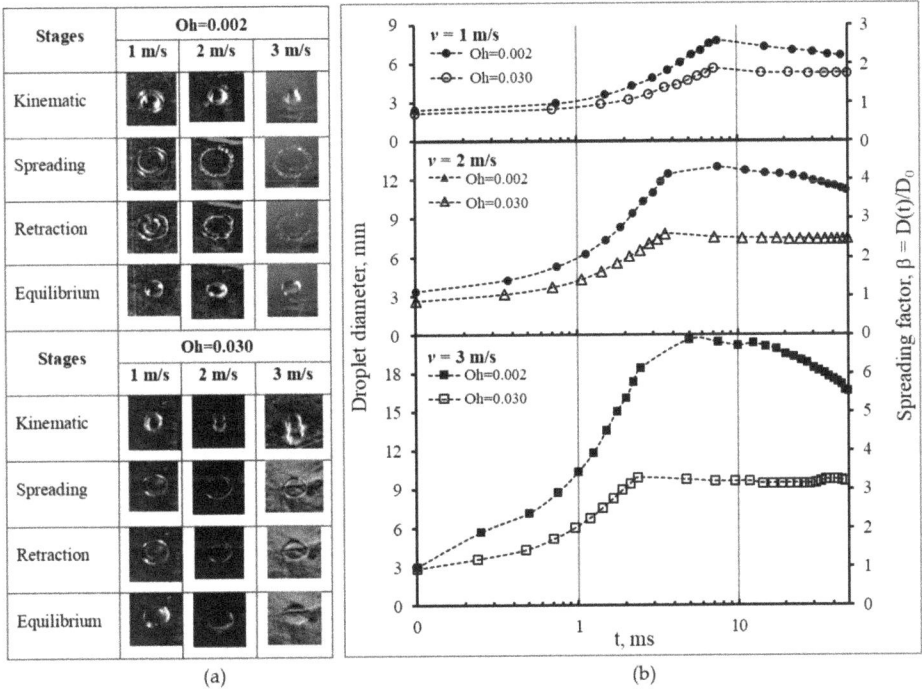

Figure 4. Evolution of droplet diameter at different impact velocities on the wetting surface: (**a**) images of spreading stages; (**b**) spreading pattern.

The impact characteristics at 1, 2, and 3 m/s were recorded by taking snapshots with the high-speed camera. Theoretically, there are four stages of the spreading following impact: kinematic, spreading, retraction, and equilibrium [24]. In general, all droplets are spherical in shape during the kinematic stage, represented by $\beta < 1$ [25]. Upon impact, the shape changes, resulting in a sudden halt in its vertical motion, causing the kinetic energy to be distributed across the liquid. The droplet continues to spread in the next stage. It spreads radially across the solid surface, driven by the remaining kinetic energy. The droplet shape becomes flatter, and its contact diameter on the surface increases and reaches maximum spreading ($1 < \beta < \beta_{max}$) [26]. As the maximum value is attained, the droplet cannot spread further. The droplets show a lamella, or a pancake form, surrounded by a periphery. Surface tension tries to minimize the surface area of the liquid droplet in the retraction stage, causing it to recoil, leading to a continuous reduction in droplet diameter and its movement back to the impact point [27]. Deposition occurs during this stage. The droplets stay in this form and reach an equilibrium stage. Figure 4b shows the spreading pattern for two substances at 1, 2, and 3 m/s. It is observed that the droplet with Oh = 0.002 spreads faster than the droplet with Oh = 0.030 at all three impact speeds. The spreading behavior was significantly altered by the impact velocity, whereby D_{max} increased with

v_0, and the rate of receding decreased. The droplet with Oh = 0.002 reached its maximum value later than the one with Oh = 0.030, and these results are summarized in Table 3.

Table 3. Maximum value of droplet spreading on the wetting surface.

Oh	v = 1 m/s			v = 2 m/s			v = 3 m/s		
	t	D_{max}	β_{max}	t	D_{max}	β_{max}	t	D_{max}	β_{max}
	ms	mm	-	ms	Mm	-	ms	mm	-
0.002	7.50	7.80	2.60	7.50	12.96	4.32	5.00	20.61	6.87
0.030	7.13	5.01	1.97	3.56	7.81	2.74	2.38	9.89	3.47

The droplets with Oh = 0.030 have smaller D_{max} of 5.01, 7.81, and 9.89 mm than the droplets with Oh = 0.002, in which D_{max} are 7.80, 12.96, and 20.61 mm at each impact speed. It can be attributed to the higher viscosity of the liquid with Oh = 0.03. As a shear-thickening fluid, the liquid acts almost like a solid when subject to rapid deformation. In this case, kinetic energy dissipates quickly upon impact, almost instantaneously converting the energy into heat or becoming stored as potential energy within the fluid structure. Internal friction limits the spread of the droplet. On the other hand, the water droplet with lower viscosity will spread out more upon impact due to a lower energy dissipation rate [28].

3.3.2. Spreading on the Non-Wetting Surface

A visual observation of the spreading from both droplets on the non-wetting surface is presented in Figure 5a.

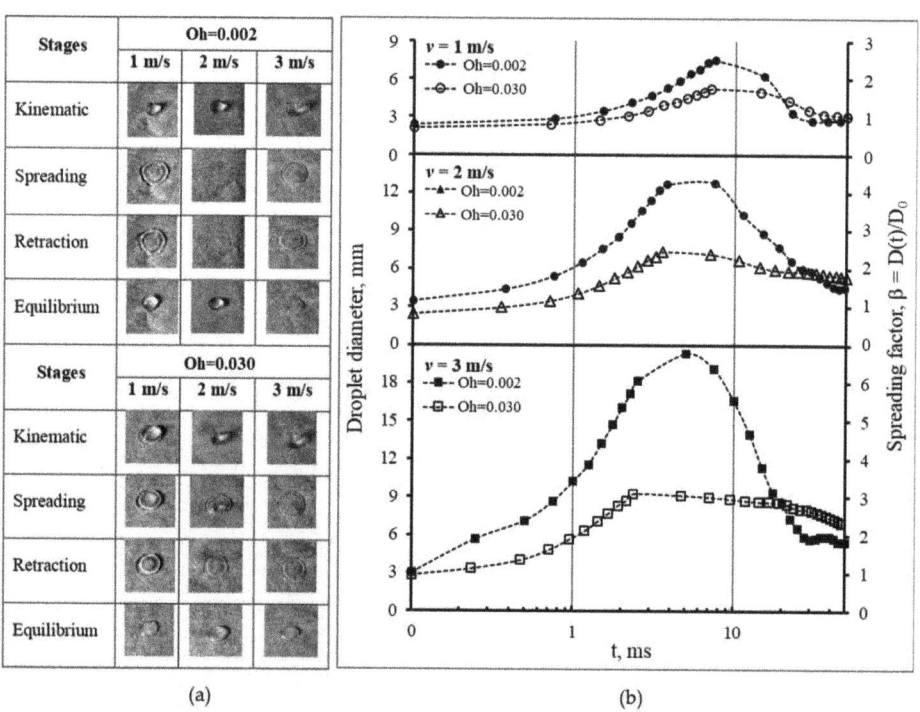

Figure 5. Evolution of droplet diameter at different impact velocities on the non-wetting surface: (a) images of spreading stages; (b) spreading pattern.

The captured images are at three impact speeds of 1, 2, and 3 m/s. All liquid droplets remain spherical at the kinematic stage. They rapidly create a thin film at the lower side while the upper side deforms into a semi-spherical form at the surface interface and no spreading lamella has been yet formed. Following the impact, the droplets expand radially, forming a flat film surrounded by a thick edge with maximum spreading. Subsequently, the edge undergoes contraction and thickening, ultimately merging with its inner boundary. The droplet with the lower Oh = 0.002 backs up to form a rounded shape, while the droplet with the higher Oh = 0.030 stays as a flat liquid layer. The droplets finally reach the equilibrium shape, with a diameter smaller than the D_{max}. Figure 5b displays the spreading pattern of liquid droplets at 1, 2, and 3 m/s speed impacts. The maximum value of the droplet spreading during the spreading stage is summarized in Table 4. The figure shows that the droplets spread, as a parabolic curve, up to 50 ms. For the droplets with the lower Oh number = 0.002, the β_{max} obtained are 2.51, 4.27, and 6.78 for each speed. With the increase of speed, the β becomes steeper, indicating faster spreading and higher contact line velocity, whereas until maximum spreading, the β_{max} for the droplets with the higher Oh = 0.030 are 1.82, 2.58, and 3.25, which tend to decrease constantly for all speed variations.

Table 4. Maximum value of droplet spreading at hydrophilic surface.

Oh	$v = 1$ m/s			$v = 2$ m/s			$v = 3$ m/s		
	t	D_{max}	β_{max}	t	D_{max}	β_{max}	t	D_{max}	β_{max}
	ms	mm	-	ms	mm	-	Ms	mm	-
0.002	7.50	7.53	2.51	7.50	12.81	4.27	5.00	20.34	6.78
0.030	7.13	5.19	1.82	3.56	7.35	2.58	2.38	9.26	3.25

The spreading phenomena of liquid droplets on the non-wetting banana show the same characteristics as on the wetting surface. The velocity affects the spreading in the non-wetting surface, in which v_0 improved D_{max} and decreased the rate of receding significantly. The spreading times of the liquid droplets with Oh number = 0.030 are shorter (7.13, 3.56, and 2.38 ms) than the ones with Oh number = 0.002 (7.5, 7.5, and 5.0 ms) for each impact velocity. Shear-thinning non-Newtonian fluids of CS/PVA exhibit lower viscosity, leading to faster spreading on unwashed surfaces.

3.4. Maximum Spreading Factor

The maximum spreading factor, β_{max}, is a parameter used in the study of spreading phenomena, which represents the ratio of the largest lamella diameter, D_{max}, over the initial one, D_0. It quantifies how much a liquid droplet spreads out when it impacts a solid surface, which generally implies better surface coverage efficiency and therefore decrease in material consumption and reduction of waste. In this work, the β_{max} in a particular range of We numbers, (40 < We < 400) for Oh = 0.002 and (60 < We < 600) for Oh = 0.030, on both hydrophilic and hydrophobic surfaces is presented in Figure 6. The β_{max} is represented as a logarithmic function of We.

For all test cases, β_{max} values are distributed as a straight-line pattern with increasing We, suggesting that inertial forces become more dominant than surface tension. Inertial forces represent the kinetic energy associated with the liquid's motion. On the other hand, surface tension is related to the cohesive forces at the liquid's interface. It is in accordance with previous studies showing that inertia regulates how a material spreads onto a surface [26,29]. A higher Weber number signifies increased kinetic energy, resulting in a more significant perturbation of the droplet, leading to more energetic spreading behavior [30]. A lower Oh = 0.002 shows a higher β_{max}. Increased viscosity leads to greater viscous friction forces in the near-wall boundary of the liquid layer. These prevent its spreading over the particle surface [31]. Two different banana surfaces have been examined, and the β_{max} data for these surfaces are nearly identical or very similar, showing that wettability minimally influences maximum spreading.

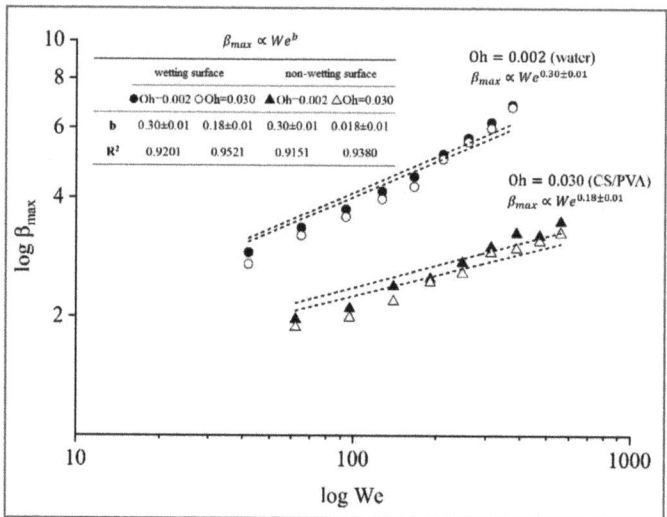

Figure 6. Maximum spreading factor, β_{max}, as a function of We number on different surfaces.

The prediction model obtained results that were in good agreement with the experimental results $\beta_{max} \propto We^b$ proposed by Clanet et al. [32], which predicted the maximum spreading for impacts on super-hydrophobic materials with a static contact angle over 150° by optimizing both surfaces, namely $\beta_{max} \propto We^{0.30\pm0.01}$ for Oh = 0.002 and $\beta_{max} \propto We^{0.18\pm0.01}$ for Oh = 0.030. To clarify the difference between the two exponents, the relationship β_{max} is dependent on the surface tension and viscosity characteristics.

3.5. Mathematical Model of Spreading

Generally, the three primary methods for predicting the β_{max} are scaling law analysis, the energy balance approach, and numerical simulation. These models are suitable for understanding the underlying physics phenomena in the spray process. They help to design process parameters to achieve desired coating thickness, distribution, and coverage. The scaling law describes how specific properties or behaviors change as a function of size or scale. It can be classified into two main categories based on the variables they use to express β_{max}: (1) Allometric Scaling Models, to express β_{max} as a power-law function of size or scale of the system, such as We, Re, and θ, where the variable θ represents either the equilibrium contact angle or the advancing contact angle, with the latter having the potential to be dynamic or static, (2) and Isometric Scaling Models, to express β_{max} as a linear function of size, without any power-law exponent [33]. Several empirical investigations have been conducted to elucidate the dynamics of spreading phenomena [11,13,34–37]. Different models have been proposed for the prediction of β_{max}. Scheller et al. [38] proposed an equation considering the correlation between the maximum spreading diameter with both Re and Oh that included two empirical coefficients, A and α. Tang et al. [26] conducted empirical investigations to determine various coefficients for five distinct surface values using the same parameters of scaling law. A similar method was utilized by Sen et al. [39] to empirically simulate the β_{max} of biofuel droplets on a stainless steel substrate. At the same time, Roisman et al. [40] proposed a semi-empirical equation that approximates the Navier–Stokes equations. This study presents a mathematical model for predicting the βmax at different Oh numbers on an organic surface using scaling law analysis. The model developed is based on the correlations found in experimental data. Figure 7 shows a non-linear regression model to fit the data into the power-law correlation. The obtained models are well fitted using correlating experimental results as a function of We and Re numbers. The fitting parameters and statistical factors, R-squared (R^2) at different Oh numbers, are indicated in Table 5.

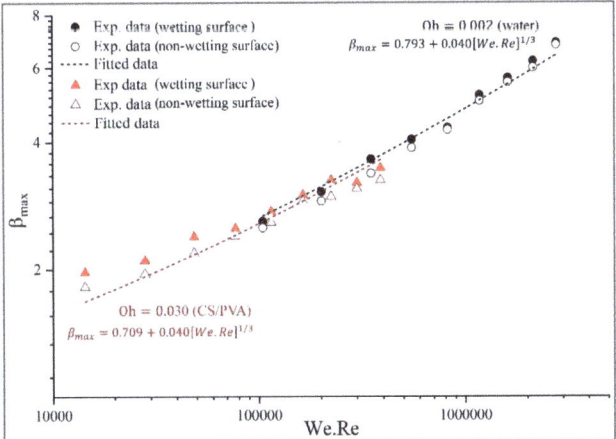

Figure 7. Reliability of model prediction for β_{max} on different wettabilities of the banana surface.

Table 5. Fitting parameters of the β_{max} on the solid surface.

Oh	Fitting Parameters			
	a	b	c	R^2
0.002	0.793	0.040	1/3	0.9623
0.030	0.709	0.040	1/3	0.9539

A comparison between previous published β_{max} data on difference surfaces and our empirical model is shown in Figure 8. The data were collected from Scheller et al. (1995) [38], Roisman et al. (2009) [40], Andrade et al. (2012) [41], Sen et al. (2014) [39], and Tang et al. (2017) [26]. The statistical values of each model are displayed in Table 6.

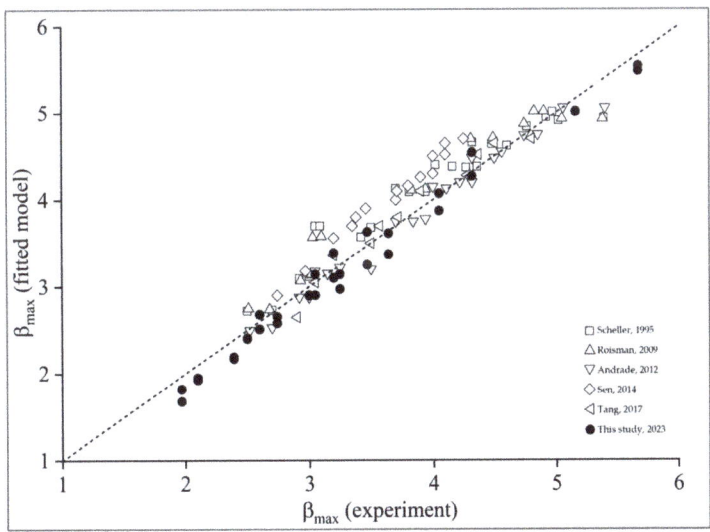

Figure 8. Comparison of model predictions and experimental data of β_{max} [26,38–41].

Table 6. Statistical values of the different models.

Literature	Fitted Model	R^2
Scheller, 1995 [38]	$\beta_{max} = 0.61\left(Re^2 Oh\right)^{0.166}$	0.941
Roisman, 2009 [40]	$\beta_{max} = 0.87 Re^{0.2} - 0.4\left(Re^{0.4}/\sqrt{We}\right)$	0.933
Andrade, 2012 [41]	$\beta_{max} = 1.28 + 0.071 We^{0.25} Re^{0.25}$	0.972
Sen, 2014 [39]	$\beta_{max} = 1.73 We^{0.14}$	0.900
Tang, 2017 [26]	$\beta_{max} = a(We/Oh)^b$, a & b depend on roughness	0.966
This study, 2023	$\beta_{max} = a + 0.04(We.Re)^{1/3}$, "a" depend on Oh	0.954

The best-fit model found by fitting experimental data is suggested:
For lower Oh (0.002):

$$\beta_{max} = 0.793 + 0.04(We.Re)^{1/3} \quad (1)$$

For higher Oh (0.030):

$$\beta_{max} = 0.709 + 0.04(We.Re)^{1/3} \quad (2)$$

The prediction model reasonably agrees with all the data in the literature. R-squared is greater than 0.958, a powerful goodness-of-fit indicator for predicting the maximum spreading factor. It scaled with $\beta_{max} = a + 0.04(We.Re)^{1/3}$, where the constants "a" depend on Oh numbers.

4. Conclusions

This work studied the droplet spreading behavior of liquid CS/PVA blends and water as a reference on fresh banana surfaces with different surface wettabilities. The Oh number of liquid droplets (0.002 for water and 0.030 for CS/PVA) is less than 1, indicating that viscous forces dominate, and surface tension effects are less significant. The We number for all tests is up to 600, meaning that the impact is sufficiently low not to allow splashing. The liquid viscosity and surface properties affect the droplets' impact behavior. Upon impact, water and CS/PVA droplets spread radially outwards on wetting surfaces, form a lamella surrounded by a periphery, and reach a maximum diameter. The surface tension causes the droplets to recoil, reduce the diameter, return to the impact point, stay in this form, and reach equilibrium. On the non-wetting surface, both liquids expand as a flat film surrounded by a thick edge and get the maximum spreading. Subsequently, the edge undergoes contraction and thickening, ultimately merging with its inner boundary. The water droplet backs up to form a rounded shape, while CS/PVA stays as a flat liquid layer. The droplets finally reach an equilibrium shape with a smaller diameter than the D_{max}. The spreading factor β, which is a function of impact velocity, demonstrates the primary role of surface tension and viscosity. The Oh numbers significantly influence the spreading dynamics. The β_{max} data for two different banana surfaces are nearly identical or very similar, indicating the banana's surface wettability minimally influences the maximum spreading. The prediction model reasonably agrees with all the data from the literature, since $R^2 = 0.958$ is a powerful goodness-of-fit indicator for predicting the maximum spreading factor. It scaled with $\beta_{max} = a + 0.04(We.Re)^{1/3}$, where the constants "a" depend on Oh numbers.

Author Contributions: E.Y.W. organized and wrote the manuscripts, proposed the subject of the review to all the other authors; N.K. and M.P.P. organized materials and methods sections; D.S. and Y.M. prepared the figures, tables and organized references; K.S. took care about results and discussions; E.G. wrote the abstract, the conclusions and supervised all the work and Finally, all the authors contributed equally to the general organization of the manuscript and its revision, with helpful suggestions about the content and the style of the text. All authors have read and agreed to the published version of the manuscript.

Funding: This work is fully funded by KEMENTERIAN PENDIDIKAN, KEBUDAYAAN, RISET, DAN TEKNOLOGI, REPUBLIK INDONESIA—DIREKTORAT JENDERAL PENDIDIKAN TINGGI, RISET, DAN TEKNOLOGI, with the contract number: 060/E5/PG.02.00.PL/2023. The authors thank Universitas Sultan Agung Tirtayasa (UNTIRTA) within the framework with Integrated Transformations of Renewable Matter Laboratory (EA TIMR 4297 UTC-ESCOM), Université de Technologie de Compiègne. Special thanks to Setyo Hanandiko, the photographer. Thank you for your talent, passion, and skill.

Conflicts of Interest: The authors declare no conflict of interest. Moreover, the funding sponsors had no role in the design of the study; in the collection, analyses, or interpretation of data; in the writing of the manuscript, or in the decision to publish the results.

References

1. Petkoska, A.T.; Daniloski, D.; D'Cunha, N.M.; Naumovski, N.; Broach, A.T. Edible packaging: Sustainable solutions and novel trends in food packaging. *Food Res. Int.* **2021**, *140*, 109981. [CrossRef]
2. Perera, K.Y.; Jaiswal, A.K.; Jaiswal, S. Biopolymer-Based Sustainable Food Packaging Materials: Challenges, Solutions, and Applications. *Foods* **2023**, *12*, 2422. [CrossRef] [PubMed]
3. Atay, H.Y. Antibacterial Activity of Chitosan-Based Systems. In *Functional Chitosan*; Springer: Berlin/Heidelberg, Germany, 2019; pp. 457–489.
4. Merzendorfer, H.; Cohen, E. Chitin/chitosan: Versatile ecological, industrial, and biomedical applications. In *Extracellular Sugar-Based Biopolymers Matrices*; Springer: Berlin/Heidelberg, Germany, 2019; pp. 541–624.
5. Basumatary, K.; Daimary, P.; Das, S.K.; Thapa, M.; Singh, M.; Mukherjee, A.; Kumar, S. Lagerstroemia speciosa fruit-mediated synthesis of silver nanoparticles and its application as filler in agar based nanocomposite films for antimicrobial food packaging. *Food Packag. Shelf Life* **2018**, *17*, 99–106. [CrossRef]
6. Werner, S.R.; Jones, J.R.; Paterson, A.H.; Archer, R.H.; Pearce, D.L. Droplet impact and spreading: Droplet formulation effects. *Chem. Eng. Sci.* **2007**, *62*, 2336–2345. [CrossRef]
7. Kalantari, D.; Tropea, C. Phase Doppler measurements of spray impact onto rigid walls. *Exp. Fluids* **2007**, *43*, 285–296. [CrossRef]
8. Igor, M.; Aidin, A.; Kyle, M.B. Magnetic Resonance Imaging measurements of a water spray upstream and downstream of a spray nozzle exit orifice. *J. Magn. Reson.* **2016**, *266*, 8–15.
9. Fansler, T.D.; Parrish, S.E. Spray measurement technology: A review. *Meas. Sci. Technol.* **2014**, *26*, 012002. [CrossRef]
10. Che, Z.; Matar, O.K. Impact of droplets on immiscible liquid films. *Soft Matter* **2018**, *14*, 1540–1551. [CrossRef]
11. Aksoy, Y.T.; Eneren, P.; Koos, E.; Vetrano, M.R. Spreading of a droplet impacting on a smooth flat surface: How liquid viscosity influences the maximum spreading time and spreading ratio. *Phys. Fluids* **2022**, *34*, 042106. [CrossRef]
12. Zhang, Y.; Zhang, G.; Han, F. The spreading and superspeading behavior of new glucosamide-based trisiloxane surfactants on hydrophobic foliage. *Colloids Surf. Physicochem. Eng. Asp.* **2006**, *276*, 100–106. [CrossRef]
13. Qin, M.; Tang, C.; Tong, S.; Zhang, P.; Huang, Z. On the role of liquid viscosity in affecting droplet spreading on a smooth solid surface. *Int. J. Multiph. Flow* **2019**, *117*, 53–63. [CrossRef]
14. Bird, J.C.; Mandre, S.; Stone, H.A. Short time dynamics of partial wetting. *Phys. Rev. Lett.* **2008**, *100*, 234501. [CrossRef] [PubMed]
15. Wang, X.; Chen, L.; Bonaccurso, E.; Venzmer, J. Dynamic wetting of hydrophobic polymers by aqueous surfactant and superspreader solutions. *Langmuir* **2013**, *29*, 14855–14864. [CrossRef] [PubMed]
16. Mohapatra, D.; Mishra, S.; Sutar, N. Banana post harvest practices: Current status and future prospects-A review. *Agric. Rev.* **2010**, *31*, 56–62.
17. Wardhono, E.Y.; Pinem, M.P.; Susilo, S.; Siom, B.J.; Sudrajad, A.; Pramono, A.; Meliana, Y.; Guénin, E. Modification of Physio-Mechanical Properties of Chitosan-Based Films via Physical Treatment Approach. *Polymers* **2022**, *14*, 5216. [CrossRef]
18. Pinem, M.P.; Wardhono, E.Y.; Clausse, D.; Saleh, K.; Guénin, E. Droplet behavior of chitosan film-forming solution on the solid surface. *South Afr. J. Chem. Eng.* **2022**, *41*, 26–33. [CrossRef]
19. *ASTM D 854-14*; Standard Test Method for Specific Gravity of Soil Solids. Annual Book of ASTM Standards. American Society for Testing and Materials: Philadelphia, PA, USA, 2014; Volume 04.08. Available online: www.astm.org (accessed on 1 October 2023).
20. Palencia, M. Surface free energy of solids by contact angle measurements. *J. Sci. Technol. Appl.* **2017**, *2*, 84. [CrossRef]
21. Koopal, L.K. Wetting of solid surfaces: Fundamentals and charge effects. *Adv. Colloid Interface Sci.* **2012**, *179*, 29–42. [CrossRef]
22. Marcotte, F.; Zaleski, S. Density contrast matters for drop fragmentation thresholds at low Ohnesorge number. *Phys. Rev. Fluids* **2019**, *4*, 103604. [CrossRef]
23. De Goede, T.; de Bruin, K.; Shahidzadeh, N.; Bonn, D. Droplet splashing on rough surfaces. *Phys. Rev. Fluids* **2021**, *6*, 043604. [CrossRef]
24. Rioboo, R.; Marengo, M.; Tropea, C. Time evolution of liquid drop impact onto solid, dry surfaces. *Exp. Fluids* **2002**, *33*, 112–124. [CrossRef]

25. Lee, J.B.; Laan, N.; de Bruin, K.G.; Skantzaris, G.; Shahidzadeh, N.; Derome, D.; Carmeliet, J.; Bonn, D. Universal rescaling of drop impact on smooth and rough surfaces. *J. Fluid Mech.* **2016**, *786*, R4. [CrossRef]
26. Tang, C.; Qin, M.; Weng, X.; Zhang, X.; Zhang, P.; Li, J.; Huang, Z. Dynamics of droplet impact on solid surface with different roughness. *Int. J. Multiph. Flow* **2017**, *96*, 56–69. [CrossRef]
27. Banks, D.; Ajawara, C.; Sanchez, R.; Surti, H.; Aguilar, G. Effects of liquid and surface characteristics on oscillation behavior of droplets upon impact. *At. Sprays* **2014**, *24*, 895–913. [CrossRef]
28. Bažant, Z.P.; Caner, F.C. Impact comminution of solids due to local kinetic energy of high shear strain rate: I. Continuum theory and turbulence analogy. *J. Mech. Phys. Solids* **2014**, *64*, 223–235. [CrossRef]
29. Bolleddula, D.A.; Berchielli, A.; Aliseda, A. Impact of a heterogeneous liquid droplet on a dry surface: Application to the pharmaceutical industry. *Adv. Colloid Interface Sci.* **2010**, *159*, 144–159. [CrossRef] [PubMed]
30. Wang, J.; Qian, J.; Chen, X.; Li, E.; Chen, Y. Numerical investigation of Weber number and gravity effects on fluid flow and heat transfer of successive droplets impacting liquid film. *Sci. China Technol. Sci.* **2023**, *66*, 548–559. [CrossRef]
31. Islamova, A.; Tkachenko, P.; Shlegel, N.; Kuznetsov, G. Effect of Liquid Properties on the Characteristics of Collisions between Droplets and Solid Particles. *Appl. Sci.* **2022**, *12*, 10747. [CrossRef]
32. Clanet, C.; Béguin, C.; Richard, D.; Quéré, D. Maximal deformation of an impacting drop. *J. Fluid Mech.* **2004**, *517*, 199–208. [CrossRef]
33. Tembely, M.; Vadillo, D.C.; Dolatabadi, A.; Soucemarianadin, A. A machine learning approach for predicting the maximum spreading factor of droplets upon impact on surfaces with various wettabilities. *Processes* **2022**, *10*, 1141. [CrossRef]
34. Laan, N.; de Bruin, K.G.; Bartolo, D.; Josserand, C.; Bonn, D. Maximum diameter of impacting liquid droplets. *Phys. Rev. Appl.* **2014**, *2*, 044018. [CrossRef]
35. Seo, J.; Lee, J.S.; Kim, H.Y.; Yoon, S.S. Empirical model for the maximum spreading diameter of low-viscosity droplets on a dry wall. *Exp. Therm. Fluid Sci.* **2015**, *61*, 121–129. [CrossRef]
36. Lee, J.B.; Derome, D.; Guyer, R.; Carmeliet, J. Modeling the maximum spreading of liquid droplets impacting wetting and nonwetting surfaces. *Langmuir* **2016**, *32*, 1299–1308. [CrossRef]
37. Wildeman, S.; Visser, C.W.; Sun, C.; Lohse, D. On the spreading of impacting drops. *J. Fluid Mech.* **2016**, *805*, 636–655. [CrossRef]
38. Scheller, B.L.; Bousfield, D.W. Newtonian drop impact with a solid surface. *AIChE J.* **1995**, *41*, 1357–1367. [CrossRef]
39. Sen, S.; Vaikuntanathan, V.; Sivakumar, D. Experimental investigation of biofuel drop impact on stainless steel surface. *Exp. Therm. Fluid Sci.* **2014**, *54*, 38–46. [CrossRef]
40. Roisman, I.V. Inertia dominated drop collisions. II. An analytical solution of the Navier–Stokes equations for a spreading viscous film. *Phys. Fluids* **2009**, *21*, 052104. [CrossRef]
41. Andrade, R.; Skurtys, O.; Osorio, F. Experimental study of drop impacts and spreading on epicarps: Effect of fluid properties. *J. Food Eng.* **2012**, *109*, 430–437. [CrossRef]

Disclaimer/Publisher's Note: The statements, opinions and data contained in all publications are solely those of the individual author(s) and contributor(s) and not of MDPI and/or the editor(s). MDPI and/or the editor(s) disclaim responsibility for any injury to people or property resulting from any ideas, methods, instructions or products referred to in the content.

Article

Native Potato Starch and Tara Gum as Polymeric Matrices to Obtain Iron-Loaded Microcapsules from Ovine and Bovine Erythrocytes

Carlos A. Ligarda-Samanez [1,2,*], Elibet Moscoso-Moscoso [1,2,*], David Choque-Quispe [1,2], Betsy S. Ramos-Pacheco [1,2], José C. Arévalo-Quijano [3], Germán De la Cruz [4], Mary L. Huamán-Carrión [1], Uriel R. Quispe-Quezada [5], Edgar Gutiérrez-Gómez [6], Domingo J. Cabel-Moscoso [7], Mauricio Muñoz-Melgarejo [8] and Wilber César Calsina Ponce [9]

1. Nutraceuticals and Biomaterials Research Group, Universidad Nacional José María Arguedas, Andahuaylas 03701, Peru; dchoque@unajma.edu.pe (D.C.-Q.); bsramos@unajma.edu.pe (B.S.R.-P.); huamancarrionmary@gmail.com (M.L.H.-C.)
2. Research Group in the Development of Advanced Materials for Water and Food Treatment, Universidad Nacional José María Arguedas, Andahuaylas 03701, Peru
3. Department of Education and Humanities, Universidad Nacional José María Arguedas, Andahuaylas 03701, Peru; jcarevalo@unajma.edu.pe
4. Agricultural Science Faculty, Universidad Nacional de San Cristobal de Huamanga, Ayacucho 05000, Peru; german.delacruz@unsch.edu.pe
5. Agricultural and Forestry Business Engineering, Universidad Nacional Autónoma de Huanta, Ayacucho 05000, Peru; uquispe@unah.edu.pe
6. Engineering and Management Faculty, Universidad Nacional Autónoma de Huanta, Ayacucho 05000, Peru; egutierrez@unah.edu.pe
7. Ambiental Engineering, Universidad Nacional San Luis Gonzaga, Ica 11001, Peru; jesus.cabel@unica.edu.pe
8. Human Medicine Faculty, Universidad Peruana los Andes, Huancayo 12006, Peru; d.mmunoz@upla.edu.pe
9. Social Sciences Faculty, Universidad Nacional del Altiplano, Puno 21001, Peru; wcalsina@unap.edu.pe
* Correspondence: caligarda@unajma.edu.pe (C.A.L.-S.); elibetmm22@gmail.com (E.M.-M.)

Citation: Ligarda-Samanez, C.A.; Moscoso-Moscoso, E.; Choque-Quispe, D.; Ramos-Pacheco, B.S.; Arévalo-Quijano, J.C.; Cruz, G.D.l.; Huamán-Carrión, M.L.; Quispe-Quezada, U.R.; Gutiérrez-Gómez, E.; Cabel-Moscoso, D.J.; et al. Native Potato Starch and Tara Gum as Polymeric Matrices to Obtain Iron-Loaded Microcapsules from Ovine and Bovine Erythrocytes. Polymers 2023, 15, 3985. https://doi.org/10.3390/polym15193985

Academic Editor: Cristina Cazan

Received: 12 September 2023
Revised: 29 September 2023
Accepted: 2 October 2023
Published: 4 October 2023

Copyright: © 2023 by the authors. Licensee MDPI, Basel, Switzerland. This article is an open access article distributed under the terms and conditions of the Creative Commons Attribution (CC BY) license (https://creativecommons.org/licenses/by/4.0/).

Abstract: Iron deficiency leads to ferropenic anemia in humans. This study aimed to encapsulate iron-rich ovine and bovine erythrocytes using tara gum and native potato starch as matrices. Solutions containing 20% erythrocytes and different proportions of encapsulants (5, 10, and 20%) were used, followed by spray drying at 120 and 140 °C. Iron content in erythrocytes ranged between 2.24 and 2.52 mg of Fe/g; microcapsules ranged from 1.54 to 2.02 mg of Fe/g. Yields varied from 50.55 to 63.40%, and temperature and encapsulant proportion affected moisture and water activity. Various red hues, sizes, and shapes were observed in the microcapsules. SEM-EDS analysis revealed the surface presence of iron in microcapsules with openings on their exterior, along with a negative zeta potential. Thermal and infrared analyses confirmed core encapsulation within the matrices. Iron release varied between 92.30 and 93.13% at 120 min. Finally, the most effective treatments were those with higher encapsulant percentages and dried at elevated temperatures, which could enable their utilization in functional food fortification to combat anemia in developing countries.

Keywords: native potato starch; tara gum; microencapsulation; erythrocytes; iron released; spray drying

1. Introduction

In recent years, new biopolymers have been studied for their use in encapsulating different bioactive compounds beneficial to human health [1,2]. The development of research on mixtures of matrices and cores is essential to understand their interactions using modern physical, chemical, and structural techniques [3,4]. The spray drying microencapsulation is used to stabilize various phytochemicals that are applied in the food and pharmaceutical industry [5,6]. The wall materials present in the microcapsules provide additional health benefits due to their protein and fiber content [7,8].

Iron is an essential micronutrient in several vital functions, such as oxygen transport, cell proliferation, immunity, deoxyribonucleic acid synthesis, and energy production. Heme iron is obtained from meat, offal, and blood containing myoglobin and hemoglobin. In contrast, non-heme iron is ferric (Fe^{+3}) or ferrous (Fe^{+2}) salt in cereals, dairy products, legumes, and other vegetables [9,10]. Iron deficiency causes anemia, mainly in children, due to the low consumption of foods containing this mineral. Heme iron is the most easily bioavailable form in the human organism, with between 15 and 40% absorption; current studies indicate that blood is one of the primary sources of iron [11,12].

Iron deficiency anemia is a global public health problem, and various iron fortification and supplementation strategies have been developed to improve its bioavailability and absorption [13]. Iron encapsulation is a promising technique that protects iron from oxidation and degradation reactions during food processing and incorporation. Iron can be explicitly released in the intestine through microencapsulation, improving absorption. Various materials have been used as encapsulants, including starches, maltodextrins, chitosan, and alginate [14]. Multiple encapsulation methods are applied in the food industry, such as emulsification, ionic gelation, extrusion, spray drying, and lyophilization [15–19]. In addition, novel chemical encapsulation techniques have been developed, such as interfacial polymerization, molecular co-crystallization, and inclusion in cyclodextrins [20,21]. Polymeric systems containing inorganic Mg and $CaCO_3$ substances that promote biodegradation, biocompatibility, and bioactivity are also being developed. The controlled incorporation of these inorganic substances into the polymers modifies the mechanical, thermal, and biological properties that could also be used for iron [22,23].

Different drawbacks were identified in fortifying foods with iron, such as adverse changes in sensory characteristics, gastrointestinal discomfort, and using non-heme iron in higher proportions. These problems have been improved by using encapsulation technology, a micro-packaging process in which a wide variety of iron compounds are protected with different polymeric matrices producing microcapsules and nanocapsules via different methodologies, among which spray drying stands out. In this way, different forms of encapsulated iron have been obtained and tested for their effectiveness in vitro and in vivo, with promising results in reducing iron deficiency anemia [9,24,25]. This paper presents an innovative methodology for obtaining iron-loaded microcapsules from sheep and cattle erythrocytes using native potato starch and tara gum as polymeric matrices. These microcapsules offer a promising application in controlled iron release, which could have significant implications for developing treatments for anemia and other related diseases.

The development of controlled release systems for bioactive compounds has become highly relevant in medicine and nutrition. Iron is essential for proper body function, and its deficiency can lead to serious health problems. Encapsulating iron in microcapsules could improve its bioavailability and enable sustained release in the body, avoiding toxicity issues associated with high doses. In this study, the use of native potato starch and tara gum as polymeric matrices to encapsulate iron from sheep and cattle erythrocytes was explored, aiming to obtain microcapsules capable of releasing the mineral in a controlled manner.

2. Materials and Methods

2.1. Materials

The blood of sheep (*Ovis orientalis aries*) and cattle (*Bos taurus*) was collected at the Municipal Slaughterhouse in San Jerónimo, province of Andahuaylas, Peru, which the National Agrarian Health Service authorizes. The blood extraction was carried out under entirely safe and aseptic conditions. Local farmers from the district of Ocobamba, province of Chincheros, Peru, kindly provided the tara. The native potato of the yanapalta variety was acquired at the central market of the district of Andahuaylas and was in optimal conditions for consumption. The research involving the use of animals was approved by the Ethics Committee of the National University José María Arguedas through Resolution N° 232-2020-CO-UNAJMA dated 22 September 2020. Sheep and cattle blood was selected because these animals are abundant in the study area, and their blood is an iron-rich by-product that is not used during the processing of these animals. In the case of the native potato, the yanapalta variety was chosen because of its good yields in the field and as it is an excellent source of native starch. As for the tara, it was chosen because it is a rich source of gum that is found in abundance on the study site.

The reagents used were hydrochloric acid reagent grade (Spectrum Chemical Mfg. Corp., Bathurst, NB, Canada), nitric acid reagent grade (Spectrum Chemical Mfg. Corp., Bathurst, NB, Canada), and absolute ethanol (Scharlau, Sentmenat, Spain).

2.2. Native Potato Starch

Around 5 kg of native potatoes from yanapalta variety were used. These were crushed using a Bosch blender (Stuttgart, Germany). Following this, several rinses were carried out with distilled water to isolate the starch by allowing it to settle. The acquired starch was then dried at 50 °C using a FED 115 BINDER forced convection oven. Subsequently, it was finely ground into powder using an agate mortar. Next, the starch underwent sieving with an analytical vibrating sieve AS 200 (Retsch, Haan, Germany) with a 45 μm mesh size. The resultant starch was collected in airtight containers and stored at 20 °C for future use [10,26,27].

2.3. Tara Gum

The germ was isolated from tara seeds, and 30 g of the germs were mixed with 800 mL of distilled water. The mixture was then stirred for 12 h at 80 degrees Celsius. Afterward, the solution was filtered through a 150 μm nylon mesh screen. It was combined with 96% ethanol at a 1:1 ratio to purify and cause the gum to precipitate. Following this, the resulting gum was diluted with distilled water until it reached a viscosity of 30 cP using a viscometer (DV-E Brookfield Engineering Laboratories, Inc., Middleboro, MA, USA). The extract was spray-dried using a mini spray dryer B-290 from BÜCHI Labortechnik AG, operating at an inlet temperature of 100 °C and an airflow rate of 650 L per minute [10].

2.4. Spray-Dried Erythrocytes

Sodium citrate was utilized as a blood anticoagulant (at a concentration of 3 g/L) to collect blood samples from sheep and cattle. The obtained blood was centrifuged at 3000 revolutions per minute for 10 min (CR4000R Centurion, Pocklington, UK). This process aimed to separate the cellular components from the rest. The resultant pellet was rinsed twice with a saline solution containing 0.15 M NaCl. After completing the washing procedures, the viscosity of the solution was adjusted to reach 30 cP (DV-E Engineering Laboratories, Inc.). Next, the material was subjected to drying using a B-290 mini spray dryer from BÜCHI Labortechnik AG, operating at an inlet temperature of 120 °C and an airflow rate of 650 L/h. The resulting atomized material was collected within low-density polyethylene bags and subsequently stored in a desiccator at 20 °C until it was ready for subsequent utilization [10].

2.5. Erythrocyte Microparticles

For the microencapsulation process, native potato starch and tara gum at a 4:1 ratio were used as the outer layer materials. These components were prepared at varying concentrations of 5%, 10%, and 20% by weight/volume (w/v). The solution for encapsulation was prepared a day in advance. As far as erythrocytes were concerned, they were made at a constant concentration of 20% (w/v). Both solutions were mixed at a 1:1 ratio and blended thoroughly using an ultraturrax device (Daihan, model HG15D, Wonju, Republic of Korea), operating at 7000 revolutions per minute for 5 min. The actual encapsulation was carried out using a B-290 mini spray (BÜCHI Labortechnik AG, Flawil, Switzerland). This process occurred at inlet temperatures of 120 °C and 140 °C and an airflow rate of 650 L/h. Following this, the encapsulated materials were collected and carefully placed into low-density bags and stored within a desiccator at a temperature of 20 °C [10].

The experimental flow diagram is shown in Figure 1, in which the abbreviations T1O, T2O, T3O, T4O, T5O, and T6O are presented to refer to the ovine erythrocyte microencapsulation treatments, and the abbreviations T1V, T2V, T3V, T4V, T5V, and T6V for the bovine erythrocyte microencapsulation treatments.

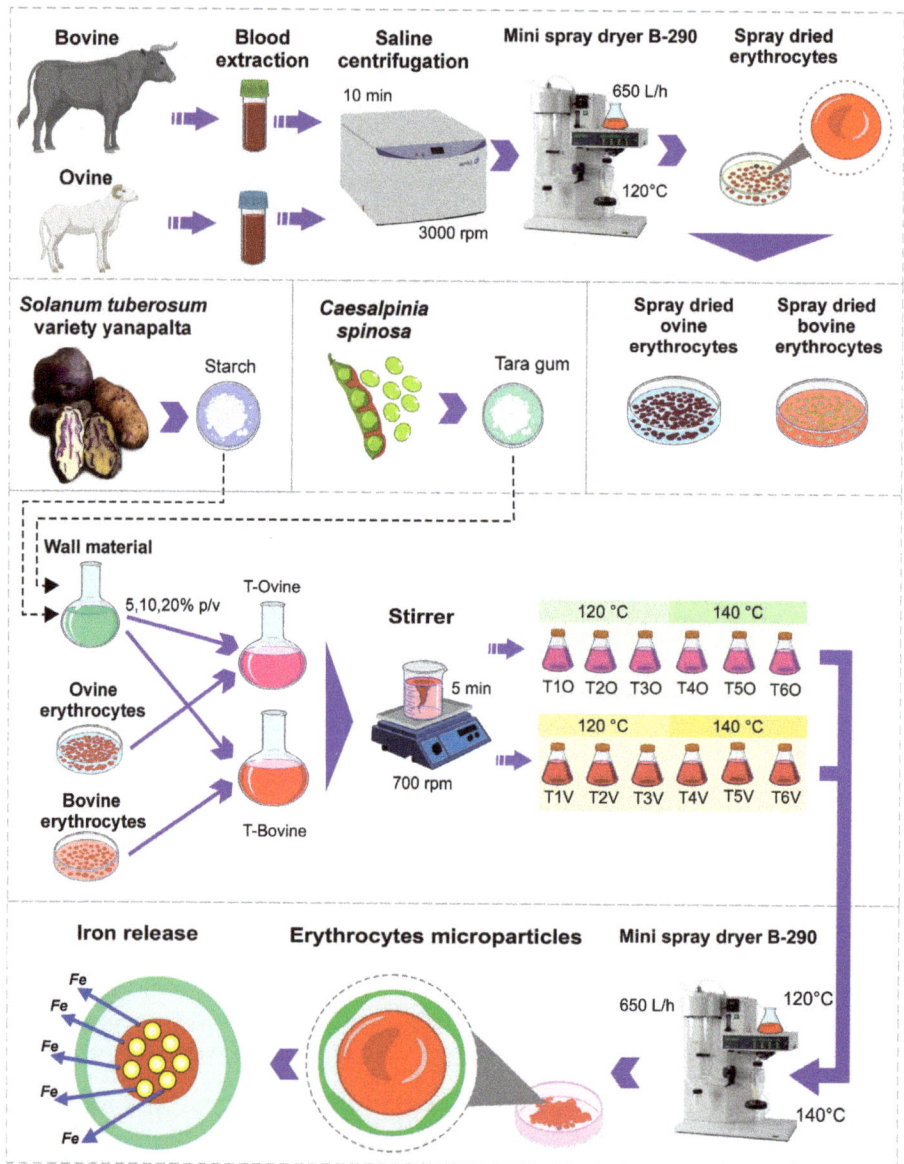

Figure 1. Experimental flow diagram.

2.6. Iron Content

A total of 200 mg of the sample underwent treatment with 3 mL of HCl and 9 mL HNO$_3$. The resulting mixture was then adjusted to a final volume of 50 mL using ultrapure water. The prepared solutions were then subjected to microwave digestion by utilizing a microwave digester (SCP Science, Miniwave, QC, Canada). An axial mode inductively coupled plasma optical emission spectrometer ICP-OES 9820 138 (Shimadzu, Kyoto, Japan) was employed to measure the iron content. Argon gas was maintained at a flow rate of 10 L/min during the measurements and readings were taken at a specific wavelength of 239.562 nm [10].

2.7. Yield, Moisture, Water Activity, and Bulk Density

The yield was calculated by considering the mass of the obtained spray-dried powder and the initial mass (wall material and core) according to the following relationship [28]:

$$Y\ (\%) = \frac{mi}{mf} \times 100 \tag{1}$$

where Y (%) represents the encapsulation yield, mi is the initial mass (g), and mf is the final mass of the spray-dried powder (g).

Moisture content was determined following the AOAC 950.10 oven drying method [29]. The water activity was assessed using a water activity meter (Rotronic, Bassersdorf, Switzerland) [30]. Bulk density was calculated by dividing the mass of the microcapsules by the volume measured using a graduated 10 mL cylinder [28].

2.8. Color Analysis

Lightness L^* and chroma a^* and b^* color attributes were ascertained using a bench-top colorimeter (CR-5, Konica Minolta, Tokyo, Japan). The degree of color change was computed using the subsequent equation [31]:

$$\Delta E_{ab}^* = \sqrt{\Delta L^{*2} + \Delta a^{*2} + \Delta b^{*2}} \tag{2}$$

where ΔE_{ab}^* is the color variation and ΔL^*, Δa^*, and Δb^* are the differences between L^*, a^*, and, b^* initials and finals.

2.9. Amylose and Amylopectin Content

The potato amylose standard (Sigma Aldrich, St. Louis, MO, USA) was used in concentrations of 0.1 to 1.0 mg/mL for the calibration curve. For amylose extraction, 20 mg of sample was taken and 0.2 mL of 95% ethanol (Scharlau, Senmanat, Spain) and 1.8 mL of 1 M NaOH (Sigma Aldrich, Darmstadt, Germany) were added and allowed to stand for 24 h at room temperature. Subsequently, the volume was adjusted to 20 mL with ultrapure water and homogenized in a vortex at 2000 RPM for minutes. For the colorimetric reaction, 0.5 mL of the extracted solution, 1 mL of 1M acetic acid (Sigma Aldrich, St. Louis, MO, USA), and 0.2 mL of lugol solution were taken, and the volume was made up to 10 mL with ultrapure water. The solution was shaken and allowed to react for 20 min, protected from light. Absorbance readings were carried out at a wavelength of 620 nm using a UV spectrophotometer (CR-5, Konica Minolta, Tokyo, Japan) [32,33].

2.10. Total Organic Carbon

A total of 0.05 g of encapsulated samples was positioned within ceramic containers to be analyzed utilizing a total organic carbon analyze TOC-L CSN-SSM 5000A (Shimadzu, Kyoto, Japan) [34,35].

2.11. SEM-EDS Analysis

The structure of erythrocytes and microcapsules was examined using a scanning electron microscope (SEM Thermo Fisher, Waltham, MA, USA) under low vacuum conditions, employing an acceleration voltage of 25 kV. Furthermore, an energy-dispersive X-ray spectroscopy (EDS) detector was utilized to perform surface chemical analysis of the samples [36].

2.12. Particle Size and ζ Potential Analysis

For particle size determination, a laser diffraction instrument, Mastersizer 3000 (Malvern Instruments, Worcestershire, UK) was employed. The samples were dissolved in isopropyl alcohol, subjected to sonication for 60 s, and measured at 600 nm. As for determining the ζ potential, 25 mg of erythrocytes and microcapsules were homogenized in 50 mL of ultra-pure water. The analysis was conducted using a dynamic light scattering (DLS) instrument, Zetasizer ZSU3100 (Malvern Instruments, Worcestershire, UK) [36].

2.13. FTIR Analysis

Fourier transform infrared spectroscopy (FTIR) was used to analyze and identify functional groups in erythrocytes and microcapsules. Pellets were prepared by combining 2 mg of the sample with 200 mg of KBr. The mixture was then pressed at a force of 10 tons to create the pellets for analysis. The FTIR measurements were conducted using the transmission module of the Nicolet IS50 FTIR (ThermoFisher, Waltham, MA, USA). The spectral range covered wavelengths from 4000 to 400 cm^{-1}. Readings were taken with a scan repetition of 32 and a resolution of 8 cm^{-1} [37].

2.14. Thermal Analysis

For the thermal stability analysis using Thermogravimetric Analysis (TGA), and 10 mg of erythrocytes and microcapsules were utilized. The measurements were conducted using a TGA 550 thermal analyzer (TA Instrument, New Castle, DE, USA) with a heating rate of 10 °C/min. Furthermore, a Differential Scanning Calorimeter (DSC2500, TA Instruments, New Castle, DE, USA) was utilized for analysis. In this process, 2 mg of microcapsules were employed. The temperature range covered was from 0 to 250 °C, employing a heating rate of 10 °C per minute. The analysis was conducted under a nitrogen atmosphere [37].

2.15. Iron Release

To conduct the in vitro release determination, uniform solutions were prepared by mixing 0.05 g of microcapsules with 500 mL of a 0.1 N HCl solution. These prepared samples were then subjected to a water bath with an agitation system (WTB 50, Memmert, Schwabach, Germany) at a temperature of 37 °C. Extractions were carried out at specific intervals: 0, 30, 60, 90, and 120 min. After extraction, the samples were analyzed on an Inductively Coupled Plasma Optical Emission Spectrometer (ICP-OES) model 9820 138 (Shimadzu, Tokyo, Japan). The resulting data were quantified in mg of Fe/g of sample and calculated using the provided relationship [10]:

$$\%L = \frac{Fe_T}{Fe_0} \times 100 \tag{3}$$

where %L is the percentage of release, Fe_T is the iron content at time t (mg/g), and Fe_0 is the initial iron content (mg/g).

2.16. Statistical Analysis

Data analysis was performed using the Origin Pro 2022 software (OriginLab Corporation, Northampton, MA, USA). The analysis involved the analysis of variance (ANOVA) along with Tukey's multiple range test, employing a significance level of 5%.

3. Results and Discussions

3.1. Instrumental Characterization of Matrices and Cores

3.1.1. SEM-EDS Analysis, Particle Size, ζ Potential, Color, and Iron Content

In Figure 2a, the characterization of native potato starch from the yanapalta variety is shown, where elliptical shapes of granules were observed with a size of approximately 32.30 μm, with a negative ζ potential, white color, and a predominant presence of carbon and oxygen on its surface, these results were similar to those reported for starch from native potatoes of the peruanita variety [36]. Regarding specifically the SEM-EDS analysis, the results were similar to the starches obtained from native potatoes grown in Cusco, Peru, which presented smooth surfaces and ellipsoidal shapes that also contained mainly carbon, oxygen, and traces of calcium, with particle sizes between 12 and 72 μm [38]. The above parameters are essential in the chemical and techno-functional properties of starch [39]. Amylose (42.97%) and amylopectin (57.03%) were also characterized, which is consistent with native potatoes of the huamantanga and qeccorani varieties [40]. The ratio of amylose to amylopectin influences the functional properties of starches [41].

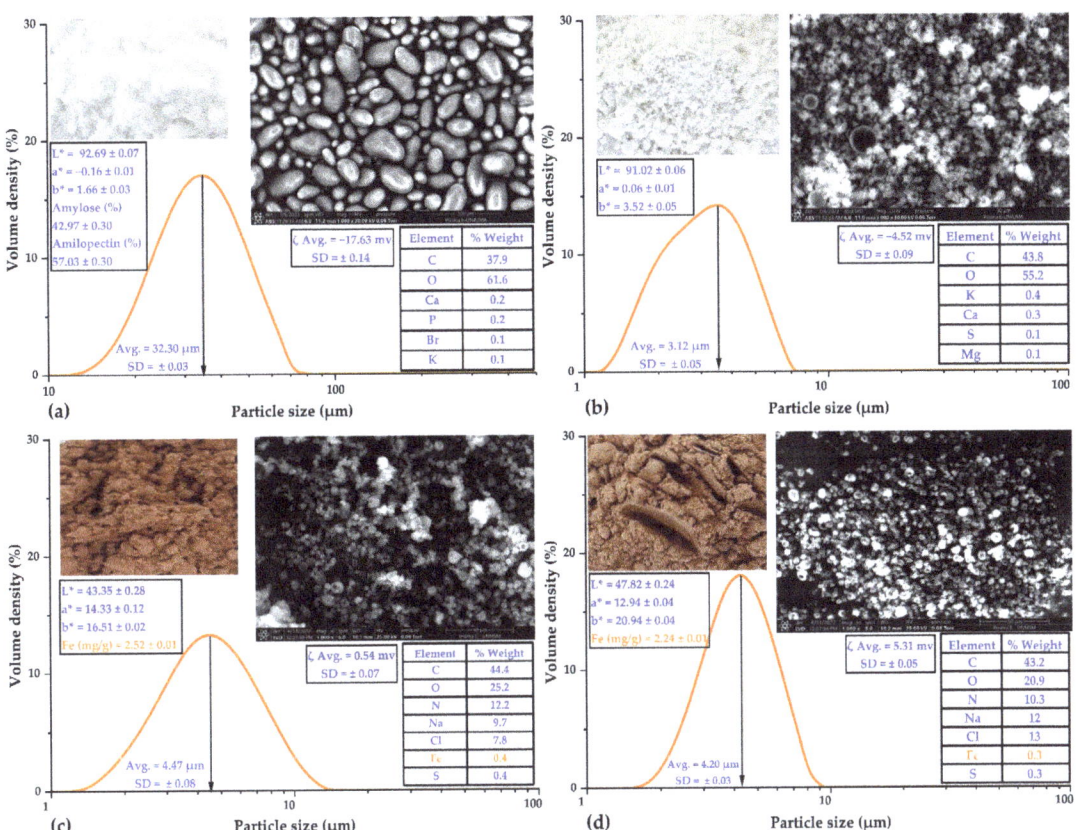

Figure 2. (**a**) Characterization of native potato starch of the yanapalta variety, (**b**) characterization of spray-dried tara gum, (**c**) characterization of spray-dried ovine erythrocytes, and (**d**) characterization of spray-dried bovine erythrocytes.

In Figure 2b, spray-dried tara gum is characterized by spherical particles obtained with an average size of 3.12 μm, negative ζ potential, white color, and a predominant presence of carbon and hydrogen on its surface. In this context, recent investigations support the notion that spray drying enables the production of spherical and uniformly sized particles [37]. The negative zeta potential could be attributed to carboxyl and hydroxyl functional groups on the particle surface, which are crucial in stabilizing colloidal dispersions [10]. The white color of the microparticles could be attributed to their size and morphology, influencing the dispersion of visible light; the predominant presence of carbon and hydrogen on the surface is likely associated with the chemical structure of tara gum, which is rich in polysaccharides [10,36].

On the other hand, Figure 2c,d depict the instrumental characterization of ovine and bovine erythrocytes, respectively, obtained through spray drying. Spherical shapes with central indentations were observed, with a size of approximately 4 μm, positive ζ potential, reddish color, and a predominant presence of carbon, oxygen, and nitrogen on their surface. Notably, the iron content of ovine and bovine erythrocytes was 2.52 and 2.24 mg of Fe/g, respectively. These values also agreed with the surface percentage characterization performed by SEM-EDS. Similar results were reported in guinea pig blood erythrocytes (3.30 mg of Fe/g) [10] and commercial bovine erythrocytes (2.49 mg of Fe/g) [42].

3.1.2. Thermal Analysis

In Figure 3a, the thermal analysis of native potato starch from the yanapalta variety and spray-dried tara gum is presented. Both materials exhibited similar thermal behaviors, and two main events were observed. The first event occurred at a temperature of 43.81 °C, marking the initiation of hydrogen bond breaking and consequent water loss, which continued until its evaporation at around 100 °C. Additionally, other volatile components were lost during this process. The second event, observed at 289.62 °C, involved the elimination of organic compounds such as carbohydrates, proteins, lipids, and fiber, which continue degrading until reaching a temperature of 600 °C.

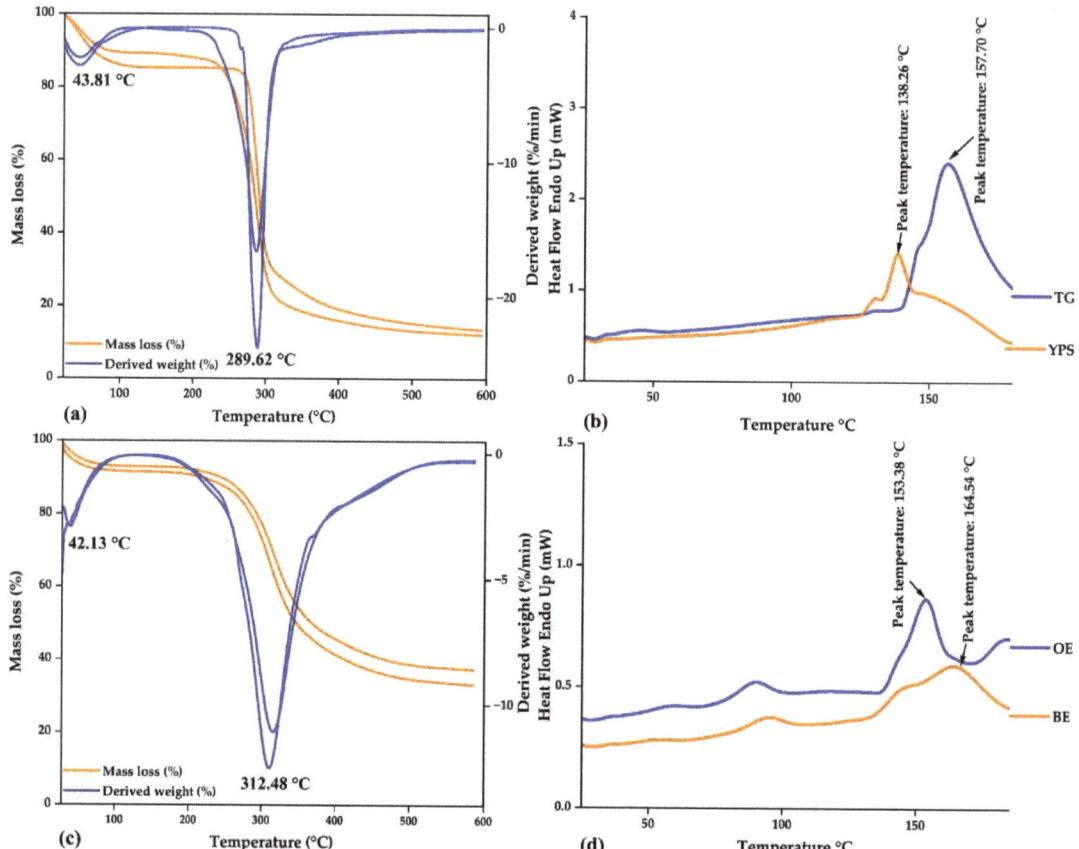

Figure 3. Thermal analysis: (**a**) DT and DTA curves in native potato starch and tara gum, (**b**) DSC curves in native potato starch (YPS) and tara gum (TG), (**c**) DT and DTA curves in ovine and bovine erythrocytes, and (**d**) DSC curves in ovine (OE) and bovine (BE) erythrocytes.

Figure 3b shows the DSC analysis for both wall materials. For the yanapalta potato starch, a glass transition temperature of 138.26 °C was found, while for the spray-dried tara gum, the value was 157.70 °C. Both values are used as references to confirm the encapsulation of the cores within the wall materials. The results in both materials are consistent with the current literature, as primarily two events were observed related to the thermal degradation of biomolecules, such as carbohydrates and complex organic compounds [10,36]. DSC analysis offers valuable insights into phase transitions and changes in the molecular structure of materials. The glass transition is a crucial feature that can impact polymeric materials' functional and encapsulation properties, holding significance for their potential applications [37,43].

On the other hand, Figure 3c,d depict the thermal analysis of spray-dried ovine and bovine erythrocytes, respectively. Two events were also observed at temperatures of 42.13 and 312.48 °C. In the first case, water loss is initiated, while in the second case, biopolymers are eliminated, ultimately resulting in the formation of ashes. The glass transition temperature was determined to be 153.38 °C for ovine erythrocytes and 164.54 °C for bovine erythrocytes. In summary, the results above provide a deeper understanding of the physical properties and their thermal behavior, holding significance for their application in various scientific and technological fields.

3.2. Characterization of the Microcapsules

3.2.1. Physical and Chemical Properties

Table 1 presents the physical and chemical properties of microcapsules obtained from ovine and bovine erythrocytes in native potato starch and tara gum matrices. The iron content varied between 1.54 and 2.02 mg of Fe/g, with similar contents observed in microcapsules of guinea pig blood erythrocytes in native potato starch and tara gum (1.32 to 2.05 mg of Fe/g) [10] and higher contents than those of commercial bovine erythrocytes microcapsules in maltodextrin (0.77 mg of Fe/g) [42]. Iron content in the microcapsules is critical [14], particularly in food fortification [13]. Iron is essential for human health, as it plays a fundamental role in oxygen transport, DNA synthesis, and immune function [44]. The encapsulation of iron within microcapsules can confer advantages in terms of stability and bioavailability [10,42].

The percentage of total organic carbon ranged from 12.80% to 14.88%, with higher proportions of the matrices used corresponding to greater TOC contents, as carbon atoms constitute an essential part of the structure of biomolecules such as carbohydrates, proteins, lipids, and fibers [35,45,46]. The percentage quantification of TOC contents confirms the successful encapsulation of the cores within the matrices, which are rich in various biopolymers [1,14,47,48].

The encapsulation efficiency ranged from 52.94% to 85.88%, while the yield fluctuated between 50.55% and 63.40%. Encapsulation yields were similar to those reported for guinea pig erythrocyte microcapsules (47.84% and 58.73%) [10] and ferrous sulfate microcapsules (47.93% and 56.26%) [49]. On the other hand, the values exceeded those obtained in commercial bovine erythrocyte microcapsules (39% and 47%) [42].

The properties studied, which are related to the shelf-life preservation of spray-dried powders, were moisture content, ranging between 4.31% and 7.49%, and water activity, oscillating between 0.36 and 0.43. Maintaining moisture content below 5% is recommended for adequately preserving dry products, and in the case of water activity, values lower than 0.6 are advised [50–54]. This way, the various reaction mechanisms that lead to food deterioration are controlled. It was also observed that moisture content and water activity decrease with increased temperature and a higher proportion of encapsulants [55,56].

Likewise, variations in luminosity were observed between 52.61 and 62.59, values that increase as the proportion of matrices is increased. Regarding the color coordinate a^*, values ranging between 7.00 and 11.57 were identified, and it was noted that the powders take on redder hues as the quantity of encapsulants is reduced. On the other hand, the color coordinate b^* exhibited variation in the range of 18.09 to 20.55. Additionally, significant differences were observed regarding the initial color, as evidenced by ΔE^*_{ab} values ranging from 5.25 to 17.84. This is attributed to non-enzymatic browning and the caramelization of carbohydrates at high temperatures [10,42].

Particle size was measured in isopropyl alcohol and using the laser diffraction technique. The values obtained ranged between 4.26 and 7.59 µm, higher than those reported for microcapsules of guinea pig erythrocytes solubilized in water and measured via the dynamic light scattering technique (817.1 and 1672.2 nm) [10]. This is attributed to the type of dispersant and the method used, with the laser diffraction technique being the most appropriate for measuring microcapsules obtained through spray drying [36,37]. On the other hand, the ζ potential presented negative values ranging from −0.11 to −3.51 mV, indicating a tendency to aggregate and precipitate due to the presence and nature of the erythrocytes used [10].

Table 1. Physical and chemical properties of microcapsules.

Microcapsules O	T1O		T2O		T3O		T4O		T5O		T6O	
Properties	$\bar{x}$ ± SD	*	$\bar{x}$ ± SD	*	$\bar{x}$ ± SD	*	$\bar{x}$ ± SD	*	$\bar{x}$ ± SD	*	$\bar{x}$ ± SD	*
Iron (mg/g)	1.99 ± 0.01	a	1.67 ± 0.01	b	1.34 ± 0.01	c	2.02 ± 0.01	d	1.87 ± 0.02	e	1.76 ± 0.02	f
TOC (%)	13.98 ± 0.01	a	14.50 ± 0.01	b	14.63 ± 0.05	b	13.78 ± 0.01	a	14.50 ± 0.03	b	14.56 ± 0.04	b
EE (%)	80.24 ± 0.21	a	66.03 ± 0.44	b	52.94 ± 0.11	c	78.72 ± 0.10	d	74.22 ± 0.73	e	69.86 ± 0.73	f
Yield (%)	53.69 ± 1.14	ab	50.55 ± 1.92	a	50.98 ± 1.91	a	56.99 ± 0.31	b	55.60 ± 0.57	ab	54.04 ± 1.46	ab
Moisture (%)	6.07 ± 0.02	a	5.07 ± 0.06	bc	4.52 ± 0.24	cd	5.27 ± 0.22	b	4.64 ± 0.12	cd	4.31 ± 0.08	e
Aw	0.43 ± 0.003	a	0.41 ± 0.004	b	0.38 ± 0.004	c	0.41 ± 0.001	b	0.40 ± 0.004	d	0.38 ± 0.002	c
L^*	54.06 ± 0.02	a	55.05 ± 0.03	b	59.16 ± 0.11	c	55.95 ± 0.28	d	58.11 ± 0.22	e	60.05 ± 0.37	f
a^*	11.34 ± 0.05	a	11.57 ± 0.01	a	9.63 ± 0.06	b	10.62 ± 0.18	c	10.06 ± 0.09	d	8.81 ± 0.17	e
b^*	19.09 ± 0.07	a	20.55 ± 0.08	b	20.06 ± 0.06	c	18.59 ± 0.18	d	20.26 ± 0.07	bc	19.46 ± 0.17	e
ΔE^*_{ab}	11.41 ± 0.30	a	12.68 ± 0.31	b	16.86 ± 0.18	c	13.30 ± 0.31	b	15.82 ± 0.25	d	17.84 ± 0.67	e
Particle size (µm)	4.26 0.13	a	6.23 ± 0.05	b	6.73 ± 0.07	c	5.46 ± 0.06	d	5.60 ± 0.12	e	6.28 ± 0.10	f
ζ potential (mV)	−0.11 ± 0.16	a	−0.98 ± 0.23	b	−2.76 ± 0.91	c	−2.90 ± 0.70	d	−3.43 ± 0.47	e	−3.51 ± 0.73	f
Microcapsules V	**T1V**		**T2V**		**T3V**		**T4V**		**T5V**		**T6V**	
Iron (mg/g)	1.88 ± 0.01	a	1.73 ± 0.01	b	1.54 ± 0.02	c	1.93 ± 0.02	d	1.77 ± 0.01	e	1.56 ± 0.01	f
TOC (%)	13.07 ± 0.02	a	13.39 ± 0.14	ab	13.97 ± 0.04	ab	12.80 ± 0.13	a	13.65 ± 0.02	ab	14.88 ± 0.09	b
EE (%)	83.95 ± 0.22	a	78.90 ± 0.14	b	69.54 ± 0.15	c	85.88 ± 0.12	d	77.12 ± 0.10	e	68.43 ± 0.82	f
Yield (%)	60.37 ± 1.11	a	55.38 ± 1.46	b	54.93 ± 0.50	b	63.40 ± 0.25	a	62.38 ± 1.40	a	61.88 ± 0.81	a
Moisture (%)	7.49 ± 0.01	a	7.27 ± 0.08	ab	7.19 ± 0.08	b	6.22 ± 0.02	c	5.80 ± 0.07	d	5.59 ± 0.01	d
Aw	0.43 ± 0.003	a	0.43 ± 0.002	a	0.40 ± 0.003	b	0.42 ± 0.002	a	0.42 ± 0.001	c	0.36 ± 0.001	d
L^*	52.61 ± 0.08	a	55.99 ± 0.14	b	61.61 ± 0.05	c	54.66 ± 0.01	d	58.20 ± 0.26	e	62.59 ± 0.78	f
a^*	11.32 ± 0.08	a	9.87 ± 0.07	b	7.17 ± 0.01	c	10.51 ± 0.02	d	9.05 ± 0.11	e	7.00 ± 0.05	c
b^*	19.51 ± 0.09	a	19.91 ± 0.08	b	18.34 ± 0.04	c	19.54 ± 0.02	a	19.49 ± 0.11	a	18.09 ± 0.02	d
ΔE^*_{ab}	5.25 ± 0.33	a	8.78 ± 0.29	b	15.17 ± 0.23	c	7.39 ± 0.24	d	11.17 ± 0.47	e	16.17 ± 0.31	f
Particle size (µm)	5.52 ± 0.03	a	5.54 ± 0.02	b	7.34 ± 0.21	c	5.97 ± 0.11	d	6.92 ± 0.09	e	7.59 ± 0.15	f
ζ potential (mV)	−0.30 ± 0.38	a	−0.40 ± 0.05	b	−1.46 ± 0.61	c	−0.51 ± 0.12	d	−1.10 ± 0.06	e	−1.89 ± 0.16	f

Where $\bar{x}$ is the arithmetic mean and SD is the standard deviation. * Different letters indicate significant difference per row evaluated through at 5% significance, for n = 3.

Figure 4 shows the principal component analysis (PCA) of the studied properties, wherein it can be observed that microcapsules T1O, T2O, T4O, T5O, T1V, T2V, T4V, and T5V (in violet color) are associated with properties such as iron content, encapsulation efficiency, encapsulation yield, humidity, water activity, and color parameters a* and b*. These correlations suggest a potential interdependence among these properties, which could influence the overall performance of the microcapsules. On the other hand, microcapsules T3V and T6V (in orange color) are more linked to luminosity and particle size, indicating that these microcapsules might possess unique optical and structural characteristics that could influence their behavior, particularly in contexts involving light dispersion. In contrast, microcapsules T3O and T6O (in green color) are more related to color variations and total organic carbon.

The PCA analysis establishes relationships among complex variables, providing a comprehensive overview of interactions between diverse properties [34]. This methodology facilitates the identification of significant trends, thereby potentially aiding the design and optimization of microcapsules with specific properties [10]. PCA is a valuable tool in studying microcapsules, with potential applications within the food and pharmaceutical industries [36].

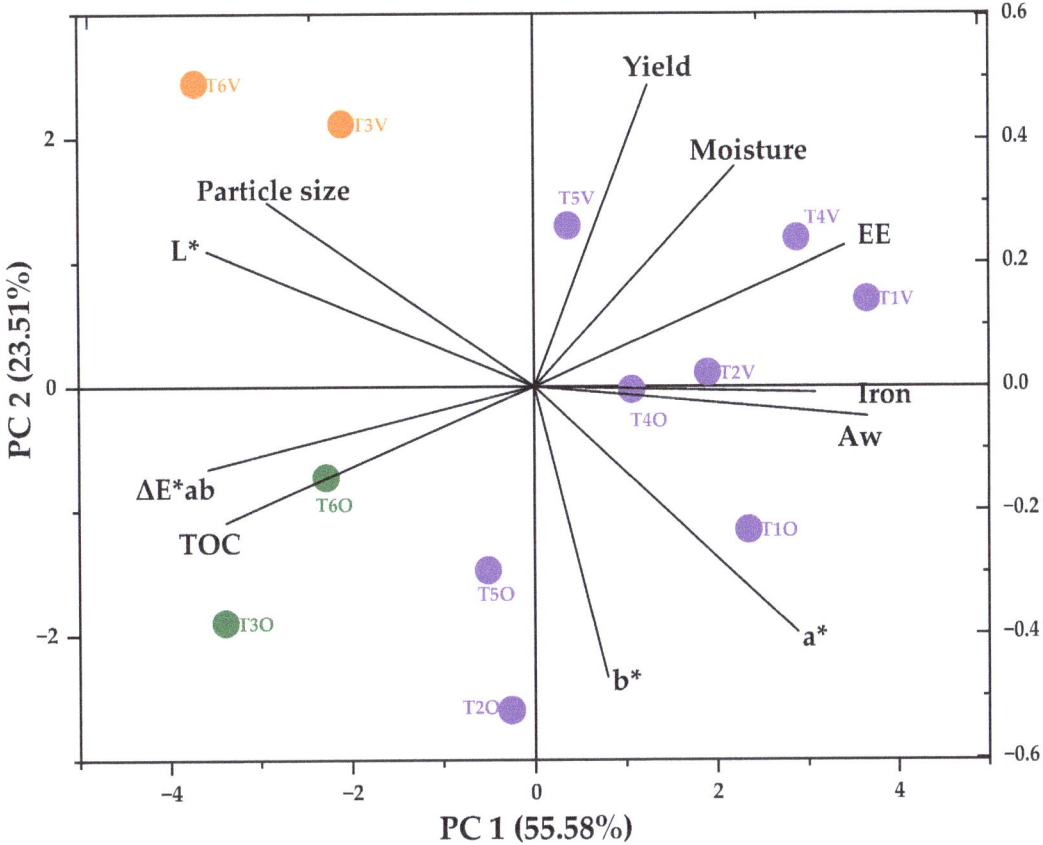

Figure 4. PCA study.

3.2.2. SEM-EDS Analysis

Figure 5 displays the SEM images of the microcapsules, revealing the formation of irregular microparticles with diverse sizes and shapes, distinctive of the spray drying process [48,57–62]. These microparticles exhibited varying dimensions and larger surface openings when a 140 °C inlet air temperature was employed. This variability appears to be influenced by feed characteristics and parameters governing the spray drying process. It is plausible that this response is linked to solvent evaporation during the drying procedure, which could have led to the contraction and loss of microcapsule sphericity [63]. The temperature increase led to a faster drying rate and the subsequent emergence of pores in the microcapsules [36]. Conversely, lower temperatures resulted in the creation of more uniform particles. However, the presence of amorphous structures in the encapsulated material is attributed to using tara gum as an encapsulating agent [10].

The results suggest that the larger size of some particles could be due to interactions between the core and matrix during microencapsulation since incorporating the core within the microcapsules may have modified the surface roughness. Iron in the microcapsules may have promoted surface collapse instead of total core retention within them [10,64].

Previous studies report on the formation of large cracks on the microparticle surface, attributing this phenomenon to the collapse of the polymeric gel network during spray drying [65]. Likewise, irradiation with electron beams by SEM can cause surface rupture of the microcapsules [66].

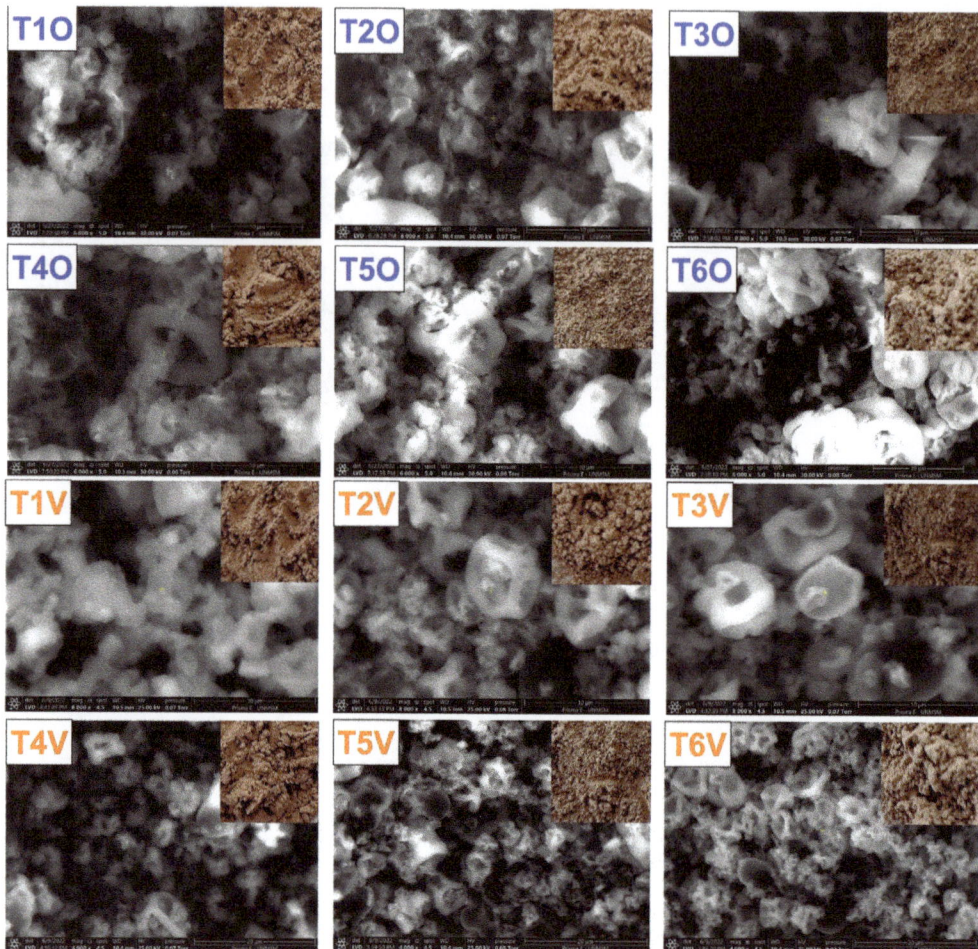

Figure 5. SEM analysis. Where T1O, T2O, T3O, T4O, T5O and T6O are the ovine erythrocyte microcapsules and T1V, T2V, T3V, T4V, T5V and T6V are the bovine erythrocyte microcapsules.

Surface chemical analysis of the microcapsules corroborated the existence of iron in the microcapsules, observing that an increase in the inlet temperature increased the amount of surface iron in the encapsulates, which varied between 0.1% and 0.3% (Table 2). Furthermore, it was verified that this temperature increase was associated with a larger particle size, potentially facilitating the encapsulation of a greater iron quantity. These findings align with the data obtained through ICP OES in this research.

On the other hand, carbon, oxygen, and nitrogen were observed, which could be attributed to biopolymers such as carbohydrates and proteins in the matrices and cores. The presence of these elements is consistent with previous observations of spray-dried erythrocyte microparticles [10,42]. The presence of sodium and chlorine was also observed due to the erythrocyte extraction being carried out in a saline solution. In addition, the presence of other chemical elements found in native potato starch, tara gum, and spray-dried erythrocytes was detected [10].

Table 2. Surface chemical analysis of microcapsules via EDS.

Element	Weight%											
	T1O	T2O	T3O	T4O	T5O	T6O	T1V	T2V	T3V	T4V	T5V	T6V
C	46.1%	43.3%	39.8%	43.5%	41.2%	39.7%	41.6%	39.8%	41.7%	39.3%	39.1%	41.7%
O	25.8%	33.8%	37.7%	33.5%	35.7%	38.6%	24.1%	33.4%	35.1%	25.8%	37.0%	34.4%
N	12.3%	10.1%	9.8%	10.1%	9.1%	8.6%	7.7%	7.5%	5.8%	7.6%	7.8%	6.1%
Na	8.3%	7.0%	6.4%	6.2%	7.0%	6.7%	15.3%	9.6%	8.8%	14.5%	7.8%	9.4%
Cl	6.7%	4.9%	5.4%	6.0%	6.1%	5.3%	10.4%	8.9%	7.8%	11.9%	7.5%	7.3%
S	0.5%	0.6%	0.6%	0.3%	0.4%	0.7%	0.4%	0.3%	0.3%	0.4%	0.3%	0.4%
Fe	0.1%	0.1%	0.1%	0.2%	0.2%	0.2%	0.2%	0.2%	0.2%	0.3%	0.3%	0.3%
P	0.1%	0.1%	0.1%	0.1%	0.1%	0.1%	0.1%	0.1%	0.1%	0.1%	0.1%	0.1%
K	0.1%	0.1%	0.1%	0.1%	0.2%	0.1%	0.2%	0.2%	0.2%	0.1%	0.1%	0.3%

3.2.3. FTIR Analysis

The analysis was conducted to verify the successful microencapsulation of iron-rich cores within the utilized matrices. To achieve this, an approved methodology for acquiring and interpreting infrared spectra was followed [67]. Figure 6 displays the characteristic spectra of the examined materials: in the case of native potato starch, tara gum, and erythrocytes, intense bands within the range of 3308–3394 cm^{-1} were identified, corresponding to the stretching vibrations of the hydroxyl group. Wavenumbers with similar intensities were detected in all microcapsules (3320 cm^{-1}). Furthermore, a prominent vibrational stretching band at 1079 cm^{-1} was observed in tara gum, indicative of a portion of the carboxylic acid structure. This same band was also evident in all microcapsules (1085 cm^{-1}) [68,69].

Likewise, additional spectral bands present in erythrocytes were also found in the microcapsules. For instance, the wavenumber around 2960 cm^{-1}, present in all microcapsules (2958 cm^{-1}), corresponds to characteristic C-H stretching vibrations. Moreover, spectral bands at 1536 and 1655 cm^{-1} were identified in erythrocytes, and these bands were also observed in the encapsulated particles, with wavenumbers ranging from 1537 to 1656 cm^{-1}, confirming the presence of amide functional groups I and II [42,49,70]. Bands around 622 cm^{-1} correspond to the pyranose ring of tara gum, a feature in all microcapsules [69,71,72].

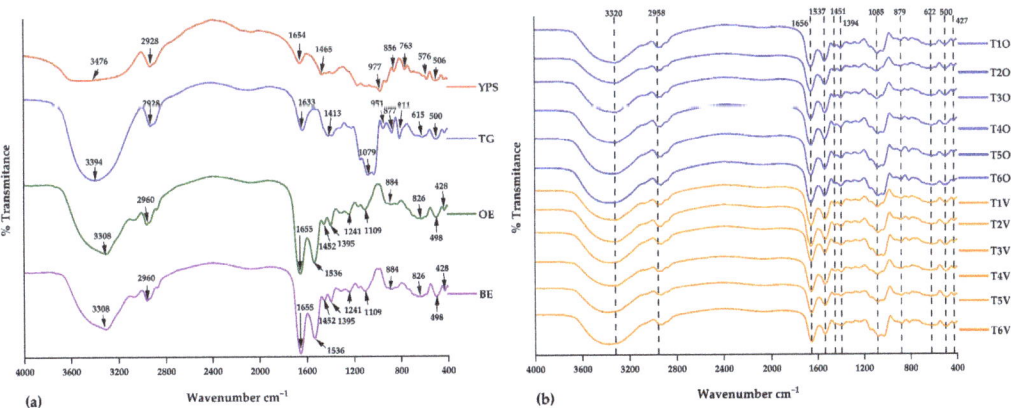

Figure 6. FTIR analysis: (a) wall materials and erythrocytes and (b) microcapsules.

The preceding results allow us to infer that the encapsulation process was successfully conducted, as the outcomes closely resembled those reported by other researchers [42,49]. Additionally, an increase in temperature from 120 °C to 140 °C induced alterations in the functional groups' intensities [10,73].

3.2.4. Thermal Analysis

The TG and DTA curves in Figure 7a,c were similar in all microcapsules. A first event was observed between 30.51 °C and 43.02 °C, resulting in approximately 30% mass loss. This mass loss can be attributed to the initiation of hydrogen bond breaking in water, which continues until temperatures close to 100 °C [37,74]. A second event occurred between 266.43 °C and 304.81 °C, with a mass loss of around 90%. This pronounced degradation rate around 300 °C can be attributed to the thermal decomposition of carbohydrates. After reaching this temperature, the thermal depolymerization of biopolymers persists until complete volatilization [37,74]. According to the obtained results, it can be appreciated that the microcapsules exhibit excellent thermal stability, attributed to the use of native potato starch and tara gum as coating materials [10].

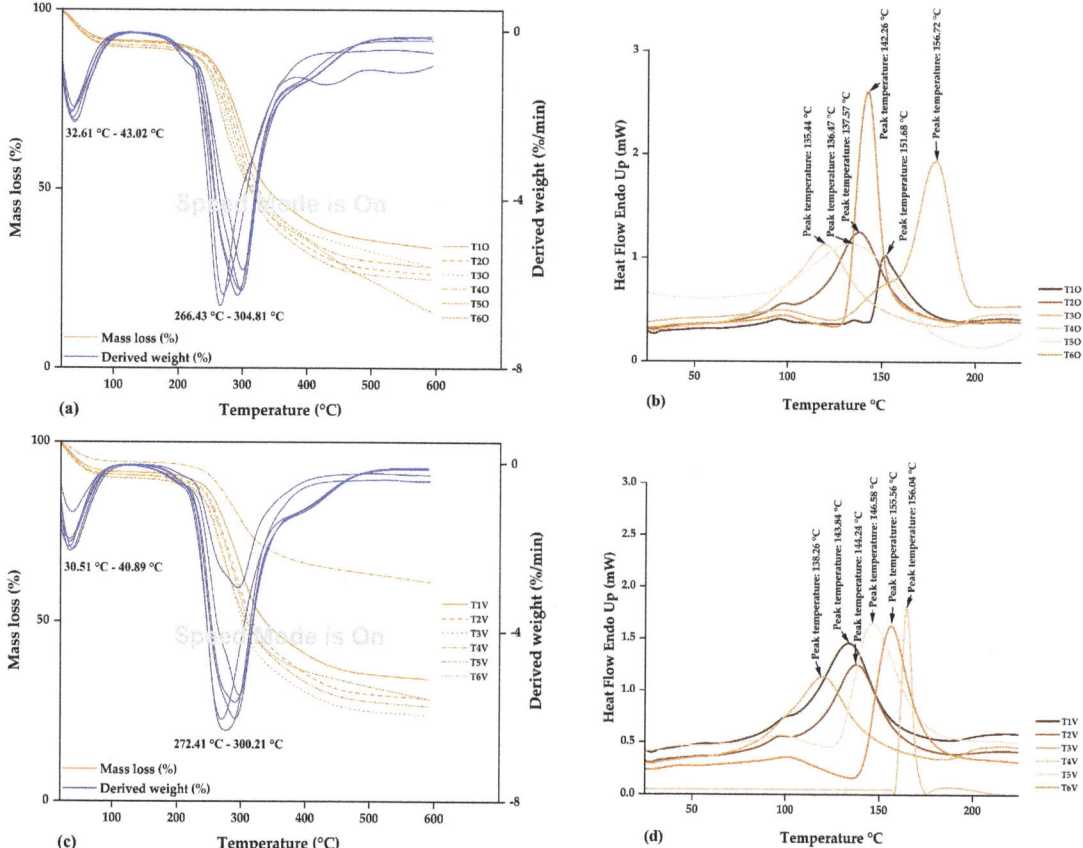

Figure 7. Thermal analysis: (**a**) DT and DTA curves in ovine erythrocyte microcapsules, (**b**) DSC curves in ovine erythrocyte microcapsules, (**c**) DT and DTA curves in bovine erythrocyte microcapsules, and (**d**) DSC curves in bovine erythrocyte microcapsules. Where T1O, T2O, T3O, T4O, T5O and T6O are the ovine erythrocyte microcapsules and T1V, T2V, T3V, T4V, T5V and T6V are the bovine erythrocyte microcapsules.

The DSC thermograms of the microcapsules are presented in Figure 7b,d, where endothermic peaks were identified at glass transition temperatures ranging from 135.44 °C to 156.72 °C. These temperatures are similar to the glass transition temperatures of native potato starch (138.26 °C) and tara gum (157.70 °C). The glass transition temperatures of the microcapsules were close to those of the used matrices. On the other hand, it is essential to mention that glass transition temperatures lower than those of the matrices would confirm the encapsulation of iron-rich erythrocyte cores [37]. These cores formed inclusive complexes with the employed encapsulants [43].

3.3. Iron Release

Figure 8a,b display the iron release profile from microcapsules derived from ovine and bovine erythrocytes. The outcomes indicate that, at 120 min, the spray-dried treatments T4O and T4V at 140 °C exhibited the highest iron release values, reaching 93.13% and 92.31%, respectively. It was demonstrated that an increase in the air temperature and a decrease in the encapsulating quantity result in a more pronounced iron release over time. Likewise, the other treatments also displayed considerable levels of iron release. This release is essential in bioavailability in the small intestine, as only 1 to 2 mg of iron are absorbed in this human body region, which is crucial for reducing the risk of anemia [13]. Thus, based on the developed in vitro iron release, the most prominent treatments would be T4O and T4V. The findings obtained in this study presented similarities with the results reported in studies involving iron encapsulation in matrices such as native potato starch and tara gum [10], potato starch and maltodextrin [75], chitosan and eudragit [76], eudraguard [77], and dextrin [78].

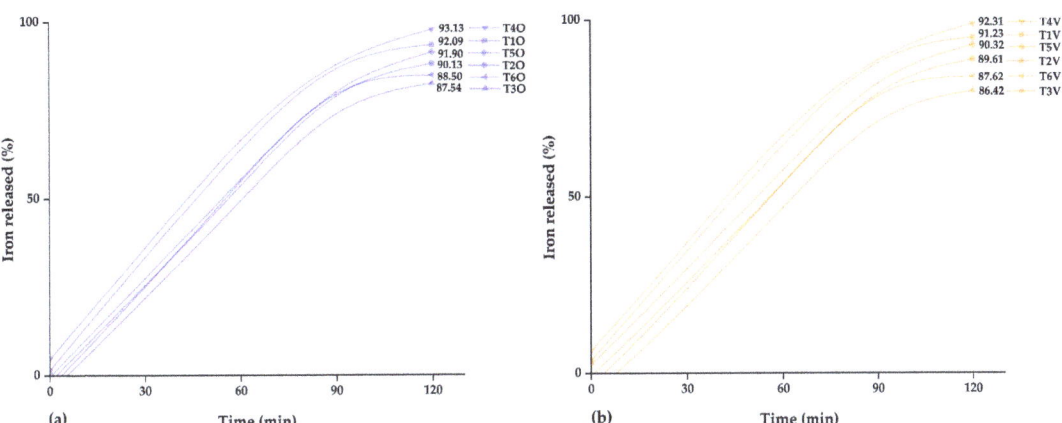

Figure 8. Iron released: (**a**) in microcapsules of the ovine erythrocytes and (**b**) in microcapsules of the bovine erythrocytes.

4. Conclusions

This study encapsulated a significant amount of heme iron extracted from ovine and bovine erythrocytes at 20% (w/v). A blend of tara gum and native potato starch was employed as a coating material at 5%, 10%, and 20% (w/v) concentrations. Encapsulation was achieved through spray drying in an aqueous environment at 120 °C and 140 °C. Consequently, elevated iron levels were achieved in erythrocytes and microcapsules, coupled with substantial in vitro bioavailability for treatments T4O and T4V and suitable physicochemical properties.

The particles exhibited micrometric dimensions and tended to aggregate in colloidal solutions. SEM-EDS analysis confirmed the presence of iron on the surface of the microcapsules. In contrast, FTIR analysis assessed the incorporation of the iron core into the polymeric matrix, substantiated by detecting functional groups within the microcapsules. Thermal analysis was also conducted, confirming the encapsulation of the cores within the matrices. In summary, the inlet temperature and the amount of coating material influenced the studied properties. Lastly, the matrix combination proved novel, and these findings pave the way for utilizing cost-effective raw materials in food fortification and combatting iron-deficiency anemia in developing nations.

Author Contributions: Conceptualization, C.A.L.-S. and E.M.-M.; methodology, D.C.-Q., J.C.A.-Q., U.R.Q.-Q. and G.D.l.C.; software, C.A.L.-S. and E.M.-M.; validation, M.L.H.-C., J.C.A.-Q., E.G.-G. and E.M.-M.; formal analysis, C.A.L.-S., D.C.-Q., E.M.-M. and M.L.H.-C.; investigation, C.A.L.-S., E.G.-G., D.J.C.-M., B.S.R.-P., D.C.-Q., W.C.C.P. and M.M.-M.; data curation, D.J.C.-M. and B.S.R.-P.; writing—original draft preparation, C.A.L.-S. and E.M.-M.; writing—review and editing, D.C.-Q., U.R.Q.-Q., W.C.C.P., G.D.l.C., M.M.-M. and B.S.R.-P.; supervision, C.A.L.-S.; project administration, W.C.C.P. All authors have read and agreed to the published version of the manuscript.

Funding: The Research Group on Nutraceuticals and Biomaterials of the Universidad Nacional José María Arguedas.

Institutional Review Board Statement: Not applicable.

Informed Consent Statement: Not applicable.

Data Availability Statement: They are available in the same article.

Acknowledgments: The authors thank the Food Nanotechnology research laboratory of the Universidad Nacional José María Arguedas.

Conflicts of Interest: The authors declare no conflict of interest.

References

1. Halahlah, A.; Piironen, V.; Mikkonen, K.S.; Ho, T.M. Polysaccharides as wall materials in spray-dried microencapsulation of bioactive compounds: Physicochemical properties and characterization. *Crit. Rev. Food Sci. Nutr.* **2023**, *63*, 6983–7015. [CrossRef] [PubMed]
2. Řepka, D.; Kurillová, A.; Murtaja, Y.; Lapčík, L. Application of Physical-Chemical Approaches for Encapsulation of Active Substances in Pharmaceutical and Food Industries. *Foods* **2023**, *12*, 2189. [CrossRef] [PubMed]
3. Saldanha Coimbra, P.P.; de Souza Neves Cardoso, F.; Branco de Andrade Goncalves, E.C. Spray-drying wall materials: Relationship with bioactive compounds. *Crit. Rev. Food Sci. Nutr.* **2021**, *61*, 2809–2826. [CrossRef] [PubMed]
4. Zhao, D.; Li, Z.; Xia, J.; Kang, Y.; Sun, P.; Xiao, Z.; Niu, Y. Research progress of starch as microencapsulated wall material. *Carbohydr. Polym.* **2023**, *318*, 121118. [CrossRef] [PubMed]
5. Akbarbaglu, Z.; Peighambardoust, S.H.; Sarabandi, K.; Jafari, S.M. Spray drying encapsulation of bioactive compounds within protein-based carriers; different options and applications. *Food Chem.* **2021**, *359*, 129965. [CrossRef]
6. Xu, Y.; Dong, M.; Xiao, H.; Young Quek, S.; Ogawa, Y.; Ma, G.; Zhang, C. Advances in spray-dried probiotic microcapsules for targeted delivery: A review. *Crit. Rev. Food Sci. Nutr.* **2023**, 1–17. [CrossRef]
7. Samborska, K.; Boostani, S.; Geranpour, M.; Hosseini, H.; Dima, C.; Khoshnoudi-Nia, S.; Rostamabadi, H.; Falsafi, S.R.; Shaddel, R.; Akbari-Alavijeh, S.; et al. Green biopolymers from by-products as wall materials for spray drying microencapsulation of phytochemicals. *Trends Food Sci. Technol.* **2021**, *108*, 297–325. [CrossRef]
8. Otálora, M.C.; Wilches-Torres, A.; Gómez Castaño, J.A. Microencapsulation of Betaxanthin Pigments from Pitahaya (*Hylocereus megalanthus*) By-Products: Characterization, Food Application, Stability, and In Vitro Gastrointestinal Digestion. *Foods* **2023**, *12*, 2700. [CrossRef]
9. Durán, E.; Villalobos, C.; Churio, O.; Pizarro, F.; Valenzuela, C. Encapsulación de hierro: Otra estrategia para la prevención o tratamiento de la anemia por deficiencia de hierro. *Rev. Chil. De Nutr.* **2017**, *44*, 234–243. [CrossRef]
10. Ligarda-Samanez, C.A.; Moscoso-Moscoso, E.; Choque-Quispe, D.; Palomino-Rincón, H.; Martínez-Huamán, E.L.; Huamán-Carrión, M.L.; Peralta-Guevara, D.E.; Aroni-Huamán, J.; Arévalo-Quijano, J.C.; Palomino-Rincón, W.; et al. Microencapsulation of Erythrocytes Extracted from Cavia porcellus Blood in Matrices of Tara Gum and Native Potato Starch. *Foods* **2022**, *11*, 2107. [CrossRef]
11. Rodríguez Cruzado, C.P.; Salcedo Robles, C.; Morán González, C.V.; Lara Sosa, I.; Sánchez Banda, L.R.; Rodríguez Mázmela, C.L.; Obregón Domínguez, J.A. Formulación de una mezcla de hierro hemínico, cacao y camu-camu en polvo instantáneo a base de quinua roja. *Rev. De Innovación Y Transf. Product.* **2021**, *2*, e004. [CrossRef]
12. Gholam Jamshidi, E.; Behzad, F.; Adabi, M.; Esnaashari, S.S. Edible Iron-Pectin Nanoparticles: Preparation, Physicochemical Characterization and Release Study. *Food Bioprocess Technol.* **2023**. [CrossRef]
13. Piskin, E.; Cianciosi, D.; Gulec, S.; Tomas, M.; Capanoglu, E. Iron Absorption: Factors, Limitations, and Improvement Methods. *ACS Omega* **2022**, *7*, 20441–20456. [CrossRef] [PubMed]
14. Kandasamy, S.; Naveen, R. A review on the encapsulation of bioactive components using spray-drying and freeze-drying techniques. *J. Food Process Eng.* **2022**, *45*, e14059. [CrossRef]
15. Wilson, R.J.; Hui, Y.; Whittaker, A.K.; Zhao, C.-X. Facile bioinspired synthesis of iron oxide encapsulating silica nanocapsules. *J. Colloid Interface Sci.* **2021**, *601*, 78–84. [CrossRef]

16. Naktinienė, M.; Eisinaitė, V.; Keršienė, M.; Jasutienė, I.; Leskauskaitė, D.J.L. Emulsification and gelation as a tool for iron encapsulation in food-grade systems. *LWT* **2021**, *149*, 111895. [CrossRef]
17. Bamidele, O.P.; Emmambux, M.N. Encapsulation of bioactive compounds by "extrusion" technologies: A review. *Crit. Rev. Food Sci. Nutr.* **2021**, *61*, 3100–3118. [CrossRef]
18. Wang, L.; Clardy, A.; Hui, D.; Wu, Y. Physiochemical properties of encapsulated bitter melon juice using spray drying. *Bioact. Carbohydr. Diet. Fibre* **2021**, *26*, 100278. [CrossRef]
19. Barick, K.; Tripathi, A.; Dutta, B.; Shelar, S.B.; Hassan, P. Curcumin encapsulated casein nanoparticles: Enhanced bioavailability and anticancer efficacy. *J. Pharm. Sci.* **2021**, *110*, 2114–2120. [CrossRef]
20. Álvarez, C.; Pando, D. Encapsulation Technologies Applied to Food Processing. In *Food Formulation*; Wiley Online Library: Hoboken, NJ, USA, 2021; pp. 121–145.
21. Parra Huertas, R.A. Food microencapsulation: A review. *Rev. Fac. Nac. De Agron. Medellín* **2010**, *63*, 5669–5684.
22. Lishchynskyi, O.; Stetsyshyn, Y.; Raczkowska, J.; Awsiuk, K.; Orzechowska, B.; Abalymov, A.; Skirtach, A.G.; Bernasik, A.; Nastyshyn, S.; Budkowski, A. Fabrication and Impact of Fouling-Reducing Temperature-Responsive POEGMA Coatings with Embedded CaCO3 Nanoparticles on Different Cell Lines. *Materials* **2021**, *14*, 1417. [CrossRef] [PubMed]
23. Ali, F.; Kalva, S.N.; Koç, M. Additive Manufacturing of Polymer/Mg-Based Composites for Porous Tissue Scaffolds. *Polymers* **2022**, *14*, 5460. [CrossRef] [PubMed]
24. Zimmermann, M.B.; Windhab, E.J. Encapsulation of Iron and Other Micronutrients for Food Fortification. In *Encapsulation Technologies for Active Food Ingredients and Food Processing*; Zuidam, N.J., Nedovic, V., Eds.; Springer New York: New York, NY, USA, 2010; pp. 187–209.
25. Li, Y.O.; González, V.P.D.; Diosady, L.L. Chapter 31—Microencapsulation of vitamins, minerals, and nutraceuticals for food applications. In *Microencapsulation in the Food Industry (Second Edition)*; Sobel, R., Ed.; Academic Press: Cambridge, MA, USA, 2023; pp. 507–528.
26. Choque, Q.D.; Ramos, P.B.S.; Ligarda, S.C.A.; Barboza, P.G.I.; Kari, F.A.; Taipe, P.F.; Choque, Q.Y. Heavy metal removal by biopolymers-based formulations with native potato starch/nopal mucilage. *Rev. Fac. De Ing. Univ. De Antioq.* **2022**, *103*, 44–50. [CrossRef]
27. Choque-Quispe, D.; Froehner, S.; Ligarda-Samanez, C.A.; Ramos-Pacheco, B.S.; Palomino-Rincón, H.; Choque-Quispe, Y.; Solano-Reynoso, A.M.; Taipe-Pardo, F.; Zamalloa-Puma, L.M.; Calla-Florez, M.; et al. Preparation and Chemical and Physical Characteristics of an Edible Film Based on Native Potato Starch and Nopal Mucilage. *Polymers* **2021**, *13*, 3719. [CrossRef] [PubMed]
28. Ligarda-Samanez, C.A.; Choque-Quispe, D.; Moscoso-Moscoso, E.; Huamán-Carrión, M.L.; Ramos-Pacheco, B.S.; Peralta-Guevara, D.E.; Cruz, G.D.; Martínez-Huamán, E.L.; Arévalo-Quijano, J.C.; Muñoz-Saenz, J.C.; et al. Obtaining and Characterizing Andean Multi-Floral Propolis Nanoencapsulates in Polymeric Matrices. *Foods* **2022**, *11*, 3153. [CrossRef]
29. Ligarda-Samanez, C.A.; Choque-Quispe, D.; Allende-Allende, L.F.; Ramos Pacheco, B.S.; Peralta-Guevara, D.E. Calidad sensorial y proximal en conservas de mondongo de res (*Bos taurus*) en salsa de ají amarillo (*Capsicum baccatum*). *Cienc. Y Tecnol. Agropecu.* **2023**, *24*, 13. [CrossRef]
30. Choque-Quispe, D.; Ligarda-Samanez, C.A.; Huamán-Rosales, E.R.; Aguirre Landa, J.P.; Agreda Cerna, H.W.; Zamalloa-Puma, M.M.; Álvarez-López, G.J.; Barboza-Palomino, G.I.; Alzamora-Flores, H.; Gamarra-Villanueva, W. Bioactive Compounds and Sensory Analysis of Freeze-Dried Prickly Pear Fruits from an Inter-Andean Valley in Peru. *Molecules* **2023**, *28*, 3862. [CrossRef]
31. Ligarda-Samanez, C.A.; Palomino-Rincón, H.; Choque-Quispe, D.; Moscoso-Moscoso, E.; Arévalo-Quijano, J.C.; Huamán-Carrión, M.L.; Quispe-Quezada, U.R.; Muñoz-Saenz, J.C.; Gutiérrez-Gómez, E.; Cabel-Moscoso, D.J.; et al. Bioactive Compounds and Sensory Quality in Chips of Native Potato Clones (*Solanum tuberosum* spp. andigena) Grown in the High Andean Region of PERU. *Foods* **2023**, *12*, 2511. [CrossRef]
32. Galicia, L.; Miranda, A.; Gutiérrez, M.G.; Custodio, O.; Rosales, A.; Ruíz, N.; Surles, R.; Palacios, N. *Laboratorio de Calidad Nutricional de Maíz y Análisis de Tejido Vegetal: Protocolos de Laboratorio*; CIMMYT: Mexico City, Mexico, 2012.
33. Jimenez, M.D.; Lobo, M.; Sammán, N. 12th IFDC 2017 Special Issue—Influence of germination of quinoa (*Chenopodium quinoa*) and amaranth (*Amaranthus*) grains on nutritional and techno-functional properties of their flours. *J. Food Compos. Anal.* **2019**, *84*, 103290. [CrossRef]
34. Ligarda-Samanez, C.A.; Choque-Quispe, D.; Palomino-Rincón, H.; Ramos-Pacheco, B.S.; Moscoso-Moscoso, E.; Huamán-Carrión, M.L.; Peralta-Guevara, D.E.; Obregón-Yupanqui, M.E.; Aroni-Huamán, J.; Bravo-Franco, E.Y.; et al. Modified Polymeric Biosorbents from Rumex acetosella for the Removal of Heavy Metals in Wastewater. *Polymers* **2022**, *14*, 2191. [CrossRef]
35. Choque-Quispe, D.; Mojo-Quisani, A.; Ligarda-Samanez, C.A.; Calla-Florez, M.; Ramos-Pacheco, B.S.; Zamalloa-Puma, L.M.; Peralta-Guevara, D.E.; Solano-Reynoso, A.M.; Choque-Quispe, Y.; Zamalloa-Puma, A. Preliminary Characterization of a Spray-Dried Hydrocolloid from a High Andean Algae (*Nostoc sphaericum*). *Foods* **2022**, *11*, 1640. [CrossRef] [PubMed]
36. Ligarda-Samanez, C.A.; Choque-Quispe, D.; Moscoso-Moscoso, E.; Huamán-Carrión, M.L.; Ramos-Pacheco, B.S.; De la Cruz, G.; Arévalo-Quijano, J.C.; Muñoz-Saenz, J.C.; Muñoz-Melgarejo, M.; Quispe-Quezada, U.R.; et al. Microencapsulation of Propolis and Honey Using Mixtures of Maltodextrin/Tara Gum and Modified Native Potato Starch/Tara Gum. *Foods* **2023**, *12*, 1873. [CrossRef] [PubMed]

37. Ligarda-Samanez, C.A.; Choque-Quispe, D.; Moscoso-Moscoso, E.; Palomino-Rincón, H.; Taipe-Pardo, F.; Aguirre Landa, J.P.; Arévalo-Quijano, J.C.; Muñoz-Saenz, J.C.; Quispe-Quezada, U.R.; Huamán-Carrión, M.L.; et al. Nanoencapsulation of Phenolic Extracts from Native Potato Clones (*Solanum tuberosum* spp. andigena) by Spray Drying. *Molecules* 2023, *28*, 4961. [CrossRef]
38. Martínez, P.; Vilcarromero, D.; Pozo, D.; Peña, F.; Manuel Cervantes-Uc, J.; Uribe-Calderon, J.; Velezmoro, C. Characterization of starches obtained from several native potato varieties grown in Cusco (Peru). *J. Food Sci.* 2021, *86*, 907–914. [CrossRef] [PubMed]
39. Kaur, L.; Singh, N.; Sodhi, N.S. Some properties of potatoes and their starches II. Morphological, thermal and rheological properties of starches. *Food Chem.* 2002, *79*, 183–192. [CrossRef]
40. Diaz Barrera, Y. Determinación de las Propiedades Físicas, Químicas, Tecnofuncionales y la Estabilidad en Congelación/Descongelación del Almidón de Cuatro Variedades de *Solanum tuberosum* ssp. *andigenum* (Papa Nativa). 2015. Available online: https://repositorio.unajma.edu.pe/handle/20.500.14168/213 (accessed on 26 September 2023).
41. Kurdziel, M.; Łabanowska, M.; Pietrzyk, S.; Pająk, P.; Królikowska, K.; Szwengiel, A. The effect of UV-B irradiation on structural and functional properties of corn and potato starches and their components. *Carbohydr. Polym.* 2022, *289*, 119439. [CrossRef]
42. Churio, O.; Valenzuela, C. Development and characterization of maltodextrin microparticles to encapsulate heme and non-heme iron. *LWT* 2018, *96*, 568–575. [CrossRef]
43. Pashazadeh, H.; Zannou, O.; Ghellam, M.; Koca, I.; Galanakis, C.M.; Aldawoud, T.M.S. Optimization and Encapsulation of Phenolic Compounds Extracted from Maize Waste by Freeze-Drying, Spray-Drying, and Microwave-Drying Using Maltodextrin. *Foods* 2021, *10*, 1396. [CrossRef]
44. Abe, E.; Fuwa, T.J.; Hoshi, K.; Saito, T.; Murakami, T.; Miyajima, M.; Ogawa, N.; Akatsu, H.; Hashizume, Y.; Hashimoto, Y. Expression of Transferrin Protein and Messenger RNA in Neural Cells from Mouse and Human Brain Tissue. *Metabolites* 2022, *12*, 594. [CrossRef]
45. Furuta, T.; Neoh, T.L. Microencapsulation of food bioactive components by spray drying: A review. *Dry. Technol.* 2021, *39*, 1800–1831. [CrossRef]
46. Rezvankhah, A.; Emam-Djomeh, Z.; Askari, G. Encapsulation and delivery of bioactive compounds using spray and freeze-drying techniques: A review. *Dry. Technol.* 2020, *38*, 235–258. [CrossRef]
47. Zhang, H.; Gong, T.; Li, J.; Pan, B.; Hu, Q.; Duan, M.; Zhang, X. Study on the Effect of Spray Drying Process on the Quality of Microalgal Biomass: A Comprehensive Biocomposition Analysis of Spray-Dried S. acuminatus Biomass. *BioEnergy Res.* 2022, *15*, 320–333. [CrossRef]
48. Samborska, K.; Poozesh, S.; Barańska, A.; Sobulska, M.; Jedlińska, A.; Arpagaus, C.; Malekjani, N.; Jafari, S.M. Innovations in spray drying process for food and pharma industries. *J. Food Eng.* 2022, *321*, 110960. [CrossRef]
49. Kaul, S.; Kaur, K.; Mehta, N.; Dhaliwal, S.S.; Kennedy, J.F. Characterization and optimization of spray dried iron and zinc nanoencapsules based on potato starch and maltodextrin. *Carbohydr. Polym.* 2022, *282*, 119107. [CrossRef] [PubMed]
50. Premi, M.; Sharma, H. Effect of different combinations of maltodextrin, gum arabic and whey protein concentrate on the encapsulation behavior and oxidative stability of spray dried drumstick (*Moringa oleifera*) oil. *Int. J. Biol. Macromol.* 2017, *105*, 1232–1240. [CrossRef] [PubMed]
51. Ruengdech, A.; Siripatrawan, U. Improving encapsulating efficiency, stability, and antioxidant activity of catechin nanoemulsion using foam mat freeze-drying: The effect of wall material types and concentrations. *LWT* 2022, *162*, 113478. [CrossRef]
52. Zotarelli, M.F.; da Silva, V.M.; Durigon, A.; Hubinger, M.D.; Laurindo, J.B. Production of mango powder by spray drying and cast-tape drying. *Powder Technol.* 2017, *305*, 447–454. [CrossRef]
53. Tuyen, C.K.; Nguyen, M.H.; Roach, P.D. Effects of spray drying conditions on the physicochemical and antioxidant properties of the Gac (*Momordica cochinchinensis*) fruit aril powder. *J. Food Eng.* 2010, *98*, 385–392. [CrossRef]
54. Ricci, A.; Mejia, J.A.A.; Versari, A.; Chiarello, E.; Bordoni, A.; Parpinello, G.P. Microencapsulation of polyphenolic compounds recovered from red wine lees: Process optimization and nutraceutical study. *Food Bioprod. Process.* 2022, *132*, 1–12. [CrossRef]
55. Chng, G.Y.V.; Chang, L.S.; Pui, L.P. Effects of maltodextrin concentration and inlet temperature on the physicochemical properties of spray-dried kuini powder. *Asia Pac. J. Mol. Biol. Biotechnol.* 2020, *28*, 113–131. [CrossRef]
56. Mahdi Jafari, S.; Masoudi, S.; Bahrami, A. A Taguchi approach production of spray-dried whey powder enriched with nanoencapsulated vitamin D3. *Dry. Technol.* 2019, *37*, 2059–2071. [CrossRef]
57. Piñón-Balderrama, C.I.; Leyva-Porras, C.; Terán-Figueroa, Y.; Espinosa-Solís, V.; Álvarez-Salas, C.; Saavedra-Leos, M.Z. Encapsulation of active ingredients in food industry by spray-drying and nano spray-drying technologies. *Processes* 2020, *8*, 889. [CrossRef]
58. Chopde, S.; Datir, R.; Deshmukh, G.; Dhotre, A.; Patil, M.; Research, F. Nanoparticle formation by nanospray drying & its application in nanoencapsulation of food bioactive ingredients. *J. Agric.* 2020, *2*, 100085. [CrossRef]
59. Baldelli, A.; Wells, S.; Pratap-Singh, A. Impact of product formulation on spray-dried microencapsulated zinc for food fortification. *Food Bioprocess Technol.* 2021, *14*, 2286–2301. [CrossRef]
60. Bordón, M.G.; Alasino, N.P.X.; Villanueva-Lazo, Á.; Carrera-Sánchez, C.; Pedroche-Jiménez, J.; del Carmen Millán-Linares, M.; Ribotta, P.D.; Martínez, M.L. Scale-up and optimization of the spray drying conditions for the development of functional microparticles based on chia oil. *Food Bioprod. Process.* 2021, *130*, 48–67. [CrossRef]
61. Caruana, R.; Montalbano, F.; Zizzo, M.G.; Puleio, R.; Caldara, G.; Cicero, L.; Cassata, G.; Licciardi, M. Enhanced anticancer effect of quercetin microparticles formulation obtained by spray drying. *Int. J. Food Sci. Technol.* 2022, *57*, 2739–2746. [CrossRef]

62. de Moura, S.C.S.R.; Schettini, G.N.; Gallina, D.A.; Dutra Alvim, I.; Hubinger, M.D. Microencapsulation of hibiscus bioactives and its application in yogurt. *J. Food Process. Preserv.* **2022**, *46*, e16468. [CrossRef]
63. Arpagaus, C.; John, P.; Collenberg, A.; Rütti, D. Nanocapsules formation by nano spray drying. In *Nanoencapsulation Technologies for the Food and Nutraceutical Industries*; Elsevier: Amsterdam, The Netherlands, 2017; pp. 346–401.
64. Moslemi, M.; Hosseini, H.; Ertan, M.; Mortazavian, A.M.; Fard, R.M.N.; Neyestani, T.R.; Komeyli, R. Characterisation of spray-dried microparticles containing iron coated by pectin/resistant starch. *Int. J. Food Sci. Technol.* **2014**, *49*, 1736–1742. [CrossRef]
65. Nayak, A.K.; Pal, D. Formulation optimization and evaluation of jackfruit seed starch–alginate mucoadhesive beads of metformin HCl. *Int. J. Biol. Macromol.* **2013**, *59*, 264–272. [CrossRef]
66. Roy, A.; Bajpai, J.; Bajpai, A.K. Dynamics of controlled release of chlorpyrifos from swelling and eroding biopolymeric microspheres of calcium alginate and starch. *Carbohydr. Polym.* **2009**, *76*, 222–231. [CrossRef]
67. Nandiyanto, A.B.D.; Oktiani, R.; Ragadhita, R. How to read and interpret FTIR spectroscope of organic material. *Indones. J. Sci. Technol.* **2019**, *4*, 97–118. [CrossRef]
68. Santiago-Adame, R.; Medina-Torres, L.; Gallegos-Infante, J.; Calderas, F.; González-Laredo, R.; Rocha-Guzmán, N.; Ochoa-Martínez, L.; Bernad-Bernad, M. Spray drying-microencapsulation of cinnamon infusions (*Cinnamomum zeylanicum*) with maltodextrin. *LWT-Food Sci. Technol.* **2015**, *64*, 571–577. [CrossRef]
69. Asghari-Varzaneh, E.; Shahedi, M.; Shekarchizadeh, H. Iron microencapsulation in gum tragacanth using solvent evaporation method. *Int. J. Biol. Macromol.* **2017**, *103*, 640–647. [CrossRef] [PubMed]
70. Valenzuela, C.; Hernández, V.; Morales, M.S.; Neira-Carrillo, A.; Pizarro, F. Preparation and characterization of heme iron-alginate beads. *LWT-Food Sci. Technol.* **2014**, *59*, 1283–1289. [CrossRef]
71. Ghayempour, S.; Montazer, M.; Rad, M.M. Tragacanth gum as a natural polymeric wall for producing antimicrobial nanocapsules loaded with plant extract. *Int. J. Biol. Macromol.* **2015**, *81*, 514–520. [CrossRef] [PubMed]
72. Santos, M.B.; Isabel, I.C.A.; Garcia-Rojas, E.E. Ultrasonic depolymerization of aqueous tara gum solutions: Kinetic, thermodynamic and physicochemical properties. *J. Sci. Food Agric.* **2022**, *102*, 4640–4646. [CrossRef]
73. Wardhani, D.H.; Wardana, I.N.; Ulya, H.N.; Cahyono, H.; Kumoro, A.C.; Aryanti, N. The effect of spray-drying inlet conditions on iron encapsulation using hydrolysed glucomannan as a matrix. *Food Bioprod. Process.* **2020**, *123*, 72–79. [CrossRef]
74. Başyiğit, B.; Sağlam, H.; Kandemir, Ş.; Karaaslan, A.; Karaaslan, M. Microencapsulation of sour cherry oil by spray drying: Evaluation of physical morphology, thermal properties, storage stability, and antimicrobial activity. *Powder Technol.* **2020**, *364*, 654–663. [CrossRef]
75. Kaur, R.; Kaur, K. Effect of processing on color, rheology and bioactive compounds of different sweet pepper purees. *Plant Foods Hum. Nutr.* **2020**, *75*, 369–375. [CrossRef]
76. Singh, A.P.; Siddiqui, J.; Diosady, L.L. Characterizing the pH-dependent release kinetics of food-grade spray drying encapsulated iron microcapsules. *Food Bioprocess Technol.* **2018**, *11*, 435–446. [CrossRef]
77. Pratap-Singh, A.; Leiva, A. Double fortified (iron and zinc) spray-dried microencapsulated premix for food fortification. *LWT* **2021**, *151*, 112189. [CrossRef]
78. Li, Y.O. *Development of Microencapsulation-Based Technologies for Micronutrient Fortification in Staple Foods for Developing Countries*; University of Toronto: Toronto, ON, Canada, 2009.

Disclaimer/Publisher's Note: The statements, opinions and data contained in all publications are solely those of the individual author(s) and contributor(s) and not of MDPI and/or the editor(s). MDPI and/or the editor(s) disclaim responsibility for any injury to people or property resulting from any ideas, methods, instructions or products referred to in the content.

Article

Preservation of Fresh-Cut 'Maradol' Papaya with Polymeric Nanocapsules of Lemon Essential Oil or Curcumin

Moises Job Galindo-Pérez [1,2], Lizbeth Martínez-Acevedo [3,4], Gustavo Vidal-Romero [2,4], Luis Eduardo Serrano-Mora [4] and María de la Luz Zambrano-Zaragoza [5,*]

1. Departamento de Procesos y Tecnología, Universidad Autónoma Metropolitana, Unidad Cuajimalpa, Av. Vasco de Quiroga 4871, Santa Fe Cuajimalpa, Ciudad de Mexico 05348, Ciudad de Mexico, Mexico; moises.galindo@zaragoza.unam.mx
2. Departamento del Área Farmacéutica, Facultad de Estudios Superiores Zaragoza, Universidad Nacional Autónoma de México, Campus II, Col. Ejército de Oriente, Iztapalapa, Ciudad de México 09230, Ciudad de Mexico, Mexico; gustavo.vidal@zaragoza.unam.mx
3. Departamento de Sistemas Biológicos, Universidad Autónoma Metropolitana, Unidad Xochimilco, Calzada del Hueso 1100, Col. Villa Quietud, Coyoacán, Ciudad de Mexico 04960, Ciudad de Mexico, Mexico; liz_martinez@comunidad.unam.mx
4. Laboratorio de Posgrado e Investigación en Tecnología Farmacéutica, Facultad de Estudios Superiores Cuautitlán, Universidad Nacional Autónoma de México, Av. 1o de Mayo s/n, Cuautitlán Izcalli 54745, Estado de Mexico, Mexico; luedserrano_h@hotmail.com
5. Laboratorio de Procesos de Transformación de Alimentos y Tecnologías Emergentes, Departamento de Ingeniería y Tecnología, Facultad de Estudios Superiores Cuautitlán, Universidad Nacional Autónoma de México, Km 2.5 Carretera Cuautitlán–Teoloyucan, San Sebastián Xhala, Cuautitlán Izcalli 54714, Estado de Mexico, Mexico
* Correspondence: luz.zambrano@unam.mx; Tel.: +52-5556231999 (ext. 39406); Fax: +52-5556232077

Citation: Galindo-Pérez, M.J.; Martínez-Acevedo, L.; Vidal-Romero, G.; Serrano-Mora, L.E.; Zambrano-Zaragoza, M.d.l.L. Preservation of Fresh-Cut 'Maradol' Papaya with Polymeric Nanocapsules of Lemon Essential Oil or Curcumin. *Polymers* 2023, *15*, 3515. https://doi.org/10.3390/polym15173515

Academic Editor: Cristina Cazan

Received: 31 July 2023
Revised: 13 August 2023
Accepted: 18 August 2023
Published: 23 August 2023

Copyright: © 2023 by the authors. Licensee MDPI, Basel, Switzerland. This article is an open access article distributed under the terms and conditions of the Creative Commons Attribution (CC BY) license (https://creativecommons.org/licenses/by/4.0/).

Abstract: Papaya is one of the most consumed fruits in the world; however, tissue damage caused by cuts quickly leads to its decay. Therefore, this study aimed to prepare and characterize lemon oil and curcumin nanocapsules to evaluate their capacity for preserving fresh-cut papaya. Lemon essential oil and curcumin nanocapsules were prepared using ethyl cellulose (EC) and poly-(ε-caprolactone) (PCL) by the emulsification–diffusion method coupled with ultrasound. The particles had sizes smaller than 120 nm, with polydispersity indices below 0.25 and zeta potentials exceeding -12 mV, as confirmed by scanning electron microscopy. The nanoparticles remained stable for 27 days, with sedimentation being the instability mechanism observed. These nanoparticles were employed to coat fresh-cut papaya, which was stored for 17 days. The results demonstrated their remarkable efficacy in reducing the respiration rate. Furthermore, nanocapsules maintained the pH and acidity levels of the papayas for an extended period. The lemon oil/EC nanocapsule treatment retained the color better. Additionally, all systems exhibited the ability to minimize texture loss associated with reduced pectin methylesterase activity. Finally, the nanocapsules showed a notable reduction in polyphenol oxidase activity correlating with preserving total phenolic compounds in the fruit. Therefore, the lemon oil and curcumin nanoparticles formed using EC and PCL demonstrated their effectiveness in preserving fresh-cut 'Maradol' papaya.

Keywords: ethyl cellulose; poly-(ε-caprolactone); polymeric nanoparticles; papaya conservation; essential oils; curcumin

1. Introduction

Sales of fresh-cut fruits have increased significantly in international markets, with a global annual increase of 6%. These products are ready to eat while maintaining freshness and nutritional quality [1]. Fresh-cut products are minimally processed fruits altered from their original form by peeling, slicing, dicing, cutting into strips, coring, or other similar methods, with or without washing or other treatments, before being packaged for consumer or retail use [2]. However, due to tissue cutting, a series of physicochemical and biochemical changes occur that promote a decrease in their shelf-life [3].

Papaya (*Carica papaya*) is a climacteric fruit rich in nutrients such as provitamin A, carotenoids, vitamin C, vitamin B, lycopene, dietary fiber, and minerals. It has laxative properties, reduces indigestion, and has been studied for its potential to prevent heart diseases and various types of cancer [4]. Consequently, papaya is the third most consumed tropical fruit in the world. It holds great economic and social importance, providing income for thousands of families and serving as a source of foreign exchange for producing countries. Mexico is the third-largest papaya producer globally [5]. However, fresh-cut papaya is highly perishable due to the side effects associated with tissue cutting, which accelerate respiration rate, ethylene production, and the overproduction of enzymes, ultimately leading to a decline in organoleptic and nutritional characteristics of the fruit [6]. Various alternatives have been proposed to increase the shelf-life of fresh-cut papaya, including chemical treatments, edible hydrocolloid-based coatings, and modified atmospheres, all of which have shown improvements in papaya quality [7–10]. Essential oils are secondary metabolites of plants and possess antioxidant, antimicrobial, and antifungal properties and can be incorporated into the treatments mentioned above to enhance the preservation of fresh-cut fruits [11].

Lemon essential oil is extracted from *Citrus lemon* and is a mixture of terpenes and terpenoids, with α-limonene being the main compound, accounting for approximately 60% of its composition. Lemon essential oil has demonstrated potent antimicrobial activity, inhibiting the growth of microorganisms such as *Aeromonas*, *Candida*, *Enterococcus*, *Escherichia*, and *Staphylococcus* [12,13]. It exhibits strong antioxidant properties [14]. Curcumin, the principal curcuminoid found in turmeric (*Curcuma longa*), possesses antioxidant, anti-inflammatory, antiviral, and antifungal properties. It demonstrates antioxidant activity comparable to that of vitamins C, E, and β-carotene, making turmeric a potential option for cancer prevention, liver protection, and the prevention of premature aging [15,16]. Due to its beneficial properties, curcumin has been utilized for the preservation of fresh-cut pineapple [17], pear [18], and apple [19].

Recently, nanotechnology has emerged as a promising tool for preserving fresh-cut products. Nanotechnology involves the application of material knowledge at the nanoscale and has found applications in various scientific fields, including the food processing chain. Inorganic nanoparticles such as ZnO in chitosan coatings [20] and montmorillonite embedded in a whitemouth croaker protein isolated matrix [21], oregano essential oil-based nanoemulsion [22], and citral-based nanoemulsion [23] have been used for the preservation of fresh-cut papaya. Polymeric nanoparticles include nanocapsules and nanospheres. Nanocapsules (NCs) are vesicular structures capable of encapsulating substances with hydrophobic or hydrophilic characteristics surrounded by a polymeric barrier. Meanwhile, nanospheres are dense polymeric matrices in which the compounds are dispersed, dissolved, or chemically bound to the polymer matrix [24]. Polymeric nanoparticles have been used to encapsulate active compounds, which protect the encapsulated components from the external environment, reducing their degradation caused by heat, light, oxygen, and pH, thereby increasing their physicochemical stability [25]. In addition, polymeric nanoparticles can enable controlled release of the encapsulated components, maximizing their functionality. Materials for nanoencapsulation of active compounds are generally biodegradable synthetic polymers or FDA-approved semisynthetic or natural biopolymers for food contact [26]. Ethyl cellulose (EC) is a polymer derived from cellulose approved by the FDA as a food additive [27]. Moreover, poly-(ε-caprolactone) (PCL) is a biodegradable polyester approved by the FDA for use in drug delivery systems and widely used in food packaging [28].

Therefore, this research aimed to develop and characterize nanocapsules of lemon essential oil and curcumin using EC and PCL as biopolymer coatings for preserving fresh-cut 'Maradol' papaya.

2. Materials and Methods

2.1. Chemical Materials

Poly-(ε-caprolactone) (Mw ≈ 80,000), polyvinyl alcohol (PVA), lemon essential oil (*Citrus lemon*), and curcumin (*Curcuma longa*) were purchased from Sigma-Aldrich®

(St. Louis, MO, USA). Analytical grade ethyl acetate was acquired from Fermont® (Mexico City, Mexico). Octenyl succinic anhydride starch (OSA-starch) was supplied from Makymat® (Naucalpan, State of Mexico, Mexico). All other used reagents were at least analytical grade.

2.2. Biological Material

Papaya (*Carica papaya*) var. 'Maradol' fruits were obtained from a fruit distribution center in the area (Cuautitln Izcalli, State of Mexico, Mexico). They were selected according to their size and shape, with a maturity level of 5 (80–90% yellow surface) and without the presence of mechanical or microbiological damage.

2.3. Polymeric Nanocapsule Preparation

The nanocapsules (NCs) were prepared by the emulsification–diffusion method using ultrasonic homogenization according to the optimized conditions by Galindo-Pérez et al. (2018) [29]. Briefly, the stabilizers (PVA and OSA-starch) were dissolved in the aqueous phase at a concentration of 30 g/L. PCL or EC (307 mg) and lemon essential oil or curcumin (237 mg) were dissolved in the organic phase (water saturated ethyl acetate). Subsequently, both solutions were emulsified with an ultrasonic homogenizer at a frequency of 26 kHz (UP200HT; Helshier; Teltow, Germany) with a sonotrode of 14 mm in diameter. The homogenization time was 4 min at an ultrasonic power of 54 W using an external ice bath as temperature control. After obtaining the emulsion, 180 mL of water was added to induce diffusion of solvent and aggregation of polymer with the formation of the nanoparticles. The diffusive process was carried out under the same conditions as the emulsion formation. Finally, the organic solvent was removed with a vacuum rotary evaporator (HB10; IKA® Works, Inc.; Wilmington, NC, USA) at 30 °C and a reduced pressure of 66.6 Pa.

2.4. Nanocapsule Characterization

2.4.1. Particle Size (Ps) and Polydispersity Index (PDI)

The dynamic light scattering technique was used for Ps and PDI measurement in a Malvern Zarasizer Nano ZS90 (Malvern Instruments Ltd.; Malvern, Worcestershire, UK) at a detection angle of 90° and a laser λ = 633 nm. One milliliter of each colloidal dispersion was diluted 10 times with distilled water. The measurements were performed in triplicate.

2.4.2. Zeta Potential (ζ) of Nanoparticles

The electrophoretic movement of particles in dispersion was measured to obtain the zeta potential in a Zarasizer Nano ZS90 using polystyrene dispersions (ζ = -55 mV) as a reference. This parameter indicates the surface charge of the particles and the degree of repulsion between adjacent particles. All measurements were carried out in triplicate.

2.4.3. Morphological Characterization of Nanocapsules

Polymeric NCs were placed on a glass slide in a refrigerated desiccator until the water evaporated entirely. The samples were coated with gold ($\approx$2 nm) using a fine coat ion sputter deposition unit (JFC-1100 fine coat ion sputter; JEOL Ltd.; Akishima, Japan) and observed under a scanning electron microscope (SEM; JSM 5600 LV-SEM® LV; JEOL Ltd.; Akishima, Japan) with a resolution of 5 nm. An electron beam of 28 kV and a chamber pressure of 12–20 Pa were the operational conditions.

2.4.4. Determination of Nanocapsule Stability

The physical stability of the nanodispersions prepared with lemon essential oil and curcumin was analyzed using a Turbiscan® Classic instrument (Toulouse, France). Five mL of each dispersion was placed in a cylindrical glass cell to ensure no air bubbles in the sample. The transmitted light was measured using a transmission detector, while the light scattered at 30° was detected using a backscattered sensor. The detection length was 55 mm, and measurements were obtained at 40 µm intervals along the sample. Measurements were carried out for four weeks. Measurements were taken at 0, 144, 193, 216, 482, 531, and 667 h (27.8 days). The Turbiscan Stability Index (TSI) was obtained from the backscattering data using Equation (1):

$$\text{TSI} = \sum_i \frac{\sum_h |\text{scan}_i(h) - \text{scan}_{i-1}(h)|}{H} \quad (1)$$

where $\text{scan}_i(h)$ is the average backscattering for each measurement time (i), $\text{scan}_{i-1}(h)$ is the average backscattering for the previous time $(i-1)$, and H is the sample height [30].

2.5. Application of Nanocoatings on Fresh-Cut Papaya

The selected papayas were washed and sanitized in a solution of iodine fruit detergent (2 g/L). Then, these were peeled, cut into approximately 1 cm cubed pieces, and immersed in a $CaCl_2$ solution (10 g/L) for 3 min. Afterward, the papaya pieces were drained for 3 min and immersed in the different nanodispersions for 3 min. The evaluated treatments were: nanocapsules of lemon essential oil/PCL (NC L/PCL), nanocapsules of lemon essential oil/EC (NC L/EC), nanocapsules of curcumin/PCL (NC C/PCL), and nanocapsules of curcumin/EC (NC C/EC). In addition, fresh-cut papaya without any treatment was considered as the control treatment. The fresh-cut papaya was packaged in crystalline polypropylene cups (approximately 100 g per cup) and stored at 4 °C for 17 days.

2.6. Respiration Rate of Fresh-Cut Papaya Treated with Nanodispersions

The respiration rate was determined using the static method reported by Wang et al. (2009) [31] and Iqbal et al. (2008) [32] in the treated and stored papaya. The CO_2 measurements were determined by measuring the headspace gas using a needle inserted through the container lid and analyzed using an O_2/CO_2 analyzer (Quantek Instruments model 905; Grafton, Massachusetts, USA) to obtain the volumetric fraction of CO_2 and O_2 inside the container. The CO_2 production was calculated based on the difference in CO_2 concentrations at different time intervals. The measurements were conducted during the storage period, in triplicate, and expressed as follows for CO_2:

$$RCO_2 = \frac{(yCO_2 - y_iCO_2)}{(t - t_i)} * \frac{V_f}{W} \quad (2)$$

where $yiCO_2$ is the initial concentration in the mixture (volumetric fraction), yCO_2 is the CO_2 concentration at any other time, t is any non-zero time expressed in hours ($t_i = 0$), RCO_2 is the CO_2 production rate, W is the mass of the product (kg), and V_f is the volume (mL) inside the container.

The O_2 consumption rate was calculated as Equation (3):

$$RO_2 = \frac{(yO_2 - y_iO_2)}{(t - t_i)} * \frac{V_f}{W} \quad (3)$$

where RO_2 is the O_2 consumption rate, yiO_2 is the initial concentration in the mixture (volumetric fraction), yO_2 is the O_2 concentration at any other time, t is any non-zero time expressed in hours ($t_i = 0$), W is the mass of the product (kg), and V_f is the volume (mL) inside the container.

2.7. Color Determination

The coloration of the treated and untreated papaya was determined by obtaining the values of L^*, a^*, and b^* using a Minolta CM-600 colorimeter calibrated with $L^* = 57.79$, $a^* = -1.09$, $b^* = 7.57$. Color measurements were taken on the cut and treated surface of the fresh-cut papaya (on any side of the cube) in triplicate during the storage period.

2.8. Determination of Firmness in Fresh-Cut Papaya

The firmness of the stored papaya was measured using a texture analyzer (CT3 Texture Analyzer; Brookfield AMETEK; Middleborough, MA, USA). The cut papaya was penetrated using a stainless steel flat-bottomed cylindrical probe with a diameter of 4 mm at a speed of 1 mm/s and a target depth of 5 mm. The measurements were performed in triplicate.

2.9. Measurement of Pectin Methylesterase (PME) Activity

The extraction of PME enzymes from papaya was performed following the protocol described by Zambrano-Zaragoza et al. (2014) [33]. Briefly, 20 g of fresh-cut papaya was homogenized with a NaCl solution (2 M). The homogenate was stirred for 10 min and then centrifuged at $7277 \times g$ for 20 min, with the supernatant containing the enzymatic extract. For the determination of PME activity, the protocol described by Hagerman and Austin (1986) [34] was followed. Briefly, in a spectrophotometric cell, 1 mL of citrus pectin (10 g/L), 580 mL of water, 200 µL of bromothymol blue (0.1 g/L), and NaCl (0.2 M) were added. Adding 200 µL of enzymatic extract initiated the reaction, and the decrease in absorbance at 640 nm (Genesys 10 UV-Vis; Thermo Fisher Scientific Inc.; Waltham, MA, USA) was measured for 3 min. The number of µmoles of released acid due to pectin methylesterase action was determined from a galacturonic acid standard curve following the treatment described by Hagerman and Austin (1986) [34].

2.10. Evolution of Polyphenol Oxidase (PPO) Activity

The polyphenol oxidase (PPO) was obtained by the methodology described by Galindo-Pérez et al. (2015) [32]. Briefly, 20 g of fresh-cut papaya was homogenized with 20 mL of a phosphate buffer solution (0.2 M; pH = 7.0). The homogenate was stirred for 10 min and then centrifuged at $7277 \times g$ for 20 min. The supernatant obtained was used as the enzymatic extract for measuring PPO activity. The PPO activity was determined by a reaction mixture of 2.8 mL citrate-phosphate buffer solution (0.2 M; pH = 6.5) containing catechol (50 mM) and 200 µL of enzymatic extract. The solution was gently stirred, and the increase in absorbance at 420 nm (Genesys 10 UV-Vis; Thermo Fisher Scientific Inc.; Waltham, MA, USA) was measured. One unit of polyphenol oxidase activity was defined as the change in absorbance per minute (0.001 Abs/min). The measurements were performed in triplicate.

2.11. Protein Determination

The analytical determination of protein in the enzymatic extracts was performed using the technique proposed by Bradford (1976) [35]. For this purpose, 100 µL of the enzymatic extract was mixed with 5 mL of Bradford reagent. The mixture was gently agitated and kept in the dark for 10 min, after which the absorbance was measured at 595 nm (Genesys 10 UV-Vis; Thermo Fisher Scientific Inc.; Waltham, MA, USA) using a bovine serum albumin standard curve under the same conditions.

2.12. Total Phenolic Compound Measurement

The extraction of total phenols was performed according to the technique reported by Waterhouse (2002) [36]. First, 20 g of fresh-cut papaya was homogenized with 20 mL of a methanol–water mixture (95%). The resulting solution was centrifuged at $7277 \times g$ for 20 min, and the supernatant contained the phenolic mixture. The quantification of total phenols was performed using the Folin–Ciocâlteu method. Briefly, 20 µL of the phenolic extract was mixed with 1.8 mL of water and 100 µL of Folin reagent and agitated for 5 min; then 300 µL of sodium carbonate was added. The mixture was kept in the dark for 1 h. The content of phenols was determined spectrophotometrically at 765 nm (Genesys 10 UV-Vis, Thermo Fisher Scientific Inc., USA) using a previously prepared gallic acid standard curve ($y = 0.0009x + 0.0153$; $R^2 = 0.996$). The results were expressed as gallic acid equivalents (GAE) per 100 g of fruit. The determinations were performed in triplicate.

2.13. Statistical Analysis

An analysis of variance (ANOVA) was conducted to analyze the effect of the different treatments on papaya preservation. The significance of the differences was determined using Tukey's multiple comparisons tests. Additionally, a Dunnett's mean comparison test was used to assess the effect of each treatment compared to the control group [37]. Differences were considered significant with p-values < 0.05.

3. Results

3.1. Characterization of Nanoparticles

3.1.1. Particle Size, Polydispersity Index, and Zeta Potential

Table 1 presents the results of NC particle size, polydispersity index, and zeta potential. The nanoparticles had particle sizes in the nanoscale range, with the NC L/PCL system having the smallest size. The PDI, with values from 0.126 to 0.246, indicates monodispersed systems with a narrow size distribution. The NC L/PCL and NC L/EC had PDI $\leq$ 0.150. The zeta potential of NCs formed ranged from -5.80 to -12.43 mV.

Table 1. Nanocapsule characterization of lemon essential oil and curcumin extract using EC and PCL as encapsulant polymers.

System	Ps (nm)	PDI	ζ (mV)
NC L/EC	116 ± 0.72	0.126 ± 0.02	−7.52 ± 0.53
NC L/PCL	87.57 ± 0.34	0.142 ± 0.01	−12.43 ± 0.98
NC C/EC	115.73 ± 0.83	0.246 ± 0.01	−6.15 ± 1.04
NC C/PCL	100.18 ± 0.98	0.167 ± 0.01	−5.80 ± 0.78

Ps = particle size; PDI = polydispersity index; ζ = zeta potential.

3.1.2. Morphology of Nanocapsules

Figure 1 shows the micrographs of the nanocapsules, showing that all systems had sizes $\leq$ 500 nm and spherical shapes. NC L/PCL and NC L/EC sizes are consistent with dynamic light scattering results (200 nm). Furthermore, NC C/PCL and NC C/EC exhibited spherical structures with Ps ranging from 150 to 200 nm.

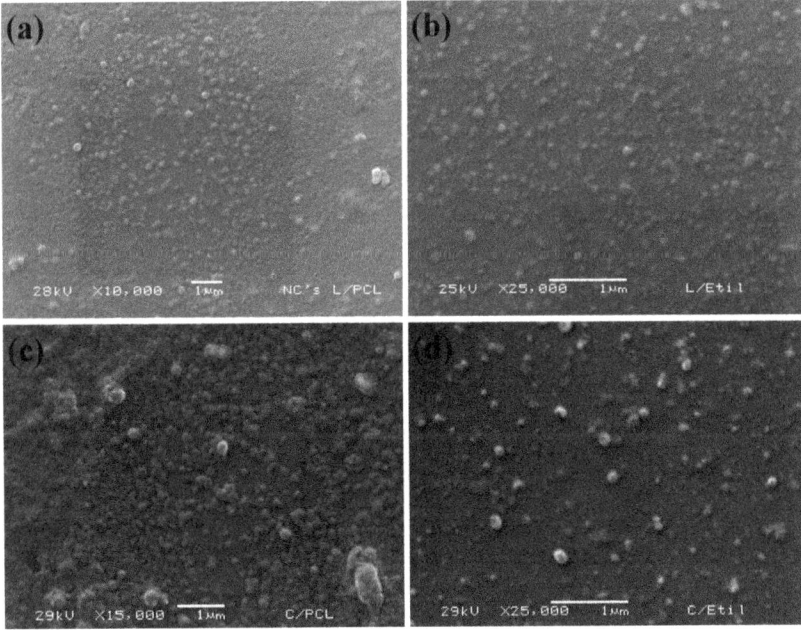

Figure 1. SEM micrographs of the different dispersions formed using the emulsification–diffusion method coupled with ultrasound: (**a**) NC L/PCL; (**b**) NC L/EC; (**c**) NC C/PCL; (**d**) NC C/EC.

3.1.3. Instability Mechanism of Nanocapsules

Table 2 presents the instability mechanism of the NCs and the migration velocities of nanoscale particles in suspension obtained by the Turbiscan® instrument. The NC L/PCL and NC C/PCL, as well as NC C/PCL and NC C/EC, were stable at room temperature during 667 h of storage. No significant changes in backscattering were observed; sedimentation was identified as the instability mechanism in all cases. This behavior is related to the calculated migration velocities of the particles. The NC L/EC and NC L/PCL had migration velocities in the bottom of 0.023 and 0.010 µm/min, respectively. The migration velocities in the top container for lemon oil NCs were 0.015 and 0.013 µm/min, demonstrating the high physical stability of the nanocapsules.

Table 2. Instability mechanism and migration velocity of lemon oil and curcumin nanocapsules manufactured with PCL or EC as barrier polymers.

System	Instability Mechanism	Average Migration Rate in the Bottom (µm/min)	Average Migration Rate in the Top (µm/min)
NC L/EC	Sedimentation	0.023	0.015
NC L/PCL	Sedimentation	0.010	0.013
NC C/EC	Sedimentation	0.011	0.012
NC C/PCL	Sedimentation	0.009	0.010

The NCs prepared with a curcumin oil core also showed good stability, as no significant changes in backscattering were observed during 4 weeks of storage at room temperature. The migration velocities in the bottom for NC C/EC and NC C/PCL were 0.011 and 0.009 µm/min, respectively.

3.1.4. TSI of the Nanocapsules

Figure 2 shows the TSI values determined from the backscattering data measured by the Turbiscan® equipment for lemon essential oil and curcumin NCs. All nanoscale systems presented TSI values lower than 1.0 after 27 days of storage at room temperature, suggesting high stability because TSI values close to zero indicate excellent stability. The TSI values for NC C/EC and NC C/PCL at the end of the measurement period were 0.10 and 0.044, respectively. Meanwhile, the NC L/EC had a TSI of 0.65 and NC L/PCL reached TSI values of 0.12.

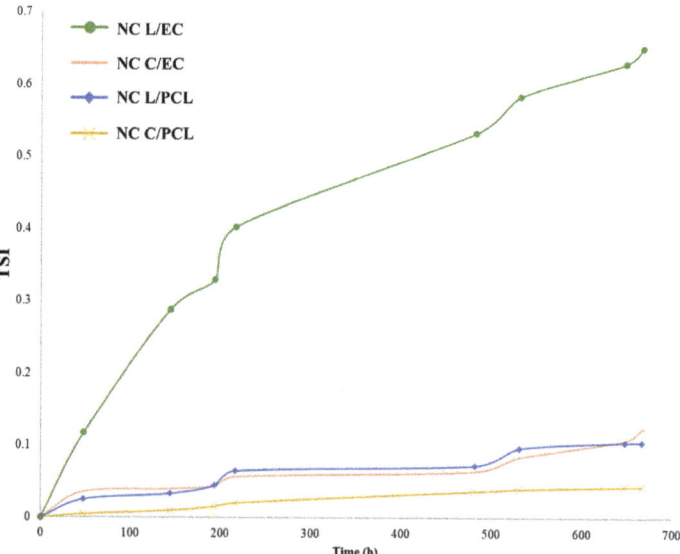

Figure 2. *Cont.*

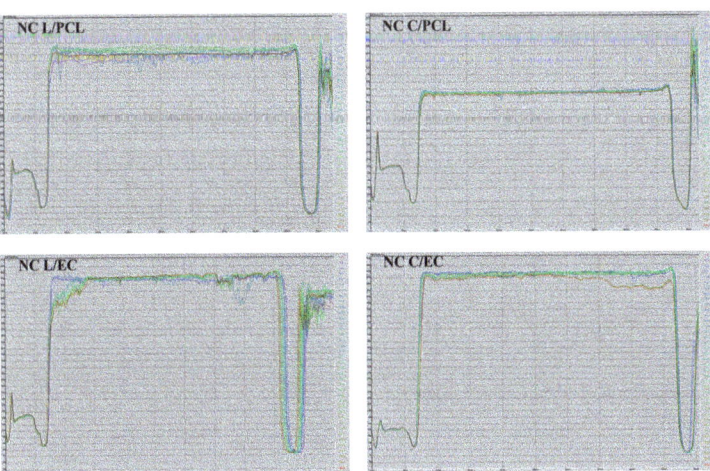

Figure 2. TSI of lemon oil and curcumin nanocapsules prepared with EC or PCL as barrier polymers.

3.2. Rate of CO_2 Production in Papayas Treated with Nanocapsules

Figure 3 presents the values for the rate of CO_2 production of papaya as a function of the treatments used. The highest rate of CO_2 production was observed in the control papaya batch, which had a significant increase during the first three days, with an average value of 3.27 mL of CO_2 kg^{-1} h^{-1}. This rate was maintained practically during the following days, and at the end of the storage period (day 14), another increase in the respiration rate was observed, reaching an average value of 5.23 mL of CO_2 kg^{-1} h^{-1}.

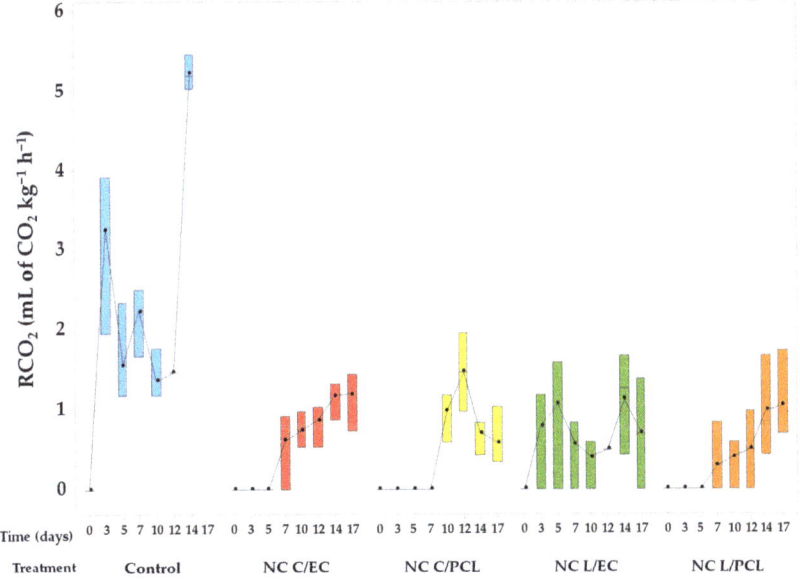

Figure 3. CO_2 production rate of papaya treated with nanocapsules containing lemon oil or curcumin and PCL or EC as biopolymers.

In contrast, in the papaya treated with NCs, a decrease in the rates of CO_2 formation was observed in all cases. The statistical analysis demonstrated significant differences ($p < 0.05$) compared to the control batches of papaya. For papaya coated with NC L/EC, it was observed that the rate of CO_2 production was slightly higher than 1.0 mL of CO_2 kg^{-1} h^{-1} on days 7 and 14. The papaya treated with NC L/PCL exhibited a CO_2 production rate of zero during the first five days of storage. However, the rate gradually increased after that, showing an upward trend in CO_2 production. It reached a maximum value of 1.04 mL of CO_2 kg^{-1} h^{-1} on day 17 of storage.

Similarly, papaya coated with NC C/EC and NC C/PCL showed no changes in CO_2 production during the initial days until 7 and 10 days, respectively. Afterward, the CO_2 production rate increased, reaching a CO_2 production of 1.19 mL of CO_2 kg^{-1} h^{-1} for papaya treated with NC C/EC and 1.45 mL of CO_2 kg^{-1} h^{-1} for papaya treated with NC C/PCL.

3.3. Effect of Nanoparticles on the Rate of O_2 Consumption

Figure 4 presents the behaviors of the rate of O_2 decrease in the headspace of containers with papaya under different treatments. It is worth noting that all systems had a significant effect ($p < 0.05$) on the rate of O_2 consumption.

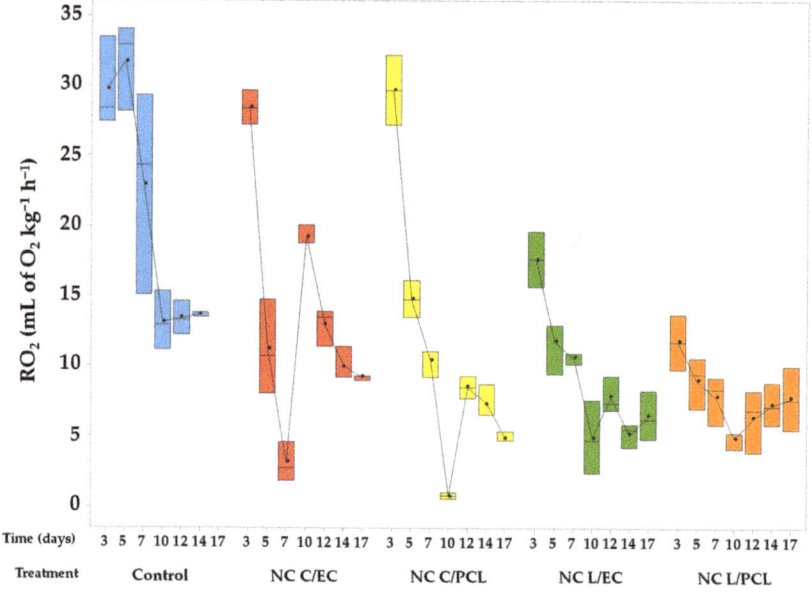

Figure 4. Oxygen consumption rate of papaya treated with nanoparticles containing lemon oil or curcumin.

The control system exhibited high O_2 consumption values during the first five days, with average rates of around 30 mL of O_2 kg^{-1} h^{-1}. The trend then decreased, and the rate of O_2 consumption dropped to 13.72 mL of O_2 kg^{-1} h^{-1} on the 14th day of testing. The nanocoatings of NC L/EC and NC L/PCL showed the lowest rates of O_2 consumption, with initial values of 17.63 and 11.75 mL of O_2 kg^{-1} h^{-1}, respectively. Moreover, the rate of O_2 consumption decreased during the remaining storage period, with O_2 consumption rates of 6.45 mL of O_2 kg^{-1} h^{-1} for NC L/EC and 7.72 mL of O_2 kg^{-1} h^{-1} for NC L/PCL. The results were statistically significant compared to the control system. Meanwhile, the O_2 consumption rates were 28.47 mL of O_2 kg^{-1} h^{-1} for the NC C/EC treatment and 29.71 mL of O_2 kg^{-1} h^{-1} for the NC C/PCL treatment during the first three days, gradually decreasing over the storage period to reach values of 9.22 and 4.91 mL of O_2 kg^{-1} h^{-1}, respectively.

3.4. pH and Acidity of Papaya Treated with Nanosystems

Table 3 presents the results obtained for the pH and acidity determinations of papaya treated with lemon oil and curcumin NCs using EC and PCL as barrier polymers during 17 days in cold storage.

Table 3. Changes in pH and acidity of papaya treated with lemon oil and curcumin nanocapsules using EC and PCL as barrier polymers.

Time		Control	NC L/EC	NC L/PCL	NC C/EC	NC C/PCL
Day 0	pH	5.20 ± 0.01	5.25 ± 0.03	5.23 ± 0.01	5.11 ± 0.04	5.20 ± 0.04
	Acidity (mg citric acid/100 g)	0.518 ± 0.05	0.532 ± 0.01	0.525 ± 0.03	0.534 ± 0.07	0.538 ± 0.06
Day 3	pH	5.21 ± 0.08	5.62 ± 0.02	5.45 ± 0.02	5.67 ± 0.02	5.33 ± 0.06
	Acidity (mg citric acid/100 g)	0.395 ± 0.02	0.529 ± 0.02	0.431 ± 0.02	0.375 ± 0.02	0.533 ± 0.02
Day 5	pH	5.30 ± 0.02	5.55 ± 0.09	5.54 ± 0.05	5.53 ± 0.04	5.59 ± 0.04
	Acidity (mg citric acid/100 g)	0.363 ± 0.05	0.497 ± 0.02	0.395 ± 0.02	0.400 ± 0.02	0.463 ± 0.02
Day 7	pH	5.23 ± 0.02	5.37 ± 0.05	5.37 ± 0.04	5.41 ± 0.05	5.38 ± 0.03
	Acidity (mg citric acid/100 g)	0.405 ± 0.02	0.568 ± 0.01	0.494 ± 0.02	0.535 ± 0.01	0.509 ± 0.03
Day 10	pH	5.35 ± 0.01	5.25 ± 0.01	5.52 ± 0.01	5.40 ± 0.07	5.48 ± 0.01
	Acidity (mg citric acid/100 g)	0.469 ± 0.04	0.520 ± 0.01	0.689 ± 0.02	0.608 ± 0.03	0.546 ± 0.03
Day 12	pH	5.16 ± 0.04	5.16 ± 0.06	5.35 ± 0.12	5.48 ± 0.03	5.46 ± 0.02
	Acidity (mg citric acid/100 g)	0.416 ± 0.03	0.560 ± 0.02	0.515 ± 0.03	0.511 ± 0.03	0.469 ± 0.02
Day 14	pH	5.08 ± 0.03	5.27 ± 0.04	4.94 ± 0.05	5.40 ± 0.02	5.30 ± 0.04
	Acidity (mg citric acid/100 g)	0.427 ± 0.05	0.417 ± 0.01	0.592 ± 0.02	0.428 ± 0.02	0.546 ± 0.03
Day 17	pH	4.97 ± 0.05	5.56 ± 0.04	4.86 ± 0.04	5.47 ± 0.04	5.29 ± 0.04
	Acidity (mg citric acid/100 g)	0.416 ± 0.03	0.480 ± 0.03	0.656 ± 0.02	0.449 ± 0.01	0.576 ± 0.03

The control samples showed a decrease in pH throughout the storage period, with an average pH value at the end of 4.97. Additionally, the acidity showed slight variations compared to its initial condition, resulting in final values of 0.416 mg citric acid/100 g of sample at the end of storage.

In papayas treated with NC L/EC, the pH barely changed, by 0.3, during storage. Similar behavior was remarked for papaya coated with NC C/EC, where the lowest pH value was found on day 7, with an average pH of 5.41. An increasing trend was observed in acidity during the first 12 days of storage, followed by a decrease. The final average value for the treatment with NC L/EC was 0.48 mg of citric acid/100 g of fruit, and for papaya coated with NC C/EC, it was 0.45 mg of citric acid/100 g. Statistical analysis showed significant differences ($p < 0.05$) between the encapsulated compounds.

The NC L/PCL and NC C/PCL showed a final value of 0.65 and 0.58 mg citric acid/100 g, respectively. In addition, the pH in papaya treated with NC L/PCL decreased drastically during the last week of storage, reaching a value of 4.8. The statistical analysis showed that pH and acidity exhibited significant differences between the untreated fresh-cut papaya, suggesting a protective effect against the deterioration of components in fresh-cut papaya.

3.5. Color Changes of Fresh-Cut Papaya Treated with Nanocapsules of Lemon Oil or Curcumin

Figure 5 shows the colorimetric values in CIELab coordinates of the surface of fresh-cut papaya treated with different nanodispersions and stored under refrigeration for 17 days. The L* value indicates the luminosity of the sample, where values close to 0 indicate black colors, while values close to 100 demonstrate white colors. This parameter in fruit preservation is related to the formation or degradation of compounds; for example, carotenoid degradation decreases the samples' luminosity [38].

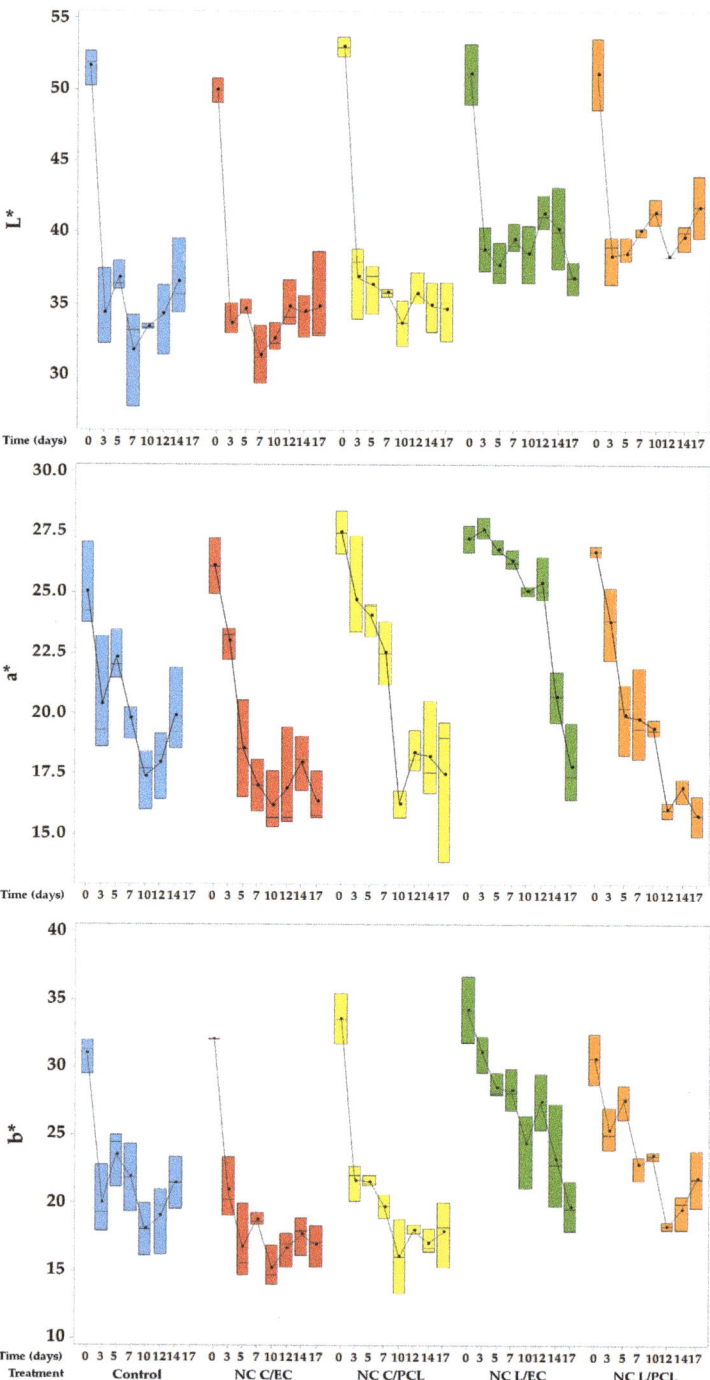

Figure 5. CIELab values (L*, a*, and b*) of the surface of fresh-cut papaya treated with different nanocapsules and storage under refrigeration.

In control papaya, the initial average luminosity value was 51.59, decreasing drastically during the first 3 days and reaching an L* value of 34.34, which remained constant in the following days of storage, with an average L* value of 36.5 on day 14, corresponding to a 30% loss of luminosity compared to its initial state. The fresh-cut papaya treated with NC C/EC and NC C/PCL showed more significant changes in the L* values, decreasing rapidly during the first days of storage and remaining constant. The final loss of luminosity in fresh-cut papayas treated with NC C/EC and NC C/PCL was 30.06% and 30.24%, respectively, compared to their initial state, with no significant differences ($p > 0.05$) between the NCs containing curcumin and the control system. In contrast, the NC L/EC and NC L/PCL treatments showed luminosity losses of 18.2% and 27.9%, respectively, compared to their initial state. Furthermore, ANOVA statistical analysis revealed significant differences ($p < 0.05$) among the encapsulated oily materials. Additionally, Dunnett's test indicated significant differences ($p < 0.05$) between the lemon oil nanoparticles and the control.

Figure 5 shows the a* value obtained from the surface of the papaya. The a* value indicates the change from red (positive) to green (negative) [38]. In the untreated control papaya, the a* value showed a decreasing trend during the storage period, with a decrease rate of −0.68 a*/day, reaching an a* value of 19.89 at the end of the storage period, representing a 30.42% loss of red coloration. The fresh-cut papaya with NC C/EC showed a change rate of −0.82 a*/day from day 0 to day 12, while for the papaya with NC C/PCL, the change rate was −0.95 a*/day in the same time interval. Dunnett's test did not show statistical differences compared to the control. In contrast, the treatment that maintained the a* value for a more extended period was composed of NC L/EC, where a smaller decrease in the a* value was observed, with a change rate of −0.33 a*/day, indicating that the treatment was able to preserve the red coloration in fresh-cut papaya. ANOVA showed statistically significant differences between NC L/EC and all other treatments.

Figure 5 shows box plots for the evolution of the b* values of fresh-cut papayas treated with nanodispersions in cold storage. The b* scale ranges from positive values, indicating yellow colors, to negative values, where colors tend to be blue [38]. In the papaya without any treatment (control), a rapid decrease in the b* value was observed during the first days of storage, with a rate of change of −1.66 b*/day. Then, the b* value remained relatively constant until day 14 of storage, with a 42.1% loss compared to its initial condition. In the papaya with NC C/EC, a rapid decrease in the b* value was observed during the first 12 days of storage, with a loss rate of −1.1 b*/day. After day 12, the rate of b* loss changed and reached a value of −0.25 b*/day.

Similarly, this behavior was noted in papaya treated with NC C/PCL, where the rate of change in the b* value during the first 12 days was −1.2 b*/day. The calculated percentage loss of the b* value on day 12 was 48.1% for NC C/EC and 46.5% for NC C/PCL. The treatments containing curcumin did not show significant differences but differed from the other tested systems.

Conversely, for the fresh-cut papaya coated with NC L/EC, a downward trend of the b* value was observed throughout the storage period, with a decrease rate of −0.72 b*/day. A final value of 19.66 was found, representing a loss of 8.78% in the b* value on day 12 and 34.5% on day 17. Also, in the fresh-cut papaya treated with NC L/PCL, a significant decrease in the b* value was seen during the first 12 days, with a change rate of −0.84 b*/day. Subsequently, the rate decreased to −0.11 b*/day with a loss of b* value of 29% compared to day 0. Both treatments showed significant differences compared to the control.

3.6. Changes in the Firmness of Fresh-Cut Papaya Treated with Different Nanosystems

The firmness of the papaya was determined during 17 days of refrigerated storage; these results are presented in Figure 6. In the control papaya, firmness decreased rapidly during the first 3 days of storage. The average firmness on day zero was 3.35 N. In contrast, on day 3, it decreased to an average firmness of 1.07 N, indicating that the control papaya lost up to 68% of its initial firmness during the early storage period.

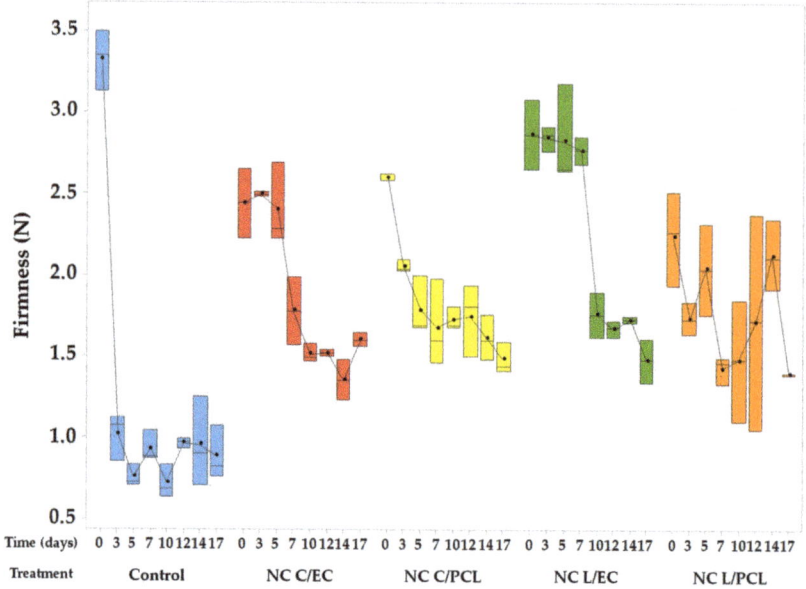

Figure 6. Firmness of papaya coated with different dispersions during 17 days of storage.

The firmness loss of fresh-cut papaya treated with NC L/EC compared to its initial value was only 9% on the 5th day. However, firmness decreased to 1.57 N on day 7, with a firmness loss percentage of 26% and reaching a final value of 36%. For the papaya treated with NC L/PCL, a firmness loss rate of 0.017 N/day was observed during the storage period, with a firmness of 1.54 N on day 17, representing a 32% decrease compared to its initial state. For the papaya coated with NC C/EC, a periodic decrease in firmness was observed, reaching a final value of 1.44 N, which represents a firmness loss of 39% compared to the initial state, reducing the percentage of loss by more than 20% compared to the untreated system. Meanwhile, in the papaya treated with NC C/PCL, firmness loss rates of 0.038 N/day were observed, resulting in a final firmness of 1.36 N with a 40% loss of firmness compared to day zero. Statistical analysis showed significant differences ($p < 0.05$) between the control system and the nanocapsule treatments containing lemon oil or curcumin, indicating firmness preservation for 17 days of refrigerated storage.

3.7. PME Activity of Fresh-Cut Papaya Treated with Different Polymeric Nanoparticles

Figure 7 shows the values of papaya coated with NC L/EC, NC L/PCL, NC C/EC, and NC C/PCL, as well as the control samples of papaya without any treatment. The control batch of papaya exhibited an increasing trend in pectin methylesterase (PME) activity during the initial days of storage, correlating with a drastic loss of firmness. The maximum PME activity of 1.44 U/mg protein was reached on the fifth day of storage, which then decreased, reaching an activity of 0.29 U/mg protein on day 17.

In papaya coated with NC L/EC, an increase in PME activity was observed, reaching a maximum value on the fifth day of storage of 0.41 U/mg of protein. This level was maintained practically throughout the rest of the testing period. In the case of NC L/PCL, the PME activity was initially low during the early days of storage, with an average activity of 0.14 U/mg of protein. However, a considerable increase in activity was noted on the seventh day, with an activity of 0.59 U/mg of protein, followed by a decrease. Papayas treated with NC C/EC showed a maximum PME activity on day 5, with an average activity of 0.72 U/mg of protein. At the same time, NC C/PCL samples reached their maximum activity on day 7, with an activity of 0.76 U/mg of protein. Statistical analysis did not show significant differences among the evaluated nanometric treatments.

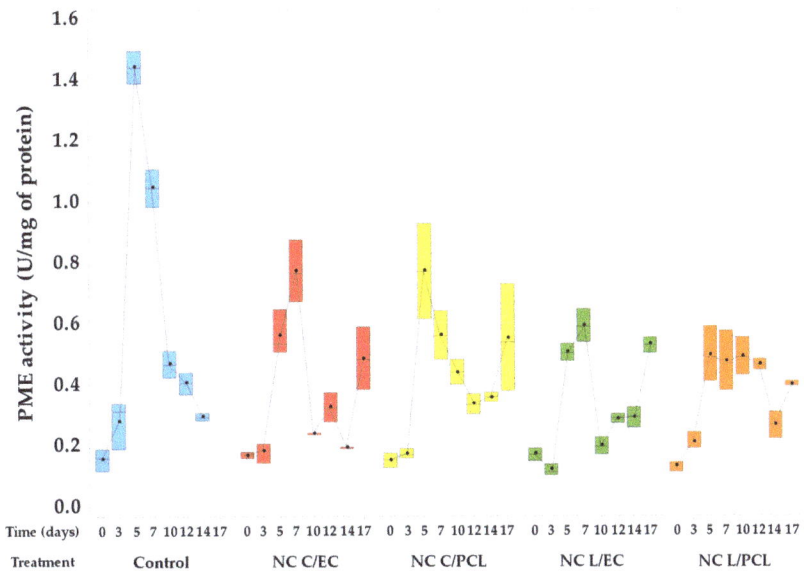

Figure 7. Changes in the activity of the PME enzyme in treated and stored papaya.

3.8. PPO Activity of Fresh-Cut Papaya Treated with Nanocapsule Systems

Figure 8 presents the results obtained for PPO enzyme activity in fresh-cut papaya treated with nanoparticles. Dunnett's multiple comparison tests revealed statistically significant differences between the nanoparticulate treatments and the control system, indicating that the treatments used decreased PPO activity. For the control system, it was found that PPO enzyme activity increased during days 5 and 7, reaching activity values of 35.53 and 32.68 U/g protein, respectively. Subsequently, PPO activity declined to values lower than 1 U/g protein. In the treatments of NC L/EC and NC L/PCL, an increase in PPO activity was observed on day 7, reaching values of activity of 17.14 and 12.35 U/g of protein, respectively, corresponding to inhibition percentages of 47.55% and 62.18%, showing statistically significant differences compared to the control system. In the case of fresh-cut papaya treated with NC C/EC and NC C/PCL, an increase in PPO activity was seen on day 7 of storage with an activity of 21.7 and 18.7 U/g of protein, with an inhibition percentage of PPO activity of 33.61% and 42.79%, respectively. The ANOVA showed significant differences compared to the control system ($p < 0.05$).

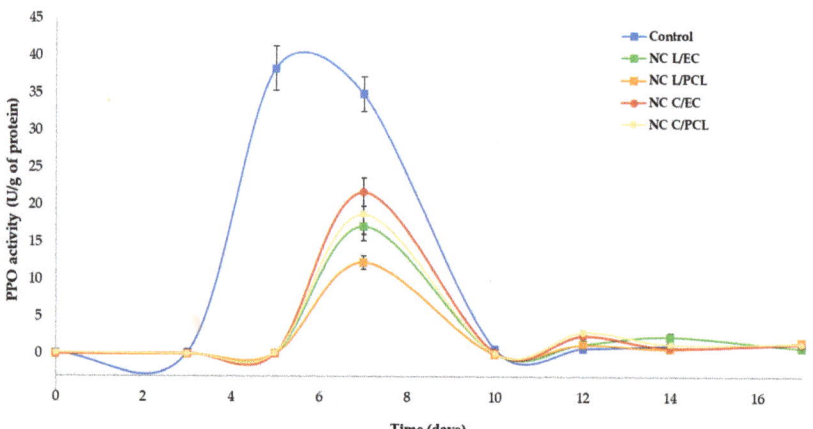

Figure 8. Evolution of PPO enzymatic activity in papaya coated with nanoparticles and stored under refrigeration.

3.9. Changes in the Total Phenolic Content in Fresh-Cut Papaya Treated with Nanodispersions

Figure 9 presents the results of total phenols in fresh-cut untreated papaya (control) and papaya coated with NCs of lemon oil and curcumin using EC or PCL as biopolymers. In the control system, an increase in the content of total phenols was observed, showing an upward trend throughout the storage period, with a production rate of phenolic compounds of 0.17 mg of gallic acid equivalents/day. The Dunnet test showed significant differences between the control papaya and the papaya coated with the nanodispersions.

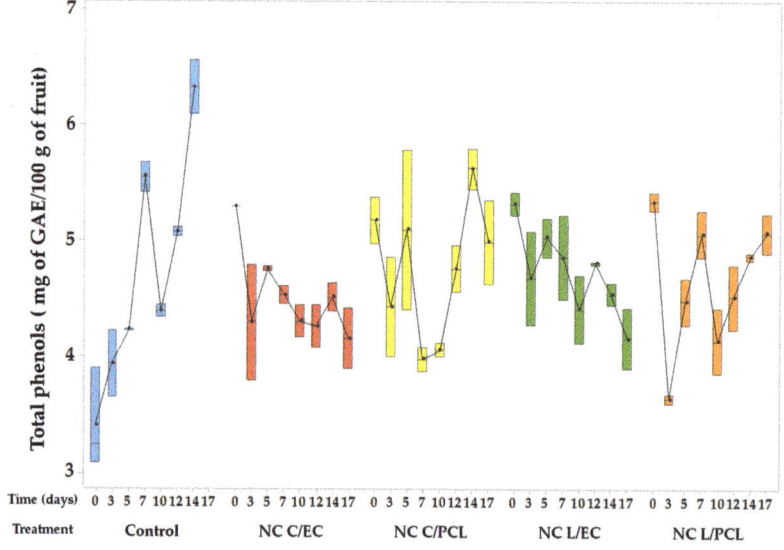

Figure 9. Total phenol content in fresh-cut papaya coated with nanodispersions and stored under refrigeration.

The trend for papaya treated with NCs formed by EC containing lemon oil or curcumin showed a slight decreased concerning phenolic compounds. In papaya treated with NC L/EC, a decrease rate in the total phenols of 0.043 mg of gallic acid equivalents/day was observed. Meanwhile, NC C/EC had a 0.047 mg gallic acid equivalents/day rate. For nanoparticles containing either lemon oil or curcumin formed with PCL, less variation in the content of phenolic compounds was observed; however, statistical analysis did not show significant differences between the NCs prepared with PCL or EC.

4. Discussion
4.1. Ps, PDI, ζ, and Morphology of Nanocapsules

Table 1 shows that the PS of lemon oil and curcumin NCs using EC and PCL ranged from 87 to 116 nm. The NC L/EC and NC L/PCL had sizes of 116 nm and 87.57 nm, respectively. Hasani et al. (2018) [39] reported Ps of 339 and 553 nm for lemon essential oil nanocapsules with chitosan and OSA-starch, respectively. Moreover, a study of nanocapsules prepared with methyl methacrylate-styrene as barrier polymer showed particle sizes of 136 nm [40], coinciding with our report. The average Ps values for the NC C/EC and NC C/PCL were 115 nm and 100 nm, respectively, falling within the nanoscale range. A study on curcumin extract NCs of sodium caseinate reported a Ps of 165 nm [41]. Similarly, the nanoencapsulation of curcumin with different oils such as castor, soybean, and Miglyol oil exhibited particle sizes of 150 nm, 142.5 nm, and 206 nm, respectively [42].

The PDI of lemon oil and curcumin NCs was lower than 0.25 (Table 1), indicating a narrow distribution of particle sizes. In lemon essential oil NCs prepared by ionic gelation and freeze-drying, a PDI of 0.424 was reported when chitosan or modified starch was used [39]. Moreover, PDIs between 0.16 and 0.29 have been reported for curcumin NCs dissolved in different oils, implying that the prepared systems have good homogeneity [43]. Then, the results of this study showed their comportment.

The ζ of NCs (Table 1) is partially correlated with stability. However, due to the characteristics of non-ionic stabilizers, low zeta potentials are generally observed, where steric effects prevent the aggregation or flocculation of the nanoparticles, and the systems remained stable for 4 weeks under ambient temperature storage conditions. For lemon oil nanocapsules prepared with chitosan and starch, ζ-values of 10.58 mV have been reported when the chitosan/starch ratio was 0.5%/9.5% [39].

The micrographs of lemon oil and curcumin NCs obtained by SEM (Figure 1) show spherical and nanometric sizes from 150 to 200 nm. Similar results have been reported for nanocapsules containing fragrances for textile scenting using lemon oil as the main compound, with spherical nanocapsules ranging from 100 to 200 nm [40]. Nanoscale spherical systems with particle sizes between 150 and 200 nm were observed with SEM for curcumin/PCL NCs obtained using solvent displacement methodology [44]. Curcumin/EC NCs prepared by dialysis have been reported as spherical structures with a particle diameter of 282.9 nm [45].

Therefore, the emulsification–diffusion process using ultrasound homogenization is an excellent tool for preparing nanoscale systems capable of encapsulating essential oils and curcumin with a good size distribution and stability, which can be helpful for food preservation purposes.

4.2. Stability of Nanocapsules

Sedimentation was the main instability mechanism for lemon oil and curcumin NCs manufactured with EC and PCL. However, the low migration velocities indicated they were highly stable at room temperature during 27 days of storage. These results correlate with TSI because values less than 1.0 were obtained (Figure 2). This pattern is due to the steric effect of the stabilizer used to prevent flocculation or aggregation of nanoparticles [46]. According to different authors [47,48], particles in dispersion with zeta potentials close to zero are considered unstable. Stabilizers like PVA can stabilize nanodispersions due to steric effects that prevent the binding of one particle to another, avoiding aggregation and promoting the stability of the NCs in suspension. In a study of curcumin nanoparticles prepared with a zein–shellac mixture (1:1), TSI values ranged from 1.54 to 8.43, with higher stability [49]. For α-limonene-based emulsions and nanoemulsions, TSI values between 2.37 and 6.01 were found after 30 days of storage [50]. D-limonene emulsions showed TSI values ranging from 1 to 15, with emulsions more stable when using Pluronic PE9400 [51]. Therefore, the fabrication of NCs increases the stability of dispersions, helping to maintain the physicochemical stability of the encapsulated active agent. In addition, the emulsification–diffusion methodology with ultrasound homogenization effectively prepares NCs with good physical stability.

The migration velocity of NC L/EC at the base of the measurement cell was two times higher than that of the other prepared systems (Table 1), with TSI values 6 to 10 times higher (Figure 2). This behavior could be attributed to the interactions between EC and fractions with high water solubility of lemon oil. The lemon essential oil obtained through steam distillation is composed of poorly water-soluble compounds, such as limonene, γ-terpinene, and α- and β-pinene. It also includes compounds with higher water solubility, such as neral, geranial, neryl acetate, and geranyl acetate [52]. Compounds with higher water solubility can interact with the hydroxyl groups of EC through hydrogen bonding, which could potentially diminish the interactions between these compounds and the hydrophilic stabilizers (PVA and OSA-starch), resulting in minor stability of the NC L/EC. According to Zhao et al. (2018) [53], citrus-based emulsions (mandarin, sweet orange, and bergamot) were more stable as the content of polar compounds increased due to hydrophilic compounds being more prone to covering the hydrophilic stabilizer (Tween 80), thereby enhancing the emulsion's interfacial properties and improving its stability.

4.3. Respiration Rate of Fresh-Cut Papaya Treated with Nanocapsules

The slicing of the tissues induces a cascade of metabolic changes in fresh-cut fruits, including an increase in respiration rate (O_2 consumption and CO_2 production). A significant reduction in respiration rate during the initial storage days contributes to the extended shelf-life of fresh-cut fruits. The reduction in respiration decreased ethylene production (a maturation hormone) and the downregulation of enzymes related to fruit degradation, such as pectin methylesterase, polygalacturonase, and polyphenol oxidase, among others. These mechanisms preserve firmness, color, and bioactive compounds like polyphenols [54]. Consequently, nanoparticle-based coatings can form a barrier that limits gas exchange, thus reducing the respiration rate in fresh-cut fruits [55].

The CO_2 production and O_2 consumption rates in the control papaya were significantly higher than in the papayas coated with the nanosystems (Figures 3 and 4). This behavior is because of cuts that increase the surface area in the control papaya. Furthermore, barriers that limit gas permeability have been eliminated, contributing to an increase in the respiration rate of the papaya [56]. The papaya utilized the O_2 inside the container to continue its metabolism, resulting in a high respiration rate and increased concentration of CO_2 in the container's headspace. Additionally, the onset of biodegradative metabolism leads to an increase in the concentration of CO_2 caused by the need to produce energy (ATP) for survival [57,58].

In contrast, the nanocoating formed by NC L/EC, NC L/PCL, NC C/EC, and NC C/PCL showed a statistically significant reduction of CO_2 production and O_2 consumption rates compared to the control system, indicating that the nanoparticles form a coating on the surface of the fresh-cut papaya, limiting the diffusion of O_2 and respiration of the product. This reduction in respiration helped to preserve the freshness, quality, and desirable attributes of the papaya, such as its texture, color, and flavor. Similar behavior was observed in chitosan nanoparticles encapsulating lemon essential oil applied as a coating on strawberries, where the O_2 consumption rate was reduced compared to chitosan nanoparticles alone and the control system [59]. In whole cucumbers coated with chitosan nanoparticles encapsulating cinnamon essential oil, the respiration rate of the cucumbers decreased significantly compared to the uncoated cucumbers, with no statistically significant differences observed compared to the chitosan treatment without nanoparticles [60]. The ability of the NCs to slow down the respiration process can be attributed to their film-forming properties and ability to create a protective layer [61]. This layer acts as a barrier against oxygen and moisture, preventing oxidative reactions and the growth of microorganisms that contribute to the deterioration of the papaya.

4.4. pH and Acidity of Fresh-Cut Papaya Treated with Nanocapsules

The pH in the control papaya had a decrescent tendency, while after day 10, the acidity experienced a sharp reduction. The decrease in pH is attributed to the formation of organic acids due to the biotransformation of carbohydrates into organic acids and the growth of microorganisms on the papaya's surface, which can produce organic acids. The reduction of acidity in control papaya is due to using organic acids such as citric acid and malic acid by enzymes during respiratory processes [62].

The NC L/EC and NC C/EC showed slightly varied pH and acidity content during storage. The slight differences in pH and acidity behavior are associated with nanosized treatments that can act as a barrier to gas exchange, reducing papaya respiration and thereby maintaining the carbohydrates that are not modified or used in respiratory processes. More minor changes in pH and fruit acidity have been observed for papaya coated with chitosan, implying that chitosan forms a barrier, reducing fruit respiration and increasing the stability of papaya components [62]. In contrast, for the NC L/PCL and NC C/PCL, an increase was observed throughout the storage period regarding acidity. While the NCs have a protective effect by reducing the respiration rate, PCL can degrade over time, forming carboxyl end groups that can decrease the acidity of the samples. The degradation of PCL occurs because the polymer in an aqueous environment can swell, causing water molecules to interact with the polymer, attacking the ester groups of PCL, leading to hydrolysis and the formation of terminal carboxyl groups. Furthermore, the hydrolysis of ester groups can be catalyzed by lipase enzymes found in papaya tissue [63].

4.5. Changes in the Color of Fresh-Cut Papaya Treated with Nanosystems

The control system showed a rapid decrease in the luminosity of the papaya surface (Figure 5). This behavior is attributed to the loss of physical barriers, allowing greater oxygen diffusion into the exposed tissue, leading to the decomposition of papaya components, especially carotenoids that give the fruit its red color. In addition, oxidation reactions of phenols increase, resulting in the darkening of the samples, as observed by González-Aguilar et al. (2009) [8] in fresh-cut 'Maradol' papaya. The systems containing lemon oil NCs maintained the papaya's luminosity, as they could limit gas transfer, especially oxygen. As a result, the oxidation of carotenoids that give the fruit its red color and the formation of dark compounds caused by polyphenol oxidase enzyme were reduced. Moreover, incorporating liposoluble compounds in edible coatings can modify the oxygen permeability properties of the formed coatings [64]. Furthermore, for fresh-cut 'Formosa' papaya coated with montmorillonite (TP = 100 nm) supported by a coating formed by a corvina protein isolate, it was observed that the use of nanoscale dispersions significantly reduced the loss of luminosity of the papaya [21].

The decrease in a* value in the control papaya (Figure 5) is attributed to the oxidation of carotenoids that give pigmentation to the fruit due to the absence of barriers that can interfere with oxygen distribution within the papaya tissues [65]. Papayas treated with NC C/EC and NC C/PCL did not differ significantly from the control. This behavior is associated with the degradation of curcumin, which leads to a loss of coloration [66]. In comparison, the NC L/EC treatment exhibited the slightest fluctuations in the a* value, suggesting that the treatment generated a barrier between the cut surface of the papaya and the environment, thereby reducing the entry of oxygen into the tissues and minimizing color changes in the product.

The rapid diminishing of the b* value in the control papaya (Figure 5) is due to the exposure of the papaya tissues and their components to oxygen, which can lead to oxidation or degradation phenomena. In papaya packaged in PVC bags with 15 microperforations, a percentage decrease in the b* value of 36.94% was noted after 9 days of storage compared to its initial state. Moreover, an increase in browning for the system with a higher number of microperforations was observed, leading to greater oxygen exposure, subsequent activation of polyphenol oxidase, and the formation of melanin [67]. In the papaya with NC C/EC and NC C/PCL, the percentage of b* value loss was higher compared to the control system; this additional decrease in b* value may be due to the decomposition of curcumin constituents that give it its characteristic yellow color, such as demethoxycurcumin, curcumin, and bisdemethoxycurcumin, which can be easily degraded by light [68], taking several days [69] and resulting in the loss of yellow coloration in the samples. Minor changes in the b* value were obtained with the NCs containing lemon oil, suggesting that nanoparticles can limit the oxygen exchange between the cut papaya and the environment, allowing the product's sensory characteristics to be maintained. A lower presence of oxygen in the cut tissues reduces the oxidation of color-giving compounds such as carotenoids. It decreases the activities of enzymes involved in forming brown compounds, such as polyphenol oxidases [58].

4.6. Firmness of Fresh-Cut Papayas Treated with Nanocapsules

Firmness is the maximum penetration force of a texturometer probe capable of breaking the surface of fresh-cut papaya. Firmness is determined by the physical anatomy of the tissue, such as size, shape, cellular arrangement, cell wall thickness, and cell cohesion degree [58].

The significant loss of firmness in the control papaya (Figure 6) is due to the higher activity of pectolytic enzymes in the control papaya that modified and hydrolyzed the pectins of the cell wall. This behavior has been widely studied in papaya. For example, in 'Sunrise Solo' papaya without any treatment, a firmness loss of approximately 50% was observed in the first two days of storage, decreasing from 4.7 to 2.5 N [70]. Likewise, González-Aguilar et al. (2009) [8] observed a 64% loss of firmness in fresh-cut 'Maradol' papaya without any treatment during the first 3 days of refrigerated storage. These are similar results to those reported in this study.

For NC treatments, up to 40% firmness loss was achieved. The coating formed on the papaya surface decreases the metabolism activity of pectic enzymes related to changes in papaya firmness. For fresh-cut 'Sinta' papaya coated with nanochitosan, it was observed that the nanosystems could reduce papaya firmness loss [71]. For fresh-cut papaya coated with nanocomposites composed of whitemouth croaker (*Micropogonias furnieri*) protein isolates and montmorillonite, a firmness loss of only 17% was found compared to untreated samples that had a firmness loss of 69.76% after 12 days of refrigerated storage [21]. Furthermore, the antioxidant capacity of lemon oil and curcumin could help maintain the structure of the cell membrane. While the rapid appearance of oxidizing species affects the structure of the cell membrane, the antioxidants contained in the NCs could limit the oxidation of phospholipids and thus maintain the physical structure of fresh-cut papaya, as has been observed by Velderrain-Rodríguez et al. (2015) [72] in papaya treated with antioxidants obtained from mango.

4.7. PME Activity in Fresh-Cut Papaya Treated with Nanoparticles

As shown in Figure 7, the control samples exhibited the highest PME activities; this behavior is attributed to the induction of ripening processes in the control papaya caused by cuts, which, in turn, increase the activities of pectolytic enzymes such as PME and polygalacturonase, leading to the modification and hydrolysis of pectins [58]. González-Aguilar et al. (2009) [8] observed that fresh-cut 'Maradol' papaya without any treatment exhibited a rapid increase in PME activity during the first 3 days of storage compared to papaya treated with chitosan. In addition, tissue damage results in the rapid release of linolenic and linoleic acids from phospholipids in cell membranes due to lipoxygenases, which act as chemical signalers in synthesizing cell wall-related enzymes. Moreover, the hydrolysis of pectins caused by the activity of pectolytic enzymes like PME and polygalacturonase results in the release of oligogalacturonides, leading to the overexpression of the PME enzyme [73].

In contrast, in the papayas treated with NCs, PME activity was partially inhibited. This behavior indicates that the NCs incorporated on the cut surface of the papaya can generate a coating that reduces the respiration rate of the papaya and the ethylene production and inhibits the signaling cascades involved in the synthesis and activation of pectinolytic enzymes such as PME [74].

4.8. PPO Activity of Fresh-Cut Papaya Treated with Nanoparticles

As shown in Figure 8, the activity of the PPO enzyme was much higher in control papayas than in the papayas treated with nanodispersions. This behavior is associated with the absence of any physical barrier that limits the transport of O_2 inside the papaya, allowing more significant interaction of O_2 with PPO and phenolic compounds found in the fruit. In untreated papaya, Arjun et al. (2015) [63] found that the PPO activity increased during the initial days and decreased after reaching peak activity. The PPO activity was significantly higher than that found in papaya treated with a chitosan–soy coating, indicating that the lower PPO activity is due to the protective effect of the coating, which limits the oxygen supply to the tissue, thereby reducing PPO activity.

In addition, the treatments with nanocapsules showed up to 63% PPO inhibition. These results are attributed to the fact that the nanoparticles can form a coating on the cut surface of the papaya. This coating dramatically limits the transfer of gases between the papaya and the environment. As a result, a modified atmosphere is generated within the fruit tissue, leading to decreased PPO activity, fewer color changes due to a decrease in the formation of dark compounds, and the preservation of biologically valuable components in fresh-cut papaya. A decrease in PPO activity has also been found for apples treated with nanochitosan, which is related to the oxygen barrier properties of chitosan [75]. The NC C/EC and NC C/PCL treatments were less effective in limiting gas transfer than the NC L/EC and NC L/PCL systems. A higher amount of oxygen in fresh-cut papaya promotes the activation of PPO enzymes, increasing browning development and decreasing the bioactive compounds. Furthermore, essential oils can inhibit PPO activity, as Eissa et al. [65] demonstrated in apple juice, where lemon grass oil extract showed inhibition of 92%. Regarding using nanostructures to decrease PPO activity, it has been found that in fresh-cut 'Red Delicious' apples treated with α-tocopherol nanocapsules, PPO enzyme activity was delayed compared to the $CaCl_2$ treatment [76].

4.9. Total Phenolic Content in Fresh-Cut Papaya Treated with Nanocapsules

The increasing trend in the total phenolic content in the control papaya (Figure 9) is attributed to the degradation of both the cell wall and the cell membrane, generating species such as oligogalacturonides from the hydrolysis of pectins or free fatty acids because of the hydrolysis of papaya cell membrane phospholipids, which serve as signals for the production and activation of new enzymes involved in protection mechanisms against pathogen attacks. Oligogalacturonides induce the formation and activation of enzymes such as phenylalanine ammonia-lyase, which is the crucial enzyme in the phenylpropanoid (or shikimic acid) pathway, producing a wide variety of phenolic compounds and lignin that obstruct pathogen attack. Likewise, the chalcone enzyme metabolizes phytoalexin production, which has antimicrobial activity [54].

Alternatively, the content of phenolic compounds in fresh-cut papaya treated with NCs had slight variations over 17 days of refrigerated storage. This behavior is attributed to the excellent interaction between the pectin in the papaya cell wall and the EC of NCs [77] that were deposited on the surface of the fruit, reducing the degradation of pectins and signaling reactions involved in the synthesis of phenolic compounds. Also, PCL NCs can integrate into the fruit surface, forming a coating that minimizes the degradation of fresh-cut papaya. Furthermore, the presence of antioxidant compounds such as lemon oil and curcumin promotes the recycling of the antioxidant activity of phenolic compounds [78]. Similar behavior has been observed in recent studies where α-tocopherol/PCL nanocapsules reduce the initial respiration rate and the enzyme activities of PME and phenylalanine ammonia-lyase in fresh-cut 'Red Delicious' apples [76,79].

5. Conclusions

The main advantage of nanoencapsulation is that it allows the incorporation of many active substances that, due to their chemical or physicochemical properties, are difficult to mix with food matrices. Also, the encapsulated active compounds are protected from degradation reactions, and controlled release of the encapsulated compounds can also be achieved. Therefore, nanocapsules manufactured with approved compounds for food use are an interesting option for preserving fresh-cut fruits.

NCs containing curcumin and lemon essential oil were obtained by the emulsification–diffusion method coupled with ultrasound and using EC or PCL as barrier biopolymers. The systems exhibited particle sizes below 150 nm, polydispersity indices below 0.2, and zeta potentials higher than −10 mV. The instability mechanism observed in the lemon oil and curcumin NCs was sedimentation. However, they remained stable for 27 days when stored at room temperature. The EC- and PCL-based NCs of lemon oil and curcumin showed a significant reduction in the respiration rate of fresh-cut papaya during 17 days of storage. The EC-based NCs displayed less variation in the acidity and pH of fresh-cut papaya. The NCs effectively mitigated physical changes associated with the degradation of fresh-cut papaya, with particular attention given to the treatment with lemon oil/EC nanocapsules, demonstrating better color and firmness retention. Furthermore, all nanosystems decreased PPO and PME enzymatic activities, which correlated with the retention of quality characteristics and total phenolic content in the fresh-cut papaya. The lemon oil nanocapsules and the curcumin-based nanocapsules employing EC and PCL as biopolymers may be extended to conserve various fresh-cut fruits and vegetables.

Author Contributions: This paper was written by all authors: conceptualization, M.d.l.L.Z.-Z.; data curation, M.J.G.-P., L.M.-A. and L.E.S.-M.; formal analysis, M.J.G.-P., L.M.-A., G.V.-R. and L.E.S.-M.; funding acquisition, M.d.l.L.Z.-Z.; investigation, M.J.G.-P.; methodology, M.J.G.-P. and M.d.l.L.Z.-Z.; project administration, M.d.l.L.Z.-Z.; software, G.V.-R.; supervision, M.d.l.L.Z.-Z.; writing—original draft, M.J.G.-P.; writing—review and editing, M.J.G.-P., L.M.-A., G.V.-R., L.E.S.-M. and M.d.l.L.Z.-Z. All authors have read and agreed to the published version of the manuscript.

Funding: The authors acknowledge the financial support to the projects PAPIIT PAPIIT IN221823 of DGAPA-UNAM, FESC UNAM CI2233.

Institutional Review Board Statement: Not applicable.

Data Availability Statement: The data presented in this paper are available upon request from the corresponding author.

Acknowledgments: Galindo-Pérez M.J. and Martínez-Acevedo L. thank the Consejo Nacional de Ciencia y Tecnología (CONACyT) of Mexico for the granted postdoctoral fellowship.

Conflicts of Interest: The authors declare no conflict of interest.

References

1. Wilson, M.D.; Stanley, R.A.; Eyles, A.; Ross, T. Innovative Processes and Technologies for Modified Atmosphere Packaging of Fresh and Fresh-Cut Fruits and Vegetables. *Crit. Rev. Food Sci. Nutr.* **2019**, *59*, 411–422. [CrossRef]
2. Hu, W.; Sarengaowa, W.; Guan, Y.; Feng, K. Biosynthesis of Phenolic Compounds and Antioxidant Activity in Fresh-Cut Fruits and Vegetables. *Front. Microbiol.* **2022**, *13*, 906069. [CrossRef] [PubMed]
3. Botondi, R.; Barone, M.; Grasso, C. A Review into the Effectiveness of Ozone Technology for Improving the Safety and Preserving the Quality of Fresh-Cut Fruits and Vegetables. *Foods* **2021**, *10*, 748. [CrossRef] [PubMed]
4. Dotto, J.M.; Abihudi, S.A. Nutraceutical Value of *Carica papaya*: A Review. *Sci. Afr.* **2021**, *13*, e00933. [CrossRef]
5. Valencia Sandoval, K.; Duana Ávila, D.; Hernández Gracia, T.J. Estudio Del Mercado de Papaya Mexicana: Un Análisis de Su Competitividad (2001–2015). *Suma Negocios* **2017**, *8*, 131–139. [CrossRef]
6. Iturralde-García, R.D.; Cinco-Moroyoqui, F.J.; Martínez-Cruz, O.; Ruiz-Cruz, S.; Wong-Corral, F.J.; Borboa-Flores, J.; Cornejo-Ramírez, Y.I.; Bernal-Mercado, A.T.; Del-Toro-Sánchez, C.L. Emerging Technologies for Prolonging Fresh-Cut Fruits' Quality and Safety during Storage. *Horticulturae* **2022**, *8*, 731. [CrossRef]

7. Ayón-Reyna, L.E.; Tamayo-Limón, R.; Cárdenas-Torres, F.; López-López, M.E.; López-Angulo, G.; López-Moreno, H.S.; López-Cervántes, J.; López-Valenzuela, J.A.; Vega-García, M.O. Effectiveness of Hydrothermal-Calcium Chloride Treatment and Chitosan on Quality Retention and Microbial Growth during Storage of Fresh-Cut Papaya. *J. Food Sci.* **2015**, *80*, C594–C601. [CrossRef]
8. González-Aguilar, G.A.; Valenzuela-Soto, E.; Lizardi-Mendoza, J.; Goycoolea, F.; Martínez-Téllez, M.A.; Villegas-Ochoa, M.A.; Monroy-García, I.N.; Ayala-Zavala, J.F. Effect of Chitosan Coating in Preventing Deterioration and Preserving the Quality of Fresh-Cut Papaya "Maradol". *J. Sci. Food Agric.* **2009**, *89*, 15–23. [CrossRef]
9. Kuwar, U.; Sharma, S.; Tadapaneni, V.R.R. Aloe Vera Gel and Honey-Based Edible Coatings Combined with Chemical Dip as a Safe Means for Quality Maintenance and Shelf Life Extension of Fresh-Cut Papaya. *J. Food Qual.* **2015**, *38*, 347–358. [CrossRef]
10. Waghmare, R.B.; Annapure, U.S. Combined Effect of Chemical Treatment and/or Modified Atmosphere Packaging (MAP) on Quality of Fresh-Cut Papaya. *Postharvest Biol. Technol.* **2013**, *85*, 147–153. [CrossRef]
11. Antunes, M.D.; Gago, C.M.; Cavaco, A.M.; Miguel, M.G. Edible Coatings Enriched with Essential Oils and Their Compounds for Fresh and Fresh-Cut Fruit. *Recent Patents Food Nutr. Agric.* **2012**, *4*, 114–122. [CrossRef] [PubMed]
12. Fisher, K.; Phillips, C.A. The Effect of Lemon, Orange and Bergamot Essential Oils and Their Components on the Survival of *Campylobacter jejuni*, *Escherichia coli* O157, *Listeria monocytogenes*, *Bacillus cereus* and *Staphylococcus aureus* In Vitro and in Food Systems. *J. Appl. Microbiol.* **2006**, *101*, 1232–1240. [CrossRef] [PubMed]
13. Ben Hsouna, A.; Ben Halima, N.; Smaoui, S.; Hamdi, N. Citrus Lemon Essential Oil: Chemical Composition, Antioxidant and Antimicrobial Activities with Its Preservative Effect against Listeria Monocytogenes Inoculated in Minced Beef Meat. *Lipids Health Dis.* **2017**, *16*, 146. [CrossRef] [PubMed]
14. Himed, L.; Merniz, S.; Monteagudo-Olivan, R.; Barkat, M.; Coronas, J. Antioxidant Activity of the Essential Oil of Citrus Limon before and after Its Encapsulation in Amorphous SiO_2. *Sci. Afr.* **2019**, *6*, e00181. [CrossRef]
15. Jyotirmayee, B.; Mahalik, G. A Review on Selected Pharmacological Activities of *Curcuma longa* L. *Int. J. Food Prop.* **2022**, *25*, 1377–1398. [CrossRef]
16. Ibáñez, M.D.; Blázquez, M.A. *Curcuma longa* L. Rhizome Essential Oil from Extraction to Its Agri-Food Applications. A Review. *Plants* **2021**, *10*, 44. [CrossRef]
17. Zou, Y.; Yu, Y.; Cheng, L.; Li, L.; Zou, B.; Wu, J.; Zhou, W.; Li, J.; Xu, Y. Effects of Curcumin-Based Photodynamic Treatment on Quality Attributes of Fresh-Cut Pineapple. *LWT* **2021**, *141*, 110902. [CrossRef]
18. Chai, Z.; Zhang, F.; Liu, B.; Chen, X.; Meng, X. Antibacterial Mechanism and Preservation Effect of Curcumin-Based Photodynamic Extends the Shelf Life of Fresh-Cut Pears. *LWT* **2021**, *142*, 110941. [CrossRef]
19. Tao, R.; Zhang, F.; Tang, Q.; Xu, C.S.; Ni, Z.J.; Meng, X. Effects of Curcumin-Based Photodynamic Treatment on the Storage Quality of Fresh-Cut Apples. *Food Chem.* **2019**, *274*, 415–421. [CrossRef]
20. Lavinia, M.; Hibaturrahman, S.N.; Harinata, H.; Wardana, A.A. Antimicrobial Activity and Application of Nanocomposite Coating from Chitosan and ZnO Nanoparticle to Inhibit Microbial Growth on Fresh-Cut Papaya. *Food Res.* **2020**, *4*, 307–311. [CrossRef]
21. Cortez-Vega, W.R.; Pizato, S.; De Souza, J.T.A.; Prentice, C. Using Edible Coatings from Whitemouth Croaker (*Micropogonias furnieri*) Protein Isolate and Organo-Clay Nanocomposite for Improve the Conservation Properties of Fresh-Cut "Formosa" Papaya. *Innov. Food Sci. Emerg. Technol.* **2014**, *22*, 197–202. [CrossRef]
22. Tabassum, N.; Aftab, R.A.; Yousuf, O.; Ahmad, S.; Zaidi, S. Application of Nanoemulsion Based Edible Coating on Fresh-Cut Papaya. *J. Food Eng.* **2023**, *355*, 111579. [CrossRef]
23. Luciano, W.A.; Pimentel, T.C.; Bezerril, F.F.; Barão, C.E.; Marcolino, V.A.; de Siqueira Ferraz Carvalho, R.; dos Santos Lima, M.; Martín-Belloso, O.; Magnani, M. Effect of Citral Nanoemulsion on the Inactivation of *Listeria monocytogenes* and Sensory Properties of Fresh-Cut Melon and Papaya during Storage. *Int. J. Food Microbiol.* **2023**, *384*, 109959. [CrossRef] [PubMed]
24. Zielińska, A.; Carreiró, F.; Oliveira, A.M.; Neves, A.; Pires, B.; Venkatesh, D.N.; Durazzo, A.; Lucarini, M.; Eder, P.; Silva, A.M.; et al. Polymeric Nanoparticles: Production, Characterization, Toxicology and Ecotoxicology. *Molecules* **2020**, *25*, 3731. [CrossRef] [PubMed]
25. Rahman, N.A. *Applications of Polymeric Nanoparticles in Food Sector*; Siddiquee, S., Hong Melvin, G.J., Rahman, M.M., Eds.; Springer Nature Switzerland AG: Cham, Switzerland, 2019; ISBN 9783319996028.
26. Jafari, S.M. *An Overview of Nanoencapsulation Techniques and Their Classification*; Elsevier Inc.: Amsterdam, The Netherlands, 2017; ISBN 9780128094365.
27. Ahmadi, P.; Jahanban-Esfahlan, A.; Ahmadi, A.; Tabibiazar, M.; Mohammadifar, M. Development of Ethyl Cellulose-Based Formulations: A Perspective on the Novel Technical Methods. *Food Rev. Int.* **2022**, *38*, 685–732. [CrossRef]
28. Din, M.I.; Ghaffar, T.; Najeeb, J.; Hussain, Z.; Khalid, R.; Zahid, H. Potential Perspectives of Biodegradable Plastics for Food Packaging Application-Review of Properties and Recent Developments. *Food Addit. Contam. Part A Chem. Anal. Control Expo. Risk Assess.* **2020**, *37*, 665–680. [CrossRef]
29. Galindo-Pérez, M.J.; Quintanar-Guerrero, D.; Cornejo-Villegas, M.d.l.Á.; Zambrano-Zaragoza, M.d.l.L. Optimization of the Emulsification-Diffusion Method Using Ultrasound to Prepare Nanocapsules of Different Food-Core Oils. *LWT* **2018**, *87*, 333–341. [CrossRef]
30. Xu, D.; Zhang, J.; Cao, Y.; Wang, J.; Xiao, J. Influence of Microcrystalline Cellulose on the Microrheological Property and Freeze-Thaw Stability of Soybean Protein Hydrolysate Stabilized Curcumin Emulsion. *LWT Food Sci. Technol.* **2016**, *66*, 590–597. [CrossRef]

31. Wang, Z.W.; Duan, H.W.; Hu, C.Y. Modelling the Respiration Rate of Guava (*Psidium guajava* L.) Fruit Using Enzyme Kinetics, Chemical Kinetics and Artificial Neural Network. *Eur. Food Res. Technol.* **2009**, *229*, 495–503. [CrossRef]
32. Iqbal, T.; Rodrigues, F.A.S.; Mahajan, P.V.; Kerry, J.P.; Gil, L.; Manso, M.C.; Cunha, L.M. Effect of Minimal Processing Conditions on Respiration Rate of Carrots. *J. Food Sci.* **2008**, *73*, 396–402. [CrossRef]
33. Zambrano-Zaragoza, M.d.l.L.; Mercado-Silva, E.; Del Real, L.A.; Gutiérrez-Cortez, E.; Cornejo-Villegas, M.A.; Quintanar-Guerrero, D. The Effect of Nano-Coatings with α-Tocopherol and Xanthan Gum on Shelf-Life and Browning Index of Fresh-Cut "Red Delicious" Apples. *Innov. Food Sci. Emerg. Technol.* **2014**, *22*, 188–196. [CrossRef]
34. Hagerman, A.E.; Austin, P.J. Continuous Spectrophotometric Assay for Plant Pectin Methyl Esterase. *J. Agric. Food Chem.* **1986**, *34*, 440–444. [CrossRef]
35. Bradford, M.M. A Rapid and Sensitive Method for the Quantitation of Microgram Quantities of Protein Utilizing the Principle of Protein-Dye Binding. *Anal. Biochem.* **1976**, *72*, 248–254. [CrossRef]
36. Waterhouse, A.L. Determination of Total Phenolics. *Curr. Protoc. Food Anal. Chem.* **2002**, *6*, I1.1.1–I1.1.8.
37. Chien, P.J.; Sheu, F.; Yang, F.H. Effects of Edible Chitosan Coating on Quality and Shelf Life of Sliced Mango Fruit. *J. Food Eng.* **2007**, *78*, 225–229. [CrossRef]
38. Sant'Anna, V.; Gurak, P.D.; Ferreira Marczak, L.D.; Tessaro, I.C. Tracking Bioactive Compounds with Colour Changes in Foods—A Review. *Dye. Pigment.* **2013**, *98*, 601–608. [CrossRef]
39. Hasani, S.; Ojagh, S.M.; Ghorbani, M. Nanoencapsulation of Lemon Essential Oil in Chitosan-Hicap System. Part 1: Study on Its Physical and Structural Characteristics. *Biol. Macromol.* **2018**, *115*, 143–151. [CrossRef]
40. Liu, C.; Liang, B.; Shi, G.; Li, Z.; Zheng, X. Preparation and Characteristics of Nanocapsules Containing Essential Oil for Textile Application. *Flavour Fragr. J.* **2015**, *30*, 295–301. [CrossRef]
41. Pan, K.; Zhong, Q.; Baek, S.J. Enhanced Dispersibility and Bioactivity of Curcumin by Encapsulation in Casein Nanocapsules. *J. Agric. Food Chem.* **2013**, *61*, 6036–6043. [CrossRef]
42. Zanotto-Filho, A.; Coradini, K.; Braganhol, E.; Schröder, R.; De Oliveira, C.M.; Simoes-Pires, A.; Battastini, A.M.O.; Pohlmann, A.R.; Guterres, S.S.; Forcelini, C.M.; et al. Curcumin-Loaded Lipid-Core Nanocapsules as a Strategy to Improve Pharmacological Efficacy of Curcumin in Glioma Treatment. *Eur. J. Pharm. Biopharm.* **2013**, *83*, 156–167. [CrossRef]
43. Klippstein, R.; Wang, J.T.; El-gogary, R.I.; Bai, J.; Mustafa, F.; Rubio, N.; Bansal, S.; Al-jamal, W.T. Passively Targeted Curcumin-Loaded PEGylated PLGA Nanocapsules for Colon Cancer Therapy In Vivo. *Small* **2015**, *11*, 4704–4722. [CrossRef]
44. Umerska, A.; Gaucher, C.; Oyarzun-Ampuero, F.; Fries-Raeth, I.; Colin, F.; Villamizar-Sarmiento, M.G.; Maincent, P.; Sapin-Minet, A. Polymeric Nanoparticles for Increasing Oral Bioavailability of Curcumin. *Antioxidants* **2018**, *7*, 46. [CrossRef] [PubMed]
45. Suwannateep, N.; Banlunara, W.; Wanichwecharungruang, S.P.; Chiablaem, K.; Lirdprapamongkol, K.; Svasti, J. Mucoadhesive Curcumin Nanospheres: Biological Activity, Adhesion to Stomach Mucosa and Release of Curcumin into the Circulation. *J. Control. Release* **2011**, *151*, 176–182. [CrossRef] [PubMed]
46. Bagherifam, S.; Griffiths, G.W.; Mælandsmo, G.M.; Nyström, B.; Hasirci, V.; Hasirci, N. Poly(Sebacic Anhydride) Nanocapsules as Carriers: Effects of Preparation Parameters on Properties and Release of Doxorubicin. *J. Microencapsul.* **2015**, *32*, 166–174. [CrossRef] [PubMed]
47. Galindo-Rodriguez, S.; Allémann, E.; Fessi, H.; Doelker, E. Physicochemical Parameters Associated with Nanoparticle Formation in the Salting-out, Emulsification-Diffusion, and Nanoprecipitation Methods. *Pharm. Res.* **2004**, *21*, 1428–1439. [CrossRef] [PubMed]
48. Piirma, I. *Polymeric Surfactants*; CRC Press: Boca Raton, FL, USA; Taylor & Francis Group: Oxford, UK, 1992; ISBN 9780824786083.
49. Sun, C.; Xu, C.; Mao, L.; Wang, D.; Yang, J.; Gao, Y. Preparation, Characterization and Stability of Curcumin-Loaded Zein-Shellac Composite Colloidal Particles. *Food Chem.* **2017**, *228*, 656–667. [CrossRef] [PubMed]
50. Trujillo-Cayado, L.A.; Alfaro, M.C.; Muñoz, J. Effects of Ethoxylated Fatty Acid Alkanolamide Concentration and Processing on D-Limonene Emulsions. *Colloids Surf. A Physicochem. Eng. Asp.* **2018**, *536*, 198–203. [CrossRef]
51. Pérez-Mosqueda, L.M.; Trujillo-Cayado, L.A.; Carrillo, F.; Ramírez, P.; Muñoz, J. Formulation and Optimization by Experimental Design of Eco-Friendly Emulsions Based on d-Limonene. *Colloids Surf. B Biointerfaces* **2015**, *128*, 127–131. [CrossRef] [PubMed]
52. Rao, J.; McClements, D.J. Impact of Lemon Oil Composition on Formation and Stability of Model Food and Beverage Emulsions. *Food Chem.* **2012**, *134*, 749–757. [CrossRef]
53. Zhao, S.; Tian, G.; Zhao, C.; Li, C.; Bao, Y.; DiMarco-Crook, C.; Tang, Z.; Li, C.; Julian McClements, D.; Xiao, H.; et al. The Stability of Three Different Citrus Oil-in-Water Emulsions Fabricated by Spontaneous Emulsification. *Food Chem.* **2018**, *269*, 577–587. [CrossRef]
54. Maringgal, B.; Hashim, N.; Mohamed Amin Tawakkal, I.S.; Muda Mohamed, M.T. Recent Advance in Edible Coating and Its Effect on Fresh/Fresh-Cut Fruits Quality. *Trends Food Sci. Technol.* **2020**, *96*, 253–267. [CrossRef]
55. Hasan, S.M.K.; Ferrentino, G.; Scampicchio, M. Nanoemulsion as Advanced Edible Coatings to Preserve the Quality of Fresh-Cut Fruits and Vegetables: A Review. *Int. J. Food Sci. Technol.* **2020**, *55*, 1–10. [CrossRef]
56. Wang, D.; Ma, Q.; Li, D.; Li, W.; Li, L.; Aalim, H.; Luo, Z. Moderation of Respiratory Cascades and Energy Metabolism of Fresh-Cut Pear Fruit in Response to High CO_2 Controlled Atmosphere. *Postharvest Biol. Technol.* **2021**, *172*, 111379. [CrossRef]
57. Baldwin, E.A.; Bai, J. Physiology of Fresh-Cut Fruits and Vegetables. In *Advances in Fresh-Cut Fruits and Vegetables Processing*; Martín-Belloso, O., Soliva-Fortuny, R., Eds.; CRC Press: Boca Raton, FL, USA; Taylor & Francis Group: Oxford, UK, 2011; pp. 87–113, ISBN 9781420031874.

58. Toivonen, P.M.A.; Brummell, D.A. Biochemical Bases of Appearance and Texture Changes in Fresh-Cut Fruit and Vegetables. *Postharvest Biol. Technol.* **2008**, *48*, 1–14. [CrossRef]
59. Perdones, A.; Sánchez-González, L.; Chiralt, A.; Vargas, M. Effect of Chitosan-Lemon Essential Oil Coatings on Storage-Keeping Quality of Strawberry. *Postharvest Biol. Technol.* **2012**, *70*, 32–41. [CrossRef]
60. Mohammadi, A.; Hashemi, M.; Hosseini, S.M. Chitosan Nanoparticles Loaded with Cinnamomum Zeylanicum Essential Oil Enhance the Shelf Life of Cucumber during Cold Storage. *Postharvest Biol. Technol.* **2015**, *110*, 203–213. [CrossRef]
61. Zambrano-Zaragoza, M.L.; González-Reza, R.; Mendoza-Muñoz, N.; Miranda-Linares, V.; Bernal-Couoh, T.F.; Mendoza-Elvira, S.; Quintanar-Guerrero, D. Nanosystems in Edible Coatings: A Novel Strategy for Food Preservation. *Int. J. Mol. Sci.* **2018**, *19*, 705. [CrossRef]
62. Ali, A.; Muhammad, M.T.M.; Sijam, K.; Siddiqui, Y. Effect of Chitosan Coatings on the Physicochemical Characteristics of Eksotika II Papaya (*Carica Papaya* L.) Fruit during Cold Storage. *Food Chem.* **2011**, *124*, 620–626. [CrossRef]
63. Mochizuki, M.; Hirano, M.; Kanmuri, Y.; Kudo, K.; Tokiwa, Y. Hydrolysis of Polycaprolactone Fibers by Lipase: Effects of Draw Ratio on Enzymatic Degradation. *J. Appl. Polym. Sci.* **1995**, *55*, 289–296. [CrossRef]
64. Perdones, Á.; Vargas, M.; Atarés, L.; Chiralt, A. Physical, Antioxidant and Antimicrobial Properties of Chitosan-Cinnamon Leaf Oil Films as Affected by Oleic Acid. *Food Hydrocoll.* **2014**, *36*, 256–264. [CrossRef]
65. Malvano, F.; Corona, O.; Pham, P.L.; Cinquanta, L.; Pollon, M.; Bambina, P.; Farina, V.; Albanese, D. Effect of Alginate-Based Coating Charged with Hydroxyapatite and Quercetin on Colour, Firmness, Sugars and Volatile Compounds of Fresh Cut Papaya during Cold Storage. *Eur. Food Res. Technol.* **2022**, *248*, 2833–2842. [CrossRef]
66. Zheng, B.; Zhang, Z.; Chen, F.; Luo, X.; McClements, D.J. Impact of Delivery System Type on Curcumin Stability: Comparison of Curcumin Degradation in Aqueous Solutions, Emulsions, and Hydrogel Beads. *Food Hydrocoll.* **2017**, *71*, 187–197. [CrossRef]
67. Jayathunge, K.G.L.R.; Gunawardhana, D.K.S.N.; Illeperuma, D.C.K.; Chandrajith, U.G.; Thilakarathne, B.M.K.S.; Fernando, M.D.; Palipane, K.B. Physico-Chemical and Sensory Quality of Fresh Cut Papaya (Carica Papaya) Packaged in Micro-Perforated Polyvinyl Chloride Containers. *J. Food Sci. Technol.* **2014**, *51*, 3918–3925. [CrossRef] [PubMed]
68. Price, L.C.; Buescher, R.W. Decomposition of Turmeric Curcuminoids as Affected by Light, Solvent and Oxygen. *J. Food Biochem.* **2007**, *20*, 125–133. [CrossRef]
69. Schneider, C.; Gordon, O.N.; Edwards, R.L.; Luis, P.B. Degradation of Curcumin: From Mechanism to Biological Implications. *J. Agric. Food Chem.* **2015**, *63*, 7606–7614. [CrossRef] [PubMed]
70. Ergun, M.; Huber, D.; Jeong, J.; Bartz, J.A. Extended Shelf Life and Quality of Fresh-Cut Papaya Derived from Ripe Fruit Treated with the Ethylene Antagonist 1-Methylcyclopropene. *J. Am. Soc. Hortic. Sci.* **2006**, *131*, 97–103. [CrossRef]
71. Allanigue, D.K.A.; Sabularse, V.C.; Hernandez, H.P.; Serrano, E.P. The Effect of Chitosan-Based Nanocomposite Coating on the Postharvest Life of Papaya (*Carica Papaya* L.) Fruits. *Philipp. Agric. Sci.* **2017**, *100*, 233–242.
72. Velderrain-Rodríguez, G.R.; Ovando-Martínez, M.; Villegas-Ochoa, M.; Ayala-Zavala, J.F.; Wall-Medrano, A.; Álvarez-Parrilla, E.; Madera-Santana, T.J.; Astiazarán-García, H.; Tortoledo-Ortiz, O.; González-Aguilar, G.A. Antioxidant Capacity and Bioaccessibility of Synergic Mango (Cv. Ataulfo) Peel Phenolic Compounds in Edible Coatings Applied to Fresh-Cut Papaya. *Food Nutr. Sci.* **2015**, *6*, 365–373. [CrossRef]
73. Karakurt, Y.; Huber, D.J. Activities of Several Membrane and Cell-Wall Hydrolases, Ethylene Biosynthetic Enzymes, and Cell Wall Polyuronide Degradation during Low-Temperature Storage of Intact and Fresh-Cut Papaya (Carica Papaya) Fruit. *Postharvest Biol. Technol.* **2003**, *28*, 219–229. [CrossRef]
74. Formiga, A.S.; Pereira, E.M.; Junior, J.S.P.; Costa, F.B.; Mattiuz, B.H. Effects of Edible Coatings on the Quality and Storage of Early Harvested Guava. *Food Chem. Adv.* **2022**, *1*, 100124. [CrossRef]
75. Pilon, L.; Spricigo, P.C.; Miranda, M.; de Moura, M.R.; Assis, O.B.G.; Mattoso, L.H.C.; Ferreira, M.D. Chitosan Nanoparticle Coatings Reduce Microbial Growth on Fresh-Cut Apples While Not Affecting Quality Attributes. *Int. J. Food Sci. Technol.* **2015**, *50*, 440–448. [CrossRef]
76. Galindo-Pérez, M.J.; Quintanar-Guerrero, D.; Mercado-Silva, E.; Real-Sandoval, S.A.; Zambrano-Zaragoza, M.L. The Effects of Tocopherol Nanocapsules/Xanthan Gum Coatings on the Preservation of Fresh-Cut Apples: Evaluation of Phenol Metabolism. *Food Bioprocess Technol.* **2015**, *8*, 1791–1799. [CrossRef]
77. Macleod, G.S.; Fell, J.T.; Collett, J.H. Studies on the Physical Properties of Mixed Pectin/Ethylcellulose Films Intended for Colonic Drug Delivery. *Int. J. Pharm.* **1997**, *157*, 53–60. [CrossRef]
78. Kagan, V.E.; Tyurina, Y.Y. Recycling and Redox Cycling of Phenolic Antioxidants. *Ann. N. Y. Acad. Sci.* **1998**, *854*, 425–434. [CrossRef] [PubMed]
79. Zambrano-Zaragoza, M.L.; Gutiérrez-Cortez, E.; Del Real, A.; González-Reza, R.M.; Galindo-Pérez, M.J.; Quintanar-Guerrero, D. Fresh-Cut Red Delicious Apples Coating Using Tocopherol/Mucilage Nanoemulsion: Effect of Coating on Polyphenol Oxidase and Pectin Methylesterase Activities. *Food Res. Int.* **2014**, *62*, 974–983. [CrossRef]

Disclaimer/Publisher's Note: The statements, opinions and data contained in all publications are solely those of the individual author(s) and contributor(s) and not of MDPI and/or the editor(s). MDPI and/or the editor(s) disclaim responsibility for any injury to people or property resulting from any ideas, methods, instructions or products referred to in the content.

Article

FDM 3D Printing and Soil-Burial-Degradation Behaviors of Residue of Astragalus Particles/Thermoplastic Starch/Poly(lactic acid) Biocomposites

Zhibing Ni [1,†], Jianan Shi [2,†], Mengya Li [2], Wen Lei [2,*] and Wangwang Yu [3,*]

1. School of Transportation Engineering, Nanjing Vocational University of Industry Technology, Nanjing 210023, China
2. College of Science, Nanjing Forestry University, Nanjing 210037, China
3. School of Mechanical Engineering, Nanjing Vocational University of Industry Technology, Nanjing 210023, China
* Correspondence: leiwen@njfu.edu.cn (W.L.); yuww@niit.edu.cn (W.Y.); Tel.: +86-25-8542-7621 (W.L.)
† These authors contributed equally to this work.

Citation: Ni, Z.; Shi, J.; Li, M.; Lei, W.; Yu, W. FDM 3D Printing and Soil-Burial-Degradation Behaviors of Residue of Astragalus Particles/ Thermoplastic Starch/Poly(lactic acid) Biocomposites. *Polymers* 2023, 15, 2382. https://doi.org/10.3390/polym15102382

Academic Editor: Cristina Cazan

Received: 26 April 2023
Revised: 10 May 2023
Accepted: 15 May 2023
Published: 19 May 2023

Copyright: © 2023 by the authors. Licensee MDPI, Basel, Switzerland. This article is an open access article distributed under the terms and conditions of the Creative Commons Attribution (CC BY) license (https://creativecommons.org/licenses/by/4.0/).

Abstract: Astragalus residue powder (ARP)/thermoplastic starch (TPS)/poly(lactic acid) (PLA) biocomposites were prepared by fused-deposition modeling (FDM) 3D-printing technology for the first time in this paper, and certain physico-mechanical properties and soil-burial-biodegradation behaviors of the biocomposites were investigated. The results showed that after raising the dosage of ARP, the tensile and flexural strengths, the elongation at break and the thermal stability of the sample decreased, while the tensile and flexural moduli increased; after raising the dosage of TPS, the tensile and flexural strengths, the elongation at break and the thermal stability all decreased. Among all of the samples, sample C—which was composed of 11 wt.% ARP, 10 wt.% TPS and 79 wt.% PLA—was the cheapest and also the most easily degraded in water. The soil-degradation-behavior analysis of sample C showed that, after being buried in soil, the surfaces of the samples became grey at first, then darkened, after which the smooth surfaces became rough and certain components were found to detach from the samples. After soil burial for 180 days, there was weight loss of 21.40%, and the flexural strength and modulus, as well as the storage modulus, reduced from 82.1 MPa, 11,922.16 MPa and 2395.3 MPa to 47.6 MPa, 6653.92 MPa and 1476.5 MPa, respectively. Soil burial had little effect on the glass transition, cold crystallization or melting temperatures, while it reduced the crystallinity of the samples. It is concluded that the FDM 3D-printed ARP/TPS/PLA biocomposites are easy to degrade in soil conditions. This study developed a new kind of thoroughly degradable biocomposite for FDM 3D printing.

Keywords: astragalus residue powder; thermoplastic starch; poly(lactic acid); biocomposite; fused deposition modeling; soil burial; mechanical property; thermal property; degradation behavior

1. Introduction

Three-dimensional (3D) printing, also known as additive manufacturing, makes it possible to design a product with accurate numerical values of the dimensions through a computer graphic program and create 3D physical objects with exact dimensions in a relatively short time [1–3]. As one of the 3D-printing technologies, fused-deposition modeling (FDM) 3D printing is gaining much attention at present. In this process, the filaments prepared from thermoplastic materials are extruded through a nozzle under some designed operating conditions and successively deposited in a melted/softened state on the print bed to form the end items [4,5].

Up to now, certain traditional polymers and their composites have been applied as feedstocks for FDM 3D printing. For example, Rahmatabadi et al. [6] investigated the FDM 3D printing of PLA-TPU compounds with different component ratios, finding that the glass transition temperatures remained almost the same for various samples; furthermore, raising the proportion of PLA in the compound would increase the loss modulus, strength and fracture toughness, while simultaneously decreasing the storage modulus and formability of the samples. Rahmatabadi et al. [7] successfully FDM-3D-printed polyvinyl chloride (PVC) samples using different printing parameters for the first time. Among all of the concerned parameters, raster angle and printing velocity had the greatest effects on the mechanical properties of the samples, whereas the nozzle diameter and layer thickness had little effect; the maximum tensile strength reached 88.55 MPa, which showed the superiority of 3D-printed PVC mechanical properties compared to other commercial filaments.

With the increasing awareness of environmental protection among people and the need for sustainable development of society, the research and application of biodegradable materials is becoming more and more important. For these reasons, biodegradable polymers, such as poly(lactic acid) (PLA), poly(butylene adipate-co-terephthalate) (PBAT) [8], butadiene styrene copolymer (PBS) [9], polycaprolactone (PCL) [10] and polyhydroxybutyrate (PHB) [11], have been chosen as the raw materials for FDM 3D printing. Among these, PLA has gained the most acceptance. PLA, synthesized from agricultural resources such as corn and tapioca, is biocompatible, compostable, recyclable, gas permeable and degradable by hydrolysis and enzymatic action. PLA is quite suitable for FDM printing because of its low melting point, low thermal expansion coefficient and lack of a pungent smell when being processed. However, its unit price is much more expensive than that of petroleum-based plastics, such as polyethylene and polypropylene. In addition, it exhibits a longer degradation time. Therefore, there is an urgent need to reduce the cost of PLA and improve its ability to degrade. The incorporation of natural fibers into the resin has proved to be an effective method to solve the aforementioned problems. As a result of this, a variety of natural fibers have been introduced into PLA for 3D printing. For example, Zhang et al. [12] printed poplar powder/PLA composites utilizing lubricant (TPW604) and polyolefin elastomer (POE) as a flexibilizer. They found that the lubricant improved the fluidity but reduced the impact strength of 3D-printing materials. The POE could improve the fluidity and toughness of the printed parts, and a higher content of POE resulted in better properties in a certain range. Aumnate et al. [13] extracted kenaf cellulose fibers (KFs) from locally grown kenaf plants, then treated them with tetraethyl orthosilicate, and prepared KFs/PLA biocomposite materials, using polyethylene glycol as a plasticizer. They found that the melt viscosities of the biocomposites increased as the fibers were loaded, but significantly decreased with the addition of plasticizers. The prototypes made from these materials could be applied in sustainable textiles and apparel, personalized prostheses and certain medical devices. Jaya et al. [14] 3D printed a continuous pineapple-leaf-fiber-reinforced PLA composite and investigated the properties of the specimens for the first time. They found that it took the same time to print the composites as to PLA, the tensile strength of the composite was increased by the application of the continuous pineapple leaf fiber, while its elongation at break was lower than that of the pure PLA part. Delphine et al. [15] produced bamboo-fiber-reinforced PLA filaments for FDM, the modulus of the filament was influenced by the length over diameter ratio of the compounded fibers, and the stiffness of the long bamboo-fiber-reinforced PLA filament could be increased by 215%. Guen et al. [16] compared the properties of the rice husk/PLA and wood/PLA filaments for FDM. The two biomasses had different effects on the rheological behavior while having similar effects on the mechanical properties of the 3D-printed samples. The complex viscosity of the compound could be increased by wood powder while being conversely decreased by rice husk powder. The mechanical properties were predominantly affected by the deposition direction.

Astragalus is a typical Chinese traditional medicinal crop; it is often cooked to extract certain bioactive polysaccharides for medical purposes. After being cooked, however, its residue, as one kind of natural fiber, is rarely re-utilized, which not only pollutes the environment but also leads to resource waste. We have conducted previous research on the

3D printing of astragalus residue powder (ARP)/PLA biocomposite. In our previous works, using ARP/PLA as the raw material for FDM 3D printing was proven to be feasible [17].

As a cheap, biodegradable, excellent renewable and easily available polysaccharide, thermoplastic starch (TPS) has been regarded as the optimal additive for blending with certain degradable polymers such as PLA, PBS and PCL. In recent years, many endeavors have been made on these blends for FDM 3D printing. Agnieszka et al. [18] developed a biodegradable and compostable TPS/PLA composite filament for FDM and investigated the properties of the filament. The incorporation of TPS not only reduces the cost of the granulate but also results in a considerable improvement in hydrophilicity and susceptibility to hydrolytic degradation. The compostability of the composite could also be enhanced in contrast to that of commercial PLA printouts. The printability for FDM and the properties of TPS/polycaprolactone composites were investigated by Zhao et al. [19], who found that the samples had the best performance in the FDM process with a starch ratio of 9 ph at 80~90 °C. The low printing temperature made it possible to introduce some bioactive components to produce antibacterial and biocompatible materials for FDM. Ju et al. [20] prepared a TPS/PLA/PBAT composite for FDM 3D printing, the filament had successful printability, and the samples could be printed accurately. Meanwhile, they found that the mechanical properties of the samples could be improved significantly when the chain extender ADR4468 was used.

Based on the above discussions, it was learned that both ARP and TPS could be used to form composites with PLA for FDM 3D printing, both the printed ARP/PLA and TPS/PLA samples had adequate properties and a much lower cost than the pure PLA. To our understanding, however, there have been no reports in the literature regarding the application of an ARP/TPS/PLA biocomposite in FDM printing.

In this work, we fabricated ecofriendly PLA composites loaded with TPS and ARP by using the FDM 3D-printing technique. The effects of the dosages of TPS and ARP on the mechanical and thermal properties of the composites, as well as their mass changes when immersed in water, were first investigated. Next, a specific ARP/TPS/PLA biocomposite was buried in soil and the biodegradation behavior of the samples was studied.

2. Experimental

2.1. Materials and Reagents

PLA (American Nature Works Co., 3052D, Minnetonka, MN, USA) in pellet form was purchased from Shanghai Xingyun International Trade Co. Ltd., China (Shanghai, China); TPS, food grade, was obtained from Shandong Hengren Industry and Trade Co., Ltd., China (Tengzhou, China); ARP, which was passed through an 120-mesh sieve, was made in our lab.

2.2. Preparation of FDM Filaments

Based on our previous work, 11 wt.% was the best proportion for ARP to add into PLA for the FDM 3D printing of ARP/PLA pieces when the cost and quality of the filament were taken into consideration simultaneously [21]. In this paper, the maximum proportion of ARP in the composite was thus controlled within 11 wt.%.

The biocomposite samples under investigation in this paper were PLA with different ARP and TPS compositions as listed in Table 1 infused into the base polymer. In each sample, 8 wt.% glycerol of the total mass of PLA, TPS and ARP was added as the plasticizing agent to lower the brittleness of the composite and guarantee the smooth production of the filament and the subsequently printed specimen.

The polymer composites using dried PLA, TPS and ARP as raw materials were first compounded with a twin screw extruder (SHJ-20, Nanjing Giant Machinery Co. Ltd., Nanjing, China) at 20 rpm and 130~160 °C, and the extrudate was granulated to make pellets; then, the filaments for FDM printing as illustrated in Figure 1a were prepared using a twin-screw extruder (KS-HXY, Kunshan Huanxinyang Electrical Equipment Co. Ltd., Suzhou, China) at 20 rpm and 170~190 °C.

Table 1. Compositions of ARP/TPS/PLA blends.

Sample ID	A	B	C	D	E
PLA (wt.%)	90	85	79	84	89
TPS (wt.%)	10	10	10	5	0
ARP (wt.%)	0	5	11	11	11

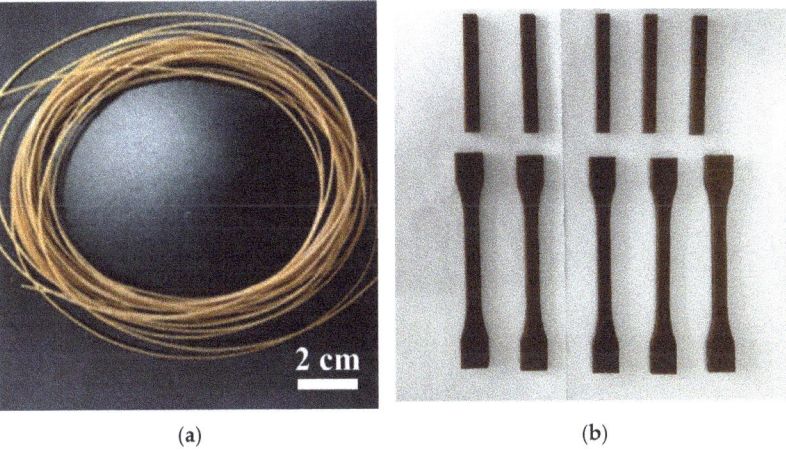

(a)   (b)

Figure 1. Physical images of the filaments and printed ARP/TPS/PLA biocomposite samples: (**a**) filaments; (**b**) printed parts.

2.3. Composite Preparation by FDM

Samples were printed with filaments prepared with the formulations listed in Table 1 on a desktop-level 3D printer (MOSHU S10, Hangzhou Shining 3D Technology Co. Ltd., Hangzhou, China) fitted with a 0.4-mm die nozzle. The printing parameters were chosen according to those for the FDM 3D printing of ARP/PLA biocomposite samples [21] and listed in Table 2.

Table 2. The parameters for the FDM 3D printing of ARP/TPS/PLA biocomposite samples.

Parameter	Print Temperature/°C	Layer Thickness/mm	Print Speed/(mm/s)	Deposition Angle/°
Data	220	0.1	50	0

The physical images of the printed samples are shown in Figure 1b.

2.4. Soil Degradation

The printed ARP/TPS/PLA parts were buried in soil collected in a 30 × 20 × 20 cm^3 paper carton, then the soil degradation test was conducted at room temperature for up to 180 days. The soil moisture during the test was controlled between 17.5% and 21.5%. At each time interval (30, 60, 90 or 180 days after burial), the parts were taken out from the soil, cleaned with water and dried thoroughly in a hot-air oven. The changes of the parts in weight, surface color, mechanical properties, thermal stability, melting and cold crystallization behavior, as well as the thermal dynamic mechanic properties, were investigated.

2.5. Testing and Characterization
2.5.1. Mechanical Testing

The samples were printed into dog-bone shapes for the tensile tests and rectangular shapes for the flexural tests, then the mechanical tests were performed using a universal testing machine (E44.304, MTS Industrial Systems (China) Co. Ltd., Shenzhen, China) with a load cell of 20 kN. The tensile and flexural tests were carried out according to the ASTM D 638 and ASTM D 790 standard testing methods with a crosshead speed of 10 mm/min and 5 mm/min, respectively.

2.5.2. Thermal Stability

The thermal stability of the samples under a nitrogen atmosphere was analyzed using a TG 209F1 thermogravimetric analyzer (NETZSCH-Gerätebau GmbH, Selb, Germany). The experiments were performed on approximately 8-mg samples from 20 °C to 550 °C at a 20 K/min heating rate to investigate changes in the initial decomposition temperature (T_i), thermal stability, and char residue of the samples.

2.5.3. Mass Change in Water

The printed dog-bone-shaped samples were chosen for water immersion experiments. The samples were dried at 80 °C for 8 h, cooled in a desiccator, and then immediately massed to the nearest 0.0001 g. Thereafter, the samples were submerged in distilled water at room temperature, removed from the water after 7 d, gently blotted with tissue paper to remove excess water from their surfaces and immediately massed to the nearest 0.0001 g again. The percentage of mass change in water was calculated by weight variation between the samples immersed in water and the dry samples using the following formula:

$$wu(\%) = \frac{w_t - w_0}{w_0} \times 100\% \tag{1}$$

where wu is the mass change rate, w_0 is the mass recorded before immersion and w_t is the mass recorded after immersion.

2.5.4. Weight Loss in Soil

The samples were taken out of the soil at every time interval, cleaned and then weighed. The weight loss rate (WL) of each sample was calculated by the weight variation between the samples before and after soil burial using the following formula:

$$wl(\%) = \frac{w_0 - w_t}{w_0} \times 100\% \tag{2}$$

where wl is the weight loss rate, w_0 is the mass recorded before soil burial and w_t is the mass recorded after soil burial.

2.5.5. Morphological Study (SEM)

The flexural fracture surfaces of the samples were first sputter-coated with a thin layer of gold to avoid any electrostatic charge and poor resolution during the scanning examination, and the SEM was operated at an accelerating voltage of 3 kV to image the samples at ×1000 magnification using a Hitachi SU 8010 field-emission scanning electron microscope (Hitachi Corporation, Tokyo, Japan).

2.5.6. Melting and Crystallization Behavior

The melting and crystallization behavior was investigated using differential scanning calorimetry (DSC). Measurements were performed using a DSC214 (NETZSCH-Gerätebau GmbH, Selb, Germany), under a nitrogen flow of 20 mL/min. Samples of approximately 5~10 mg were cut from the 3D-printed samples and sealed in aluminum pans. The samples were heated from the ambient temperature to 220 °C at a ramp rate of 10 °C/min, and held in an isothermal state for 5 min, then cooled down to room temperature at a rate of 10 °C/min and subsequently reheated to 220 °C at a rate of 10 °C. The enthalpies of melting (ΔH_m) and cold crystallization (ΔH_{cc}) were evaluated using the NETZSCH analysis software by integrating the areas of the melting and cold crystallization peaks. The T_g, T_m

and T_{CC} values were taken from the second heating curves. The crystallization percentage of each piece was obtained via the following equation:

$$x_c = \frac{|\Delta H_m - \Delta H_{cc}|}{\omega \Delta H^*} \times 100\% \quad (3)$$

where ω is the mass fraction of PLA in the sample, ΔH_m is the melting enthalpy, ΔH_{cc} is the cold crystallization enthalpy and $\Delta H^* = 93.6$ J/g is the melting enthalpy of 100% crystalline PLA [22].

2.5.7. Thermal Dynamic Mechanic Testing

A dynamic mechanical analyzer (DMA 242C, Netzsch, Bavaria, Germany) was employed to investigate the dynamic mechanical properties of the FDM-3D-printed ARP/TPS/PLA parts at different soil burial durations. The test was carried out under a nitrogen atmosphere and at a sinusoidal frequency of 3.33 Hz, over a temperature range of 30~120 °C at a heating rate of 5 °C/min. During the test, the sample was held by a dual cantilever fixture.

3. Results and Discussion

3.1. Effects of Compositions on Properties of the Biocomposites

3.1.1. Mechanical Properties

The results of the tensile and bending tests conducted on the printed parts are presented in Figure 2. As the ARP loading increased (samples A, B and C), the values of the tensile and flexural moduli of the composites increased, while the corresponding strengths decreased slightly. For sample A, the tensile strength, the flexural strength, the tensile modulus and the flexural modulus were 18.73 MPa, 84.28 MPa, 353.61 MPa and 11,066.99 MPa, respectively; the corresponding strengths for sample C decreased to 17.46 MPa and 82.06 MPa, while the moduli increased to 397.48 MPa and 11,922.16 MPa, accordingly. The decrease in strength with the loading of the filler has been observed for several other natural fiber/polymer composites, such as peanut husk/poly(butylene adipate-co-terephthalate) (PBAT) [23], wood flour/PBAT [24] and sisal fiber/polypropylene [25]; this was attributed to the higher number of voids with higher filler content, as well as the poor dispersion of the hydrophilic natural fiber in the hydrophobic polymer matrix. According to the mixing rule, however, the modulus of the composite material was decided by the contribution of each component, the modulus of ARP/TPS/PLA was thus gradually increased when more ARP was used because of its greater modulus as a natural fiber than that of PLA. When the mechanical properties of samples C, D and E were taken into consideration, it was found that both the tensile and flexural strengths decreased with the increased substitution of PLA by TPS, and the elongation at break also decreased gradually, indicating that the incorporation of TPS would reduce the strengths of ARP/PLA composites and make the composite more fragile; thus, the content of TPS in the ARP/TPS/PLA composites should be controlled within a reasonable range. As an example, Figure 2d illustrates the stress–strain curve of sample E during the tensile test, a yield point appeared on the curve, meaning that this sample would show ductile fracture but not obviously. As reported, a complex thermal and mechanical history would be induced on the printed part during the FDM process; whether the yield point would appear on the stress–strain curve of the sample or not during the tensile test was directly decided by the free volume portion and its size distribution [26]. When ARP was incorporated into PLA, the free volume portion and its size distribution in the matrix changed. Meanwhile, the tensile strength of ARP is greater than the strength and adhesion between them; with the loading of the specimen, the fibers change direction in the loading direction [27], and thus behave as a ductile break, different from the commonly recognized fragile break of pure PLA [28].

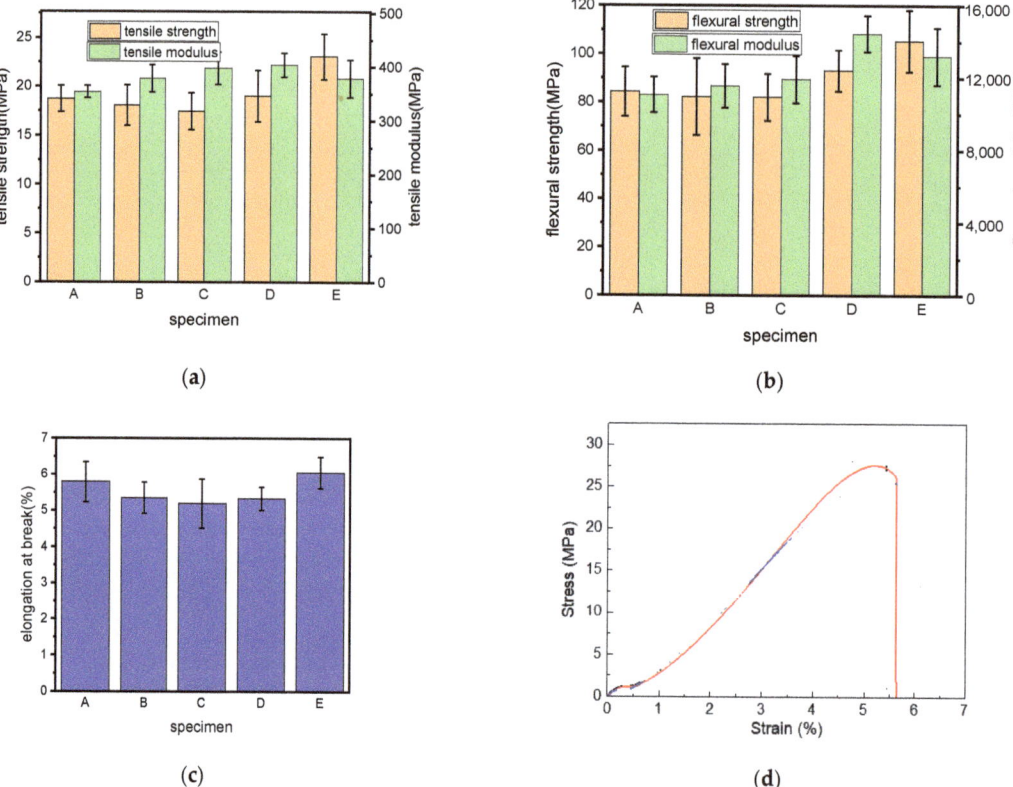

Figure 2. Mechanical properties of ARP/TPS/PLA biocomposite samples: (**a**) tensile strength and modulus, (**b**) flexural strength and modulus, (**c**) elongation at break, (**d**) a typical stress–strain curve of sample E during the tensile test.

3.1.2. Thermal Properties

Figure 3 presents the TGA curves of the composites and their raw materials. It shows that all of the composites exhibited similar thermal degradation curves under identical conditions. Figure 3 further indicates that all of the composites were more easily thermally degraded than PLA while being much more difficult to thermally degrade than TPS and ARP. The main decomposition happened between 330 °C and 370 °C. Table 3 lists the characteristic thermal degradation parameters of the thermogravimetric curves of various composites, such as the initial temperature at which the sample began to decompose (T_i) and the temperature at the maximum decomposition rate (T_p). For sample A, T_i was 335.8 °C and T_p was 370.3 °C. For samples B and C, the T_i values decreased to 334.1 °C and 326.4 °C. Meanwhile, the T_p values decreased to 363.8 °C and 354.9 °C, respectively, indicating that the thermal stability became poorer with the replacement of more PLA by ARP. This is in accordance with the conclusions reported in many studies that the incorporation of natural fiber would worsen the thermal stability of PLA [28–30]. The reason is that the thermal stability of ARP is much poorer than PLA, as evidenced in Figure 3a.

Where samples C, D and E were concerned, it was found that upon increasing the percentage of TPS in the biocomposite, the sample would become more thermally unstable, which was mainly caused by the easier decomposition of TPS than PLA as depicted in Figure 3b.

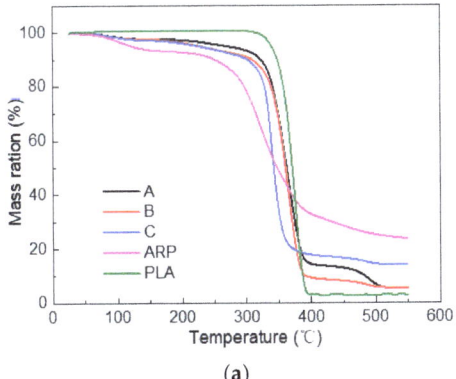
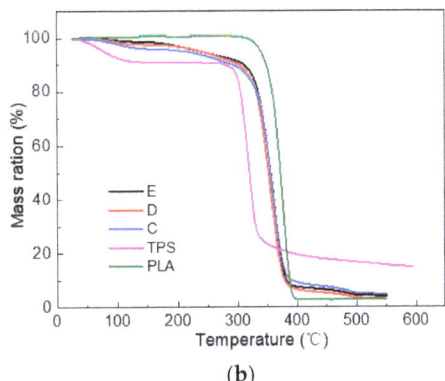

Figure 3. The mass loss curve of different specimens: (**a**) PLA, ARP and biocomposites with different ARP content; (**b**) PLA, TPS and biocomposites with different TPS content.

Table 3. Thermogravimetric analysis information table of TPS/ARP/PLA ternary composites at different formulations.

Sample	Ti/°C	Tp/°C	W/% (550 °C)
A	335.8	370.3	5.92
B	334.1	363.8	5.33
C	326.4	354.9	5.27
D	327.6	356.3	3.68
E	331.3	362.4	3.93

3.1.3. Mass Change in Water

The mass change rate in water of each sample was demonstrated in Figure 4. It can be found that the mass of sample A was actually reduced after immersion in water for 7 d, the mass change rate was a negative value. When the dosage of ARP was increased, i.e., for samples B and C, the mass change rate of the samples increased gradually to positive values. When the content of ARP was kept constant in the specimen, the mass change rate was monotonically reduced with the increase in the amount of TPS. When the ARP/TPS/PLA sample was immersed in water, TPS was easily dissolved, resulting in mass reduction in the sample. The calculated mass change rate was accordingly decreased even though all of the components may absorb some water, and the sample containing more TPS lost its mass more heavily; this is why the total mass of the biocomposite sample reduced gradually with the increase in TPS. As one kind of natural fiber, ARP has many hydrophilic groups such as hydroxyl on its molecular structure, which combine with water molecules once they contact water. For the ARP/TPS/PLA composite with little ARP, the mass loss caused by the dissolvement of TPS held the dominant position. As a result, the total mass decreased, and the calculated mass change rate by Equation (1) showed a negative value. When more ARP was used, the mass of water absorbed exceeded that of TPS dissolved; consequently, the total mass change of the sample became positive; thus, sample C had a much greater positive mass change rate than sample B. With the invasion of water molecules, the internal structure of the biocomposites may be expanded and finally destroyed. The dissolvement of TPS and the damage to the internal structure of the biocomposite illustrated that sample C should be the easiest to degrade in wet conditions among all of the samples.

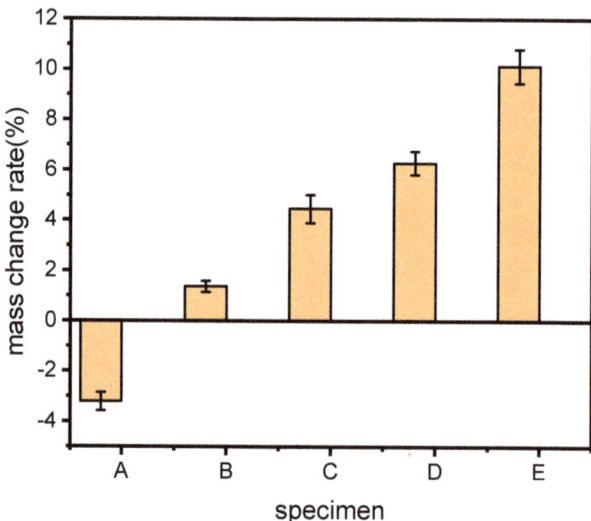

Figure 4. Mass change rate of the samples.

To sum up, increasing the dosage of ARP would reduce the strengths and thermal stability but improve the moduli of ARP/TPS/PLA biocomposites, while increasing the dosage of TPS would reduce the strengths and thermal stability and enhance the brittleness of ARP/TPS/PLA biocomposites. Even so, the mechanical and thermal properties of all of the samples could still meet the requirement for application. Among all of the samples, sample C was the easiest to degrade in wet conditions.

As a biodegradable composite, its cost must be taken into consideration for its wider application, in addition to possessing an excellent comprehensive performance. For ARP/TPS/PLA, the prices of each raw material differ greatly from one another, the prices of PLA and TPS are about 4500 USD/t and 500 USD/t in China, respectively. ARP is not available on the market as a kind of natural fiber; its price shall be about 80 USD/t when referring to the price of wood flour. The material cost of sample C will thus be about 3613.8 USD/t, 11.86% and 9.00% lower than those of samples A and E, respectively.

Taking into account the above factors comprehensively, sample C has adequate mechanical properties and thermal stability, though a little poorer than the other samples. Meanwhile, it is the most easily degraded in wet conditions and has a low cost; it was thus chosen as a research object in the following section, wherein the degradation behavior of sample C under soil burial was investigated.

3.2. Biodegradation Behavior of the Biocomposites

3.2.1. Visual Appearance

Figure 5 demonstrates the visual appearances of the printed samples after soil burial for different periods. All of the unburied samples had smooth and uniformly colored surfaces; after soil burial for 30 days, all the surfaces became gray with some black spots; with the continuation of soil burial, however, the surfaces of the samples became darkened and the surface color became more and more severely uneven. Meanwhile, certain components were found to detach from the samples, leaving some micro holes on the bodies. All the phenomena mentioned showed that degradation happened to the samples, and this degradation became more serious when the samples were buried for a longer time.

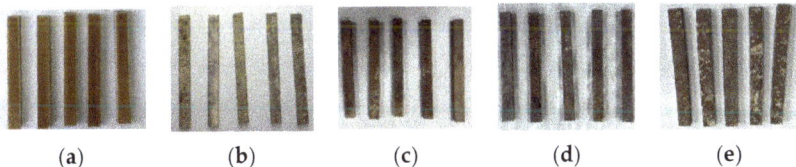

Figure 5. Visual appearances of ARP/TPS/PLA samples at different soil-burial days: (**a**) 0; (**b**) 30; (**c**) 60; (**d**) 90; (**e**) 180.

3.2.2. Weight Loss

Figure 6 shows the percentage weight change as a function of soil-burial time for ARP/TPS/PLA composites. It was found that the sample lost its weight obviously with time: after being buried in soil for 180 days, the weight loss was 21.40%, indicating that the extension of soil-burial time would promote the biodegradation of ARP/TPS/PLA, which was consistent with the results from the visual appearance observation. Comparing the weight with that of pure PLA [21], it was found that ARP and TPS greatly accelerated the degradation of PLA.

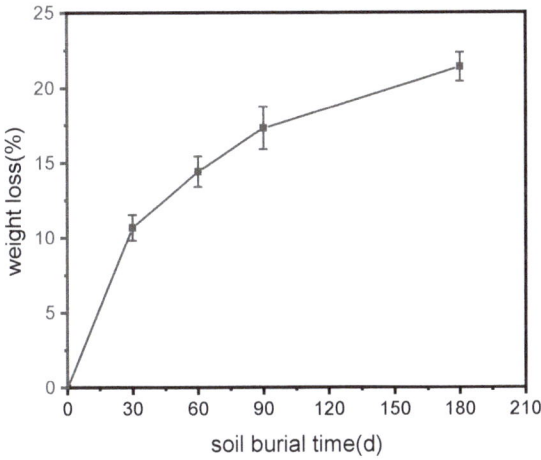

Figure 6. Weight losses of ARP/TPS/PLA samples with soil burial days.

3.2.3. Prediction of Flexural Properties

The effects of soil burial on the strength and modulus of ARP/TPS/PLA composites were investigated by bending tests. Figure 7 shows the trends of the properties with the soil-burial time. The flexural properties indicated that both the strength and modulus of ARP/TPS/PLA reduced gradually and almost linearly with time: after soil burial for 180 days, the flexural strength and modulus decreased to 47.6 MPa and 6653.92 MPa, reduced dramatically by 42.02% and 44.19% from those before soil burial. The fitting equations of the flexural strength and modulus could be expressed as $FS = 80.69 - 0.19t$, and $FM = 11329.76 - 26.96t$, respectively. Here, FS and FM represent the flexural strength and modulus, respectively, and t represents the soil burial days. It was derived from these two equations that the samples were speculated to lose their flexural strength and modulus thoroughly after soil burial for 376 d and 480 d, respectively. Although these were the results in the extreme cases predicted from the models, it could be expected that the samples would degrade greatly within 500 d, indicating that ARP/TPS/PLA was an easily biodegradable composite.

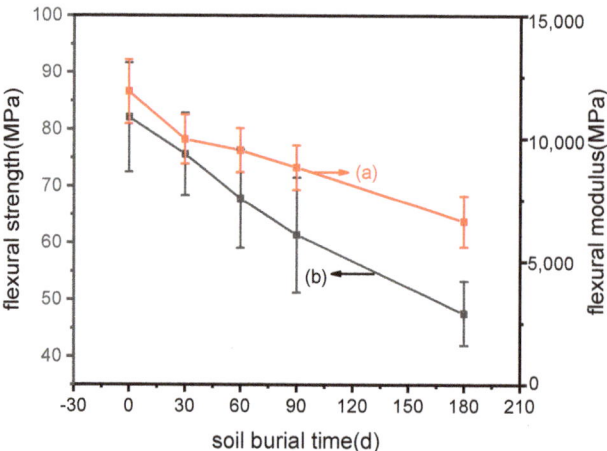

Figure 7. Bending properties' changes of ARP/TPS/PLA samples with soil burial days: (**a**) flexural modulus: (**b**) flexural strength.

3.2.4. Cross-Sectional Morphologies

The SEM images at ×1000 magnification after fracture obtained from 3D printing materials at different soil burial stages were shown in Figure 8. The pristine ARP/TPS/PLA had a relatively smooth fracture surface, showing that both ARP and TPS were enclosed well by PLA; however, in the case of the soil-buried samples, the fracture surface became much rougher and more cracks or holes appeared with the prolonging of the soil burial, and this phenomena was the most obvious for the sample after being buried in soil for 180 days. The existence of the cracks or holes made it easier for water to be absorbed by and transported in the sample; consequently, the internal structure of the sample would be destroyed more easily and the composite would degrade more heavily.

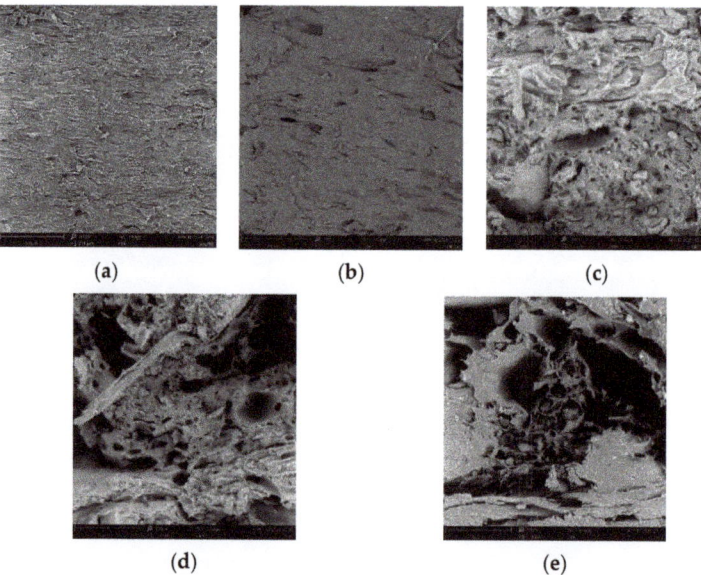

Figure 8. Cross-section morphology of ARP/TPS/PLA at different soil burial days: (**a**) 0; (**b**) 30; (**c**) 60; (**d**) 90; (**e**) 180.

3.2.5. Thermogravimetric Analysis

The thermal decomposition of the samples at different soil burial stages was investigated under a nitrogen atmosphere. Figure 9a,b illustrates the TGA and DTG thermograms during the decomposition of the printed samples in the temperature range of 20~550 °C, respectively.

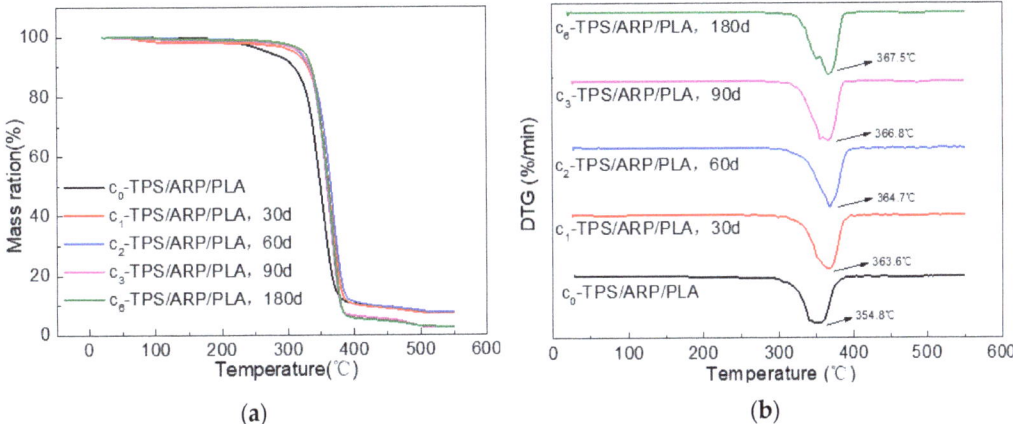

Figure 9. (a) TGA curves and (b) their derivatives of ARP/TPS/PLA biocomposites.

The derivative of TGA (DTG) curves in Figure 8b and the corresponding testing values in Table 4 indicate that the main decomposition of each sample fell into the range between 300 °C and 400 °C. Both the initial decomposition and the maximum degradation temperatures of the samples increased with the extension of soil-burial time. After being buried in soil for 180 d, the T_i and T_p values of the samples were 340.0 °C and 367.5 °C, increased by 10.6 °C and 12.7 °C from those of the unburied samples, respectively, meaning that the soil burial improved the thermal stability of the samples. This trend was similar with that of rice straw powder/PLA biocomposites [22].

Table 4. Thermogravimetric analysis information of different specimens.

Soil-Burial Time/Day	T_i/°C	T_p/°C	Char Residue/% (550 °C)
0	329.4	354.8	7.72
30	336.3	363.6	7.28
60	337.5	364.7	7.79
90	338.9	366.8	2.43
180	340.0	367.5	2.72

The possible reasons for the improved thermal stability of ARP/TPS/PLA biocomposites might come from three aspects. Firstly, PLA, as a polymer, was composed of crystal and amorphous domains; after being buried in soil, the amorphous domains would be destroyed by microorganisms, leaving the more thermally stable crystal domains, PLA itself may behave as more thermally stable in this situation. Secondly, TPS, as a polysaccharide, was easily degraded by microorganisms in the soil; the reduced amount of TPS in the soil was helpful for the improvement in the thermal stability of the samples because of its much poorer thermal stability than PLA as shown in Figure 3. Thirdly, ARP, as one kind of natural fiber, may easily rot in soil by itself; meanwhile, it was also much more thermally unstable than PLA and the biocomposite as shown in Figure 3. With the proceeding of soil burial, more and more ARP rotted and its content in the biocomposite reduced further and further, leading to the enhanced thermal stability of the whole sample.

3.2.6. DSC Thermal Analysis

The glass transition temperature (T_g), melting point (T_m) and cold crystallization temperature (T_{cc}) are the important parameters for the composites. These parameters were investigated by a two-step heating cycle using DSC operating under a nitrogen atmosphere. The secondary heating curves in the DSC traces of the samples at different soil burial stages are shown in Figure 10, and the corresponding properties of the biocomposites are presented in Table 5. The T_g, T_{cc} and T_m were 62.7 °C, 121.2 °C and 150.9 °C, respectively, for unburied ARP/TPS/PLA. The T_g decreased, while both the T_{cc} and T_m increased with the prolonging of the soil burial; however, these changes were not statistically significant. However, the degree of crystallinity showed a small decrease with the prolonging of the soil burial. This may be due to the following two reasons, one is that ARP, TPS and the amorphous areas in PLA were gradually destroyed during degradation, and a bigger free volume appeared in the matrix; as a result, the chain segment motion ability was enhanced, leading to a gradually reduced glassy transition temperature; meanwhile, the cold crystallization and melting would happen at higher temperatures. The other is that ARP, as a natural fiber, could act as a nucleating agent for the polymer matrix. When it was gradually degraded in the soil, the degree of crystallinity would be reduced accordingly.

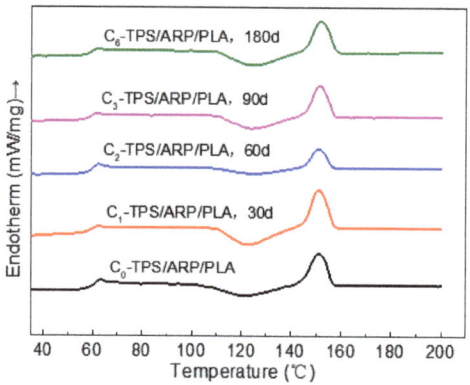

Figure 10. The secondary heating curves of the samples at different degradation periods.

Table 5. DSC thermal information of ARP/TPS/PLA samples at different soil-burial durations.

Soil-Burial Time/d	T_g/°C	T_{cc}/°C	T_m/°C	ΔH_{cc}/(J/g)	ΔH_m/(J/g)	X_c/%
0	62.7	121.2	150.9	−13.70	16.76	4.2
30	62.2	122.2	151.0	−18.38	19.26	1.2
60	62.0	124.3	151.1	−7.83	8.67	1.1
90	61.9	124.4	151.2	−14.81	14.52	0.4
120	61.7	124.6	151.4	−15.87	16.13	0.3

3.2.7. Thermo-Dynamic Mechanical Properties

Figure 11 shows the DMA results of the soil-buried samples. The storage modulus (E′) (30 °C) shown in the figure was 2395.3 MPa, 2284.9 MPa, 1921.2 MPa, 1866.4 MPa and 1476.5 MPa for the samples buried in soil for 0, 30, 60, 90 and 180 days, respectively. Generally, the extension of soil-burial time decreased the E′ value gradually, and the value was significantly decreased by 38.4% after 180 d of soil burial. The existence of the hydrophilic natural fiber and starch made the sample more easily invaded and destroyed by water and microorganisms, accelerating the breaking of the molecular chains; as a result, the storage modulus of the sample was greatly reduced.

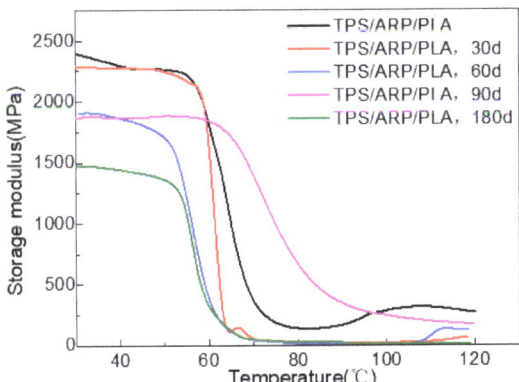

Figure 11. Storage-modulus curves of the samples at different degradation periods.

4. Conclusions

This study first investigated the effect of the dosages of ARP and TPS on the mechanical properties, thermal stability and mass changes in water of FDM 3D-printed ARP/TPS/PLA biocomposite samples, then studied the effects of soil-burial duration on the degradation behavior of the printed samples containing 11 wt.% ARP, 10 wt.% TPS and 79 wt.% PLA. The following conclusions can be drawn from the results of this study:

1. Raising the dosage of ARP or TPS decreased the strengths of the biocomposites. The tensile and flexural strengths of the samples containing 0 wt.% ARP, 10 wt.% TPS, and 90 wt.% PLA were 18.73 MPa and 84.28 MPa, respectively, The strengths of the samples containing 11 wt.% ARP, 0 wt.% TPS and 89 wt.% PLA were 23.07 MPa and 105.39 MPa, respectively, while those of the samples containing 11 wt.% ARP, 10 wt.% TPS and 79 wt.% PLA dropped to 17.46 MPa and 82.06 MPa, respectively.
2. All of the composites were more easily thermally decomposed than PLA. Increasing the percentage of TPS or ARP in the biocomposites resulted in the samples becoming more thermally unstable.
3. After immersion in water for 7 days, the masses of all of the samples would be changed due to the dissolution of TPS and the water absorption by the samples. For sample A containing 90 wt.% PLA and 10 wt.% TPS, the mass change rate was negative. With the increase in the dosage of ARP, the mass change rate of the samples increased gradually to positive values. When the content of ARP was kept constant in the specimens, the mass change rate was monotonically reduced with the increase in the amount of TPS.
4. Soil burial altered the surfaces and fracture surfaces of the samples. After soil burial, the surface color became uneven, some components detached from the samples, leaving some micro holes on the bodies. The fracture surfaces became much rougher, and more cracks or holes appeared with the prolonging of the soil-burial time.
5. Extending the soil-burial time resulted in the samples' extensive mass loss and reduction in storage modulus and flexural properties. After soil burial for 180 d, the weight, the storage modulus at 30 °C, as well as the flexural strength and modulus, were greatly reduced by 21.40%, 38.36%, 42.02% and 44.19%, respectively, when compared with those before soil burial.
6. Extending soil-burial time increased the thermal stability and decreased the crystallinity of the samples gradually, but had little effect on the glass transition temperature, cold crystallization temperature or melting point of the samples.

Through experimentation, it is confirmed that ARP/TPS/PLA can be subjected to degradation in soil and the incorporation of ARP and TPS can accelerate the degradation of PLA.

Author Contributions: Conceptualization, W.L. and W.Y.; methodology and formal analysis, J.S. and W.L.; investigation, J.S., Z.N. and M.L.; data curation, J.S., W.L. and M.L.; writing—original draft preparation, J.S., Z.N. and W.Y.; writing—review and editing, W.L. and W.Y. All authors have read and agreed to the published version of the manuscript.

Funding: This research was funded by the Natural Science Foundation of the Jiangsu Higher Education Institutions of China [22KJA430012].

Institutional Review Board Statement: Not applicable.

Informed Consent Statement: Not applicable.

Data Availability Statement: Not applicable.

Conflicts of Interest: The authors declare no conflict of interest.

References

1. Khorasani, M.; Ghasemi, H.; Rolfe, B.; Gibson, I. Additive manufacturing a powerful tool for the aerospace industry. *Rapid Prototyp. J.* **2022**, *28*, 87–100. [CrossRef]
2. Beg, M.D.H.; Pickering, K.L.; Gauss, C. The effects of alkaline digestion, bleaching and ultrasonication treatment of fibre on 3D printed harakeke fibre reinforced polylactic acid composites. *Compos. Part A* **2023**, *166*, 107384. [CrossRef]
3. Choo1, Y.J.; Boudier-Revéret, M.; Chang, M.C. 3D printing technology applied to orthosis manufacturing: Narrative review. *Ann. Palliat. Med.* **2020**, *9*, 4262–4270. [CrossRef] [PubMed]
4. Melocchi, A.; Parietti, F.; Loreti, G.; Maroni, A.; Gazzaniga, A. Lucia Zema. 3D printing by fused deposition modeling (FDM) of a swellable/erodible capsular device for oral pulsatile release of drugs. *J. Drug Deliv. Sci. Technol.* **2015**, *30*, 1–8.
5. Raja, S.; Agrawal, A.P.; Patil, P.; Thimothy, P.; Capangpangan, R.Y.; Singhal, P.; Wotango, M.T. Optimization of 3D printing process parameters of polylactic acid filament based on the mechanical test. *Int. J. Chem. Eng.* **2022**, *2022*, 5830869. [CrossRef]
6. Rahmatabadi, D.; Ghasemi, I.; Baniassadi, M.; Abrinia, K.; Baghani, M. 3D printing of PLA-TPU with different component ratios: Fracture toughness, mechanical properties, and morphology. *J. Mater. Res. Technol.* **2022**, *21*, 3970–3981. [CrossRef]
7. Davood, R.; Kianoosh, S.; Mohammad, A.; Elyas, S.; Ismaeil, G.; Majid, B.; Karen, A.; Mahdi, B.; Mostafa, B. Development of Pure Poly Vinyl Chloride (PVC) with Excellent 3D Printability and Macro- and Micro-Structural Properties. *Macromol. Mater. Eng.* **2022**, *308*, 2200568. [CrossRef]
8. Badouard, C.; Traon, F.; Denoual, C.; Mayer-Laigle, C.; Paës, G.; Bourmaud, A. Exploring mechanical properties of fully compostable flax reinforced composite fifilaments for 3D printing applications. *Ind. Crops Prod.* **2019**, *135*, 246–250. [CrossRef]
9. Abdullah, T.; Qurban, R.O.; Abdel-Wahab, M.S.; Salah, N.A.; Melaibari, A.A.; Zamzami, M.A.; Memić, A. Development of nanocoated filaments for 3D fused deposition modeling of antibacterial and antioxidant materials. *Polymers* **2022**, *14*, 2645. [CrossRef]
10. Tran, T.N.; Bayer, I.S.; Heredia-Guerrero, J.A.; Frugone, M.; Lagomarsino, M.; Maggio, F.; Athanassiou, A. Cocoa shell waste biofilaments for 3D printing applications. *Macromol. Mater. Eng.* **2017**, *302*, 1700219. [CrossRef]
11. Vaidya, A.A.; Collet, C.; Gaugler, M.; Lloyd-Jones, G. Integrating softwood biorefinery lignin into polyhydroxybutyrate composites and application in 3D printing. *Mater. Today Commun.* **2019**, *19*, 286–296. [CrossRef]
12. Zhang, Q.F.; Cai, H.Z.; Zhang, A.D.; Lin, X.N.; Yi, W.M.; Zhang, J.B. Effects of lubricant and toughening agent on the fluidity and toughness of poplar powder-reinforced polylactic acid 3D printing materials. *Polymers* **2018**, *10*, 932. [CrossRef] [PubMed]
13. Aumnate, C.; Soatthiyanon, N.; Makmoon, T.; Potiyaraj, P. Polylactic acid/kenaf cellulose biocomposite filaments for melt extrusion based-3D printing. *Cellulose* **2021**, *28*, 8509–8525. [CrossRef]
14. Suteja, J.; Firmanto, H.; Soesanti, A.; Christian, C. Properties investigation of 3D printed continuous pineapple leaf fiber reinforced PLA composite. *J. Thermoplast. Compos. Mater.* **2020**, *35*, 2052–2061. [CrossRef]
15. Depuydt, D.; Balthazar, M.; Hendrickx, K.; Six, W.; Ferraris, E.; Desplentere, F.; Ivens, J.; Vuure, V.A.W. Production and characterization of bamboo and flax fiber reinforced polylactic acid filaments for fused deposition modeling (FDM). *Polym. Compos.* **2019**, *40*, 1951–1963. [CrossRef]
16. Le Guen, M.J.; Hill, S.; Smith, D.; Theobald, B.; Gaugler, E.; Barakat, A.; Mayer-Laigle, C. Influence of rice husk and wood biomass properties on the manufacture of filaments for Fused Deposition Modelling. *Front. Chem.* **2019**, *7*, 735. [CrossRef] [PubMed]
17. Yu, W.W.; Shi, J.N.; Sun, L.W.; Lei, W. Effects of printing parameters on pProperties of FDM 3D printed residue of astragalus/polylactic acid biomass composites. *Molecules* **2022**, *27*, 7373. [CrossRef]
18. Harynska, A.; Janik, H.; Sienkiewicz, M.; Mikolaszek, B.; Kucinska-Lipka, J. PLA—potato thermoplastic starch filament as a sustainable alternative to the conventional PLA filament: Processing, characterization, and FFF 3D printing. *ACS Sustain. Chem. Eng.* **2021**, *9*, 6923–6938. [CrossRef]
19. Zhao, Y.Q.; Yang, J.H.; Ding, X.K.; Ding, X.J.; Duan, S.; Xu, F.J. Polycaprolactone/polysaccharide functional composites for low-temperature fused deposition modelling. *Bioact. Mater.* **2020**, *5*, 185–191. [CrossRef]
20. Ju, Q.; Tang, Z.P.; Shi, H.D.; Zhu, Y.F.; Shen, Y.C.; Wang, T.W. Thermoplastic starch based blends as a highly renewable filament for fused deposition modeling 3D printing. *Int. J. Biol. Macromol.* **2022**, *219*, 175–184. [CrossRef]

21. Yu, W.W.; Shi, J.; Qiu, R.; Lei, W. Degradation behavior of 3D-printed residue of astragalus particle/poly(Lactic Acid) biocomposites under soil conditions. *Polymers* **2023**, *15*, 1477. [CrossRef] [PubMed]
22. Yu, W.W.; Dong, L.L.; Lei, W.; Shi, J.N. Rice straw powder/polylactic acid biocomposites for three-dimensional printing. *Adv. Compos. Lett.* **2020**, *29*, 1–8. [CrossRef]
23. Wu, C.S. Utilization of peanut husks as a fifiller in aliphaticearomatic polyesters: Preparation, characterization, and biodegradability. *Polym. Degrad. Stab.* **2012**, *97*, 2388–2395. [CrossRef]
24. Singamneni, S.; Smith, D.; LeGuen, M.J.; Truong, D. Extrusion 3D printing of polybutyrate-adipate-terephthalate-polymer composites in the pellet form. *Polymers* **2018**, *10*, 922. [CrossRef]
25. Prakash, K.; Chantara, T.; Sivakumar, M. Enhancements in crystallinity, thermal stability, tensile modulus and strength of sisal fifibres and their PP composites induced by the synergistic effects of alkali and high intensity ultrasound (HIU) treatments. *Ultrason. Sonochem.* **2017**, *34*, 729–742.
26. Aberoumand, M.; Soltanmohammadi, K.; Soleyman, E.; Rahmatabadi, D.; Ghasemi, I.; Baniassadi, M.; Abrinia, K.; Baghani, M. A comprehensive experimental investigation on 4D printing of PET-G under bending. *J. Mater. Res. Technol.* **2022**, *18*, 2552–2569. [CrossRef]
27. Moradi, M.; Aminzadeh, A.; Rahmatabadi, D.; Rasouli, S.A. Statistical and Experimental Analysis of Process Parameters of 3D Nylon Printed Parts by Fused Deposition Modeling: Response Surface Modeling and Optimization. *J. Mater. Eng. Perform.* **2021**, *30*, 5441–5454. [CrossRef]
28. Daver, F.; Lee, M.; Brandt, M.; Shanks, R. Cork–PLA composite filaments for fused deposition modelling. *Compos. Sci. Technol.* **2018**, *168*, 230–237. [CrossRef]
29. Manshor, R.; Anuar, H.; Aimi, N.; Fitrie, I.; Nazri, B.; Sapuan, M.; Shekeil, A.; Wahit, U. Mechanical, thermal and morphological properties of durian skin fibre reinforced PLA biocomposites. *Mater. Des.* **2014**, *59*, 279–286. [CrossRef]
30. Sun, Y.; Wang, Y.; Mu, W.; Zheng, Z.; Yang, B.; Wang, J.; Zhang, R.; Zhou, K.; Chen, L.; Ying, J.; et al. Mechanical properties of 3D printed micro-nano rice husk/polylactic acid filaments. *J. Appl. Polym. Sci.* **2022**, *139*, e52619. [CrossRef]

Disclaimer/Publisher's Note: The statements, opinions and data contained in all publications are solely those of the individual author(s) and contributor(s) and not of MDPI and/or the editor(s). MDPI and/or the editor(s) disclaim responsibility for any injury to people or property resulting from any ideas, methods, instructions or products referred to in the content.

Article

Furan as Impurity in Green Ethylene and Its Effects on the Productivity of Random Ethylene–Propylene Copolymer Synthesis and Its Thermal and Mechanical Properties

Joaquín Hernández-Fernández [1,2,3,*], Esneyder Puello-Polo [4] and Edgar Márquez [5,*]

1. Chemistry Program, Department of Natural and Exact Sciences, San Pablo Campus, University of Cartagena, Cartagena 130015, Colombia
2. Chemical Engineering Program, School of Engineering, Universidad Tecnológica de Bolívar, Parque Industrial y Tecnológico Carlos Vélez Pombo Km 1 Vía Turbaco, Cartagena 130001, Colombia
3. Department of Natural and Exact Science, Universidad de la Costa, Barranquilla 080002, Colombia
4. Group de Investigación en Oxi/Hidrotratamiento Catalítico Y Nuevos Materiales, Programa de Química-Ciencias Básicas, Universidad del Atlántico, Puerto Colombia 081001, Colombia
5. Grupo de Investigaciones en Química Y Biología, Departamento de Química Y Biología, Facultad de Ciencias Básicas, Universidad del Norte, Carrera 51B, Km 5, Vía Puerto Colombia, Barranquilla 081007, Colombia
* Correspondence: jhernandezf@unicartagena.edu.co (J.H.-F.); ebrazon@uninorte.edu.co (E.M.); Tel.: +57-301-562-4990 (J.H.-F.)

Citation: Hernández-Fernández, J.; Puello-Polo, E.; Márquez, E. Furan as Impurity in Green Ethylene and Its Effects on the Productivity of Random Ethylene–Propylene Copolymer Synthesis and Its Thermal and Mechanical Properties. *Polymers* 2023, 15, 2264. https://doi.org/10.3390/polym15102264

Academic Editor: Cristina Cazan

Received: 15 February 2023
Revised: 29 April 2023
Accepted: 4 May 2023
Published: 11 May 2023

Copyright: © 2023 by the authors. Licensee MDPI, Basel, Switzerland. This article is an open access article distributed under the terms and conditions of the Creative Commons Attribution (CC BY) license (https:// creativecommons.org/licenses/by/ 4.0/).

Abstract: The presence of impurities such as H_2S, thiols, ketones, and permanent gases in propylene of fossil origin and their use in the polypropylene production process affect the efficiency of the synthesis and the mechanical properties of the polymer and generate millions of losses worldwide. This creates an urgent need to know the families of inhibitors and their concentration levels. This article uses ethylene green to synthesize an ethylene–propylene copolymer. It describes the impact of trace impurities of furan in ethylene green and how this furan influences the loss of properties such as thermal and mechanical properties of the random copolymer. For the development of the investigation, 12 runs were carried out, each in triplicate. The results show an evident influence of furan on the productivity of the Ziegler–Natta catalyst (ZN); productivity losses of 10, 20, and 41% were obtained for the copolymers synthesized with ethylene rich in 6, 12, and 25 ppm of furan, respectively. PP0 (without furan) did not present losses. Likewise, as the concentration of furan increased, it was observed that the melt flow index (MFI), thermal (TGA), and mechanical properties (tensile, bending, and impact) decreased significantly. Therefore, it can be affirmed that furan should be a substance to be controlled in the purification processes of green ethylene.

Keywords: furan; green ethylene; Ziegler–Natta catalyst; random copolymer; catalyst; mechanical properties; melt flow index

1. Introduction

Currently, the chemical, petrochemical, and agri-food industries and the research community in general aim to mitigate the environmental impact these sectors have generated over the years [1]. For this reason, one of the strategies is to optimize the production processes of polymeric substances of fossil origin that function as raw materials to obtain products of great commercial utility [2,3]. However, the poor disposal of fossil substances and their derivatives causes damage to the health and environment of the communities near the plants that produce these substances [1–4]. This has led to the implementation of new ideas, such as using raw materials that allow the production of polymers through processes free of fossil fuels [1–7].

For the industrial synthesis of polymers, it is common to use the raw material ethylene from fossil fuels and endothermic processes [1–9]. The green ethylene obtained from the fermentation of corn, glucose, and starch has been implemented to meet the global demand for ethylene of fossil origin and thus reduce the environmental impact [1]. One route for obtaining green ethylene is from the pyrolysis of methanol [9]. Regardless of the ethylene and propylene source, these hydrocarbons are used to synthesize random or impact copolymers since they are of great academic and industrial interest for better physical and mechanical properties [10–14]. One of the limitations of ethylene of fossil origin is the presence of impurities of sulfur-containing chemical compounds, ketones, alcohols, arsine, phosphine, carboxylic acids, and thiols that act as aggressive inhibitors of the Ziegler–Natta catalytic systems [2], which act as a catalytic agent in the synthesis of copolymers. It is necessary to highlight that these impurities can affect other properties of the polymer such as the melt flow index (MFI), thermal degradation, molecular weight, and mechanical properties [2], which affect the productivity of the catalysts and the physico-chemical properties of the synthesized copolymers. Since green ethylene is of natural origin, it is understood that it is not in the presence of these contaminants, thus having another competitive advantage. Green ethylene may have another type of contaminant present in trace concentrations since it is obtained from the transformation of natural products and methanol. In this research, we focus on heterocyclic compounds such as furan. Furan is the name by which a five-membered aromatic heterocyclic compound is recognized [14]; it is soluble in organic media, transparent and colorless with a high degree of flammability, and very volatile [14,15]. The constant study of the structure and properties of this compound has allowed us to affirm that around 135 different isomers of it are known to date [16], and many of them generate negative impacts both on health and on the production costs of certain substances and raw materials in the industrial sector. The slightest impurities in a raw material can affect the final stages of the production of copolymers [1]. In the identification of furan in renewable and non-renewable sources, techniques such as infrared radiation [1,11] and gas chromatography [2] have been used. To date, very little information is available on how compounds such as furan can influence the copolymer, how it affects the physicochemical and mechanical properties of this product, and how it can affect its production rate.

Since furan can be present as an impurity in green ethylene and can also be formed during the pyrolysis of natural waste, we will develop this research and propose reaction mechanism pathways that explain furan formation, the reaction of furan with the Ziegler–Natta catalyst, and the effect of different concentrations of furan on the efficiency of copolymer synthesis, and we will use other instrumental techniques to determine how the melt index, molecular weight distribution, thermal stability, and mechanical properties of the copolymer are affected.

2. Materials and Methods

2.1. Polymerization Process: Synthesis of Random Ethylene–Propylene Copolymer

Table 1 shows the list of reagents used to obtain polypropylene. For the development of this research, a fourth-generation spherical Ziegler–Natta catalyst was used, with $MgCl_2$ support with 3.6 wt.% Ti; diisobutyl phthalate (DIBP) was used as an internal donor with a purity degree of 99.99% from Sudchemie, Germany. As a co-catalyst, triethylaluminium (TEAL) with a purity of 98%, obtained from Merck, Germany, was used as well as cyclohexyl methyl dimethoxy silane, the latter of which had a purity of 99.9%. Hydrogen and nitrogen gases obtained by Lynde were used. Finally, the propylene used was obtained from Airgas with a purity of 99.999%.

Table 1. List of reagents for obtaining polypropylene.

Materials	Supplier Used	Purity
Diisobutyl phthalate (DIBP)	(In-house donor) Sudchemie, Germany	99.99%
Triethylaluminium	(Co-catalyst) Merck, Germany	98%
Cyclohexyl methyl dimethoxysilane (CMDS)	(External donor) Merck, Germany	99.9%
Hydrogen	Lynde	99.999%
Nitrogen	Lynde	99.999%
Ethylene	Airgas	99.999%
Propylene	Airgas	99.999%

For the synthesis of polypropylene, the methodology proposed in [1,12] was followed, which suggests obtaining polypropylene with the help of a Ziegler–Natta catalyst [13–18]. The process starts with the use of a fluidized bed reactor which is purged with nitrogen; after this, the system is fed with hydrogen and propylene, which provide fluidization in addition to absorbing heat from the reaction. A Ziegler-Nata catalyst, TEAL, and a selectivity control agent were also incorporated along with nitrogen. The quantities are shown in Table 2. The process was carried out at 70 °C and 27 bar pressure in a discontinuous condition; the reactor product gases were captured with a compressor and then taken to a heat exchanger, where they were cooled and recirculated to the system. To obtain virgin resin, the method proposed by [19–34] was followed. The resin obtained from the reactor was taken to a purge tower fed with nitrogen to eliminate the hydrocarbons leaving the system.

Table 2. Identification of samples and sampling points.

Materials	Run											
	PP0	PP0-1	PP0-2	PP1	PP1-1	PP1-2	PP2	PP2-1	PP2-2	PP3	PP3-1	PP3-2
Catalyst, kg/h	5	5	5	5	5	5	5	5	5	5	5	5
Propylene TM/h	1.2	1.2	1.2	1.2	1.2	1.2	1.2	1.2	1.2	1.2	1.2	1.2
Green ethylene	0.6	0.6	0.6	0.6	0.6	0.6	0.6	0.6	0.6	0.6	0.6	0.6
TEAL, kg/h	0.25	0.25	0.25	0.25	0.25	0.25	0.25	0.25	0.25	0.25	0.25	0.25
Hydrogen, g/h	30	30	30	30	30	30	30	30	30	30	30	30
Furan (ppm)	0	0	0	6	5.8	6.2	12	12.2	12.5	25	25.5	24.6
Selectivity control agent, mol/h	1	1	1	1	1	1	1	1	1	1	1	1
T, °C	70	70	70	70	70	70	70	70	70	70	70	70
Pressure, bar	27	27	27	27	27	27	27	27	27	27	27	27

2.2. Gas Chromatography Analysis with Selective Mass Detector (GC-FID)

The data obtained to express the quantification of furan in the green ethylene samples were obtained with the aid of a gas chromatograph (Agilent 7890B), with a front and rear injector of (250 °C, 7.88 psi, 33 mL min^{-1}) and (250 °C, 11.73 psi, 13 mL min^{-1}), respectively. The volume used ranged from 0.25 to 1.0 mL. The oven was started at 40 °C × 3 min, which increased to 60 °C–10 °C min^{-1} × 4 min and then increased to 170 °C at 35 °C min^{-1}.

2.3. Melt Flow Index—MFI

The flow index was determined using the methodology proposed by [12], in which a Tinius Olsen MP1200 plastometer was used. The temperature inside the equipment was 230 °C, and a 2.16 kg piston was used to displace the molten material.

2.4. Thermogravimetric Analysis—TGA

For the development of the thermogravimetric analysis (TGA), a TGA Q500 thermal analyzer was used to achieve better results and continued with the methodology established by [12], where they proposed a heating rate of 10 °C min^{-1} from 40 to 800 °C in an air atmosphere (50 cm^3 min^{-1}). Therefore, the equipment was calibrated for temperature and weight via standard methods.

2.5. Analysis of Mechanical Properties

2.5.1. Injection Molding

Several authors propose that materials obtained from thermoplastics should be processed with a twin-screw extruder and then subjected to an injection molding process [32]. These composites can also be molded via compression molding, among other techniques. For the present study, the samples were obtained via injection molding; to achieve better results when molding the samples, variables such as temperature and pressure were controlled, keeping the former at 50 °C and having strict control over the latter.

2.5.2. Test Specimen Preparation

ASTM standards were followed to prepare the specimens. Two molds were used to design the tensile (ASTMD638) and flexural (ASTMD790) test samples to be stored for 48 h at the same temperature as the surrounding environment and then used for each of their respective tests.

2.5.3. Tensile Test

To verify the resistance of thermoplastic composites to axial tensile forces and to determine the ability of the composite to elongate before cracking and failure, a tensile strength test was performed. The tensile test was performed according to the ASTMMD638 standard on a computerized H50KL universal testing machine (TiniusOlsen). The samples were clamped at both ends and then subjected to uniaxial tensile force. Since the samples were obtained via injection molding, they were classified as TYPE 1. The dimensions of the pieces to be studied were followed according to the methodology proposed by Suraj and Sanyay, 2022 [32], where the value of the gauge length (G) was 50 mm, the width of the narrow section (W) was 12.7 mm, and the thickness (T) was 3.4 mm.

2.5.4. Flexural Test

Flexural strength is considered to be the capacity of a composite to withstand the bending force to which it is subjected, which is applied transversally to the shaft. For the development of this research, this resistance was evaluated by taking into account the ASTMD790 standard and using the H50KL universal testing machine (TiniusOlsen).

2.5.5. Impact Test

Using the model IT504 pendulum impact tester (TiniusOlsen), the Izod impact test according to ASTMD256 was performed. Each specimen is assigned a 45°, 2.5 mm deep AV notch. One end of the notched specimen was fixed using a cantilevered vice.

3. Results

3.1. Effects of Furan on the Polymerization Process of Green Random Copolymer Rat

Figure 1 shows that the presence of furan was inversely proportional to the productivity of the Ziegler–Natta catalyst since the higher the furan concentration (ppm), the lower the catalyst's effectiveness. In the absence of furan, the productivity of the ZN catalyst was 47 MT/kg. The average concentration of 6, 12, and 25 ppm of furan in ethylene affected productivity by 10, 20, and 41%, respectively.

The standard error for each of the variables was calculated based on the central limit theorem (see Equations (1) and (2)).

$$Typical\ error = \frac{\sigma}{\sqrt{n}} \quad (1)$$

$$\sigma = \sqrt{\frac{\sum_{i=1}^{N}(x_i - \bar{x})^2}{N}} \quad (2)$$

It is essential to point out that all the groups studied presented significant differences ($p < 0.05$) when varying the concentration of furan from one group to another. This result is because furan acted as an inhibitor of the reaction since it reacted with the active center of titanium, preventing the propylene from polymerizing in the vibrant center of titanium. It should be noted that the inhibitory behavior of impurities on the productivity of the Ziegler–Natta catalyst has been previously reported [1,3,12,35–40], showing that the presence of impurities such as H_2S, oxygenated compounds, and thiol, among others, affects the efficiency in the production of polypropylene on an industrial scale.

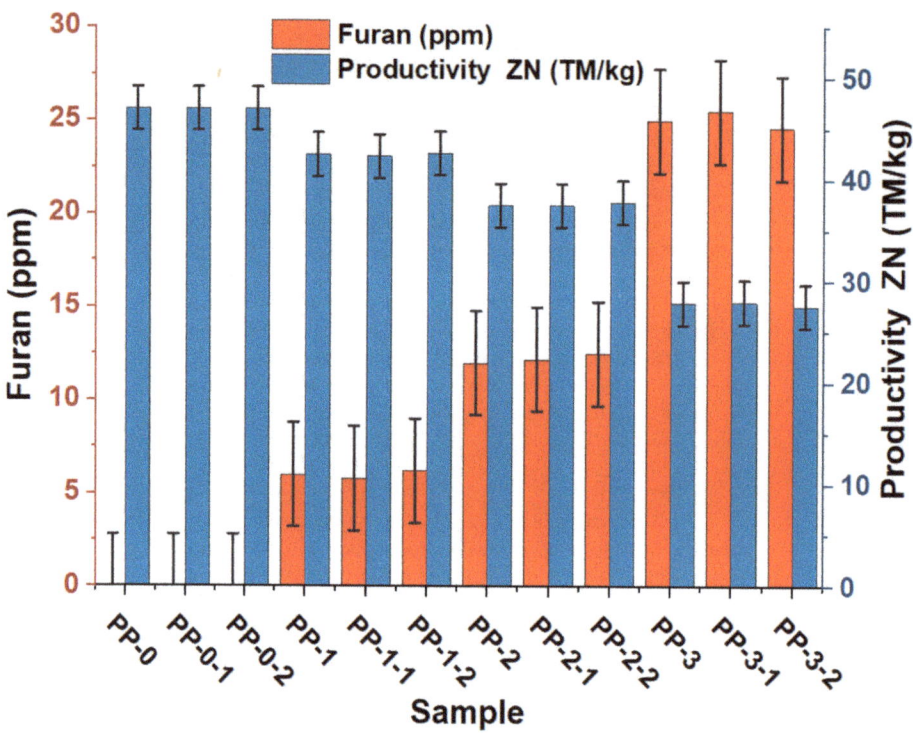

Figure 1. Productivity lost in the polymerization process of the green random copolymer.

Proposed Mechanism of Action of Furan on the ZN Catalyst

Figure 2 shows the mechanism of furan formation in corn residues and other natural products. As mentioned in this study, two of the sources of green ethylene are corn and starch, which have glucose in their matrix. From this fact, Figure 2 shows the reaction mechanism for furan formation while obtaining ethylene from the fermentation of corn and starch. Figure 2 indicates that the furan formation process begins with four intramolecular rearrangements in three equilibria of the glucose molecule. The first structure moves an electron pair (double bond) from the carbonyl group to the alpha–beta carbon region to form an alpha–beta carbon–carbon double bond. In the second structure of Figure 2, the electronic pair transfers to form the carbonyl bond (C=O) on the beta carbon. The third equilibrium is the transfer of the electron pair to form a carbon–carbon beta–gamma double bond. Molecules 2 and 4 of the three proposed equilibriums undergo successive dehydration, losing hydroxyl groups and hydrogen atoms, with which the conditions for closing the ring between the beta–carbon carbonyl group and the epsilon carbon are obtained. In this way, the furan ring is thus formed. Molecule 2 additionally forms an equilibrium in which two electronic pairs are transferred from the carbonyl groups of the alpha and beta carbons to form alkene bonds between the alpha and beta and gamma and delta carbons; here, in the same way, dehydration occurs until the furan ring is formed, with the difference that in this way, a molecule of glycolic acid first detaches from the structure before starting the furan ring.

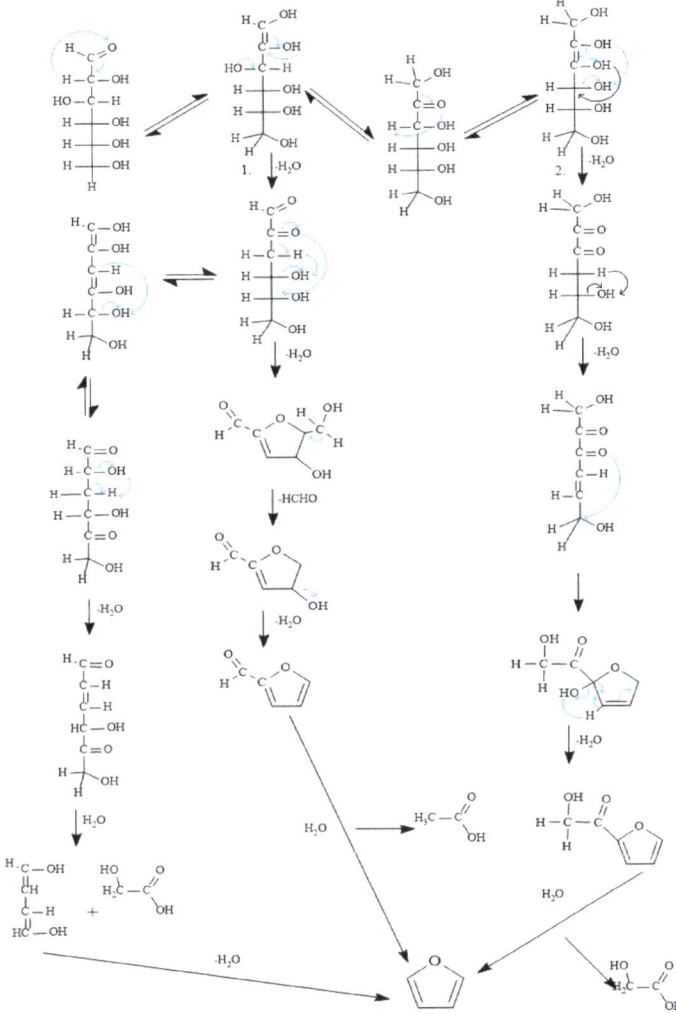

Figure 2. Proposal for the mechanism of furan biogenesis.

The furan originated in the pyrolysis of natural residues, which traces of unfermented glucose can generate. These glucose molecules can undergo a dehydration process in the first two steps of the mechanism presented in Figure 3, causing the characteristic and propitious condition at carbon 1,4 for the formation of the furanoid ring; later, this ring undergoes further dehydration, followed by dehydrogenation and the removal of the carbonyl group that combines with the ring, leaving hydrogen to form formaldehyde. This mechanism is because the pyrolysis conditions destabilize the glucose molecule, which decomposes to create more stable species under such conditions, such as furan.

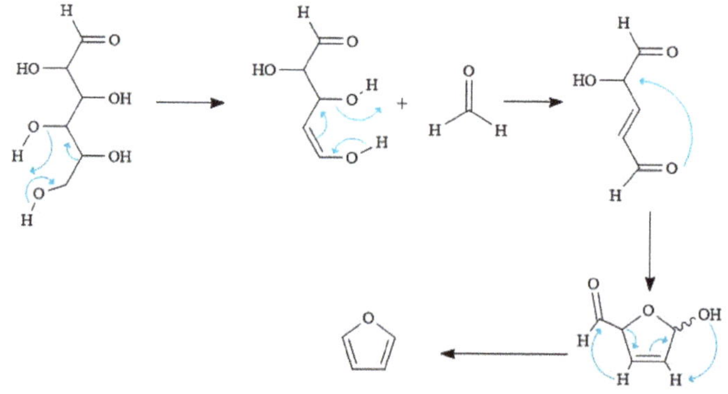

Figure 3. Proposal of the mechanism of furan formation during the pyrolysis of natural residues.

In the mechanism proposed in Figure 4, furan competes with propylene monomer for the Ti-active site. First, a π complex is formed by coordinating the furan with the Ti of the TiCl$_4$/MgCl$_2$ complex. The furan-Ti interaction is carried out through the interaction of the electropositive Ti with the lone pair of electrons of the oxygen atom of the furan heterocyclic structure. This interaction is proposed to have a higher energy gain than that of a π complex in Ti-propylene. Therefore, the furan-Ti reaction predominates. In other investigations, it has been shown that to synthesize this family of polymers, there is a propylene insertion barrier that varies between 6 and 12 kcal mol^{-1} because the probability of the occurrence of the propylene insertion reaction in Ti is supported by thermodynamics since a favoring of approximately 20 kcal mol^{-1} is observed. For the efficiency of the reaction to be affected, the growth of the PP chain will be affected when the active site of Ti reacts with inhibitors of different polarities. Since furan exhibits intermediate polarities to other poisons such as H$_2$S, their energy values are expected to be within the ranges of H$_2$S. As shown in Figure 4, furan interferes with the formation of propylene complexes and their insertion.

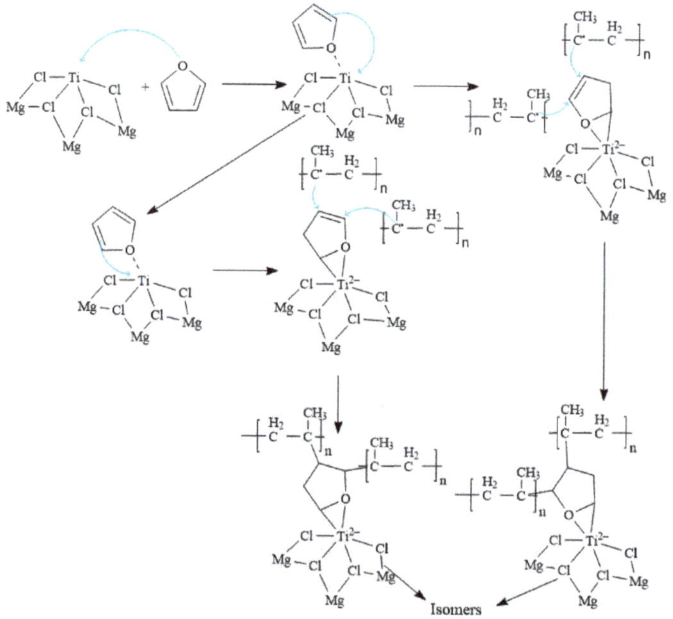

Figure 4. Proposal of the reaction mechanism between furan and the Ziegler–Natta catalyst.

3.2. Effects of Furan on the TGA of the Random Copolymer

TGA determined the thermal degradation of the copolymer; the curves are shown in Figure 2. In these, it can be seen that as the concentration of furan increases, its thermal stability decreases. This is due to the inhibitory capacity of furan in the polymerization process [12] and the favoring of the formation of new functional groups as a consequence of incomplete polymerization. This generates an alteration in the behavior and structure of the copolymer at the micromolecular and macro-molecular levels. When evaluating thermal degradation, it is evident that samples PP0 and PP1 have similar behavior in terms of weight loss, which is 5% by weight at 390 °C. For PP3 and PP4, there is loss of 5% at 370 °C and 320 °C, respectively as shown in Figure 5.

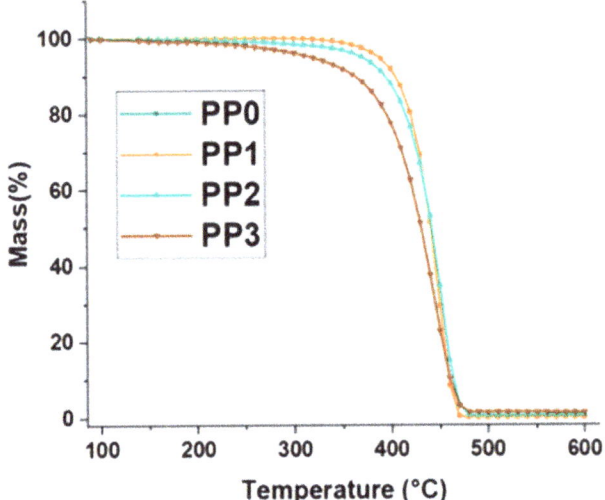

Figure 5. Effects of furan on thermal degradation of the random copolymer.

3.3. Effects of Furan on the MFI of the Green Random Copolymer

Figure 6 shows furan's effect on the copolymer's flow rate. Significant differences ($p < 0.05$) were observed when varying the furan concentration, observing a proportional relationship between the concentration of furan in the copolymer and the MFI index of the samples PP1, PP2, and PP3. The MFI of the copolymer without furan was 20 and increased to approximately 21, 23, and 27 g/10 min due to 6, 12, and 25 ppm furan, respectively. This corroborates what was observed in Figure 1, which shows that furan has an adverse action on the catalytic activity of ZN and, therefore, on the polymerization reaction. The chemical structure of furan may allow its oxygen atom in its heterocyclic ring to react with the active center of the titanium of the ZN catalyst to form a new stable complex that prevents the growth of the chain length of the obtained copolymer [1,12,22,29,30], its increase in fluidity, and the decrease in the molecular weight of the polymer. This can be seen in Figure 7, where the rise in furan concentration increases the MFI and decreases its molecular weight. This inhibitory behavior caused by polluting chemicals in the production of resins of industrial interest has been previously reported (Hernandez et al., 2022) and demonstrated that compounds such as Arsenia in concentrations of 0.001 to 4.32 ppm affect the MFI, and consequently, the molecular weight of the polymer.

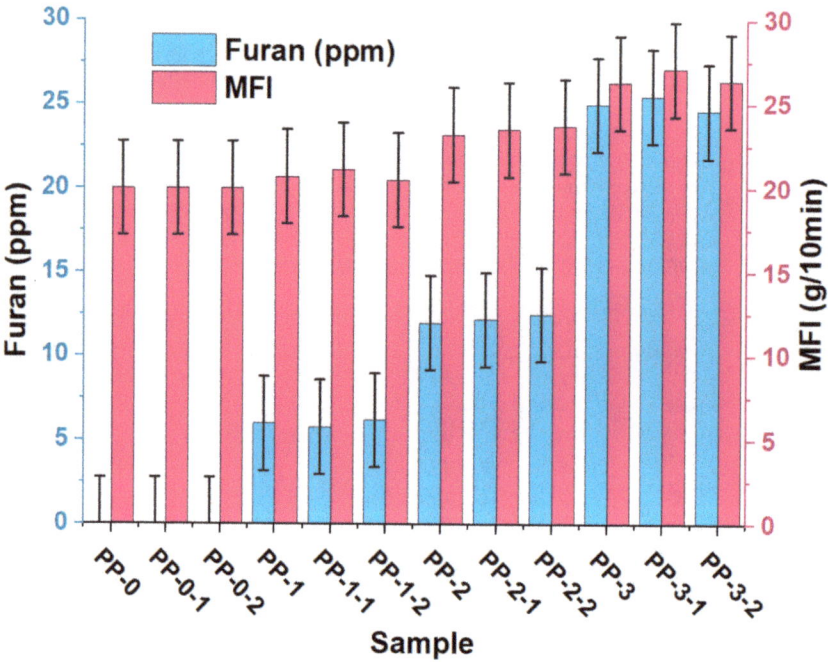

Figure 6. Effects of furan on the MFI of the random copolymer.

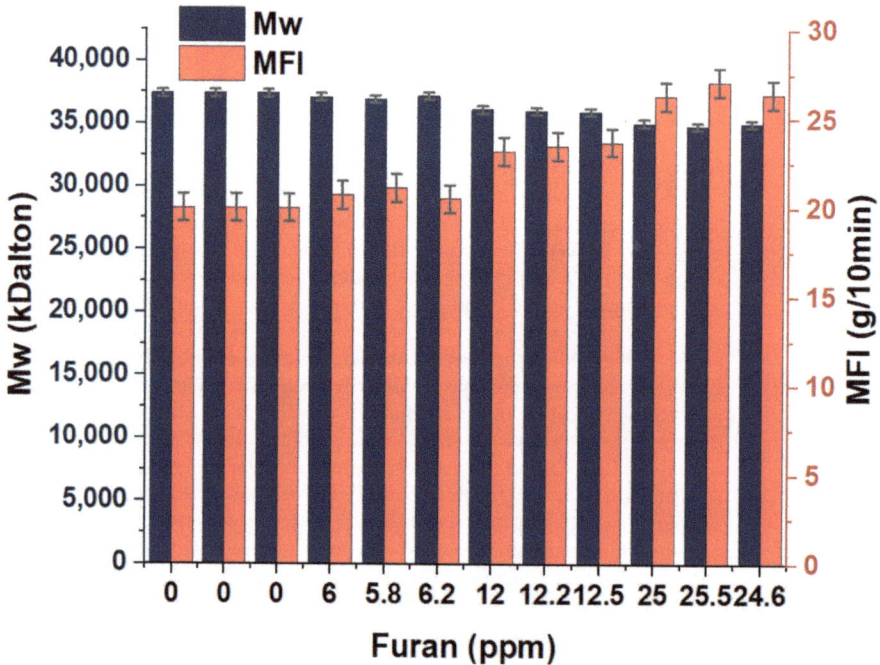

Figure 7. Effects of furan on the Mw and MFI of the random copolymer.

3.4. Effects of Furan on the Mechanical Properties (Tensile, Flexural, and Impact) of the Random Green Copolymer

The influence of the presence of furan on the mechanical properties of the copolymer is shown in Figures 8–10. When evaluating the mechanical properties of the obtained product, a notable difference is observed in the PP0 copolymer compared to PP1, PP2, and PP3, which allows us to affirm that the presence of furan can directly affect these properties of the copolymer since as the concentration of the component increases, the values of tension, bending, and impact decrease. PP0 presented average bending, tensile, and Izod values of 211,160 psi, 411,833 psi, and 11.6 ft-Lb*in, respectively. Increasing furan concentrations by an average of 6, 12, and 25 ppm caused flex decreases of 1, 11, and 18%, respectively. The tensile decreases were 4, 13, and 18%, respectively. The Izod impact trend was also inversely proportional to the furan concentration, showing percentage decreases of 9, 18, and 22%. This variation in the data is mainly due to the changes generated at the structural level that occur in the copolymer, a product of the formation of new compounds with different functional groups, in addition to incomplete polymerization and the presence of an oxygen atom in the furan, directly influence the mechanical properties of the copolymer. This behavior is associated with low flow rates and molecular weight distribution, which directly affect the mechanical properties of polypropylene [24].

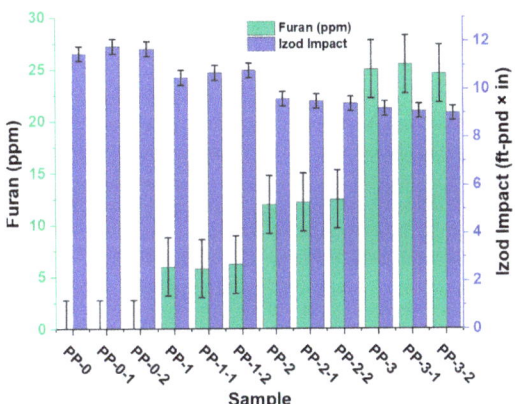

Figure 8. Effects of furan on the impact of the random copolymer.

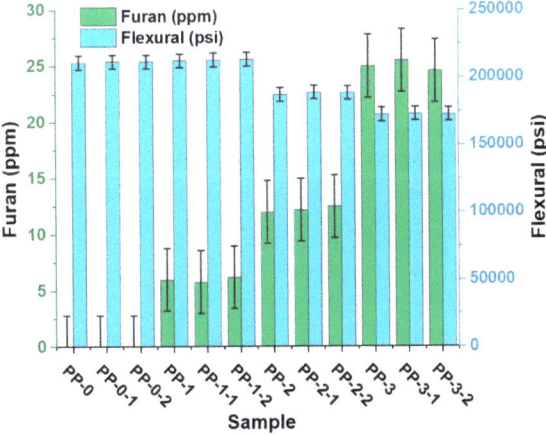

Figure 9. Effects of furan on the flexural of the random copolymer.

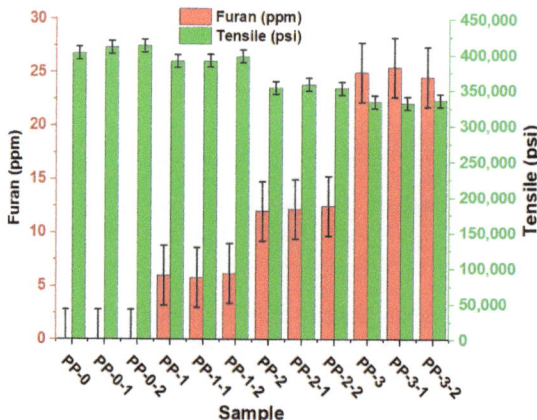

Figure 10. Effects of furan on the tensile of the random copolyme.

4. Conclusions

In the present investigation, an analysis of a random green copolymer was carried out to determine the influence of the presence of furan on the mechanical properties, MFI, productivity, and TGA of the copolymer. The results showed an evident effect of furan on the random copolymer polymerization process, inhibiting the Ziegler–Natta catalyst's capacity in this process, which generated losses in the productivity of this catalyst. A notable decrease in the thermal capacity of copolymer was evidenced, and the results were supported by the TGA levels obtained, as well as an increase in the MFI levels, which may have been associated with a decrease in the molecular weight values due to the degradation of the chemical structure of copolymer. Finally, a remarkable dominance of furan on the mechanical properties of the random copolymer was observed due to the changes generated in the levels of MFI, thermal degradation, and loss of the productivity of the catalyst.

Author Contributions: Conceptualization, J.H.-F.; methodology, J.H.-F. and E.P.-P.; validation, E.P.-P. and J.H.-F.; formal analysis, J.H.-F.; investigation, E.P.-P., E.M. and J.H.-F.; writing—original draft preparation, E.M.; writing—review and editing, E.M., E.P.-P. and J.H.-F. supervision, J.H.-F. and E.M.; project administration, J.H.-F. All authors have read and agreed to the published version of the manuscript.

Funding: This research received no external funding.

Institutional Review Board Statement: Not applicable.

Informed Consent Statement: Not applicable.

Data Availability Statement: The data presented in this study are available on request from the corresponding author.

Acknowledgments: We acknowledge Dalila Fernandez Viloria.

Conflicts of Interest: The authors declare no conflict of interest.

References

1. Hernández, J.; Guerra, Y.; Espinosa, E. Development and Application of a Principal Component Analysis Model to Quantify the Green Ethylene Content in Virgin Impact Copolymer Resins During Their Synthesis on an Industrial Scale. *J. Polym. Environ.* **2022**, *30*, 4800–4808. [CrossRef]
2. Hernández, J.; Rodríguez, E. Determination of phenolic antioxidants additives in industrial wastewater from polypropylene production using solid phase extraction with high-performance liquid chromatography. *J. Chromatogr. A* **2019**, *1607*, 460442. [CrossRef] [PubMed]

3. Hernández, J.; Lopez, J. Quantifcation of poisons for Ziegler Natta catalysts and efects on the production of polypropylene by gas chromatographic with simultaneous detection: Pulsed discharge helium ionization, mass spectrometry and fame ionization. *J. Chromatogr. A* **2020**, *1614*, 460–736. [CrossRef]
4. Hernández, J.; Cano, H.; Guerra, Y.; Polo, E.; Ríos, J.; Vivas, R.; Oviedo, J. Identifcation and quantifcation of microplastics in effluents of wastewater treatment plant by differential scanning calorimetry (DSC). *Sustainability* **2022**, *14*, 4920. [CrossRef]
5. Picó, Y.; Soursou, V.; Alfarhan, A.; El-Sheikh, M.; Barceló, D. First evidence of microplastics occurrence in mixed surface and treated wastewater from two major Saudi Arabian cities and assessment of their ecological risk. *J. Hazard. Mater.* **2021**, *416*, 125–747. [CrossRef]
6. Mallow, O.; Spacek, S.; Schwarzböck, T.; Fellner, J.; Rechberger, H. A new thermoanalytical method for the quantifcation of microplastics in industrial wastewater. *Environ. Pollut.* **2020**, *259*, 113–862. [CrossRef] [PubMed]
7. Vajravel, S.; Sirin, S.; Kosourov, S.; Allahverdiyeva, Y. Towards sustainable ethylene production with cyanobacterial artifcial biofilms. *Green Chem* **2020**, *22*, 6404–6414. [CrossRef]
8. Wang, Z.; Shi, R.; Zhang, T. Three-phase electrochemistry for green ethylene production. *Curr. Opin. Electrochem.* **2021**, *30*, 100–789. [CrossRef]
9. Penteado, A.; Kim, M.; Godini, H.; Esche, E.; Repke, J. Biogas as a renewable feedstock for green ethylene production via oxidative coupling of methane: Preliminary feasibility study. *Chem. Eng. Trans.* **2017**, *61*, 589–594. [CrossRef]
10. Du, Z.; Xu, J.; Dong, Q.; Fan, Z. Thermal fractionation and efect of comonomer distribution on the crystal structure of ethylene-propylene copolymers. *Polymer* **2009**, *50*, 2510–2515. [CrossRef]
11. Rani, M.; Marchesi, C.; Federici, S.; Rovelli, G.; Alessandri, I.; Vassalini, I.; Ducoli, S.; Borgese, L.; Zacco, A.; Bilo, F.; et al. Miniaturized near-infrared (MicroNIR) spectrometer in plastic waste sorting. *Materials* **2019**, *12*, 2740. [CrossRef] [PubMed]
12. Hernández, J.; Guerra, Y.; Puello, E.; Marquez, E. Effects of Different Concentrations of Arsine on the Synthesis and Final Properties of Polypropylene. *Polymers* **2022**, *14*, 3123. [CrossRef] [PubMed]
13. Hernández, J.; López, J. Experimental study of the auto-catalytic effect of triethylaluminum and TiCl4 residuals at the onset of non-additive polypropylene degradation and their impact on thermo-oxidative degradation and pyrolysis. *J. Anal. Appl. Pyrolysis* **2021**, *155*, 105052. [CrossRef]
14. Mardueño, S.; Barron, J.; Verdin, B.; Mendeleev, E.; Montalvo, R. *Química Orgánica Introducción a la Química Heterocíclica*; Universidad Autónoma de Nayarit, México: Tepic, Mexico, 2013; p. 181.
15. Ruiz, M.L. *Tesis Sobre Determinación y Evaluación de las Emisiones de Dioxinas y Furanos en la Producción de Cemento en España*; Universidad Complutense de Madrid (UCM): Madrid, Spain, 2007.
16. Sarra, A.J. Dioxinas y furanos derivados de la combustión. *Rev. Científica De La Univ. De Belgrano* **2018**, *1*, 143–157.
17. González, P. *Tesis Sobre Contribución al Control de la Contaminación por Cloropropanoles en el Ámbito Alimentario Mediante el Desarrollo de Nuevas Metodologías Analíticas*; Universidad de Santiago de Compostela: Santiago, Spain, 2015.
18. Pedreschi, F. *Tesis Technologies for Furan Mitigation in Highly Consumed Chilean Foods Processed at High Temperatures*; Pontifica Universidad Católica de Chile: Santiago, Chile, 2015.
19. Cahuaya, M.A. *Tesis Sobre la Determinación de Dioxinas y Furanos en Harina de Pescado por Cromatografía de Gases Acoplada a Espectrometría de Masas Triple Cuadrupolo*; Universidad Peruana Cayetano Heredia: San Martín de Porres District, Peru, 2021.
20. Naccha, L.R. *Tesis Sobre Cuantificación de Dioxinas por Cromatografía de Gases*; Universidad Autónoma de Nuevo León: San Nicolás de los Garza, México, 2010.
21. Cobo, M.I.; Hoyos, A.E.; Aristizábal, B.; de Correa, C.M. Dioxinas y furanos en cenizas de incineración. *Rev. Fac. De Ing. Univ. De Antioq.* **2004**, *32*, 26–38.
22. Shubhra, Q.T.H.; Alam, A.K.M.M.; Quaiyyum, M.A. Mechanical properties of polypropylene composites: A review. *J. Thermoplast. Compos. Mater.* **2013**, *26*, 362–391. [CrossRef]
23. Karol, F.J.; Jacobson, F.I. Catalysis and the Unipol Process. *Stud. Surf. Sci. Catal.* **1986**, *25*, 323–337.
24. Hernández, J.A. *Tesis Sobre el Uso de Aditivos Sostenibles en la Estabilización Térmica del Polipropileno en su Proceso de Síntesis*; Universidad Politecnica de Valencia: Valencia, Spain, 2018.
25. Quinteros, J.G. *Tesis sobre Síntesis y Aplicaciones de Ligandos Arsinas. Estudios de Sistemas Catalíticos de Pd y Au*; Universidad Nacional de Córdoba: Córdoba, Spain, 2016.
26. Treichel, P.M. A Review of: "Phosphine, Arsine, and Stibine Complexes of The Transition Elements. C. A. McAuliffe and W. Levason, Elsevier Scientific Publishing Co., Amsterdam, The Netherlands, 1979, 546 pp, $84.50". *Synth. React. Inorganic Met. Nano-Metal Chem.* **1979**, *9*, 507–508. [CrossRef]
27. Alsabri, A.; Tahir, F.; Al-Ghamdi, S.G. Life-Cycle Assessment of Polypropylene Production in the Gulf Cooperation Council (GCC) Region. *Polymers* **2021**, *13*, 3793. [CrossRef]
28. Caicedo, C.; Crespo, L.M.; Cruz, H.D.L.; Álvarez, N.A. Propiedades termo-mecánicas del Polipropileno: Efectos durante el reprocesamiento. *Ing. Investig. Y Tecnol.* **2017**, *18*, 245–252.
29. Caceres, C.A.; Canevarolo, S.V. Correlação entre o Índice de Fluxo à Fusão ea Função da Distribuição de Cisão de Cadeia durante a degradação termo-mecânica do polipropileno. *Polimeros* **2006**, *16*, 294–298. [CrossRef]
30. Ferg, E.E.; Bolo, L.L. A correlation between the variable melt flow index and the molecular mass distribution of virgin and recycled polypropylene used in the manufacturing of battery cases. *Polym. Test.* **2013**, *32*, 1452–1459. [CrossRef]

31. Yan, H.; Hui-li, Y.; Gui-bao, Z.; Hui-liang, Z.; Ge, G.; Li-song, D. Rheological, Thermal and Mechanical Properties of Biodegradable Poly(propylene carbonate)/Polylactide/Poly(1,2-propylene glycol adipate) Blown Films. *Chin. J. Polym. Sci.* **2015**, *12*, 1702–1712. [CrossRef]
32. Suraj, K.; Sanjay, L. Evaluation of mechanical properties of polypropylene—Acrylonitrile butadiene styrene blend reinforced with cowrie shell [Cypraeidae] powder. *Mater. Proc.* **2022**, *50*, 1644–1652. [CrossRef]
33. Chacon, H.; Cano, H.; Fernández, J.H.; Guerra, Y.; Puello-Polo, E.; Ríos-Rojas, J.F.; Ruiz, Y. Effect of Addition of Polyurea as an Aggregate in Mortars: Analysis of Microstructure and Strength. *Polymers* **2022**, *14*, 1753. [CrossRef]
34. Hernández-Fernández, J.; Castro-Suarez, J.R.; Toloza, C.A.T. Iron Oxide Powder as Responsible for the Generation of Industrial Polypropylene Waste and as a Co-Catalyst for the Pyrolysis of Non-Additive Resins. *Int. J. Mol. Sci.* **2022**, *23*, 1708. [CrossRef]
35. Hernández-Fernández, J.; Rayón, E.; López, J.; Arrieta, M.P. Enhancing the Thermal Stability of Polypropylene by Blending with Low Amounts of Natural Antioxidants. *Macromol. Mater. Eng.* **2019**, *304*, 1900379. [CrossRef]
36. Hernández-Fernández, J. Quantification of oxygenates, sulphides, thiols and permanent gases in propylene. A multiple linear regression model to predict the loss of efficiency in polypropylene production on an industrial scale. *J. Chromatogr. A* **2020**, *1628*, 461478. [CrossRef]
37. Joaquin, H.F.; Juan, L.M. Autocatalytic influence of different levels of arsine on the thermal stability and pyrolysis of polypropylene. *J. Anal. Appl. Pyrolysis* **2022**, *161*, 105385. [CrossRef]
38. Hernández-Fernández, J.; Cano, H.; Aldas, M. Impact of Traces of Hydrogen Sulfide on the Efficiency of Ziegler–Natta Catalyst on the Final Properties of Polypropylene. *Polymers* **2022**, *14*, 3910. [CrossRef]
39. Hernández-Fernández, J.; Ortega-Toro, R.; Castro-Suarez, J.R. Theoretical–Experimental Study of the Action of Trace Amounts of Formaldehyde, Propionaldehyde, and Butyraldehyde as Inhibitors of the Ziegler–Natta Catalyst and the Synthesis of an Ethylene–Propylene Copolymer. *Polymers* **2023**, *15*, 1098. [CrossRef] [PubMed]
40. Hernández-Fernández, J.; Vivas-Reyes, R.; Toloza, C.A.T. Experimental Study of the Impact of Trace Amounts of Acetylene and Methylacetylene on the Synthesis, Mechanical and Thermal Properties of Polypropylene. *Int. J. Mol. Sci.* **2022**, *23*, 2148. [CrossRef] [PubMed]

Disclaimer/Publisher's Note: The statements, opinions and data contained in all publications are solely those of the individual author(s) and contributor(s) and not of MDPI and/or the editor(s). MDPI and/or the editor(s) disclaim responsibility for any injury to people or property resulting from any ideas, methods, instructions or products referred to in the content.

Article

The Contribution of BaTiO₃ to the Stability Improvement of Ethylene–Propylene–Diene Rubber: Part I—Pristine Filler

Tunde Borbath [1], Nicoleta Nicula [2,*], Traian Zaharescu [1,2,*], Istvan Borbath [1] and Tiberiu Francisc Boros [1]

1. ROSEAL SA, 5 A Nicolae Bălcescu, Odorheiu Secuiesc, 535600 Harghita, Romania
2. INCDIE ICPE CA, 313 Splaiul Unirii, 030138 Bucharest, Romania
* Correspondence: nicoleta.nicula@icpe-ca.ro (N.N.); traian.zaharescu@icpe-ca.ro (T.Z.)

Abstract: This study presents the functional effects of BaTiO₃ powder loaded in ethylene–propylene–diene rubber (EPDM) in three concentrations: 0, 1, and 2.5 phr. The characterization of mechanical properties, oxidation strength, and biological vulnerability is achieved on these materials subjected to an accelerated degradation stimulated by their γ-irradiation at 50 and 100 kGy. The thermal performances of these materials are improved when the content of filler becomes higher. The results obtained by chemiluminescence, FTIR-ATR, and mechanical testing indicate that the loading of 2.5 phr is the most proper composition that resists for a long time after it is γ-irradiated at a high dose. If the oxidation starts at 176 °C in the pristine polymer, it becomes significant at 188 and 210 °C in the case of composites containing 1 and 2.5 phr of filler, respectively. The radiation treatment induces a significant stability improvement measured by the enlargement of temperature range by more than 1.5 times, which explains the durability growth for the radiation-processed studied composites. The extension of the stability period is also based on the interaction between degrading polymer substrate and particle surface in the composite richest in titanate fraction when the exposure is 100 kGy was analyzed. The mechanical testing as well as the FTIR investigation clearly delimits the positive effects of carbon black on the functionality of EPDM/BaTiO₃ composites. The contribution of carbon black is a defining feature of the studied composites based on the nucleation of the host matrix by which the polymer properties are effectively ameliorated.

Keywords: ethylene–propylene–diene rubber-based composites; barium titanate as oxidation protector filler; radiation stability; chemiluminescence; mechanical testing; antifungal strength

Citation: Borbath, T.; Nicula, N.; Zaharescu, T.; Borbath, I.; Boros, T.F. The Contribution of BaTiO₃ to the Stability Improvement of Ethylene–Propylene–Diene Rubber: Part I—Pristine Filler. *Polymers* **2023**, *15*, 2190. https://doi.org/10.3390/polym15092190

Academic Editor: Cristina Cazan

Received: 11 March 2023
Revised: 26 April 2023
Accepted: 3 May 2023
Published: 5 May 2023

Copyright: © 2023 by the authors. Licensee MDPI, Basel, Switzerland. This article is an open access article distributed under the terms and conditions of the Creative Commons Attribution (CC BY) license (https://creativecommons.org/licenses/by/4.0/).

1. Introduction

The production of various polymers and their composite items requires detailed and accurate investigations of the functional characteristics, such as mechanical strengths, long-term stability, and biological solution to the fungi attack that describe complementary global durability under hazardous conditions [1]. The analysis of operational performance is the main goal of material qualification when the technical products are destined for nuclear power plants [2]. For the keystone analysis of radiation effects on polymers, a comprehensive book chapter was published [3]. The most important aspect that reveals the radiochemical stability and the durability of polymer materials is the value of the ratio between the radiochemical yields of scission and crosslinking, G(S)/G(X). Its superunitary value, resulting from the three times higher ethylene segments with respect to the propylene content, is proof for an appropriate radiation processing or long-term operation in nuclear applications. The individual values of G(S) and G(X) for ethylene–propylene–diene rubber (EPDM) are placed between the material characteristics of polyethylene (G(X) = 0.8–1.1 and (G(S) = 0.4–0.5) and polypropylene (G(X) = 0.8–1.1 and (G(S) = 0.4–0.5) [3]. The consumption of diene (ethylene–norbornene) with a radiochemical yield of 32.1 [4] and the formation of various oxygenated degradation products (ketones—G = 13.9, acids—G = 4.4, alcohols—G = 4.1 and peroxides—G = 0.3) depict the behavior of this elastomer under the action of high energy radiation when the protector is not technologically added.

The thermal and radiochemical stabilization effects in the studied elastomer correspond to the technical requirements for various applications such as sealants, vibration buffers, automotive items, medical wear, corrosion protection layers, hydrophobic impregnations, and shoe soles. There are many alternatives through which EPDM may gain improved durability by the addition of phenolic antioxidants [5–7], inorganic fillers [8–10], and wood wastes [11]. If the evolution of radiation-processed EPDM was already reported [12], the stabilization mechanisms promoted by the inorganic materials based on the surface interactions through which the free radicals born during the oxidative degradation are still elaborating [13].

The inhibition of degradation is effectively accomplished by the appropriate compounds, which are able to break the oxidation chain. On the propagation stage, the conversion of peroxyl radicals into final stabile products would be efficiently interrupted if the filler does not act on the free radicals by bonding or the active component exceeds a certain critical threshold. The hazardous conditions existing in nuclear power plants require a certain advanced stability by which the rubber products are able to resist with proper functional parameters between the successive maintenance stops [14]. Under various circumstances, the nature and loading degree of filler [15], as well as the content of diene composing the elastomer structure [16], are determining factors that characterize the material strength against oxidation [17]. The high-performance materials that are radio-oxidation resistant may be obtained either by the incorporation of another suitable polymer to promote crosslinking [18,19], the addition of a crosslinker [20], or the compatibilization of an efficient filler [21]. Accordingly, high-tech materials may be produced by the inspired combination between the compositional features and the foreseen technological parameters [22].

Crosslinked EPDM is a suitable material for several operation areas where the degradation factors do not alter the functional features. The technical concern of stability characteristics is related to the double role of the contained diene: a connecting element between the molecular chains (Figure 1) and the weaker spot that is facilely broken.

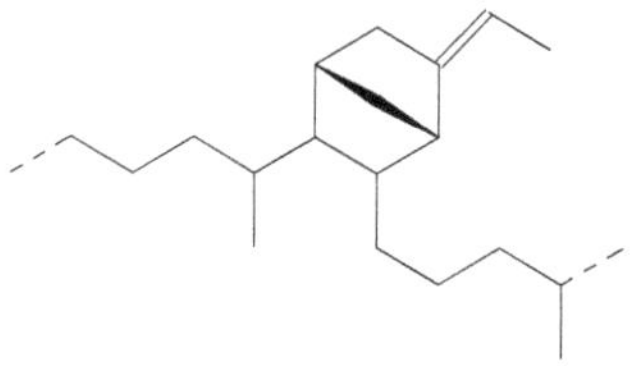

Figure 1. The structure of ethylene-propylene-diene-monomer.

During the exposure of this polymer to a certain γ-dose not greater than and especially less than 100 kGy, the scission of norbornene and the bonds of tertiary carbon atoms from propylene moieties create radicals that are involved in crosslinking and oxidation [4,23]. The presence of any oxidation protector increases the crosslinking degree due to the turning of radicals onto the reaction with another analogous fragment [24].

The stability characterization of EPDM is a good opportunity for the presentation of various faces of material qualification when different approaches reveal the peculiar features of a product in connection with its structural understanding. The thermal analysis assay [25], the improvement of ablation resistance [26], the explanation of dielectric properties under various environmental conditions [27], the preparation of new structural materials by the blending EPDM with unlike products such as NBR [28], PP and PA6 [29] or Kevlar fibers [30], the increase in crosslinking degree [31–33], the anti-migration effects of graphene oxide [34], and many other practical aspects asked by the market are the study themes that become important for certain applications. One interesting aspect that is related to the valorization of EPDM by converting it into composites or blends is the recycling of its wastes [35,36], the main reason that it may be considered as an example for many other polymer materials.

The modification of the functional performances of EPDM-based products by the addition of an advisable blending component represents an attractive goal on which the radiation processing may create new material structures with enlarged application areas. The investigations on the stability effects of aluminum trioxide on fire retardancy [37], improvement of mechanical characteristics required by electrical applications [38], assay on radiation shielding for source depositing [39], changes in electrical properties of cables for nuclear power plants [40], and applications in the production of commodities [41] are the topics investigated, by which the selected compositions may offer the desired versions of focused interest.

The present assay is addressed to the manufacturers on long-life polymer items, which may resist certain energetic efforts. As it was previously demonstrated [9], the presence of titanate filler is a good solution for the production of high-tech polymer composites. If this early manuscript concerns the pristine barium titanate, the second part of the study will report the stabilization effects brought about by doped barium titanates, which act more efficiently on the degradation rate of EPDM by radical scavenging.

2. Materials and Methods

In the present paper, the composite materials based on EPDM are tested for their application in various environments that promote accelerated oxidation.

2.1. Materials

The polymer matrix was manufactured by ethylene–propylene–diene terpolymer (EPDM) used in the DSM Elastomers (Heerlen, The Netherlands) as KELTAN 8340. The filler, barium titanate, was purchased from Thermo Scientific (Shandong Deshang Chemical Co., Ltd., Jining, China). The pristine polymer material contained ethylene and propylene moieties in the proportion of 3:1 and 5 wt% 5-ethylidene 2-norbornene (EBN). While the preparation of sheets for the mechanical testing of samples was performed according to an appropriate procedure, the thermal and radiation stabilities were checked on the thin pellicles obtained by the solvent removal. This raw polymer was not subjected to any purification process before the present investigations. Barium titanate has the following characteristics: purity > 99% and an average particle size of 300 μm.

2.2. Sample Preparation

For mechanical testing, the plate samples were obtained by means of the thermal pressing process achieved at 180 °C for 15 min. The equipment was produced by Nicovala (Sighisoara, Romania), whose features (plate dimensions: 400 mm × 450 mm, charge: 160 tf, electrical heating) allow the optimal preparation of EPDM plates. The vulcanization pressure was established at 100 bars. From these plates, standard dumbbells according to ISO 527 were obtained. The black samples contain a significant amount of carbon content (36 phr).

For chemiluminescence investigation, thin films were obtained by solvent removal from aluminum round caps (diameter 5 mm) where the liquids were poured. Three concentrations of barium titanate (0, 1, and 2.5 phr) were prepared by the addition of appropriate amounts of powder followed by vigorous shaking for homogenization. Finally, when the dry samples were obtained, they were weighted for the determination of specimen masses. These mass values were used for the normalization of measured CL intensities to the standard unit allowing reliable results for their comparison. Photon counting is achieved by the connection between the experimental device and the attached computer. The specific program for the addition of counting information allows the conversion of results into the Origin representations, which are presented herein.

2.3. γ-Irradiation

The radiation treatment was achieved in air at room temperature in an Ob Servo Sanguis irradiation machine (Budapest, Hungary) provided with a ^{60}Co source. Three doses were selected (0, 50, and 100 kGy) for the radiation processing of specimens. The dose rate was 0.5 kGy h^{-1}. All measurements were performed immediately after the end of exposures, avoiding any alteration of results due to the decay of short-life radicals.

2.4. Characterization of Filler

Several attempts for the identification of any differences between the XRD, Raman, and FTIR spectra as well as the SEM images of crystalline pristine and γ-irradiated (100 kGy) barium titanate powders failed. The overlapping of the recorded results did not allow the detection of any radiation effect on this inorganic phase. These details prove that there were no radiolysis effects on the inorganic blending component.

2.5. Characterization of Composites
2.5.1. Mechanical Testing

Mechanical testing was achieved using the ZMR 250 equipment (VEB THURINGER, Raunestein, Germany) applying the testing standard ISO 37/2012. The evaluation of Shore A hardness was based on ISO 7619/1 (2011) with testing unit equipment purchased from STENDAL, Stendal, Germany).

2.5.2. Chemiluminescence (CL)

The stability assay was achieved by means of a LUMIPOL 3 device produced by the Institute of Polymers, Slovak Academy of Science, Bratislava (Slovakia). The two available investigation procedures, isothermal and nonisothermal, were used for the complementary assay. For nonisothermal determinations, there were four selected heating rates: 5, 10, 15, and 20 °C min^{-1}. The isothermal measurements were achieved at 160, 170, and 180 °C. The CL spectra were electronically transferred from the measurement device onto the computing system, which allows the graphical approach.

2.5.3. Spectral Assay by FTIR-ATR

A Fourier transform infrared (FTIR) spectrometer (Interspectrum, INTERSPEC 200-X, Tõravere, Estonia) equipped with a device for attenuated total reflectance (ATR) was used in the wavenumber range between 4000 cm^{-1}–750 cm^{-1} at room temperature for evaluation of samples spectra. Each ATR–FTIR spectrum is an average of over 10 scans, using air as a reference, and 4 cm^{-1} as the nominal spectral resolution.

2.5.4. Antifungal Properties

The antifungal properties of the samples consisting of EPDM and barium titanate in various proportions were inspected according to ASTM-G21-09 standard [42]. Briefly, small square samples (2.5 cm × 2.5 cm) with a smooth surface and free of defects were placed in Petri dishes containing a layer of Chapex-Dox agar. A 0.1 mL of fungal spore suspension (consisting of a mixture of *Aspergillus brasiliensis*—ATCC 9642, *Penicillium funiculosum*—ATCC 11797, *Chaetomium globosum*—ATCC 6205, *Trichoderma virens*—ATCC 9645, and *Aureobasidium pullulans*—ATCC 15233) was poured onto the sample surface. These samples were incubated for 28 days at 27 ± 2 °C and 95% relative humidity. After the incubation period, the samples were visually evaluated for fungal growth. The extent of growth was rated on a scale of 0–4. Number 0 indicates the lack of growth, while Number 4 indicates intense growth covering the entire sample surface. The evaluation of fungal growth extent is useful for the nomination of inspected materials that are "resistant" or "non-resistant" products. Materials with a rating of 0 or 1 are considered resistant materials, while the other ones presenting a rating of 2, 3, or 4 are considered non-resistant products. The assay was conducted in triplicates.

3. Results

The thermal stability is directly related to the breadth of degradation that is initiated by the formation of peroxyl radicals [4] The oxidation is propagated by the further reactions of $RO_2\cdot$, whose generating rate depends on the free radical abundance and the diffusion rate of molecular oxygen. In the case of EPDM, the most vulnerable spots are the position of methyl in the polypropylene moieties and the unsaturation of diene (EBN) [43]. The evolution of oxidation in the presence of stabilizers during the radiation processing of EPDM products occurs differently, because the antioxidants may turn on the mechanisms in peculiar ways [44]. Even though the γ-irradiation is an accelerated process of degradation, it may be delayed in two different manners: crosslinking [45] and/or stabilization [46].

The progress of oxidative degradation caused by accidental or current energy depositing onto polymeric items may be slowed down by the effect of inorganic structures and determines the diffusion and retention of free radicals inside the traps of stabilizer [9,47]. The delay of aging is always correlated with the stabilization efficiency that characterizes the strength of scavenging and the size of available free space inside the structured oxidation protector [48]. The γ-processing of polymers that causes profound damage to macromolecules may be judged as a proper way of material improvement when the stabilization action of filler is accompanied by radiation crosslinking [49]. In the case of EPDM, this assumption is valid, because this polymer belongs to the polymer class of radiation crosslinking materials [50].

3.1. Chemiluminescence

The emission of photons during polymer oxidation from the reacting entities, according to the deexcitation of the triplet state of ketones onto the background, allows the observation of the evolution of degradation. It is obvious that the progress of oxidation involves the counting of the quantum number that is proportional to the concentration of peroxyl radicals [51]. The shapes of CL curves offer real indications concerning the fate of radicals and their decay in relation to the presence of oxidation protectors [52].

3.1.1. Nonisothermal Chemiluminescence

Certain formulations of material composition lead to peculiar thermal behavior, which is the overall result involving the contributions of blended components. Accordingly, the temperature when the oxidation starts is determined by the easiness of splitting of the weakest existing bond. The influence of composition on the thermal stability of EPDM is demonstrated by the evolution of oxidation with and without carbon black (Figure S1). Even though the addition of carbon black initiates a crosslinking effect [53], it also provides an effect of stabilization. This feature was previously reported [54,55]. Fortunately, photon counting is possible with both groups of samples (with and without carbon black). However, the application of isothermal determinations as the measurement procedure failed with the black samples, this is because the long-time counting involves the self-absorption of CL photons over the first part of the experiment.

The nonisothermal CL measurements of the EPDM-based composites reveal the influence of filler concentration on the progress of oxidation when the testing temperature monotonically increases (Figure 2). The evident effect is the increase of starting oxidation temperature as the filler loading becomes higher. The presence of carbon black in the sample formulations causes a diminution of CL emission. It may be assumed that the oxidation is inhibited by carbon black as it was also previously reported [38].

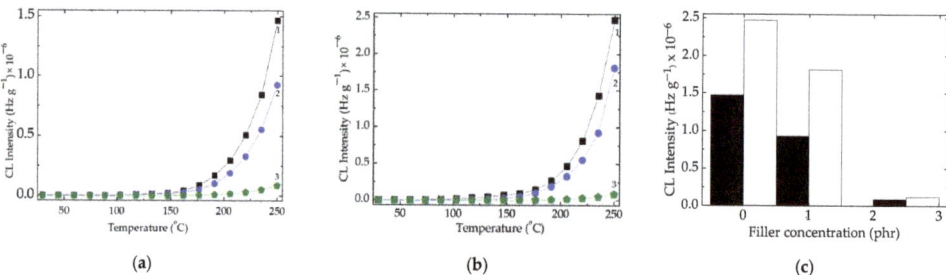

Figure 2. Nonisothermal CL spectra recorded on (**a**) polymer with carbon black and (**b**) polymer without carbon black. Filler concentration: (1) 0 phr, (2) 1 phr, and (3) 2.5 phr. Heating rate: 15 °C min^{-1}. (**c**) Histogram illustrating the CL intensities at 250 °C. (black) polymer with carbon black; (white) polymer without carbon black.

The increase in the filler loading determines an evident improvement in the polymer stability, which reaches a good degree when the barium titanate has a concentration of 2.5 phr (Figure 2). This behavior is in good agreement with the previous results obtained by the preparation of nanocomposites [56]. The comparison of CL intensities measured at 250 °C provides obvious proof for the contribution of carbon black on the material stabilization during thermal oxidation.

The influence of filler concentration on the development of oxidation is illustrated in Figure 3. The consequences of γ-irradiation on the progress of oxidation are related to the availability of EPDM for crosslinking, being known that the dose range up to 100 kGy is proper for the increase of its insoluble fraction [18,57]. This trend of amelioration in the polymer stability is more evident at a higher dose (100 kGy) when the involvement of free radicals in the oxidative degradation is really prevented by the presence of inorganic filler. The extension of the stability range of temperature as the fillet content and processing γ-dose growth is a main advantage; therefore, these compositions are recommended as the proper formulations for their applications in degradation-providing environments.

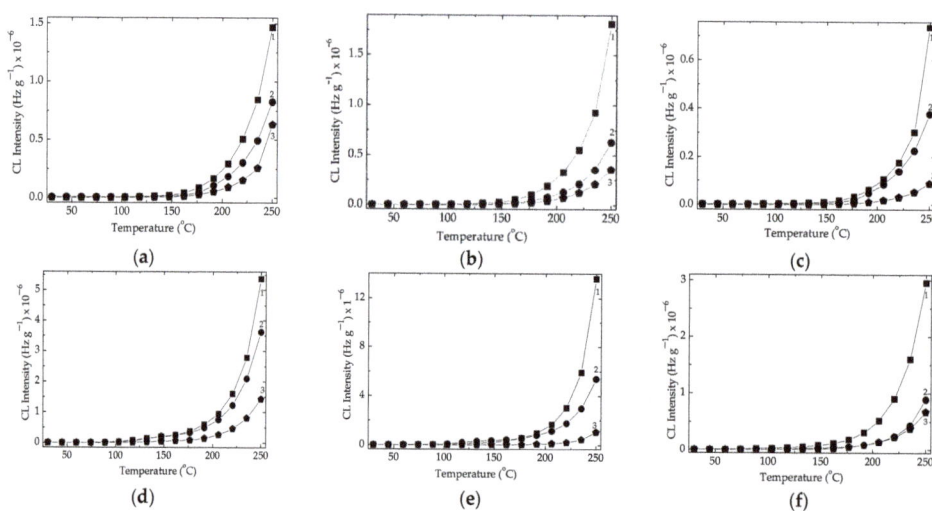

Figure 3. Nonisothermal CL spectra recorded on composite EPDM/BaTiO$_3$ probes; filler contents: (1) 0 phr; (2) 1 phr; (3) 2.5 phr. Samples with carbon black: (**a**) D 0 kGy, (**b**) D 50 kGy (**c**) D 100 kGy. Samples without carbon black: (**d**) D 0 kGy, (**e**) D 50 kGy, (**f**) D 100 kGy. Heating rate: 15 °C min^{-1}.

3.1.2. Isothermal Chemiluminescence

The evolution of degradation at a constant temperature (170 °C) reveals the high stability degree in the presence of barium titanate (Figure 4). The effects on the oxidation rates, the maximum period of degradation, and the height of emission intensity obtained for the investigated EPDM/BaTiO$_3$ composites are relevant for the protective action of filler during the accelerated degradation occurring during γ-irradiation. The chemiluminescence measurements on the EPDM-based composite samples containing carbon black prove (Figure S1) that technical items are more resistant when this additive is incorporated.

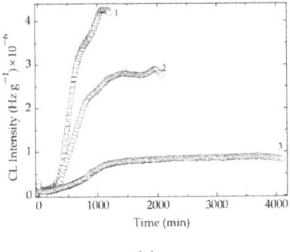
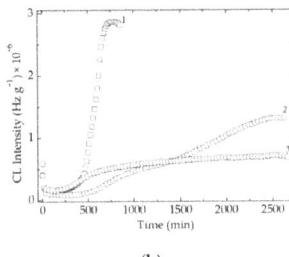
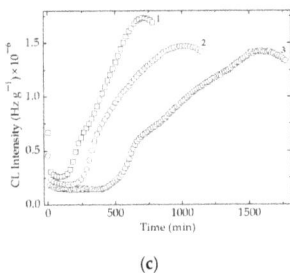

(a) (b) (c)

Figure 4. Isothermal CL spectra recorded on composites containing EPDM and BaTiO$_3$ without carbon black exposed to various γ-doses. Testing temperature: 170 °C. Filler content: (1) 0 phr, (2) 1 phr, (3) 2.5 phr. (**a**) 0 kGy, (**b**) 50 kGy, (**c**) 100 kGy.

The unusual long periods of degradation at 170 °C, even at technological γ-irradiation doses (50 and 100 kGy), are reliable evidence that incorporated filler is able to delay oxidation under hard energetic conditions of operation. The amplitudes of CL emission intensities are low for composite formulations with respect to pristine material, which depicts the jointing tightness of molecular fragments withdrawn from their reaction with diffused oxygen. The differences that exist between the values of oxidation rates, maximum Cl intensities, and oxidation induction times are the relevant indices for the scavenging action of filler particles that superficially takes place before the reactions between free radicals and penetrating gaseous oxygen. The progress of aging is efficiently delayed by the interaction between energetic gaps existing in the outer layers of particle surface and the hydrocarbon moieties resulting from bond scissions. For the low irradiation stages (0 and 50 kGy), the local concentration of oxidation intermediates is not too high. Consequently, the pseudoplateau parts of CL curves are obtained, indicating the efficacious ability of filler grains to catch reacting units. They also indicate the lack of desorption. At 100 kGy, the ascendant curves show the generation of high amounts of oxidation initiators. The gaps become highly populated, and an important fraction of radicals are oxidized prior to their structural blocking. The defects existing in the lattice of titanate may also contribute to the behavior amelioration of EPDM-based materials.

3.2. Mechanical Testing

The expected behavior regarding the mechanical properties of the studied formulations is based on the correlation between the effect of filler that inhibits oxidative degradation and the contribution of γ-treatment on the yielded crosslinking degree. As it was previously reported [58], EPDM presents a scission yield of 5.2×10^{-7} mol J^{-1}, by which this polymer provides free radicals for crosslinking according to the mechanism based on radical recombination [48,59]. The structural dependency of crosslink density is the consequence of the involvement of diene, namely 2-ethylidene-5-norbornene [60], whose contribution to the material curing characterizes the modification that occurred in the mechanical properties of EPDM composites. The aging changes undergone in γ-irradiated EPDM are influenced by the existence of filler, which interacts with basic material by the mechanistic modification of the fate of radicals [61]. This approach was also reported, when the ethylene-propylene-diene elastomer material was modified by the presence of lead vanadate [62] or titania [63].

The degradation of plastics occurs as a fragmentation process [64], when the smaller molecular parts composing the processed materials may be separated. The mechanical properties of these studied samples are tightly related to the level of γ-irradiation [43], whose values are the combination between this scission process and a certain degree of crosslinking [65]. The illustration of this overlapping is based on the extended durability that occurs in nuclear power plants [40,66].

The performances of irradiated composites based on elastomeric materials are the overall effects by which free radicals undergo the reactions that may partially restore the initial characteristics [67]. Because there are differences between the values of mechanical properties (Figure 5), their explanation is based on the evolution of oxidation states whose ways are preceded according to the protective action of filler [68]. The deterioration of the mechanical behavior in EPDM leads to a reduction of molecular mobility by crosslinking. This proof is indicated by the shift of the glass transition relaxation temperature towards higher temperatures.

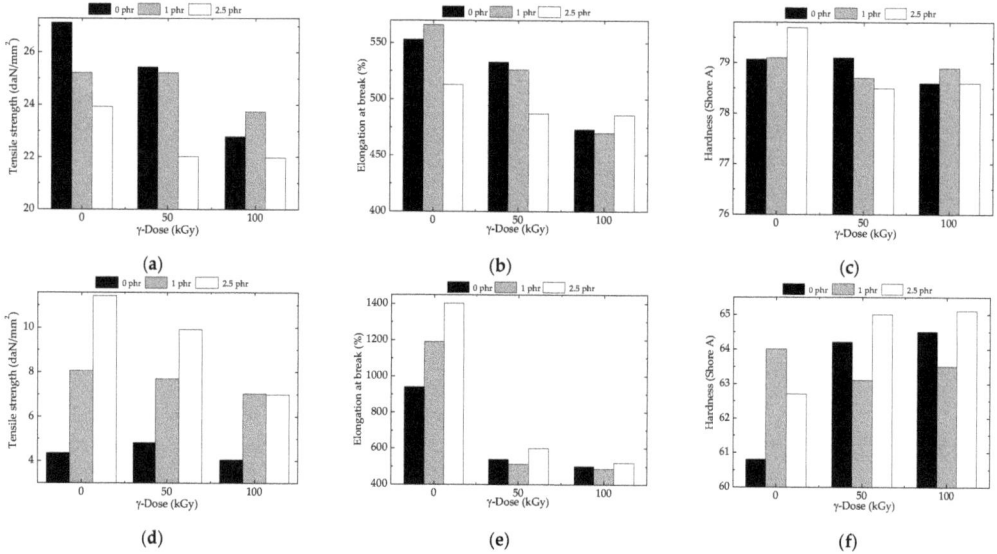

Figure 5. Mechanical behavior of EPDM/BaTiO$_3$ composites subjected to γ-irradiation. Samples with carbon black: (**a**) tensile strength, (**b**) elongation at break, (**c**) hardness; samples without carbon black: (**d**) tensile strength, (**e**) elongation at break, (**f**) hardness.

The reinforcement of EPDM containing relatively low amounts of barium titanate (1 and 2.5 phr) and enough high loading of carbon black (36 phr) is observed after the application of γ-radiation treatment. This means that the fragments that result from the splitting of macromolecules are agglomerated around fillers. They promote nucleation as it occurs in other EPDM composites [69,70]. Of course, the compositions with carbon black are suitable for several applications (O rings, anticorrosive protection, phonic insulations, vibration attenuators, impregnation agents for hydrophobic layers of ceramics, and many other opportunities.

3.3. FTIR for the Oxidation of Composites

The course of oxidation progresses during the propagation stage of degradation, when free radicals are attacked by molecular oxygen. It grows the content of oxygenated products, whose evolution may be well illustrated by the modifications occurring in the carbonyl band (Figure 6). There are great differences between the samples containing carbon black and the probes free of it [71].

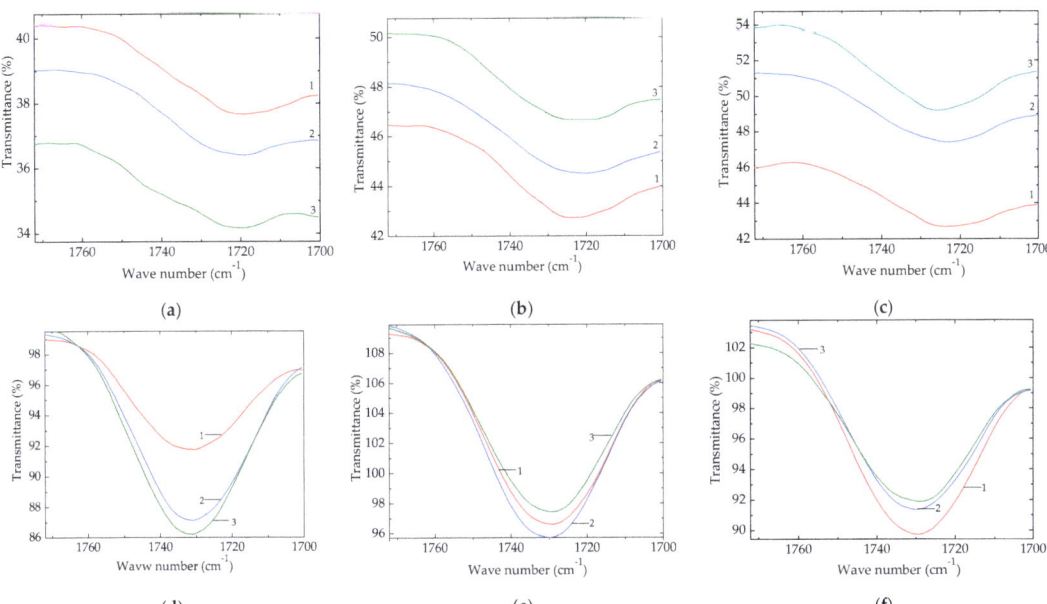

Figure 6. FTIR spectra for EPDM/BaTiO$_3$ composites subjected to oxidative degradation under γ-irradiation. Samples with carbon black: (**a**) 0 kGy; (**b**) 50 kGy; (**c**) 100 kGy. Samples free of carbon black: (**d**) 0 kGy; (**e**) 50 kGy; (**f**) 100 kGy. Filler loadings: (1) 0 phr; (2) 1 phr; (3) 2.5 phr.

The modification of absorbance values follows the contributions of various intermediates that produce the oxygenated stabile structures accumulating in the final state of the material. The presence of BaTiO$_3$ filler determines a dissimilar modification in the transmittance values for carbonyl vibrations (Figure S2) that provide proof for the protection activity of the inorganic phase by means of its capacity to inactivate radicals in respect to their oxidation. It is well known that EPDM composites are efficiently protected by inorganic fillers [72,73] due to the elemental composition comprising oxygen. However, the structure of the host polymer plays an important role in the concentration, distribution, and inner movement of oxidation promoters [74].

The relative order of peaks (Figure 6) that reflects the involvement of filler particles in polymer aging is relevant for the industrial applications of processed EPDM when the diminution of oxidation rate is the main factor for the extension of material durability. The concentration of 2.5 phr of barium titanate provides evident chemical stability, which is proven by the slower accumulation of carbonyl structures. Though the samples containing carbon black keep a relatively constant oxidation level, the compositions free of carbon black present a relatively linear level of oxidation that signifies an active implication of it in the progress of aging [75]. The interaction between elastomer and CB particle surface was reported [76], demonstrating the formation of slight chemical bonds between the polymer macromolecules and some superficial carbon atoms.

A previously published report on the effects of inorganic fillers, such as carbon black, silica, and nanoclay on ethylene–propylene–diene elastomer [77] stated that the best features are obtained when the inorganic component reaches high amounts. Another proof of the improved strength of EPDM composites is shown by the elastomer fibers containing POSS, a three-dimensional silicon-oxygen polyhedral configuration, which presents a good effect on the extension of durability [78]. The attaching of molecular moieties on the particle surface may be considered an electronegativity action of oxygen atoms contained in titanate molecules. The electrostatic attraction promotes the tight joint of radicals on the inorganic phase, where hydrocarbon fragments are electrically blocked vs the oxidation of polymer.

The studied composites may be considered as good examples for other inorganic compounds that are required for certain special applications.

3.4. Resistance against Fungi Actions

Table 1 presents the images of the samples exposed to fungi mixture for 28 days, and Table 2 shows the interpretation of the results according to ASTM G21-09 [42] after 7, 14, 21, and 28 days.

Ethylene–propylene–diene monomer (EPDM) rubber is a synthetic rubber material that has good resistance to heat, weather, and chemicals, which makes it suitable for various industrial and automotive applications. When it comes to microbial behavior, EPDM rubber is generally considered to be resistant to microbial growth. EPDM rubber is non-porous and has a smooth surface, which makes it difficult for microorganisms to adhere to and grow on the surface. Additionally, EPDM rubber is resistant to moisture and water absorption, which further reduces the likelihood of microbial growth [79]. However, it is important to note that under certain conditions, such as exposure to various environment stressors (humidity, moisture, organic compounds, UV and gamma radiations) or their modification with different additives, the microbial behavior of EPDM can be changed, and the microorganisms can grow on EPDM rubber surfaces [80].

While there is some research on the antimicrobial properties of barium titanate (most of them being reported on its nano form), the results are not conclusive, and more research is needed to fully understand its potential as an antimicrobial agent [81]. Some studies have shown that barium titanate nanoparticles can inhibit the growth of certain bacteria and fungi [81–83]. One proposed mechanism for this antimicrobial activity is the production of reactive oxygen species (ROS) when the nanoparticles come into contact with water, which can damage the cell membranes of microorganisms [84].

As can be seen from Tables 1 and 2, all types of non-irradiated EPDM samples show high resistance to fungi (i.e., 0 and 1 ratings). The resistance to fungi for the EPDM samples with barium titanate content seems to be influenced by the irradiation dose, thus at 50 kGy, the resistance to fungi is lower than at 100 kGy for both concentrations of $BaTiO_3$, even after the first 7 days of exposure. This can be due either to the higher degree of radio-induced crosslinking of the material at 100 kGy than at 50 kGy, and therefore the enzymatic degradation of the material becomes more difficult, or to a stabilizing effect of the $BaTiO_3$ that prevents the degradation of the material. In reality, it is possible to have a synergistic effect of the two types of reticulation-stabilization mechanisms. In the case of EPDM samples with carbon black (CB) and barium titanate, the resistance to fungi is higher than in the samples without CB, even after 28 days of exposure. Although there are not many studies that attest to the antifungal activity of BC, it is possible that it functions as a radio-induced free radical trap or as a decomposer of peroxides to form stable products [85].

This present interpretation is correlated with the results obtained by chemiluminescence and FTIR measurements.

Table 1. Sample images after 28 days of fungal exposure.

Sample	Dose (kGy)		
	0	50	100
EPDM			
EPDM + 1% BaTiO$_3$			
EPDM + 2.5% BaTiO$_3$			
EPDM + carbon black			
EPDM + 1% BaTiO$_3$ + carbon black			
EPDM + 2.5% BaTiO$_3$ + carbon black			

Table 2. Sample ratings according to ASTM G21-9.

Sample	7 Days			14 Days			21 Days			28 Days		
	0 kGy	50 kGy	100 kGy	0 kGy	50 kGy	100 kGy	0 kGy	50 kGy	100 kGy	0 kGy	50 kGy	100 kGy
EPDM	0	1	1	0	2	1	0	2	1	1	3	2
EPDM + 1% BaTiO$_3$	0	2	1	0	3	1	0	3	1	0	3	2
EPDM + 2.5% BaTiO$_3$	0	2	1	0	2	1	0	2	1	0	2	1
EPDM + CB *	0	0	0	0	0	0	0	0	0	0	1	1
EPDM + CB + 1% BaTiO$_3$	0	0	0	0	0	0	0	0	0	0	1	1
EPDM + CB + 2.5% BaTiO$_3$	0	0	0	0	0	0	0	0	0	0	1	1

* CB means carbon black powder.

4. Discussion

The behavior analysis of all studied formulations reveals the decisive role of fillers that efficiently hinder the oxidation of EPDM as it happens in many polymeric systems with special availabilities like poly (vinylidene fluoride) [86], polyketone [87], polysiloxane [88], polyurethane [89], ultrahigh molecular weight polyethylene [90], and Nafion [91]. The way of polymer stabilization by inorganic compounds becomes open for certain salt structures, which directly act on the breaking degradation chain [9,92]. The initiation of stabilization activity is often related to the capacity of filler or additive for the assistance of structural nonconformity like lattice holes and low electron density [9,93]. The stabilization systems based on structured carbon are appropriate examples for the restructuring of polymer networks [94].

The characterization of the stabilization contribution of barium titanate is suggested by the interaction between the surface of filler particles and available free radicals that may be placed in their neighborhood. A possible explanation is the differences that exist in the electron charge distribution near the lattice gaps. The reliable support is the increase of protection effects as the filler concentration becomes greater (Figure 4). The increase in the temperature values for inspected formulations, when the oxidation effectively starts (onset oxidation temperatures, Figure 2a,b), supports the explanation of protective effects by the superficial concentration of defects. More than that, the decrease in CL intensities either for the samples containing carbon black or for the probes free of it explains the involvement of electronic interaction by which free radicals are scavenged and withdrawn from the degradation chains.

The dissimilar mechanism of stabilization promoted by barium titanate with respect to the classical mechanism ascribed to classical antioxidants concerns the strong physical interactions, which are possible due to the unpaired electron that is present in scission fragments. The exposure of EPDM/BaTiO$_3$ composite to the action of γ-rays amplifies the stabilization effects when the irradiation dose is not too high for the initiation of oxidative degradation (Figure 3). The availability of free radicals for their trend of recombination instead of their oxidation is revealed by the evolution of carbonyl vibration (1720 cm^{-1}) during the FTIR investigation (Figure 6). The increase in the transmittance values is correlated with the fate of intermediates when they are decayed according to the reactions through which they are combined.

The mechanical properties are also based on the evidence of the anti-aging effects of barium titanate, as it protects oxidation and allows the improvement of material elasticity. The most relevant aspect concerning the amelioration of the mechanical behavior of composites is the dissimilarity between the tensile strength and elongation at break values as well as their evolution by the modification of irradiation dose (Figure 5). Striking differences appear for the sample subjected to 50 kGy, where the increase in received energy reached the gelation dose (15 kGy) [95], and the crosslinking exceeds the degradation.

The material properties at 100 kGy become more convenient, but the content of titanate influences the stabilization degree.

Starting with the result analysis from a previous report [96], it is easy to consider that irradiation processing is a welcome procedure for the manufacture of O-rings in the presence of carbon black. However, the favorable course of degradation is found only for the elastomer containing 2.5 phr of barium titanate. Under these circumstances, the nonirradiated materials offer an excellent answer to the fungi attack (Tables 1 and 2). The radiation processing decreases the resistance at the fungi attack, but it becomes relevant at low filler loading (1 phr). The selection of these formulations for the applications under outdoor conditions, namely electrical cable insulations, is a suitable decision, because these materials present appropriate mechanical properties and oxidation strength, with their irradiation being included in the manufacturing procedure.

The inclusion of titanate into the composition of the EPDM structure is accompanied by the restrictive condition of movement for free radicals. The higher content of titanate determines the hindering of fragments to meet oxygen for their conversion into degradation products. If the initiation stage does not depend on the formulation, the propagation step complicates the mechanism by the overlapping of superficial scavenging and inactivation [97].

The formation of free radicals during the γ-irradiation of EPDM indicates the possibility of crosslinking [18,98]. This feature explains the modification noticed in the values of mechanical properties, which are tightly dependent on the number of intermolecular bridges that appear by irradiation curing. The radiation vulcanization of EPDM [99] may provide a resistant material whose utilization in an energetic hazardous environment would be recommended [100]. The hybrid-structured EPDM compounds with barium titanate allow radiation processing as a suitable version of fabrication for many industrially applied items [101,102].

The presence of some appropriate compounds whose structures recommend them as suitable stabilizers may increase the period of safe usage of EPDM compounds explained by the effective inhibition of oxidation based on strong catching by stable chemical bonds [103,104]. The delay of degradation evaluated by the evolution of FTIR spectroscopy is a fast and convenient effect brought about by certain fillers and additives that extends the material durability without significant influence of mechanical properties. Accordingly, the improvement of barium titanate on the material lifetime is mandatory and required by the optimization of functional properties and the minimizing the destructive effects of degrading stressors. The essential requirement for the inclusion of an oxidation protector is its participation in the overall amelioration of material behavior simultaneously accompanied by the lack of alterations [105].

5. Conclusions

The present study reports the conditioned behavior of EPDM substrate by the presence of various amounts of barium titanate particles. The radiation degradation of polymer fraction is delayed by the scavenging action of fillers when the composites are processed by γ-irradiation. The compositional feature individualized by the presence of carbon black powder may be considered as an appropriate characteristic, which leads to high oxidation resistance, good mechanical properties, and slow advancing oxidation states. The significant increase in onset oxidation temperature (the temperature that indicates the start of oxidation) of 10–15 °C for each additional 50 kGy is a main benefit of long-term applications that expends either the operation ranges or the warranty conditions. The proposed composite formulations containing higher titanate loadings may be suitable for the manufacturing of high-tech products capable of keeping their oxidation states at a low level, as it is demonstrated by spectroscopic investigation. The mechanical properties are relevant for safe usage when the content of barium titanate is 2.5 phr. It is well corroborated with the longest oxidation induction times (around 500 min) presented by these γ-processed composites. The chemiluminescence results prove the efficient protection activity of $BaTiO_3$

filler when γ-irradiation is applied and the thermal aging in the measuring furnace stimulates lower CL emission from the polymer samples with higher filler concentrations (1 and 2.5 phr). The improved functional characteristics of EPDM composites including barium titanate make the manufacture of a large series of materials possible, which can protect surfaces against corrosion, eliminate fluid loss by perfect and long-term sealing, assure the attenuation of vibrations in the jointing part of bridges or buildings, and protect outdoor devices against the wrong effects of forecast factors.

Supplementary Materials: The following supporting information can be downloaded at: https://www.mdpi.com/article/10.3390/polym15092190/s1, Figure S1: The isothermal CL spectra recorded on pristine EPDM/carbon black samples; Figure S2. The FTIR spectra were recorded on EPDM/BaTiO$_3$ composites exposed to a γ-dose of 100 kGy.

Author Contributions: Conceptualization, T.B., N.N. and T.Z.; methodology, T.Z.; software, I.B.; validation, T.B., N.N., T.Z., I.B. and T.F.B.; formal analysis, T.Z.; investigation, T.B., N.N., T.Z., I.B. and T.F.B.; data curation, I.B. and T.F.B.; writing—original draft preparation, T.B. and T.Z.; writing—review and editing, T.B., N.N., T.Z. and I.B.; visualization, I.B.; supervision, T.Z. All authors have read and agreed to the published version of the manuscript.

Funding: Financial support was provided by the Romanian Ministry of research, Innovation and Digitalization through con-tracts PN 23140201-42N/2023 and 25PFE/2021(between National R&D Institute of Electrical Engineering, ICPE-CA and Romanian Ministry of Research, Innovation and Digitalization.

Institutional Review Board Statement: Not applicable.

Data Availability Statement: The data presented in this study are available on request from the corresponding authors.

Conflicts of Interest: The authors declare no conflict of interest.

References

1. Shanmugam, V.; Johnson Rajendran, D.J.; Babu, K.; Rajendran, S.; Veerasimman, A.; Marimuthu, U.; Singh, S.; Das, O.; Neisiany, R.E.; Hedenqvist, M.S.; et al. The mechanical testing and performance analysis of polymer-fibre composites prepared through the additive manufacturing. *Polym. Test.* **2021**, *93*, 106925. [CrossRef]
2. Ferry, M.; Roma, G.; Cochin, F.; Esnouf, S.; Dauvois, V.; Nizeyimana, F.; Gervais, B.; Ngono-Ravache, Y. Polymers in the nuclear power industry. In *Comprehensive Nuclear Materials*, 2nd ed.; Konings, R.J.K., Stoller, R.E., Eds.; Elsevier: New York, NY, USA, 2020; Volume 3, pp. 545–580.
3. Zaharescu, T.; Jipa, S. Radiochemical Modifications in Polymers. In *Landolt-Börnstein Numerical Data and Functional Relationships in Science and Technology. Group VIII, Arndt, K.-F., Lechner, M.D., Eds.*; Springer: Berlin/Heidelberg, Germany, 2013; Volume 6, pp. 93–184.
4. Rivaton, A.; Cambon, S.; Gardette, J.-L. Radiochemical yields of EPDM elastomers. 2. Identification and quantification of chemical changes in EPDM and EPR films γ-irradiated under oxygen atmosphere. *Nucl. Instrum. Meth. Phys. Res. B* **2005**, *227*, 343–356. [CrossRef]
5. Wong, W.-K.; Hsuan, Y.G. Interaction of antioxidants with carbon black in polyethylene using oxidative induction time methods. *Geotext. Geomembr.* **2014**, *42*, 641–647. [CrossRef]
6. Rivaton, A.; Cambon, S.; Gardette, J.-L. Radiochemical yields of EPDM elastomers. 4. Evaluation of some anti-oxidants. *Polym. Degrad. Stab.* **2006**, *91*, 136–143. [CrossRef]
7. Liu, Q.; Wei, P.; Cong, C.; Meng, X.; Zhou, Q. Synthesis and antioxidation behavior in EPDM of novel macromolecular antioxidants with crosslinking and antioxidation effects. *Polym. Degrad. Stab.* **2022**, *205*, 110155. [CrossRef]
8. Morlat-Therias, S.; Fanton, E.; Tomer, N.S.; Rana, S.; Singh, R.P.; Gardette, J.-L. Photooxidation of vulcanized EPDM/montmorillonite nanocomposites. *Polym. Degrad. Stab.* **2006**, *91*, 3033–3039. [CrossRef]
9. Zaharescu, T. Stabilization effects of doped inorganic filler on EPDM for space and terrestrial applications. *Mater. Chem. Phys.* **2019**, *234*, 102–109. [CrossRef]
10. Han, S.-W.; Choi, N.-S.; Ryu, S.-R.; Lee, D.-J. Mechanical property behavior and aging mechanism of carbon-black-filled EPDM rubber reinforced by carbon nano-tubes subjected to electro-chemical and thermal degradation. *J. Mech. Sci. Technol.* **2017**, *31*, 4073–4078. [CrossRef]
11. Craciun, G.; Manaila, E.; Ighigeanu, D.; Stelescu, M.D. A method to improve the characteristics of EPDM rubber based eco-composites with electron beam. *Polymers* **2020**, *12*, 215. [CrossRef]

12. Basfar, A.A.; Abdel-Aziz, M.M.; Mofti, S. Stabilization of γ-radiation vulcanized EPDM rubber against accelerated aging. *Polym. Degead. Stab.* **1999**, *66*, 191–197. [CrossRef]
13. Cai, Y.; Zheng, J.; Hu, Y.; Wei, J.; Fan, H. The preparation of polyolefin elastomer functionalized with polysiloxane and its effect in ethylene-propylene-diene monomer/silicon rubber blends. *Eur. Polym. J.* **2022**, *177*, 111468. [CrossRef]
14. Ohki, Y.; Hirai, N.; Okada, S. Penetration routes of oxygen and moisture into the insulation of FR-EPDM cables for nuclear power plants. *Polymers* **2022**, *14*, 5318. [CrossRef] [PubMed]
15. Lee, S.-H.; Park, G.-W.; Kim, H.-J.; Chung, K.; Jang, K.-S. Effects of filler functionalization on filler-embedded natural rubber/ethylene-propylene-diene monomer composites. *Polymers* **2022**, *14*, 3502. [CrossRef]
16. De Almeida, A.; Chazeau, L.; Vigier, G.; Marque, G.; Goutille, Y. Influence of PE/PP ratio and EBN content on the degradation kinetics of γ-irradiated EPDM. *Polym. Degrad. Stab.* **2014**, *110*, 175–183. [CrossRef]
17. Chen, B.; Dai, J.; Song, T.; Guan, Q. Research and development of high-performance high-damping rubber Materials for high-damping rubber isolation bearings: A review. *Polymers* **2022**, *14*, 2427. [CrossRef] [PubMed]
18. Balaji, A.B.; Ratnam, T.C.; Khalid, M.; Walvekar, R. Effect of electron beam irradiation on thermal and crystallization behavior of PP/EPDM blend. *Radiat. Phys. Chem.* **2017**, *141*, 179–189. [CrossRef]
19. Lipińska, M.; Imiela, M. Morphology, rheology and curing of (ethylene-propylene elastomer/ hydrogenate acrylonitrile-butadiene rubber) blends reinforced by POSS and organoclay. *Polym. Test.* **2019**, *75*, 26–37. [CrossRef]
20. Yasin, T.; Khan, S.; Nho, Y.-C.; Ahmad, R. Effect of polyfunctional monomers on properties of radiation crosslinked EPDM/waste tire dust blend. *Radiat. Phys. Chem.* **2012**, *81*, 421–425. [CrossRef]
21. Balachandran Nair, A.; Nandakumar, N.; Ayswarya, E.P.; Resmi, V.C.; Francis, V.; Varghese, N.; Nelson Joseph, P.; Joseph, R. Ethylene-propylene-diene (5-ethylidene-2-norbornene) terpolymer/aluminium hydroxide nanocomposites: Thermal, mechanical and flame retardant. *Mater. Today. Proc.* **2023**, *72*, 3093–3099. [CrossRef]
22. Taguet, A.; Cassagnau, P.; Lopez-Cuesta, J.-M. Structuration, selective dispersion and compatibilizing effect of (nano)fillers in polymer blends. *Prog. Polym. Sci.* **2014**, *39*, 1526–1563. [CrossRef]
23. Özdemir, T. Gamma irradiation degradation/modification of 5-ethylidene 2-norbornene (ENB)-based ethylene propylene diene rubber (EPDM) depending on ENB content of EPDM and type/content of peroxides used in vulcanization. *Radiat. Phys. Chem.* **2008**, *77*, 787–793. [CrossRef]
24. Le Lay, F. Study on the lifetime of EPDM seals in nuclear-powered vessels. *Radiat. Phys. Chem.* **2013**, *84*, 210–217. [CrossRef]
25. Mazhar, H.; Shehzad, F.; Hong, S.-G.; Al-Harthi, M.A. Thermal Degradation Kinetics Analysis of Ethylene-Propylene Copolymer and EP-1-Hexene Terpolymer. *Polymers* **2022**, *14*, 634. [CrossRef] [PubMed]
26. Ma, X.; Ji, T.; Zhang, J.; Shen, S.; Wang, S.; Wang, J.; Hou, X.; Yang, S.; Ma, X. A double-decker silsesquioxane of norbornene and performance of crosslinking reactive modified EPDM ablation resistance composites. *Compos. Part A* **2023**, *32*, 107370. [CrossRef]
27. Rizwan, M.; Chandan, M.R. Mechanistic insights into the ageing of EPDM micro/hybrid composites for high voltage insulation application. *Polym. Degrad. Stab.* **2022**, *204*, 110114. [CrossRef]
28. Valentini, F.; Dorigato, A.; Fambri, L.; Bersani, M.; Grigiante, M.; Pegoretti, A. Production and characterization of novel EPDM/NBR panels with paraffin for potential thermal energy storage applications. *Therm. Sci. Eng. Prog.* **2022**, *32*, 101309. [CrossRef]
29. Hasanpour, M.; Mehrabi Mazidi, M.; Razavi Aghjeh, M.K. The effect of rubber functionality on the phase morphology, mechanical performance and toughening mechanisms of highly toughened PP/PA6/ EPDM ternary blends. *Polym. Test.* **2019**, *79*, 106018. [CrossRef]
30. George, K.; Biswal, M.; Mohanty, S.; Nayak, S.K.; Panda, B.P. Nanosilica filled EPDM/Kevlar fiber hybrid nanocomposites: Mechanical and thermal properties. *Mater. Today. Proc.* **2021**, *41*, 983–986. [CrossRef]
31. Samaržija-Jovanović, S.; Jovanović, V.; Marković, G.; Marinović-Cincović, M.; Budinski-Simendić, J.; Janković, B. Ethylene–propylene–diene rubber-based nanoblends: Preparation, characterization and applications. In *Rubber Nano Blends. Preparation, Characterization and Applications*; Markovic, G., Visakh, P.M., Eds.; Springer Series on Polymer and Composite Materials; Springer: Cham, Switzerland, 2017; pp. 281–349.
32. Abdel-Hakim, A.; El-Gamal, A.A.; El-Zayat, M.M.; Sadek, A.M. Effect of novel sucrose based polyfunctional monomer on physico-mechanical and electrical properties of irradiated EPDM. *Radiat. Phys. Chem.* **2021**, *189*, 109729. [CrossRef]
33. Yasin, T.; Khan, S.; Shafiq, M.; Gill, R. Radiation crosslinking of styrene–butadiene rubber containing waste tire rubber and polyfunctional monomers. *Radiat. Phys. Chem.* **2015**, *106*, 343–346. [CrossRef]
34. Lu, Z.; Hu, Y.; Zhang, B.; Zhang, G.; Guo, F.; Jiang, W. EPDM/GO composite insulation for anti-migration of plasticizers. *J. Polym. Res.* **2022**, *29*, 385. [CrossRef]
35. Gong, C.; Cao, J.; Guo, M.; Cai, S.; Xu, P.; Lv, J.; Li, C. A facile strategy for high mechanical performance and recyclable EPDM rubber enabled by exchangeable ion crosslinking. *Eur. Polym. J.* **2022**, *175*, 111339. [CrossRef]
36. Ismail, H.; Ishak, S.; Hamid, Z.A.A. Effect of blend ratio on cure characteristics, tensile properties, thermal and swelling properties of mica-filled (ethylene-propylene-diene monomer)/(recycled ethylene-propylene-diene monomer) (EPDM/r-EPDM) blends. *J. Vinyl Addit. Techn.* **2015**, *21*, 1–6. [CrossRef]
37. Planes, E.; Chazeau, L.; Vigier, G.; Fournier, J.; Stevenson-Royaud, I. Influence of fillers on mechanical properties of ATH filled EPDM during ageing by gamma irradiation. *Polym. Degrad. Stab.* **2010**, *95*, 1029–1038. [CrossRef]

38. Pourmand, P.; Hedenqvist, L.; Pourrahimi, A.M.; Furó, I.; Reitberger, T.; Gedde, U.W.; Hedenqvista, M.S. Effect of gamma radiation on carbon-black-filled EPDM seals in water and air. *Polym. Degrad. Stab.* **2017**, *146*, 184–191. [CrossRef]
39. Özdemir, T.; Güngöra, A.; Akbaya, I.K.; Uzuna, H.; Babucçuoglu, Y. Nano lead oxide and epdm composite for development of polymer based radiation shielding material: Gamma irradiation and attenuation tests. *Radiat. Phys. Chem.* **2018**, *144*, 248–255. [CrossRef]
40. Šarac, T.; Devaux, J.; Quiévy, N.; Gusarov, A.; Konstantinović, M.J. The correlation between elongation at break and thermal decomposition of aged EPDM cable polymer. *Radiat. Phys. Chem.* **2017**, *132*, 8–12. [CrossRef]
41. Stelescu, M.D.; Airinei, A.; Manaila, E.; Fifere, N.; Craciun, G.; Varganici, C.; Doroftei, F. Exploring the effect of electron beam irradiation on the properties of some EPDM-flax fiber composites. *Polym. Compos.* **2019**, *40*, 315–327. [CrossRef]
42. ASTM G21-09; Standard Practice for Determining Resistance of Synthetic Polymeric Materials to Fungi. American Society for Testing and Materials: West Conshohocken, PA, USA, 2009. [CrossRef]
43. Assink, R.A.; Celina, M.; Gillen, K.T.; Clough, R.L.; Alam, T.M. Morphology changes during radiation-thermal degradation of polyethylene and an EPDM copolymer by 13C NMR spectroscopy. *Polym. Degrad. Stab.* **2001**, *73*, 355–362. [CrossRef]
44. Zaharescu, T.; Giurginca, M.; Jipa, S. Radiochemical oxidation of ethylene–propylene elastomers in the presence of some phenolic antioxidants. *Polym. Degrad. Stab.* **1999**, *63*, 245–251. [CrossRef]
45. Wang, Y.; Liu, H.; Li, P.; Wang, L. The effect of cross-linking type on EPDM elastomer dynamics and mechanical properties: A molecular dynamics simulation study. *Polymers* **2022**, *14*, 1308. [CrossRef] [PubMed]
46. Blanco, I.; Abate, L.; Bottino, F.A.; Bottino, P. Thermal behaviour of a series of novel aliphatic bridged polyhedral oligomeric silsesquioxanes (POSSs)/polystyrene (PS) nanocomposites: The influence of the bridge length on the resistance to thermal degradation. *Polym. Degrad. Stab.* **2014**, *102*, 132–137. [CrossRef]
47. Ma, C.; Sánchez-Rodríguez, D.; Kamo, T. Influence of thermal treatment on the properties of carbon fiber reinforced plastics under various conditions. *Polym. Degrad. Stab.* **2020**, *178*, 109199. [CrossRef]
48. Manaila, E.; Stelescu, M.D.; Craciun, G. Aspects regarding radiation crosslinking of elastomers. In *Advanced Elastomers*; Boczkowska, A., Ed.; IntechOpen: London, UK, 2012; pp. 3–34.
49. Pagacz, J.; Hebda, E.; Michałowski, S.; Ozimek, J.; Sternik, D.; Pielichowski, K. Polyurethane foams chemically reinforced with POSS—Thermal degradation studies. *Thermochim. Acta* **2016**, *642*, 5–104. [CrossRef]
50. Makuuchi, K.; Cheng, S. Fundamentals of radiation crosslinking. In *Radiation Processing of Polymer Materials and Its Industrial Applications*; Makuuchi, K., Cheng, S., Eds.; Wiley: New York, NY, USA, 2012; pp. 26–70.
51. Naikwadi, A.T.; Kumar Sharma, B.; Bhatt, K.D.; Mahanwar, P.A. Gamma radiation processed polymeric materials for high performance applications: A review. *Front Chem.* **2022**, *10*, 837111. [CrossRef]
52. Matisová-Rychlá, L.; Rychlý, J. Inherent relations of chemiluminescence and thermooxidation of polymers. In *Polymer Durability: Degradation, Stabilization and Lifetime Prediction*; Clough, R.L., Billingham, N.C., Gillen, K.T., Eds.; Advances in Chemistry Series; American Chemical Society: Washington, DC, USA, 1996; Volume 249, pp. 175–193.
53. Richaud, E.; Fayolle, B.; Verdu, J.; Rychlý, J. Co-oxidation kinetic model for the thermal oxidation of polyethylene-unsaturated substrate systems. *Polym. Degrad. Stab.* **2013**, *98*, 1081–1088. [CrossRef]
54. Le Hel, C.; Alcouffe, P.; Lucas, A.; Cassagnau, P.; Bounor Legaré, V. Curing agent-dependent localization of carbon black in thermoplastic vulcanizates. *Mater. Phys. Chem.* **2022**, *282*, 125926. [CrossRef]
55. Delor-Jestin, F.; Lacoste, J.; Barrois-Oudin, N.; Cardinet, C.; Lemaire, J. Photo-, thermal and natural ageing of ethylene–propylene–diene monomer (EPDM) rubber used in automotive applications. Influence of carbon black, crosslinking and stabilizing agents. *Polym. Degrad. Stab.* **2000**, *67*, 467–477. [CrossRef]
56. Nabil, H.; Ismail, H.; Azura, A.R. Comparison of thermo-oxidative ageing and thermal analysis of carbon black-filled NR/Virgin EPDM and NR/Recycled EPDM blends. *Polym. Test.* **2013**, *32*, 631–639. [CrossRef]
57. Pielichowski, K.; Njuguna, J.; Majka, T.M. Thermal degradation of polymer (nano)composites. In *Thermal Degradation of Polymer Materials*; Pielichowski, K., Njuguna, J., Majka, T.M., Eds.; Elsevier: London, UK, 2023; pp. 251–286.
58. Clough, R.L. High-energy radiation and polymers: A review of commercial processes and emerging applications. *Nucl. Instrum. Meth. Phys. Res. B* **2001**, *185*, 8. [CrossRef]
59. Planes, E.; Chazeau, L.; Vigier, G.; Fournier, J. Evolution of EPDM networks aged by gamma irradiation—Consequences on the mechanical properties. *Polymers* **2009**, *50*, 4028–4038. [CrossRef]
60. Decker, C.; Mayo, F.R.; Richardson, H. Aging and degradation of polyolefins. III. Polyethylene and ethylene–propylene copolymers. *J. Polym. Sci. Polym. Chem.* **1973**, *11*, 2879–2898. [CrossRef]
61. Perejón, A.; Sánchez-Jiménez, P.E.; Gil-González, E.; Pérez-Maqueda, L.A.; Criado, J.M. Pyrolysis kinetics of ethylene–propylene (EPM) and ethylene–propylene–diene (EPDM). *Polym. Degrad. Stab.* **2013**, *98*, 1571–1577. [CrossRef]
62. Zeid, M.M.A. Radiation effect on properties of carbon black filled NBR/EPDM rubber blends. *Eur. Polym. J.* **2007**, *43*, 4415–4422. [CrossRef]
63. Huang, W.; Yang, W.; Ma, Q.; Wu, J.; Fan, J.; Zhang, K. Preparation and characterization of γ-ray radiation shielding PbWO$_4$/EPDM composite. *J. Radioanal. Nucl. Chem.* **2016**, *309*, 1097–1103. [CrossRef]
64. Sarangapani, V.; Rajamanickam, D. Effect of gamma irradiation on titanium dioxide-filled polymer composites in cable insulation applications. *Iran. Polym. J.* **2022**, *31*, 809–820. [CrossRef]
65. Singh, B.; Sharma, N. Mechanistic implications of plastic degradation. *Polym. Degrad. Stab.* **2008**, *93*, 561–584. [CrossRef]

66. Ferrari, M.; Pandini, S.; Zenoni, A.; Donzella, G.; Battini, D.; Avanzini, A.; Salvini, A.; Zelaschi, F.; Andrighetto, A.; Bignotti, F. Degradation of EPDM and FPM elastomers irradiated at very high dose rates in mixed gamma and neutron fields. *Polym. Eng. Sci.* **2019**, *59*, 2522–2532. [CrossRef]
67. Manaila, E.; Airinei, A.; Stelescu, M.D.; Sonmez, M.; Alexandrescu, L.; Craciun, G.; Pamfil, D.; Fifere, N.; Varganici, C.-D.; Doroftei, F.; et al. Radiation processing and characterization of some ethylene-propylene-diene terpolymer/butyl (halobutyl) rubber/nanosilica composites. *Polymers* **2020**, *12*, 2431. [CrossRef]
68. Davenas, J.; Stevenson, I.; Celette, N.; Vigier, G.; David, L. Influence of the molecular modifications on the properties of EPDM elastomers under irradiation. *Nucl. Instrum. Meth. Phys. Res. B* **2003**, *208*, 461–465. [CrossRef]
69. Barala, S.S.; Manda, V.; Singh Jodha, A.; Ajay, C.; Gopalani, D. Thermal stability of gamma irradiated ethylene propylene diene monomer composites for shielding applications. *J. Appl. Polym. Sci.* **2022**, *139*, 52975. [CrossRef]
70. Chea, S.; Luengchavanon, M.; Anancharoenwong, E.; Techato, K.-A.; Jutidamrongphan, W.; Chaiprapat, S.; Niyomwas, S.; Marthosa, S. Development of an O-ring from NR/EPDM filled silica/CB hybrid filler for use in a solid oxide fuel cell testing system. *Polym. Test.* **2020**, *88*, 106568. [CrossRef]
71. Zagórski, Z.P.; Kornacka, E.M. Radiation processing of elastomers. In *Advances in Elastomers*; Advanced Structured Materials; Visakh, P., Thomas, S., Chandra, A., Mathew, A., Eds.; Springer: Berlin/Heidelberg, Germany, 2013; Volume 11, pp. 375–452.
72. Samaržija-Jovanović, S.; Jovanović, V.; Marković, G.; Konstantinović, S.; Marinović-Cincović, M. Nanocomposites based on silica-reinforced ethylene–propylene–diene–monomer/acrylonitrile–butadiene rubber blends. *Compos. B* **2011**, *41*, 1244–1250. [CrossRef]
73. Mokhothu, T.H.; Luyt, A.S.; Messuri, M. Preparation and characterization of EPDM/silica nanocomposites prepared through non-hydrolytic sol-gel method in the absence and presence of a coupling agent. *eXPRESS Polym. Lett.* **2014**, *8*, 809–822. [CrossRef]
74. Sidi, A.; Colombani, J.; Larché, J.-F.; Rivaton, A. Multiscale analysis of the radiooxidative degradation of EVA/EPDM composites. ATH filler and dose rate effect. *Radiat. Phys. Chem.* **2018**, *142*, 14–22. [CrossRef]
75. Yang, H.; Gong, J.; Wen, X.; Xue, J.; Chen, Q.; Jiang, Z.; Tian, N.; Tang, T. Effect of carbon black on improving thermal stability, flame retardancy and electrical conductivity of polypropylene/carbon fiber composites. *Compos. Sci. Technol.* **2015**, *113*, 31–37. [CrossRef]
76. Wilke, L.A.; Robertson, C.G.; Karsten, D.A.; Hardman, N.J. Detailed understanding of the carbon black–polymer interface in filled rubber composites. *Carbon* **2023**, *201*, 520–528. [CrossRef]
77. Tan, H.; Isayev, A.I. Comparative study of silica-, nanoclay and carbon black-filled EPDM rubbers. *J. Appl. Polym. Sci.* **2008**, *109*, 767–774. [CrossRef]
78. Xue, M.; Zhang, X.; Ma, L.; Gu, Z.; Lin, Y.; Bao, C.; Tian, X. Structure and thermal behavior of EPDM/POSS composite fibers prepared by electrospinning. *J. Appl. Polym. Sci.* **2013**, *128*, 2395–2401. [CrossRef]
79. Przybyłek, M.; Bakar, M.; Mendrycka, M.; Kosikowska, U.; Malm, A.; Worzakowska, M.; Szymborski, T.; Kędra-Królik, K. Rubber elastomeric nanocomposites with antimicrobial properties. *Mater. Sci. Eng. C* **2017**, *76*, 269–277. [CrossRef]
80. Basik, A.A.; Sanglier, J.J.; Yeo, C.T.; Sudesh, K. Microbial degradation of rubber: Actinobacteria. *Polymers* **2021**, *13*, 1989. [CrossRef] [PubMed]
81. Shah, A.A.; Khan, A.; Dwivedi, S.; Musarrat, J.; Azam, A. Antibacterial and antibiofilm activity of barium titanate nanoparticles. *Mater. Lett.* **2018**, *229*, 130–133. [CrossRef]
82. Boschetto, F.; Doan, H.N.; Vo, P.P.; Zanocco, M.; Yamamoto, K.; Zhu, W.; Adachi, T.; Kinashi, K.; Marin, E.; Pezzotti, G. Bacteriostatic behavior of PLA-BaTiO$_3$ composite fibers synthesized by centrifugal spinning and jubjected to aging test. *Molecules* **2021**, *26*, 2918. [CrossRef]
83. Fouda, S.M.; Gad, M.M.; Ellakany, P.; Al-Thobity, A.M.; Al-Harbi, F.A.; Virtanen, J.I.; Raustia, A. The effect of nanodiamonds on candida albicans adhesion and surface characteristics of PMMA denture base material-an in vitro study. *J. Appl. Oral Sci.* **2019**, *27*, e20180779. [CrossRef]
84. Ahamed, M.; Akhtar, M.J.; Khan, M.A.M.; Alhadlaq, H.A.; Alshamsan, A. Barium titanate (BaTiO$_3$) nanoparticles exert cytotoxicity through oxidative stress in human lung carcinoma (A549) cells. *Nanomaterials* **2020**, *10*, 2309. [CrossRef]
85. Mwila, J.; Miraftab, M.; Horrocks, A.R. Effect of carbon black on the oxidation of polyolefins. An overview. *Polym. Degrad. Stab.* **1994**, *44*, 351–356. [CrossRef]
86. Tang, X.; Pionteck, J.; Pötschke, P. Improved piezoresistive sensing behavior of poly(vinylidene fluoride)/carbon black composites by blending with a second polymer. *Polymer* **2023**, *268*, 125702. [CrossRef]
87. Orozco, F.; Salvatore, A.; Sakulmankongsuk, A.; Ribas Gomes, D.; Pei, Y.; Hermosilla, E.A.; Pucci, A.; Moreno-Villoslada, I.; Picchioni, F.; Bose, R.K. Electroactive performance and cost evaluation of carbon nanotubes and carbon black as conductive fillers in self-healing shape memory polymers and other composites. *Polymer* **2022**, *260*, 125365. [CrossRef]
88. Zhai, W.; Xia, Q.; Zhou, K.; Yue, X.; Ren, M.; Zheng, G.; Dai, K.; Liu, C.; Shen, C. Multifunctional flexible carbon black/polydimethylsiloxane piezoresistive sensor with ultrahigh linear range, excellent durability and oil/water separation capability. *Chem. Eng. J.* **2019**, *372*, 373–382. [CrossRef]
89. Wang, W.; Pan, H.; Yu, B.; Pan, Y.; Song, L.; Liew, K.M.; Hu, Y. Fabrication of carbon black coated flexible polyurethane foam for significantly improved fire safety. *RSC Adv.* **2015**, *5*, 55870–55878. [CrossRef]

90. Cao, X.; Li, C.; He, G.; Tong, Y.; Yang, Z. Composite phase change materials of ultra-high molecular weight. polyethylene/paraffin wax/carbon nanotubes with high performance and excellent shape stability for energy storage. *J. Energy Stor.* **2021**, *44*, 103460. [CrossRef]
91. Zhang, X.; Yu, S.; Li, M.; Zhang, M.; Zhang, C.; Wang, M. Enhanced performance of IPMC actuator based on macroporous multilayer MCNTs/Nafion polymer. *Sens. Actuators A* **2022**, *339*, 113489. [CrossRef]
92. Mishra, T.; Mandal, P.; Kumar Rout, A.; Sahoo, D. A state-of-the-art review on potential applications of natural fiber-reinforced polymer composite filled with inorganic nanoparticle. *Compos. C* **2022**, *9*, 100298. [CrossRef]
93. Tayouri, M.I.; Estaji, S.; Mousavi, S.R.; Khasraghi, S.S.; Jahanmardi, R.; Nouranian, S.; Arjmand, M.; Khonakdar, H.A. Degradation of polymer nanocomposites filled with graphene oxide and reduced graphene oxide nanoparticles: A review of current status. *Polym. Degrad. Stab.* **2022**, *206*, 110179. [CrossRef]
94. Li, H.-X.; Zare, Y.; Rhee, K.Y. The percolation threshold for tensile strength of polymer/CNT nanocomposites assuming filler network and interphase regions. *Mater. Chem. Phys.* **2018**, *207*, 76–83. [CrossRef]
95. Basfar, A.A.; Abdel-Aziz, M.M.; Mofti, S. Accelerated aging and stabilization of radiation-vulcanized EPDM rubber. *Radiat. Phys. Chem.* **2000**, *57*, 405–409. [CrossRef]
96. Müller, M.; Šleger, V.; Čedík, J.; Pexa, M. Research on the material compatibility of elastomer sealing O-rings. *Polymers* **2022**, *14*, 3323. [CrossRef]
97. Kornacka, E.M. Radiation-induced oxidation in polymers. In *Applications of Ionizing Radiation in Materials Processing*; Chmielewski, A., Sun, Y., Eds.; Institute of Nuclear Chemistry and Technology: Warsaw, Poland, 2017; pp. 185–192.
98. Deepalaxmi, R.; Rajini, V. Performance evaluation of gamma irradiated SiR-EPDM blends. *Nucl. Eng. Design* **2014**, *273*, 602–614. [CrossRef]
99. Scagliusi, S.R.; Cardoso, E.C.L.; Lugao, A.B. Radiation-induced degradation of butyl rubber vulcanized by three different crosslinking systems. *Radiat. Phys. Chem.* **2012**, *81*, 991–994. [CrossRef]
100. Haji-Saeid, M.; Sampa, M.H.O.; Chmielewski, A.G. Radiation treatment for sterilization of packaging materials. *Radiat. Phys. Chem.* **2007**, *76*, 1535–1541. [CrossRef]
101. Abdel-Aziz, M.M.; Amer, H.A.; Atia, M.K.; Rabie, A.M. Effect of gamma radiation on the physicomechanical characters of EPDM rubber/modified additives nanocomposites. *J. Vinyl Addit. Technol.* **2017**, *23*, E188–E200. [CrossRef]
102. El-Nemr, K.F.; Ali, M.A.M.; Hassan, M.M.; Hamed, H.E. Features of the structure and properties of radiation vulcanizates based on blends of polybutadiene and ethylene-propylene diene rubber. *J. Vinyl Addit. Technol.* **2019**, *25*, E64–E72. [CrossRef]
103. Ning, N.; Ma, Q.; Zhang, Y.; Zhang, L.; Wu, H.; Tian, M. Enhanced thermo-oxidative aging resistance of EPDM at high temperature by using synergistic antioxidants. *Polym. Degrad. Stab.* **2014**, *102*, 1–8. [CrossRef]
104. Zhao, W.; He, J.; Yu, P.; Jiang, X.; Zhang, L. Recent progress in the rubber antioxidants: A review. *Polym. Degrad. Stab.* **2023**, *207*, 110223. [CrossRef]
105. Yuan, S.; Li, S.; Zhu, J.; Tang, Y. Additive manufacturing of polymeric composites from material processing to structural design. *Compos. B* **2021**, *219*, 108903. [CrossRef]

Disclaimer/Publisher's Note: The statements, opinions and data contained in all publications are solely those of the individual author(s) and contributor(s) and not of MDPI and/or the editor(s). MDPI and/or the editor(s) disclaim responsibility for any injury to people or property resulting from any ideas, methods, instructions or products referred to in the content.

Article

Reinforcement Behavior of Chemically Unmodified Cellulose Nanofiber in Natural Rubber Nanocomposites

Bunsita Wongvasana [1], Bencha Thongnuanchan [1], Abdulhakim Masa [2], Hiromu Saito [3,*], Tadamoto Sakai [4] and Natinee Lopattananon [1,*]

1. Department of Rubber Technology and Polymer Science, Faculty of Science and Technology, Prince of Songkla University, Pattani 94000, Thailand
2. Rubber Engineering & Technology Program, International College, Prince of Songkla University, Songkhla 90110, Thailand
3. Department of Organic and Polymer Materials Chemistry, Tokyo University of Agriculture and Technology, Tokyo 184-8588, Japan
4. Organization for Innovation & Social Collaboration, Shizuoka University, Shizuoka 432-8011, Japan
* Correspondence: hsaitou@cc.tuat.ac.jp (H.S.); natinee.l@psu.ac.th (N.L.)

Citation: Wongvasana, B.; Thongnuanchan, B.; Masa, A.; Saito, H.; Sakai, T.; Lopattananon, N. Reinforcement Behavior of Chemically Unmodified Cellulose Nanofiber in Natural Rubber Nanocomposites. *Polymers* **2023**, *15*, 1274. https://doi.org/10.3390/polym15051274

Academic Editor: Cristina Cazan

Received: 7 February 2023
Revised: 24 February 2023
Accepted: 1 March 2023
Published: 2 March 2023

Copyright: © 2023 by the authors. Licensee MDPI, Basel, Switzerland. This article is an open access article distributed under the terms and conditions of the Creative Commons Attribution (CC BY) license (https://creativecommons.org/licenses/by/4.0/).

Abstract: We investigated the reinforcement behavior of small amounts of chemically unmodified cellulose nanofiber (CNF) in eco-friendly natural rubber (NR) nanocomposites. For this purpose, NR nanocomposites filled with 1, 3, and 5 parts per hundred rubber (phr) of cellulose nanofiber (CNF) were prepared by a latex mixing method. By using TEM, a tensile test, DMA, WAXD, a bound rubber test, and gel content measurements, the effect of CNF concentration on the structure–property relationship and reinforcing mechanism of the CNF/NR nanocomposite was revealed. Increasing the content of CNF resulted in decreased dispersibility of the nanofiber in the NR matrix. It was found that the stress upturn in the stress–strain curves was remarkably enhanced when the NR was combined with 1–3 phr CNF, and a noticeable increase in tensile strength (an approximately 122% increase in tensile strength over that of NR) was observed without sacrificing the flexibility of the NR in the NR filled with 1 phr CNF, though no acceleration in their strain-induced crystallization was observed. Since the NR chains were not inserted in the uniformly dispersed CNF bundles, the reinforcement behavior by the small content of CNF might be attributed to the shear stress transfer at the CNF/NR interface through the interfacial interaction (i.e., physical entanglement) between the nano-dispersed CNFs and the NR chains. However, at a higher CNF filling content (5 phr), the CNFs formed micron-sized aggregates in the NR matrix, which significantly induced the local stress concentration and promoted strain-induced crystallization, causing a substantially increased modulus but reduced the strain at the rupture of the NR.

Keywords: natural rubber; nanocomposites; cellulose nanofibers; mechanical property; reinforcement

1. Introduction

Natural rubber (NR), a natural polymer of cis-1,4-polyisoprene obtained from natural sources, is an important raw material in the rubber industry. NR is known to have excellent mechanical properties due to its stretchable nature and its ability to crystallize after stretching [1,2]. It is, therefore, widely used in the rubber industry to manufacture rubber products, specifically, automobile tires, vibration insulators, and surgical gloves [3–6]. In the manufacturing process, NR is often added with fillers to achieve a desirable reinforcement, lower its price, and improve processability.

Currently, the addition of cellulose nanofibers (CNFs) as a load-bearing filler has received significant attention for the formulation of high-performance polymer nanocomposites due to the outstanding mechanical properties presented by these CNFs. The CNFs were reported to exhibit high Young's moduli (~100–160 GPa) [7–11] and high strength (~1.6–3 GPa) [12,13]. Due to the impressive mechanical properties of CNF, along with its inherent biodegradability, abundant availability, renewability, and low density, several research groups have investigated the use of CNF in a wide variety of polymers, such as thermosets/thermoplastics [14–16], biodegradable polymers [17–19], and synthetic rubbers [20–23]. Over the years, CNFs have become a potential nano-filler candidate to be combined with NR. Abraham et al. [24] dispersed CNF together with sulfur and zinc-based crosslinking agents in an NR matrix using an NR latex mixing method, and they found that the introduction of increasing CNF contents (1–10 wt% based on the weight of the dried NR) markedly increased the tensile modulus and strength of the NR. Similar observations have also been presented by other authors [25–28]. These authors have ascribed the properties increase in the NR to the establishment of a chemical network of Zn/cellulose nanofiber complex in the NR. Kato et al. [28] reported a great increase in the reinforcing efficiency of pristine CNFs in NR with increasing the filling level from 1 to 5 wt%. The use of chemically modified CNFs further increased the stiffness and reduced the thermal expansion of the NR nanocomposite due to the finely dispersed CNF and the formation of chemical crosslinks between the CNF and the NR. Owing to the above observations, the findings have clearly shown that both CNF dispersion and bonding strength at the interface between the CNF and the NR were the main reasons for the rise in mechanical and thermal properties at low levels of addition.

Due to the stereoregularity of NR, the crystallization in NR under deformation, called strain-induced crystallization (SIC), presents a major interest in rubber technology. The formation of crystallites in a natural rubber network leads to a strengthening of this material, providing NR with a self-reinforcement character [29–31]. Generally, it is well-established that the strain-induced crystallization of NR is sensitive to the microstructure of the NR network and its changes during deformation [1,32]. Furthermore, the presence of popular nano-fillers such as nanoclay, silica, carbon black, carbon nanotubes, and graphene was found to activate an early crystallization, as well as promote the overall crystallization of NR during uniaxial deformation [4,5,33–36]. Recently, Wongvasana and co-workers [37] was the first group to compare the structure–property relationship of NR nanocomposites reinforced with nanoclay and CNF at a filling level of 5 phr. The results from this study showed clear distinctions between the nanoclay and the CNF in terms of their reinforcing effects and mechanisms. The nanoclays were found to finely disperse in the NR, and they effectively increased the crystalline phase in the NR due to the orientation of the NR chains introduced by the cooperation of the clay rotation and crosslinking in the NR network during stretching. As a consequence, the 5 phr nanoclay/NR nanocomposite exhibited high tensile strength and breaking strain. On the contrary, the CNF at a content of 5 phr formed an aggregated structure consisting of entangled nanofibers dispersed in the NR. The CNF aggregates were shown to impart high stiffness to the NR, with a low breaking strain. Interestingly, the ability of the aggregated CNFs to induce the NR crystallization upon stretching was also noted, even at low strain of approximately 150%.

NR has shown different mechanical properties when combined with different loadings of fillers [4,38,39]. Previous works [4] have shown that the microstructure of NR was changed by the dispersed fillers and their contents, and the NR microstructure strongly affected the strain-induced crystallization and mechanical properties of the NR nanocomposites. Up to now, studies on the strain-induced crystallization of NR reinforced with CNF have been very limited, and therefore, information on the mechanistic reinforcement is not adequate for the development of eco-friendly and sustainable materials which require the effective use of CNFs.

In this study, we aimed to explore CNF's effects and the structures they form at different contents on the properties of NR. Pristine CNFs were used at concentrations of 1, 3, and 5 phr. The use of CNF without chemical modification is of benefit to manufacturing from an economical and environmental perspective. The CNFs were mixed with NR using a latex mixing method, as previously outlined in the literature [37], and crosslinked with dicumyl peroxide (DCP) to obtain CNF/NR nanocomposites. The neat NR was prepared and used as a control. To clarify the CNF's effects on the mechanistic reinforcement of the NR at different contents, we investigated the microstructures, mechanical properties, bound rubber contents, crosslink densities, and strain-induced crystallization levels of the CNF/NR nanocomposites by transmission electron microscopy (TEM), tensile tests, dynamic mechanical analyses, measurements of bound rubber, solvent-induced swelling, and gel contents, and wide-angle X-ray diffraction (WAXD), respectively.

2. Materials and Methods

2.1. Materials

High ammonia (HA) concentrated natural rubber (NR) latex containing a dry rubber content (DRC) of 60% was supplied by Yala Latex Co., Ltd. (Yala, Thailand). Cellulose nanofibers (CNF, Nanoforest-S) made from wood pulp using the aqueous counter collision (ACC) method were kindly supplied by Chuetsu Pulp and Paper Co., Ltd. (Tokyo, Japan). Dicumyl peroxide (DCP) was manufactured by Wuzhou International Co., Ltd. (Shenzhen, China), and 2, 2, 4-trimethyl-1,2-dihydroquinone (TMQ) was supplied by Lanxess AG (Cologne, Germany). Paraffinic oil (white oil grade A, no. 15) was provided by China Petrochemical International Co., Ltd. (Shanghai, China).

2.2. Preparation of CNF/NR Nanocomposites

The CNF/NR nanocomposites were prepared through the latex mixing method schematically shown in Figure 1. In the latex mixing method, the aqueous CNF suspension (1 wt%), obtained by mixing the CNFs in water, as outlined in the literature [37], was firstly mixed with NR latex under vigorous stirring (600 rpm) at room temperature for 30 min using an IKA® RW 20 digital mixer (IKA®-Werke, Staufen, Germany). The obtainable CNF/NR mixtures having amounts of CNF of 1, 3, and 5 phr were then dried at 50 °C for 2 days. The dried CNF/NR masterbatches were later compounded with the rubber additives in a Hakke internal mixer (Thermo Electron Corporation, Karlsruhe, Germany) at a temperature and rotor speed of 50 °C and 60 rpm, respectively, for 12 min. The compositions of the CNF/NR nanocomposite compounds are listed in Table 1. The compounded CNF/NR nanocomposites were crosslinked with DCP in a hot-pressing machine at 160 °C for 10 min. The neat NR used as a reference specimen was also prepared using the same procedure as described above. Photographs of the NR and NR nanocomposite samples are shown in Figure 1. The chemically unmodified CNF-reinforced NR was visibly transparent at CNF filling levels of 1–5 wt%. In this study, the DCP-crosslinked NR nanocomposites with 1, 3, and 5 phr CNF were designated CNF1/NR, CNF3/NR, and CNF5/NR, respectively.

Table 1. Formulation of the NR and the CNF/NR nanocomposites.

Ingredients	Parts per Hundred Rubber (phr)			
	NR	CNF1/NR	CNF3/NR	CNF5/NR
NR	100	100	100	100
CNF	-	1	3	5
Paraffinic oil	20	20	20	20
TMQ	2	2	2	2
DCP	1	1	1	1

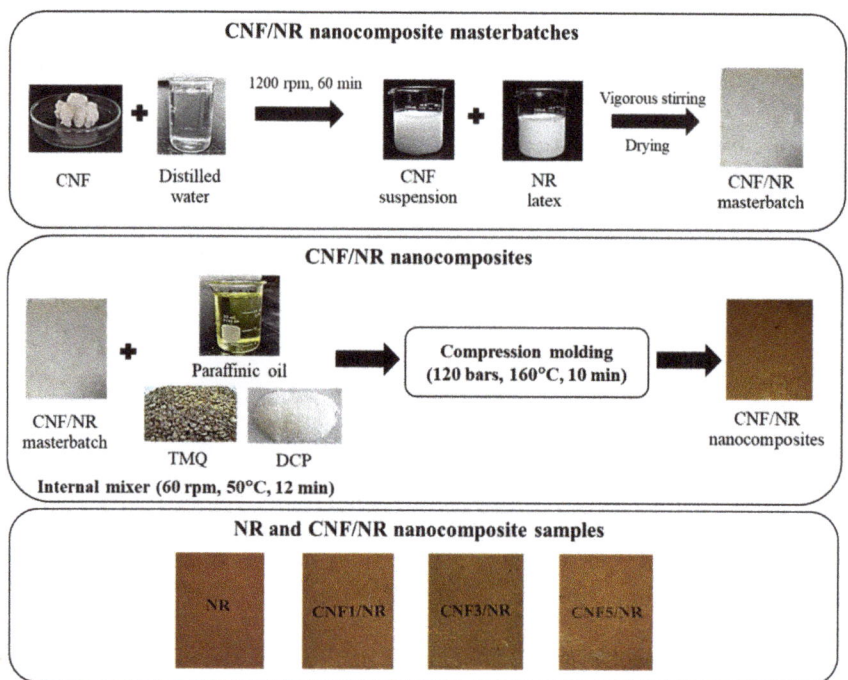

Figure 1. Schematic diagram of the preparation of the NR and the CNF/NR nanocomposites.

2.3. Characterization

2.3.1. Transmission Electron Microscopy (TEM)

TEM was used to study the dispersion of the CNFs in the CNF/NR nanocomposites. TEM imaging was conducted using a JEOL JEM 2010 (JEOL Co., Tokyo, Japan). Ultrathin sections (approximately 100 nm) were cut with a diamond knife at a temperature of −120 °C using an ultramicrotome (RMC MT-XL, RMC Products Group, Ventana Medical System, Inc., Oro Valley, AZ, USA).

2.3.2. Wide-Angle X-ray Diffraction (WAXD) Measurements

The degree of crystallinity in the NR and the CNF/NR nanocomposites during tensile stretching was assessed by wide-angle X-ray diffraction (WAXD) using a NANO-Viewer system (Rigaku Co., Ltd., Tokyo, Japan). Cu-Kα radiation with a wavelength of 0.154 nm was generated at an accelerated voltage of 46 kV and a target current of 60 mA. The sample-to-detector distance was 15 mm. An imaging plate (IP) (Fujifilm BAS-SR 127) was used as a two-dimensional detector and an IP reading device (R-AXIS Ds3, Rigaku Co., Japan) was used to transform the obtained image to text data. The sample was stretched in steps after WAXD measurements at a fixed strain using a miniature tensile machine (Imoto Machinery Co., Ltd., Kyoto, Japan). The exposure time was 15 min at room temperature (20 °C). The scattering intensity was corrected with respect to the exposure time, the sample thickness, and the transmittance.

The area of the crystalline diffraction peaks assigned to the (200) and (120) planes and the area of the amorphous halo were fitted using Origin®9.1 software. The value of X_c was calculated using Equation (1):

$$X_c = \frac{A_c}{A_c + A_a} \times 100\,\%, \tag{1}$$

where A_c represents the areas of the crystalline region and A_a corresponds to the amorphous region.

2.3.3. Mechanical Property Measurements

The mechanical properties were measured on a Hounsfield Tensometer (H10KS, Hounsfield Test Equipment Co., Ltd., Surrey, UK) at a temperature of 25 ± 2 °C with an extension rate of 500 ± 50 mm/min by ASTM D412. The dumb-bell-shaped specimens were cut from the crosslinked rubber films. An average of ten specimens was considered for the tensile test.

2.3.4. Dynamic Mechanical Analysis (DMA)

The dynamic mechanical properties of the NR and the CNF/NR nanocomposites were measured using an advanced rheometric expansion system rheometer (model ARES-RDA W/FCO, TA Instruments Ltd., New Castle, DE, USA). The storage modulus (E′) and loss factor or damping factor (tan δ = E″/E′, where E″ is a loss modulus) were determined with the tension mode at temperatures ranging from −95 °C to 80 °C using a heating rate of 2 °C/min, a frequency of 1.0 Hz, and a dynamic strain amplitude of 0.5%.

2.3.5. Bound Rubber

Bound rubber measurements were performed to determine the physical linkages between the rubber and the CNF. Approximately 0.2 g (g) of uncured rubber compounds contained in a metal cage were immersed in 20 mL of toluene at room temperature for 3 days, with the solvent replaced every day. Then, the samples were removed from the toluene solvent and dried at 105 °C until they reached a consistent weight. The bound rubber content was estimated using the following equation [40]:

$$Bound\ rubber\ (\%) = \frac{W_{fg} - W_f}{W_p}, \qquad (2)$$

where W_{fg} represents to the weighted sample after immersion, W_f is the weight of the CNF in the specimen, and W_p refers to the weight of the NR in the specimen.

2.3.6. Gel Content

Gel content measurements were performed to measure the extent of the crosslinking of the NR phase in the NR and the CNF/NR nanocomposites. Specimens weighing between 0.17 and 0.20 g were cut into small pieces and directly immersed in a 250 mL round bottom boiling flask containing ~100 mL of toluene and attached to a condenser. The gel content determination was carried out for 8 h. The insoluble residues were taken out and dried at room temperature for 48 h prior to weighting. The gel content was calculated using the following equation [41]:

$$Gel\ content = 100 - \left[\left(\frac{W_{final}}{(1-F)\ W_{rubb}}\right) \times 100\right], \qquad (3)$$

where W_{final} is the weight of the sample after extraction, W_{rubb} is the initial weight of the rubber in the sample, and F is the volume fraction of the filler.

3. Results and Discussion

3.1. Dispersion of CNF in the CNF/NR Nanocomposites

The effect of the CNF content on the filler dispersion state in the NR matrix was examined by the TEM technique, and the results are shown in Figures 2 and 3. Figure 2 shows TEM photomicrographs of thin sections of the CNF/NRs containing 1, 3, and 5 phr CNF taken at low magnification levels. In the early work of Thomas et al. [25], in a TEM photograph of NR without filler, the absence of fillers was apparent. However, the obtained TEM images of the NR nanocomposites concerning the dispersion of the CNF showed the CNF structure in the NR matrix. The sizes of the CNFs in the various CNF/NR samples were measured from the TEM images using Image J software, and their sizes were represented by the thicknesses. The results are given in Table 2. From Figure 2A–C, it can be seen that different grades of CNF dispersion were formed in the NR matrixes, depending on the content of CNF. It has been reported that individual CNFs obtained from wood sources had thicknesses of approximately 3–5 nm [13,42,43]. Based on the measured sizes of the nanofibers shown in Figure 2 and Table 2, it was clear that the CNF1/NR consisted of CNFs which were separate from the nanofiber and bundles of nanofibers due to high extent of CNF-CNF interactions via the hydrogen bonding of the active hydroxyl group (-OH) on the CNF surfaces [14,22,44]. When the addition of the CNFs was increased to 3 phr, the nanofibers were held together to form fiber bundles, and their thicknesses were apparently increased (Figure 2B and Table 2). With further addition of CNFs of up to 5 phr, the CNFs were mostly aggregated, and the aggregated dimensions were approximately 1–3 µm (Figure 2C and Table 2). At higher magnification, as shown in Figure 3, the TEM images clearly displayed the nanofiber structure in the CNF1/NR sample and the aggregated structure composed of highly entangled nanofibers in the CNF5/NR sample. In an early work by Fiorote et al. [45], the effect of CNF content (0.5, 1, 2.5, and 5 phr) on the morphology of CNF/NR nanocomposites was investigated. The results showed that the degree of nanofiber dispersion decreased with increasing contents of CNF. Similarly, Zhang et al. [46] incorporated CNFs of different contents (1–10 phr) in NR nanocomposites, and they demonstrated that poor nanofiber dispersion was observed for the nanocomposites loaded with CNF in the amounts of 5 and 10 phr. In this study, the findings from the TEM analysis led to the conclusion that there was a homogeneously dispersed, nano-sized CNF in the CNF1/NR sample and a micro-sized domain of aggregated nanofiber in the CNF5/NR sample.

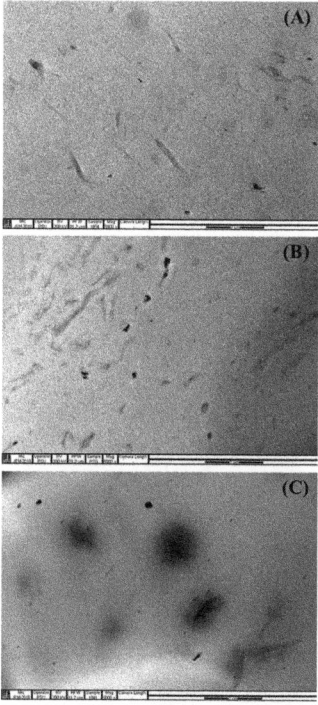

Figure 2. TEM images of (**A**) CNF1/NR, (**B**) CNF3/NR, and (**C**) CNF5/NR at low magnification (X5,000).

Table 2. Dimensions of the dispersed CNFs in the CNF/NR nanocomposites.

Samples	Dimension Range of CNFs (nm)	Average Thickness of the CNFs (nm)
CNF1/NR	3–184	65 ± 63
CNF3/NR	30–345	140 ± 99
CNF5/NR	1000–3000	1700 ± 700

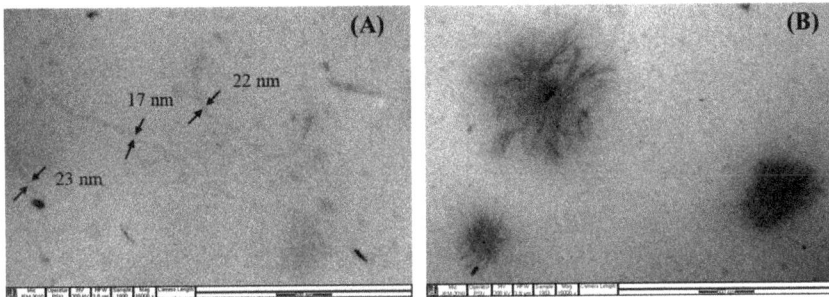

Figure 3. TEM images of (A) CNF1/NR and (B) CNF5/NR at high magnification (X15,000).

3.2. Stress-Strain Behavior of NR and CNF/NR Nanocomposites

Figure 4 shows the representative stress–strain behavior of the CNF/NRs filled with different CNF contents. As can be seen in Figure 4, it was obvious that the characteristic stress–strain curves of the NR, CNF1/NR, and CNF3/NR samples, but not that of the CNF5/NR sample, were very similar; that is, their stresses gradually increased as a function of the applied strain and turned upward sharply beyond a certain strain, as indicated by the arrows. It was also interesting to see that the upward turn was pronounced upon the addition of the CNFs into the NR. In the unfilled NR, the abrupt upturn of stress at high strains was generally assigned to the strain-induced crystallization (SIC) process [4,47,48]. Conversely, the CNF5/NR sample showed a different stress–strain behavior. The tensile stress exerted on this sample was dramatically raised upon stretching until it reached the rupture stress at low applied strain (~300%), where the abrupt upturn in stress was about to occur. As we clearly demonstrated that the CNFs in the CNF5/NR sample were inhomogeneously dispersed in the NR (Figures 2 and 3), the aggregated nanofibers in the CNF5/NR sample could have acted as crack precursors that reduced the breaking strain of the NR.

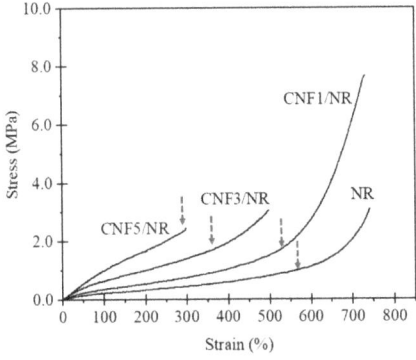

Figure 4. Stress–strain curves of the NR and the CNF/NR nanocomposites.

The tensile moduli at 50%, 100%, and 300%, as well as the tensile strength, strain at break of the NR, and various CNF/NRs, were also compared, as shown in Table 3. These results clearly showed the influence of the different CNF addition levels on the mechanical properties of the NR nanocomposites. The tensile moduli at 50%, 100%, and 300% strains obviously increased with the increasing CNF content. Several authors have reported a dependence of the modulus of a polymer on the filler content [49–51]. In Table 3, it is seen that the increases in the moduli at the 50%, 100%, and 300% strains of the NR were significant in the CNF5/NR sample (the increases were 110%, 304%, and 420% for the 50% modulus, 100% modulus, and 300% modulus, respectively). The tensile strength of the CNF/NR samples increased when CNFs were incorporated at 1 phr, and then they leveled off as the CNF contents of 3–5 phr were added. For the CNF1/NR sample, it was seen that the tensile strength of the CNF1/NR sample was remarkably improved by approximately 122% over that of the NR, and its strain at break was approximately 757% comparable to that of the NR (which had a breaking strain of approximately 759%). The high tensile strength and good flexibility may be ascribed to the well-dispersed CNFs in the CNF1/NR sample. The crosslink density determined from the equilibrium swelling measurement is also included in Table 3. In general, the crosslink density of a composite material is a measure of the filler–rubber interaction [27,39]. Based on the data, it was clear that increments in overall crosslink density resulted from more interaction between the CNF and the NR. Therefore, the addition of more CNF caused higher restricted NR chain mobility, which accounted for the increase in the tensile modulus and the decrease in the rubber flexibility. However, the tensile strength was inconsistently increased with the increasing crosslink density.

Based on these observations, a noteworthy result obtained was that the characteristic stress–strain behaviors of the NR and the NR nanocomposites with lower CNF contents (1–3 phr CNF) were clearly distinguishable from those of the high CNF content samples (5 phr CNF). Furthermore, the tensile properties of the NR nanocomposites changed in variation with the incorporated CNF contents. To explain these observations, a study on the microstructural evolution of NR networks in various CNF/NR samples using WAXD analysis was carried out, and their features of strain-induced crystallization were compared and are discussed in the next section.

Table 3. Summary of the mechanical properties of the NR and the CNF/NR nanocomposites.

Samples	50% Modulus (MPa)	100% Modulus (MPa)	300% Modulus (MPa)	Tensile Strength (MPa)	Elongation at Break (%)	Crosslink Density ($\times 10^{-5}$ mol/g)
NR	0.22 ± 0.02	0.24 ± 0.03	0.49 ± 0.06	3.26 ± 0.66	759 ± 20	3.22 ± 0.20
CNF1/NR	0.23 ± 0.03	0.35 ± 0.02	0.70 ± 0.06	7.26 ± 1.03	757 ± 38	3.67 ± 0.14
CNF3/NR	0.42 ± 0.05	0.76 ± 0.08	1.56 ± 0.16	3.08 ± 0.47	470 ± 43	4.79 ± 0.20
CNF5/NR	0.50 ± 0.06	0.90 ± 0.03	2.55 ± 0.43	2.56 ± 0.31	302 ± 21	4.92 ± 0.11

3.3. Strain-Induced Crystallization of the NR and the CNF/NR Nanocomposites

Figure 5 displays two-dimensional (2D) WAXD images of the NR and the CNF/NR samples containing 1, 3, and 5 phr CNF at various applied strains.

Figure 5 shows that the different positions of the reflection spots seen in these photographs were assigned to different crystallographic planes, and the crystallographic planes that corresponded to (200) and (120) were of interest. It was clear that the applied strain had a significant impact on the patterns in the WAXD images. At strains of 0 and 150%, no reflection spots were observed in these images due to the fact that no crystallization had occurred. On the other hand, several reflection spots belonging to different crystallographic planes appeared when the samples were stretched up to strains of approximately 175–300%. These reflection spots became more pronounced, with increasing deformations, suggesting that the strain promoted crystallization and molecular chain orientation [3].

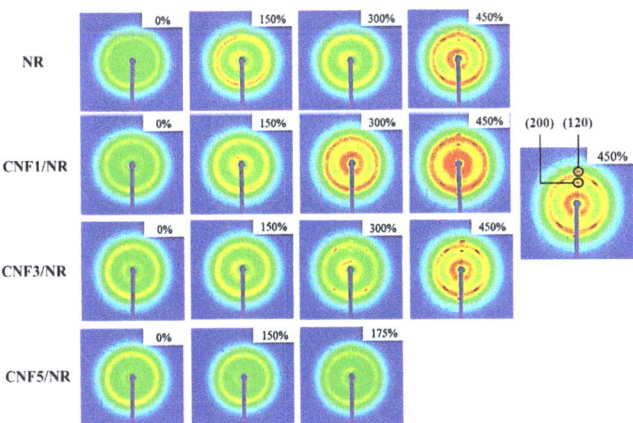

Figure 5. Typical two-dimensional WAXD images as a function of the applied strain for the NR and the CNF/NR nanocomposites.

To obtain clear information about strain-induced crystallization in the CNF/NR samples, the 2D WAXD data were transformed into 1D data, and the results are shown in Figure 6. Figure 6 shows the 1D WAXD patterns of the NR and the various CNF/NR samples selected at strain levels of 200%, 300%, and 450%. The diffraction peaks observed at 2θ of approximately 16° and 24° corresponded to the (200) and (120) planes [52,53]. No crystal peaks were observed at 200% strains for the NR, CNF1/NR, and CNF3/NR samples, indicating crystallization had not occurred in these samples. The crystallization in the NR, CNF1/NR, and CNF3/NR samples was initially seen at a strain of 300%, in which the two diffraction peaks at 2θ of approximately 16° and 24° were observed. These two peaks became more pronounced with further deformation, implying the enhancement of the crystallinity with the strain. Unlike the NR and CNF/NR samples with 1–3 phr CNF, the diffraction peaks corresponding to the (200) and (120) planes in the CNF5/NR sample were observed at a low strain of 200%, suggesting an early crystallization process in this sample. Since the CNF5/NR sample was broken at strain of approximately 300%, no further enhancement of crystallinity was observed in this sample.

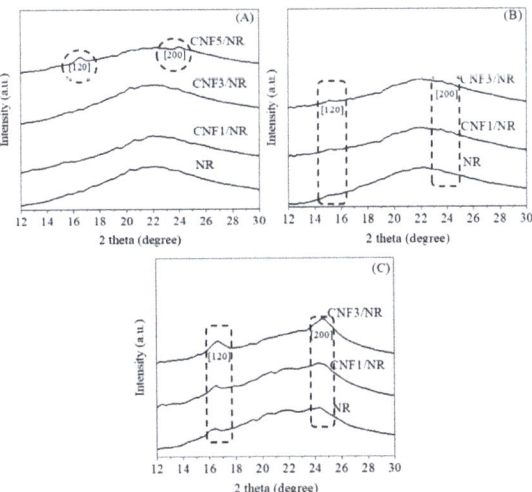

Figure 6. The 1D WAXD patterns of the NR and the CNF/NR nanocomposites measured at strains of (**A**) 200%, (**B**) 300%, and (**C**) 450%.

Based on the 1D WAXD images, the crystallinity (X_c) of the stretched NR and different CNF/NR samples could be estimated using Equation (1). The X_c results are shown in Figure 7.

Figure 7 shows the change in crystallinity degree (X_c) as a function of the applied strain for the NR and the various CNF/NR samples filled with different amounts of CNFs. It was obvious that the X_c of all samples increased with the increasing strain, indicating that the crystallization of the NR and the nanocomposites was caused by tensile deformations. The X_c values of the NR and the CNF1/NR and CNF3/NR samples were initially seen at a strain of approximately 300%. This implied that the onset strains of the strain-induced crystallization in these three samples were similar. The variation in X_c upon stretching and at the same strain levels was also comparable among these samples, suggesting that the crystallization process that took place in the NR was similar to those of the CNF1/NR and CNF3/NR samples, even though the latter contained CNF as reinforcement. Therefore, the characteristic patterns of the stress–strain curves of the NR and the CNF/NR samples containing 1 and 3 phr CNF were very similar, as discussed earlier (Figure 4). On the other hand, the CNF5/NR sample showed a dramatic decrease in strain value (175%) at the onset of crystallization and a progression of crystallization with increasing the applied strain from 175% to 225%. No further deformation and crystallization developed because the sample had failed (~300% strain). It was proposed that the immobilized NR chains at the surface of the aggregated CNF contributed to the local stress concentration and the strain-induced crystallization behavior in the CNF5/NR sample [37], and thereby, they significantly increased the moduli at different strains (Figure 4 and Table 3). As the CNF5/NR sample was strained up to approximately 300%, the amount of local stress concentration was significantly high, which resulted in the quick failure of the CNF5/NR sample.

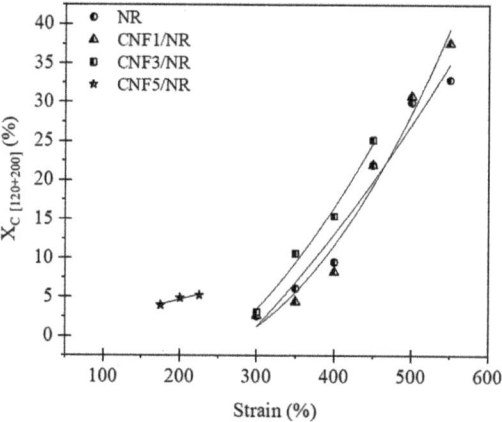

Figure 7. Degree of crystallinity (X_c) as a function of the applied strain for the NR and the CNF/NR nanocomposites.

The most surprising aspect of the above observations was that the accelerated strain-induced crystallization was not detected in the CNF/NR samples with comparatively lower CNF contents (1–3 phr), and their degrees of crystallization upon stretching did not depend on their CNF content, though the tensile properties showed different variations. Thus, further investigations to reveal the influence of CNF concentration on the nanocomposite structure and their reinforcement effects through DMA analysis, bound rubber formation, and gel content measurement were performed.

3.4. Dynamic Mechanical Properties of the CNF/NR Nanocomposites

Figure 8 shows the correlation between the storage modulus (E′) and the damping factor (Tan δ) as a function of the temperature for the NR and the CNF/NR samples containing 1, 3, and 5 phr CNF. Generally, the addition of CNF significantly enhanced the E′ in a rubbery state and decreased the tan δ, reflecting the influence of CNF on the reinforcement of the NR. The values of E′ at 25 °C, the tan $δ_{max}$ of the NR (the height of the tan δ peak), and the glass transition temperature (T_g) of the NR and the CNF/NR samples are also listed for comparison in Table 4.

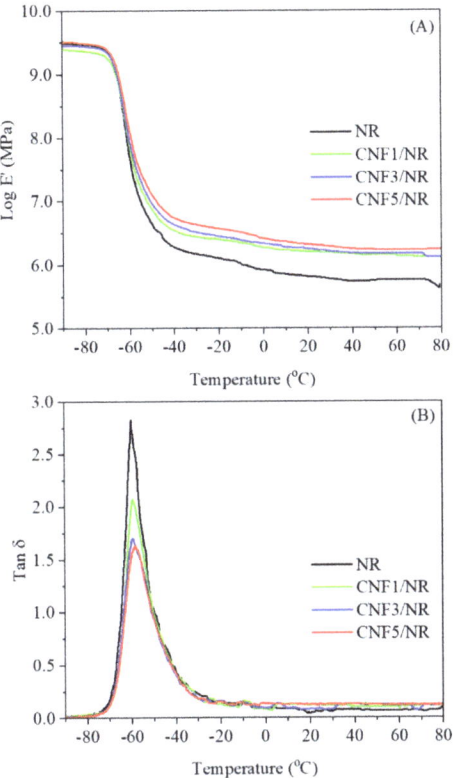

Figure 8. Variations in (**A**) storage modulus (log E′) and (**B**) tan δ as a function of temperature for the NR and the CNF/NR nanocomposites.

As can be seen from Figure 8A and Table 4, the inclusion of CNF improved the E′ of the NR at 25 °C, and the magnitude of the increment increased with increasing CNF contents. This resulted from the rubber being more rigid as a result of the higher filling levels of CNF [37]. The rigidity of the pristine CNF could impede the movement of the chain segment of the NR through the filler–rubber interfacial actions [14,54]. Thus, in our study, it was likely that that the improvement in the E′ at 25 °C could mainly attributed to the physical interaction or entanglement between the pristine CNFs and the NR chains in the CNF/NR samples. Moreover, it was seen that the pristine CNFs reduced the tan $δ_{max}$ of the NR depending on the amount of CNF. The reduction in the tan $δ_{max}$ with the increasing CNF contents indicated the higher restricted movement of the NR chain segments at the interface of the CNF and the NR [37,55–57]. The glass transition temperature illustrated by the tan δ peak temperature of the NR (−60.1 °C) was systematically shifted to higher temperature as the CNF content was increased. When the NR chains adhered to the surfaces of CNFs via interfacial interactions, as discussed previously, a higher energy was required to achieve the same level of chain segment movement in the CNF/NR samples than in the neat NR. Similar results have been found in CNF-reinforced polyethylene oxide

(PEO) [14] and styrene-butadiene (SBR) nanocomposites [22]. Therefore, the lowering of the tan δ_{max} and the increment of the T_g with the incorporated CNF further substantiated the interfacial interaction between the nanofibers and the NR at the interface of the CNF/NR samples. Owing to the results demonstrated by the DMA technique, the CNF-reinforced NR nanocomposites showed better dynamic properties than the NR due to the interfacial reinforcement in the CNF/NR nanocomposites.

Table 4. Storage moduli (log E') at 25 °C, maximum tan δ peaks (tan δ_{max}), and glass transition temperatures (T_g) of the NR and the CNF/NR nanocomposites.

Samples	Log E' at 25 °C (MPa)	Tan δ $_{max}$	T_g (°C)
NR	5.79	2.85	−60.1
CNF1/NR	6.18	2.55	−59.2
CNF3/NR	6.21	1.71	−58.9
CNF5/NR	6.30	1.64	−58.1

3.5. Bound Rubber and Gel Content of the CNF/NR Nanocomposites

Table 5 shows the effect of CNF concentration on bound rubber and gel content formation. The bound rubber is a measure of the elastomer adsorption onto the filler surface [40,58], while the gel content reveals information about the chemical crosslink density in the NR network [59].

It was seen that bound rubber was not detected in the CNF1/NR and CNF3/NR samples. This implied that the NR molecules did not interact chemically with the reinforcing nanofibers and they could be readily removed from the unreacted CNF1/NR and CNF3/NR compounds after being immersed in toluene for a given period of time. On the other hand, the CNF5/NR sample in which the nanofibers were mostly aggregated (Figures 2C and 3B) showed a significant bound rubber content of approximately 9.06%. It was shown that the non-extractable NR observed in the CNF5/NR sample was formed by the insertion of NR chains into the aggregated CNFs. These inserted NR chains led to a number of immobilized NR chains and a significant local stress concentration, which had a large influence on the tensile properties and crystalline formation in the CNF5/NR sample, as discussed earlier in our previous work [37]. These results suggested that the NR chains were not inserted into the CNF bundles of the CNF1/NR and CNF3/NR samples.

Table 5. Bound rubber contents and gel contents of the NR and the CNF/NR nanocomposites.

Samples	Bound Rubber Content (%)	Gel Content (%)
NR	N/A	80.12 ± 0.11
CNF1/NR	N/A	80.24 ± 0.32
CNF3/NR	N/A	80.38 ± 0.08
CNF5/NR	9.06 ± 1.18	80.43 ± 0.73

Considering the data of gel content measurements in Table 5, it was clearly seen that each gel content of the NR and the CNF/NR samples filled with 1, 3, and 5 CNF phr was not different, meaning that the incorporation of CNF did not change the degree of chemical crosslinking in the NR by the peroxide vulcanization. Therefore, the changes in the mechanical properties of the CNF/NR nanocomposites were largely governed by the CNFs' dispersibility and their microstructure formations. Unlike the CNF5/NR sample, the NR nanocomposites reinforced with relatively lower CNF contents, particularly the CNF1/NR sample, showed high levels of improvement in the tensile strength of the NR, with good flexibility, even though the acceleration of the strain-induced crystallization by the CNF incorporation and the bound rubber in this sample were not observed. These results may interestingly suggest a different reinforcement mechanism of the CNFs in the NR nanocomposites with relatively low (1 phr) and high (5 phr) CNF contents.

3.6. Model of Reinforcement Mechanism

Based on the observations mentioned above, we proposed a mechanistic model explaining the reinforcement of the CNF/NR nanocomposites with low CNF contents, as depicted in Figure 9. The focus was on the NR nanocomposites filled with 1 phr CNF, as the reinforcement mechanism of the NR nanocomposites containing high CNF loading (5 phr) was well-described in our earlier publication [37]. It should be noted here that the CNF1/NR sample exhibited separate nanofibers and small bundles of a nano-sized scale (Figure 2(A)), implying that the surface area of the CNF for the interaction with the NR in this sample was relatively high.

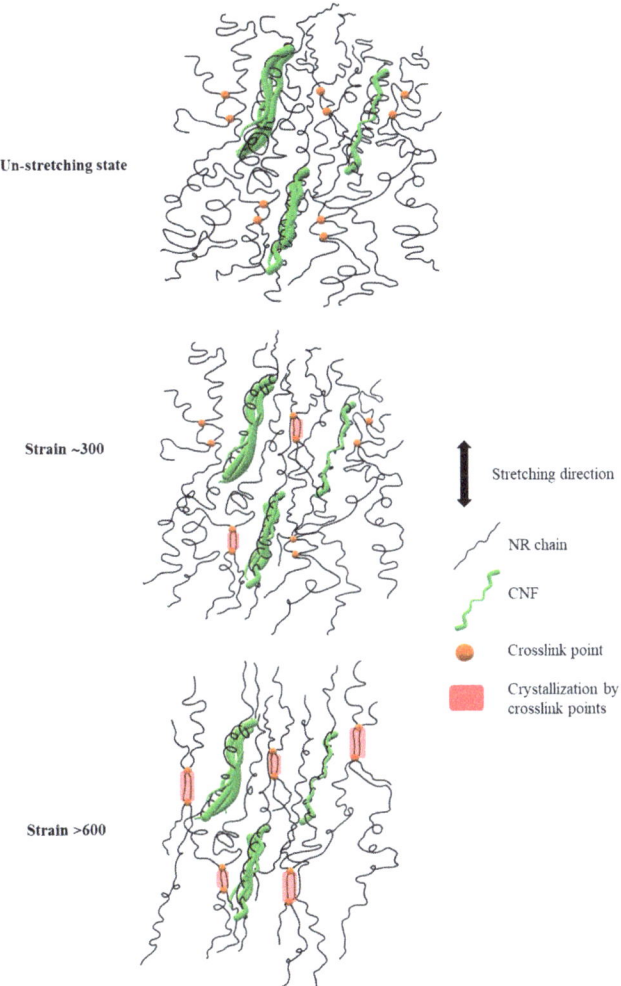

Figure 9. Proposed model for the strain-induced crystallization mechanism of the CNF/NR nanocomposite with 1 phr CNF.

In an unstretched state, the long chains of the NR molecules would most likely interact with the single CNF and bundled CNFs through physical entanglement, as shown in Figure 9. Upon tensile stretching, the NR network was deformed, whereas the stiff CNF was not deformed. Theoretically, in a classical model of short-fiber composites, the reinforcement of rigid fiber occurs through the transfer of tensile stress from the matrix to the fiber by means of interfacial shear stress [60,61]. By this mechanism, the tensile stress in the NR was built up by the transfer of the shear stress from the NR to the CNF across the CNF/NR interface. Therefore, the CNF in the CNF1/NR would contribute to carry more tensile stress upon deformation, owing to relatively large interfacial area for the stress transfer from the CNF to the NR. However, the nano-sized CNFs prevented the NR chains from aligning and crystallizing because of the lack of stress concentration at the interface between the CNF and the NR chains in the CNF1/NR sample. As a result, enhancement of the strain-induced crystallization caused by the nanofiber was not observed in the NR nanocomposites containing small amounts of CNF. On the other hand, the presence of the local stress concentration at the interface between the aggregated CNF and the NR caused by the mutually entangled structure of the CNF aggregates and the NR chains, as demonstrated by the bound rubber measurements (Table 5), was the main factor for the acceleration of the strain-induced crystallization at the low strain in the CNF5/NR sample (Figures 5–7). When the tensile deformation reached a strain of 300%, crystallization was observed in the CNF1/NR sample, which was due to the strain-induced crystallization by the short NR chains around the dense crosslinking points. The crystallization of the NR matrix progressively increased with the applied strains because the strain caused the orientation and alignment of the NR chains. At a large tensile deformation (>600% strain), the interfacial shear stress at the interface region between the CNF and the NR was significantly high, leading to a large increase in load bearing in the CNF and, thus, a significant enhancement of the NR reinforcement. The breaking strain of CNF1/NR was also comparable to the neat NR owing to the stretching without debonding at the CNF1/NR interface by the interaction through the physical entanglement.

4. Conclusions

We found the reinforcement behaviors of small amounts of chemically unmodified cellulose nanofiber (CNF) in eco-friendly natural rubber (NR). The tensile modulus and the storage modulus of the CNF-reinforced NR increased with increasing CNF concentrations. The NR nanocomposite with 1 phr CNF showed the maximum tensile strength, which was an approximate 122% increase over that of the NR, together with a large strain at break (757%). The CNF in amounts of 1–3 phr were well-dispersed in the NR matrixes, without microscaled aggregation, leading to significant enhancements in stress upturn during stretching. However, it was observed that the addition of CNF at low concentrations (1–3 phr) did not participate in the strain-induced crystallization process of the NR, and their degree of crystallinity was not dependent on the CNF filling contents. Therefore, the high tensile strength for the 1 phr CNF-filled NR nanocomposite was based on the increase in the dispersion state of the CNF, which, in turn, increased the CNF/NR interaction for the effective stress transfer capability from the NR to the embedded CNF. On the other hand, at a filling content of 5 phr CNF, the nanofibers were aggregated, resulting in a local stress concentration and accelerated strain-induced crystallization. This contributed to a high tensile modulus but low tensile strength and strain at break. Thus, this study revealed that the effects of CNF on the mechanistic reinforcement of NR varied depending on the different CNF filling concentrations.

Author Contributions: Conceptualization, B.W., H.S. and N.L.; methodology, formal analysis, and investigation, B.W. and N.L.; validation, B.T., H.S., A.M. and T.S.; writing—original draft preparation, B.W., A.M. and N.L.; writing—review and editing and supervision, H.S., A.M., T.S. and N.L.; project administration, N.L.; funding acquisition, N.L. All authors have read and agreed to the published version of the manuscript.

Funding: This research was funded by The Thailand Research Fund (TRF) through The Royal Golden Jubilee Ph.D. Program (grant no. PHD/0156/2560), The Thailand Science Research and Innovation (TSRI), and The National Research Council of Thailand (NRCT).

Institutional Review Board Statement: Not applicable.

Informed Consent Statement: Not applicable.

Data Availability Statement: The data presented in this study are available on request from the corresponding author.

Acknowledgments: The authors would like to acknowledge The Thailand Research Fund (TRF) through The Royal Golden Jubilee Ph.D. Program, The Thailand Science Research and Innovation (TSRI), and The National Research Council of Thailand (NRCT).

Conflicts of Interest: The authors declare no conflict of interest.

References

1. Toki, S.; Sics, I.; Ran, S.; Liu, L.; Hsiao, B.S. Molecular orientation and structural development in vulcanized polyisoprene rubbers during uniaxial deformation by in situ synchrotron X-ray diffraction. *Polymer* **2003**, *44*, 6003–6011. [CrossRef]
2. Trabelsi, S.; Albouy, P.-A.; Rault, J. Crystallization and Melting Processes in Vulcanized Stretched Natural Rubber. *Macromolecules* **2003**, *36*, 7624–7639. [CrossRef]
3. Masa, A.; Iimori, S.; Saito, R.; Saito, H.; Sakai, T.; Kaesaman, A.; Lopattananon, N. Strain-induced crystallization behavior of phenolic resin crosslinked natural rubber/clay nanocomposites. *J. Appl. Polym. Sci.* **2015**, *132*, 42580. [CrossRef]
4. Masa, A.; Saito, R.; Saito, H.; Sakai, T.; Kaesaman, A.; Lopattananon, N. Phenolic resin-crosslinked natural rubber/clay nanocomposites: Influence of clay loading and interfacial adhesion on strain-induced crystallization behavior. *J. Appl. Polym. Sci.* **2016**, *133*, 43214. [CrossRef]
5. Masa, A.; Saito, H.; Sakai, T.; Kaesaman, A.; Lopattananon, N. Morphological evolution and mechanical property enhancement of natural rubber/polypropylene blend through compatibilization by nanoclay. *J. Appl. Polym. Sci.* **2017**, *134*, 44574. [CrossRef]
6. Nie, Y.; Qu, L.; Huang, G.; Wang, X.; Weng, G.; Wu, J. Homogenization of Natural Rubber Network Induced by Nanoclay. *J. Appl. Polym. Sci.* **2014**, *131*, 40324. [CrossRef]
7. Hsieh, Y.-C.; Yano, H.; Nogi, M.; Eichhorn, S.J. An estimation of the Young's modulus of bacterial cellulose filaments. *Cellulose* **2008**, *15*, 507–513. [CrossRef]
8. Rusli, R.; Eichhorn, S.J. Determination of the stiffness of cellulose nanowhiskers and the fiber-matrix interface in a nanocomposite using Raman spectroscopy. *Appl. Phys. Lett.* **2008**, *93*, 033111. [CrossRef]
9. Matsuo, M.; Sawatari, C.; Iwai, Y.; Ozaki, F. Effect of Orientation Distribution and Crystallinity on the Measurement by X-ray Diffraction of the Crystal Lattice Moduli of Cellulose I and II. *Macromolecules* **1990**, *23*, 3266–3275. [CrossRef]
10. Sakurada, I.; Nukushina, Y.; Ito, T. Experimental determination of the elastic modulus of crystalline regions in oriented polymers. *J. Polym. Sci.* **1962**, *57*, 651–660. [CrossRef]
11. Šturcová, A.; Davies, G.R.; Eichhorn, S.J. Elastic Modulus and Stress-Transfer Properties of Tunicate Cellulose Whiskers. *Biomacromolecules* **2005**, *6*, 1055–1061. [CrossRef]
12. Saito, T.; Kuramae, R.; Wohlert, J.; Berglund, L.-A.; Isogai, A. An Ultrastrong Nanofibrillar Biomaterial: The Strength of Single Cellulose Nanofibrils Revealed via Sonication-Induced Fragmentation. *Biomacromolecules* **2013**, *14*, 248–253. [CrossRef]
13. Nechyporchuk, O.; Belgacem, M.N.; Bras, J. Production of cellulose nanofibrils: A review of recent advances. *Ind. Crop. Prod.* **2016**, *93*, 2–25. [CrossRef]
14. Xu, X.; Liu, F.; Jiang, L.; Zhu, J.Y.; Haagenson, D.; Wiesenborn, D.P. Cellulose Nanocrystals vs. Cellulose Nanofibrils: A Comparative Study on Their Microstructures and Effects as Polymer Reinforcing Agents. *ACS Appl. Mater. Interfaces* **2013**, *5*, 2999–3009. [CrossRef] [PubMed]
15. Wang, G.; Yang, X.; Wang, W. Reinforcing Linear Low-Density Polyethylene with Surfactant-Treated Microfibrillated Cellulose. *Polymers* **2019**, *11*, 441. [CrossRef]
16. Yasim-Anuar, T.A.T.; Arin, H.; Norrrahim, M.N.F.; Hassan, M.A.; Andou, Y.; Tsukegi, T.; Nishida, H. Well-Dispersed Cellulose Nanofiber in Low Density Polyethylene Nanocomposite by Liquid-Assisted Extrusion. *Polymers* **2020**, *12*, 927. [CrossRef] [PubMed]
17. Siqueira, G.; Bras, J.; Dufresne, A. Cellulose whiskers versus microfibrils: Influence of the nature of the nanoparticle and its surface functionalization on the thermal and mechanical properties of nanocomposites. *Biomacromolecules* **2009**, *10*, 425–432. [CrossRef] [PubMed]
18. Herrera, N.; Mathew, A.P.; Oksman, K. Plasticized polylactic acid/cellulose nanocomposites prepared using melt-extrusion and liquid feeding: Mechanical, thermal and optical properties. *Compos. Sci. Technol.* **2015**, *106*, 149–155. [CrossRef]
19. Lo Re, G.; Engström, J.; Wu, Q.; Malmström, E.; Gedde, U.W.; Olsson, R.T.; Berglund, L. Improved Cellulose Nanofibril Dispersion in Melt-Processed Polycaprolactone Nanocomposites by a Latex-Mediated Interphase and Wet Feeding as LDPE Alternative. *ACS Appl. Nano Mater.* **2018**, *1*, 2669–2677. [CrossRef]

20. Fumagalli, M.; Berriot, J.; Gaudemaris, B.; Veyland, A.; Putaux, J.-L.; Molina-Boisseau, S.; Heux, L. Rubber Materials from Elastomers and Nanocellulose Powders: Filler Dispersion and Mechanical Reinforcement. *Soft Matter* **2018**, *14*, 2638–2648. [CrossRef] [PubMed]
21. Fukui, S.; Ito, T.; Saito, T.; Noguchi, T.; Isogai, A. Surface-hydrophobized TEMPO-nanocellulose/rubber composite films prepared in heterogeneous and homogeneous systems. *Cellulose* **2019**, *26*, 463–473. [CrossRef]
22. Sinclair, A.; Zhou, X.; Tangpong, S.; Bajwa, D.S.; Quadir, M.; Jiang, L. High-Performance Styrene-Butadiene Rubber Nanocomposites Reinforced by Surface-Modified Cellulose Nanofibers. *ACS Omega* **2019**, *4*, 13189–13199. [CrossRef] [PubMed]
23. Balachandrakurup, V.; Gopalakrishnan, J. Enhanced performance of cellulose nanofibre reinforced styrene butadiene rubber nanocomposites modified with epoxidised natural rubber. *Ind. Crop. Prod.* **2022**, *183*, 114935. [CrossRef]
24. Abraham, E.; Deepa, B.; Pothan, L.A.; John, M.; Narine, S.S.; Thomas, S.; Anandjiwala, R. Physicomechanical properties of nanocomposites based on cellulose nanofibre and natural rubber latex. *Cellulose* **2013**, *20*, 417–427. [CrossRef]
25. Thomas, M.G.; Abraham, E.; Jyotishkumar, P.; Maria, H.J.; Pothan, L.A.; Thomas, S. Nanocelluloses from jute fibres and their nanocomposites with natural rubber: Preparation and characterization. *Int. J. Biol. Macromol.* **2015**, *81*, 768–777. [CrossRef]
26. Kumagai, A.; Tajima, N.; Iwamoto, S.; Morimoto, T.; Nagatani, A.; Okazaki, T.; Endo, T. Properties of natural rubber reinforced with cellulose nanofibers based on fiber diameter distribution as estimated by differential centrifugal sedimentation. *Int. J. Biol. Macromol.* **2019**, *121*, 989–995. [CrossRef]
27. Dominic, M.; Joseph, R.; Begum, P.M.S.; Joseph, M.; Padmanabhan, D.; Morris, L.A.; Kumar, A.S.; Formela, K. Cellulose Nanofibers Isolated from the Cuscuta Reflexa Plant as a Green Reinforcement of Natural Rubber. *Polymers* **2020**, *12*, 814. [CrossRef]
28. Kato, H.; Nakatsubo, F.; Abe, K.; Yano, H. Crosslinking via sulfur vulcanization of natural rubber and cellulose nanofibers incorporating unsaturated fatty acids. *RSC Adv.* **2015**, *5*, 29814–29819. [CrossRef]
29. Chenal, J.-M.; Gauthier, C.; Chazeau, L.; Guy, L.; Bomal, Y. Parameters governing strain induced crystallization in filled natural rubber. *Polymer* **2007**, *48*, 6893–6901. [CrossRef]
30. Laghmach, R.; Biben, T.; Chazeau, L.; Chenal, J.M.; Munch, E.; Gauthier, C. Strain-induced crystallization in natural rubber: A model for the microstructural evolution. In *Constitutive Models for Rubber VIII*, 1st ed.; Gil-Negrete, N., Alonso, A., Eds.; CRC Press: London, UK, 2013.
31. Candau, N.; Laghmach, R.; Chazeau, L.; Chenal, J.-M.; Gauthier, C.; Biben, T.; Munch, E. Strain-Induced Crystallization of Natural Rubber and Cross-Link Densities Heterogeneities. *Macromolecules* **2014**, *47*, 5815–5824. [CrossRef]
32. Masa, A.; Hayeemasae, N.; Soontaranon, S.; Mohd Pisal, M.H.; Mohamad Rasidi, M.S. Effect of Stretching Rate on Tensile Response and Crystallization Behavior of Crosslinked Natural Rubber. *Malays. J. Fundam. Appl. Sci.* **2021**, *17*, 217–225. [CrossRef]
33. Fu, X.; Huang, G.; Xie, Z.; Xing, W. New insights into reinforcement mechanism of nanoclay-filled isoprene rubber during uniaxial deformation by in situ synchrotron X-ray diffraction. *RSC Adv.* **2015**, *5*, 25171–25182. [CrossRef]
34. Ozbas, B.; Toki, S.; Hsiao, B.S.; Chu, B.; Register, R.A.; Aksay, I.A.; Prud'homme, R.K.; Adamson, D.H. Strain-Induced Crystallization and Mechanical Properties of Functionalized Graphene Sheet-Filled Natural Rubber. *J. Polym. Sci. B Polym. Phys.* **2012**, *50*, 718–723. [CrossRef]
35. Beurrot-Borgarino, S.; Huneau, B.; Verron, E.; Rublon, P. Strain-induced crystallization of carbon black-filled natural rubber during fatigue measured by in situ synchrotron X-ray diffraction. *Int. J. Fatigue* **2013**, *47*, 1–7. [CrossRef]
36. Weng, G.; Huang, G.; Qu, L.; Nie, Y.; Wu, J. Large-Scale Orientation in a Vulcanized Stretched Natural Rubber Network: Proved by In Situ Synchrotron X-ray Diffraction Characterization. *J. Phys. Chem. B* **2010**, *114*, 7179–7188. [CrossRef] [PubMed]
37. Wongvasana, B.; Thongnuanchan, B.; Masa, A.; Saito, H.; Sakai, T.; Lopattananon, N. Comparative Structure–Property Relationship between Nanoclay and Cellulose Nanofiber Reinforced Natural Rubber Nanocomposites. *Polymers* **2022**, *14*, 3747. [CrossRef]
38. Arroyo, M.; Lo'pez-Manchado, M.A.; Herrero, B. Organo-montmorillonite as substitute of carbon black in natural rubber compounds. *Polymer* **2003**, *44*, 2447–2453. [CrossRef]
39. Qu, L.; Huang, G.; Liu, Z.; Zhang, P.; Weng, G.; Nie, Y. Remarkable reinforcement of natural rubber by deformation-induced crystallization in the presence of organophilic montmorillonite. *Acta Mater.* **2009**, *57*, 5053–5060. [CrossRef]
40. Dannenberg, E.M. Bound Rubber and Carbon Black Reinforcement. *Rubber Chem. Technol.* **1986**, *59*, 512–524. [CrossRef]
41. Lopattananon, N.; Tanglakwaraskul, S.; Kaesaman, A.; Seadan, M.; Sakai, T. Effect of Nanoclay Addition on Morphology and Elastomeric Properties of Dynamically Vulcanized Natural Rubber/Polypropylene Nanocomposites. *Int. Polym. Process.* **2014**, *29*, 332–341. [CrossRef]
42. Siró, I.; Plackett, D. Microfibrillated cellulose and new nanocomposite materials: A review. *Cellulose* **2010**, *17*, 459–494. [CrossRef]
43. Mishra, R.K.; Sabu, A.; Tiwari, S.K. Materials chemistry and the futurist eco-friendly applications of nanocellulose: Status and prospect. *J. Saudi Chem. Soc.* **2018**, *22*, 949–978. [CrossRef]
44. Kargarzadeh, H.; Mariano, M.; Gopakumar, D.; Ahmad, I.; Thomas, S.; Dufresne, A.; Huang, J.; Lin, N. Advances in cellulose nanomaterials. *Cellulose* **2018**, *25*, 2151–2189. [CrossRef]
45. Fiorote, J.A.; Freire, A.P.; Rodrigues, D.D.S.; Martins, M.A.; Andreani, L.; Valadares, L.F. Preparation of composites from natural rubber and oil palm empty fruit bunch cellulose: Effect of cellulose morphology on properties. *Bioresources* **2019**, *14*, 3168–3181. [CrossRef]
46. Zhang, C.; Zhai, T.; Sabo, R.; Clemons, C.; Dan, Y.; Turng, L.-S. Reinforcing Natural Rubber with Cellulose Nanofibrils Extracted from Bleached Eucalyptus Kraft Pulp. *J. Biobased Mater. Bioenergy* **2014**, *8*, 317–324. [CrossRef]
47. Thomas, S.; Stephen, R. *Rubber Nanocomposites: Preparation, Properties and Applications*, 1st ed.; Wiley: Singapore, 2010; pp. 291–330.

48. Karino, T.; Ikeda, Y.; Yasuda, Y.; Kohjiya, S.; Shibayama, M. Nonuniformity in Natural Rubber As Revealed by Small-Angle Neutron Scattering, Small-Angle X-ray Scattering, and Atomic Force Microscopy. *Biomacromolecules* **2007**, *8*, 693–699. [CrossRef]
49. Dalmas, F.; Chazeau, L.; Gauthier, C.; Cavaillé, J.-Y.; Dendievel, R. Large deformation mechanical behavior of flexible nanofiber filled polymer nanocomposites. *Polymer* **2006**, *47*, 2802–2812. [CrossRef]
50. Kristo, E.; Biliaderis, C.G. Physical properties of starch nanocrystal-reinforced pullulan films. *Carbohydr. Polym.* **2007**, *68*, 146–158. [CrossRef]
51. Georgopoulos, S.; Tarantili, P.A.; Avgerinos, E.; Andreopoulos, A.G.; Koukios, E.G. Thermoplastic polymers reinforced with fibrous agricultural residues. *Polym. Degrad. Stab.* **2005**, *90*, 303–312. [CrossRef]
52. Beurrot-Borgarino, S.; Huneau, B.; Verron, E.; Thiaudière, D.; Mocuta, C.; Zozulya, A. Characteristics of Strain-Induced Crystallization in Natural Rubber During Fatigue Testing: In situ Wide-Angle X-ray Diffraction Measurements Using Synchrotron Radiation. *Rubb. Chem. Technol.* **2014**, *87*, 184–196. [CrossRef]
53. French, A.D. Idealized powder diffraction patterns for cellulose polymorphs. *Cellulose* **2014**, *21*, 885–896. [CrossRef]
54. Peng, S.; Iroh, J.O. Dependence of the Dynamic Mechanical Properties and Structure of Polyurethane-Clay Nanocomposites on the Weight Fraction of Clay. *Compos. Sci.* **2022**, *6*, 173. [CrossRef]
55. Visakh, P.M.; Thomas, S.; Oksman, K.; Mathew, A.P. Effect of cellulose nanofibers isolated from bamboo pulp residue on vulcanized natural rubber. *BioRes* **2012**, *7*, 2156–2168. [CrossRef]
56. Ikeda, Y.; Phakkeeree, T.; Junkong, P.; Yokohama, H.; Phinyocheep, P.; Kitano, R.; Kato, A. Reinforcing biofiller "Lignin" for high performance green natural rubber nanocomposites. *RSC Adv.* **2017**, *7*, 5222–5231. [CrossRef]
57. Kumar, V.; Alam, M.N.; Manikkavel, A.; Song, M.; Lee, D.-J.; Park, S.-S. Silicone Rubber Composites Reinforced by Carbon Nanofillers and Their Hybrids for Various Applications: A Review. *Polymers* **2021**, *13*, 2322. [CrossRef] [PubMed]
58. Robertson, C.G.; Hardman, N.J. Nature of Carbon Black Reinforcement of Rubber: Perspective on the Original Polymer Nanocomposite. *Polymers* **2021**, *13*, 538. [CrossRef]
59. Huang, Y.; Gohs, U.; Müller, M.T.; Zschech, C.; Wießner, S. Evaluation of Electron Induced Crosslinking of Masticated Natural Rubber at Different Temperatures. *Polymers* **2019**, *11*, 1279. [CrossRef]
60. Cox, H.L. The elasticity and strength of paper and other fibrous materials. *J. Appl. Phys.* **1952**, *3*, 72–79. [CrossRef]
61. Hull, D.; Clyne, T.W. *An Introduction to Composite Materials*, 2nd ed.; Cambridge University Press: New York, NY, USA, 1996.

Disclaimer/Publisher's Note: The statements, opinions and data contained in all publications are solely those of the individual author(s) and contributor(s) and not of MDPI and/or the editor(s). MDPI and/or the editor(s) disclaim responsibility for any injury to people or property resulting from any ideas, methods, instructions or products referred to in the content.

Review

Recent Advances in Lignocellulose-Based Monomers and Their Polymerization

Fuyun Pei [1], Lijuan Liu [1], Huie Zhu [2,*] and Haixin Guo [3,*]

[1] CECEP Techand Ecology & Environment Co., Ltd., Shenzhen 518004, China
[2] Graduate School of Engineering, Tohoku University, Sendai 980-8579, Japan
[3] Agro-Environmental Protection Institute, Ministry of Agriculture and Rural Affairs, Tianjin 300191, China
* Correspondence: zhuhuie@tohoku.ac.jp (H.Z.); haixin_g@126.com (H.G.)

Abstract: Replacing fossil-based polymers with renewable bio-based polymers is one of the most promising ways to solve the environmental issues and climate change we human beings are facing. The production of new lignocellulose-based polymers involves five steps, including (1) fractionation of lignocellulose into cellulose, hemicellulose, and lignin; (2) depolymerization of the fractionated cellulose, hemicellulose, and lignin into carbohydrates and aromatic compounds; (3) catalytic or thermal conversion of the depolymerized carbohydrates and aromatic compounds to platform chemicals; (4) further conversion of the platform chemicals to the desired bio-based monomers; (5) polymerization of the above monomers to bio-based polymers by suitable polymerization methods. This review article will focus on the progress of bio-based monomers derived from lignocellulose, in particular the preparation of bio-based monomers from 5-hydroxymethylfurfural (5-HMF) and vanillin, and their polymerization methods. The latest research progress and application scenarios of related bio-based polymeric materials will be also discussed, as well as future trends in bio-based polymers.

Keywords: cellulose; HMF; vanillin; bio-based polymers; biopolyester

Citation: Pei, F.; Liu, L.; Zhu, H.; Guo, H. Recent Advances in Lignocellulose-Based Monomers and Their Polymerization. *Polymers* **2023**, *15*, 829. https://doi.org/10.3390/polym15040829

Academic Editor: Jean-Marie Raquez

Received: 19 December 2022
Revised: 26 January 2023
Accepted: 3 February 2023
Published: 7 February 2023

Copyright: © 2023 by the authors. Licensee MDPI, Basel, Switzerland. This article is an open access article distributed under the terms and conditions of the Creative Commons Attribution (CC BY) license (https://creativecommons.org/licenses/by/4.0/).

1. Introduction

Since the industrial revolution, large-scale use of fossil resources has led to serious environmental problems such as the greenhouse effect, atmospheric pollution, soil, and marine pollution. Among these, cheap, lightweight, and versatile polymers are essential for all aspects of daily life, but most polymers are still highly dependent on chemicals derived from fossil fuels for their production [1]. Not only is the polymer industry facing a depletion of raw materials, but the production process also has an irreversible negative impact on the environment. For example, studies have shown that the equivalent CO_2 emissions per kilogram of polypropylene and polyethylene produced are 1.34 and 1.48 kg, respectively [2]. In addition to this, only 14% of plastic packaging is collected for recycling after use and a significant amount escapes into the environment [3]. The resulting white pollution and greenhouse effect also pose an unprecedented challenge for industrial development and materials research and development. For this issue, mass production and applications of bio-based polymer materials through the efficient use of green and renewable plant resources are the most promising means for solving the problems [4]. The research and development of bio-based polymer materials have been well developed in recent years in terms of bio-based monomer preparation with high efficiency, novel polymerization methods of bio-based monomers, and the preparation of novel catalysts. Depending on the synthetic methods, bio-based polymers include two main categories: natural polymers and bio-based synthetic polymers from bio-based monomers [5]. Natural polymers are mainly represented by cellulose and lignin, etc., which are widely found in plants. For the latter category, the mainly used raw materials for bio-based synthetic polymers are lignocellulose, cellulose, hemicellulose, and lignin from biomass, which can be converted to polymerizable monomers through various chemical processes.

As the most abundant renewable resource on earth, lignocellulose is widely found in hardwoods (e.g., aspen), softwoods (e.g., pine needles), agricultural waste (e.g., straw), and herbaceous plants [6–9]. Its main components are cellulose, hemicellulose, and lignin (Figure 1) [10–14]. Depending on the type of biomass, the content of the individual components varies [4]. In wood raw materials such as hardwoods and softwoods, the cellulose content is high, up to more than 50%, while in agricultural waste such as straw, the cellulose content varies from 28% to 40%. Due to the importance of wood raw materials in structural materials and in order to reduce the diversion of food resources from biomass development, the conversion of non-edible agricultural waste into high-value-added chemicals is one of the most effective tools that have been investigated in recent years. The production of new lignocellulose-based polymers includes five steps: (1) fractionation of lignocellulose into cellulose, hemicellulose, and lignin; (2) depolymerization of the fractionated cellulose, hemicellulose, and lignin into carbohydrates and aromatic compounds; (3) catalytic or thermal conversion of the depolymerized carbohydrates and aromatic compounds to platform chemicals; (4) further conversion of the platform chemicals to the desired bio-based monomers; (5) polymerization of the above monomers to bio-based polymers by suitable polymerization methods [15,16]. This review article will focus on the progress of bio-based monomers derived from lignocellulose, in particular the preparation of bio-based monomers from 5-hydroxymethylfurfural (5-HMF) and vanillin, and their polymerization methods (Figure 2). The latest research progress and application scenarios of related bio-based polymeric materials will be also discussed, as well as future trends in bio-based polymers.

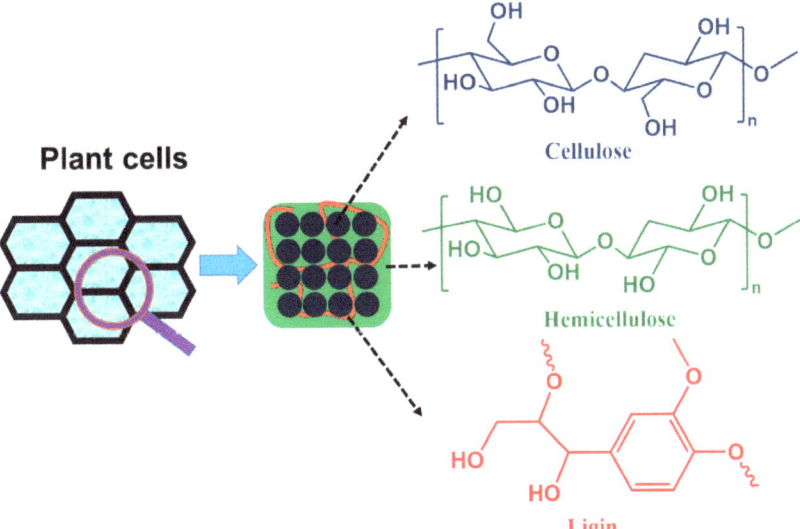

Figure 1. The three most common components contained in plant cell walls: cellulose, hemicellulose, and lignin, collectively known as lignocellulose.

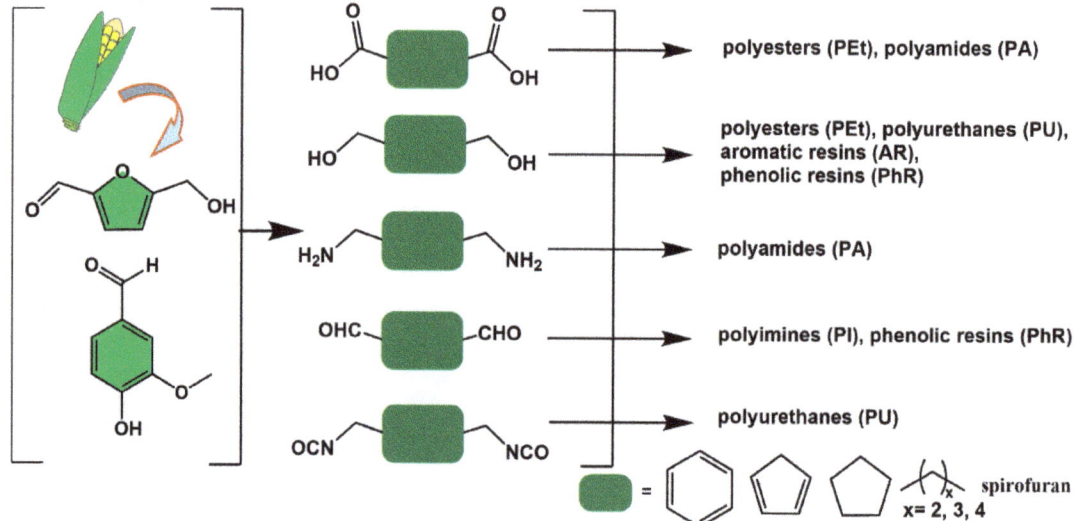

Figure 2. The preparation of bio-based polymers from biomass resources.

2. Bio-Based Monomers from Lignocellulose

2.1. C5/C6 Sugar Platform Chemicals Based on Cellulose/Hemicellulose

Cellulose is a homopolymer of glucose and is mainly found in the cell walls of plants. The hydrolysis products of cellulose are mainly the six-membered sugar glucose (C6 sugars), and their chemical structures are shown in Figure 3a. The hydrolysis products of hemicellulose include not only glucose but also other five-membered sugars (C5: xylose, arabinose) and six-membered sugars (C6: mannose, galactose, rhamnose). These sugar intermediates can be used to prepare various first-generation furan derivatives, with furfural and 5-hydroxymethylfurfural (5-HMF) being the most common platform chemicals used to prepare polymerizable monomers (Figure 3b).

Most of the bio-based monomers are prepared based on 5-HMF, furfural, and furfuryl alcohol (FA). Of these, 5-HMF is one of the twelve core chemicals selected by the U.S. Department of Energy Biomass Program (2004) and is the most versatile intermediate. 5-HMF can be prepared highly selectively from hexoses such as fructose, and all the six carbon atoms initially in the C6 sugars are retained in the molecular structure of 5-HMF. In 2009, Binder et al. reported that a solvent of N, N-dimethylacetamide (DMAc) containing lithium chloride (LiCl)) (DMAc/LiCl) can be used to efficiently prepare 5-HMF in a single-step reaction at low temperatures ($\leq$140 °C) [17]. Reaction mechanism studies have shown that the loosely bound halide ions in the DMAc-LiCl ion pair play a crucial role in the reaction process. In addition, Dumesic et al. studied the impact of solvent choice on HMF yield by adding inorganic salts (e.g., NaCl) to a concentrated fructose aqueous solution (30 wt%) in a biphasic system containing an organic phase. The 5-HMF yield could be increased by increasing the partitioning of 5-HMF in the extraction phase [18]. The results indicated when tetrahydrofuran (THF) was used as the organic extraction phase, high selectivity (83%) and high extracting power (R = 7.1) at 150 °C were achieved. A research team from Ehime University developed a high-temperature, high-pressure continuous flow microwave organic reactor that can achieve a special reaction operation: rapid rising of reaction temperature from room temperature to 400 °C within 0.01 s and rapid reduction of temperature after the reaction. The process can effectively suppress the generation of side reactions and increase the yield of 5-HMF to 70%. The obtained 5-HMF as a platform chemical can be converted into a variety of derivatives, which are important raw materials for bio-based polymers [19–24]. The reported types of derivatives can be divided into two categories according to whether the furan ring structure remains intact in the derivatives: (1) 5-HMF-derived monomers without furan ring structure and (2) 5-HMF-derived monomers with a furan ring structure.

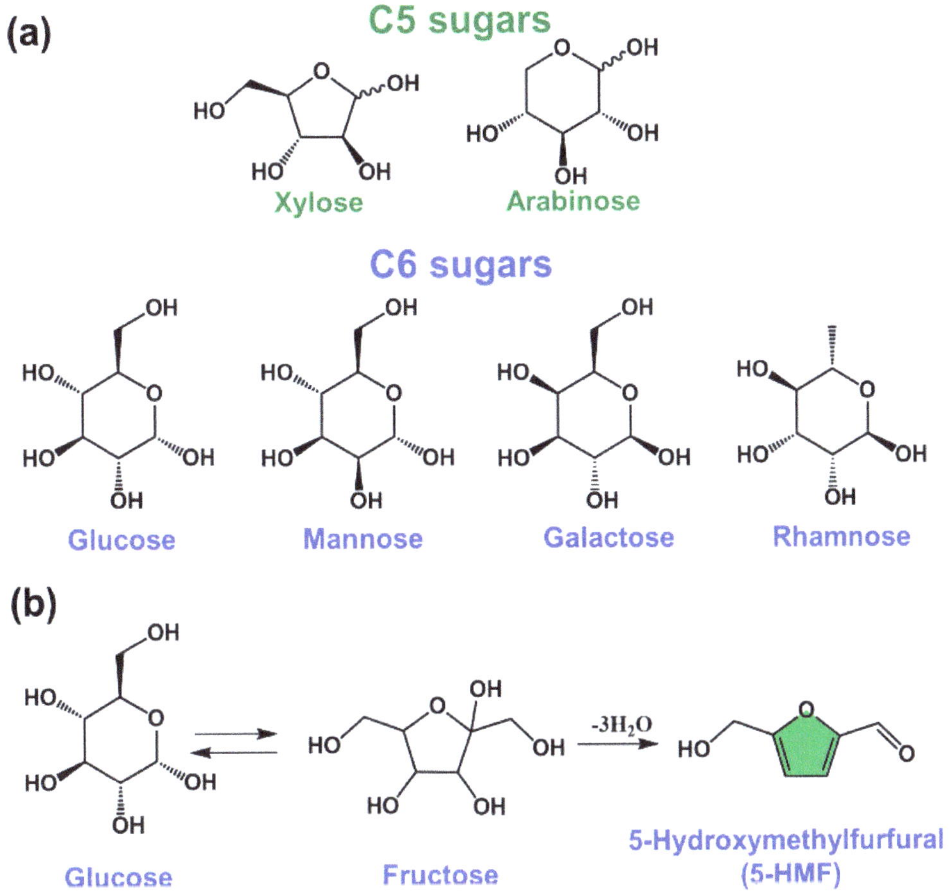

Figure 3. (a) Chemical structures of C5/C6 sugars, and (b) a general reaction pathway for the preparation of 5–hydroxymethylfurfural by the dehydration reaction of glucose.

2.2. Bio-Based Monomers without Furan Ring Structures Derived from 5-HMF

The furan ring in the 5-HMF structure can undergo various reactions such as with electrophilic reagents, nucleophilic reagents, oxidizing agents, reducing agents, cycloadditions, and metals and metal derivatives to synthesize various bio-based monomers as well as precursors (Figure 4a) [25]. Production of 2,5-hexanedione from 5-HMF was achieved us high temperature water (HTW) as a reaction medium and Zn as a catalyst [26]. The HTW at 250 °C was desirable for acid/base catalyzed HMF conversion. Maleic anhydride is a versatile chemical intermediate used in the production of various compounds and polymers, such as 1,4-butanediol, fumaric acid, tetrahydrofuran, and polyester resins. It was abandoned to use benzene as starting chemicals for synthesis of maleic anhydride in the industry because of the toxicity and undesirable loss of two carbons for the structure during reactions. The replacement of a benzene-based process with a biomass-based process is promising. Xu et al. obtained a from-5-HMF-to-maleic anhydride yield of 52% using VO(acac)$_2$ as a catalyst after a reaction at 90 °C for 4 h in acetonitrile through the simultaneous oxidation of hydroxymethyl group and C–C bond cleavage of 5-HMF [27]. Levulinic acid is an important biomass-derived platform chemical as an intermediate for the preparation of polymer monomers such as succinic acid [28–30]. The acid-catalyzed decomposition of 5-HMF for the preparation of levulinic acid has been widely reported, but most of them were limited by low yields and deficient economic efficiency. Recently, a new strategy for the efficient green preparation of levulinic acid using bifunctional Brønsted Lewis acids (HScCl$_4$) as catalysts has been proposed by Yu et al. [31]. In addition, the catalytic hydrogenation of 5-HMF using Pd/Al$_2$O$_3$ catalysts afforded 2,5-tetrahydrofuran dimethanol (THFDM) with a 9:1 ratio of cis to trans structure [32]. The oxidation of THFDM in water using a hydrotalcite-supported gold nanoparticle (AuNP) catalyst (~2 wt%) gave rise to 2,5-tetrahydrofuran dicarboxylic acid (THFDCA) with a high yield of 91%, which is a compound with potential applications in the biopolymer industry (Figure 4b) [33]. The AuNP size and basicity of hydrotalcite (HT) significantly impact the performance of the catalyst and that sintering of AuNPs was the main pathway for catalyst deactivation. 1,6-Hexanediol (HDO) is a highly valuable and important compound with two hydroxyl groups at the end of the molecule. This structure makes it an ideal monomer for the synthesis of polymers such as polyesters, polyurethanes, adhesives, and unsaturated polyesters. Bio-based HDO can also be prepared by directly catalytic conversion of 5-HMF using a Pd/ZrP catalyst and formic acid as a hydrogen source in 43% yield after 21 h of reaction at 413 K [34]. HDO was also prepared by direct catalytic conversion of 5-HMF in a fixed-bed reactor using a Pd/SiO$_2$ + Ir-ReOx/SiO$_2$ bilayer catalyst. Under optimal reaction conditions (373 K, 7.0 MPa H$_2$, mixed solvent of water (40%), and tetrahydrofuran (60%)), the yield of HDO was 57.8% [35]. The direct conversion of 5-HMF to caprolactone is difficult and therefore an indirect method is used. HDO is one of the most important intermediates and can also be prepared by oxidative cyclization of caprolactone [36]. The conversion of caprolactone to caprolactam through reacting with ammonia is a commonly used industrial process.

2.3. Bio-Based Monomers with Furan Ring Structures Derived from 5-HMF

The chemical modification of the hydroxyl and aldehyde groups in the 5-HMF molecular structure can also be selectively reacted to prepare bio-based monomers with the furan ring structure remaining unchanged. The monomers can be used for the synthesis of polyamides, polyesters, polyurethanes, polycarbonates, polyimides, and polyureas. Figure 5 illustrates various monomers obtained from the 5-HMF side group reactions [25]. 2,5-Furandicarboxylic acid (FDCA), 2,5-diformylfuran (DFF), and 2,5-bis(hydroxymethyl)furan (BHF) are the most common 5-HMF derived monomers with symmetrical chemical structures. These chemicals can also be used as intermediates to design the synthesis of bio-based monomers with reactive groups such as vinyl, acetylene, amine, and epoxy groups. FDCA is one of the most important near-market platform chemicals with an estimated value of USD 50.5 billion and is considered a promising alternative to petroleum-based terephthalic acid (TPA) for the production of green polymers such as polyethylene 2,5-furandicarboxylate (PEF) [37–39].

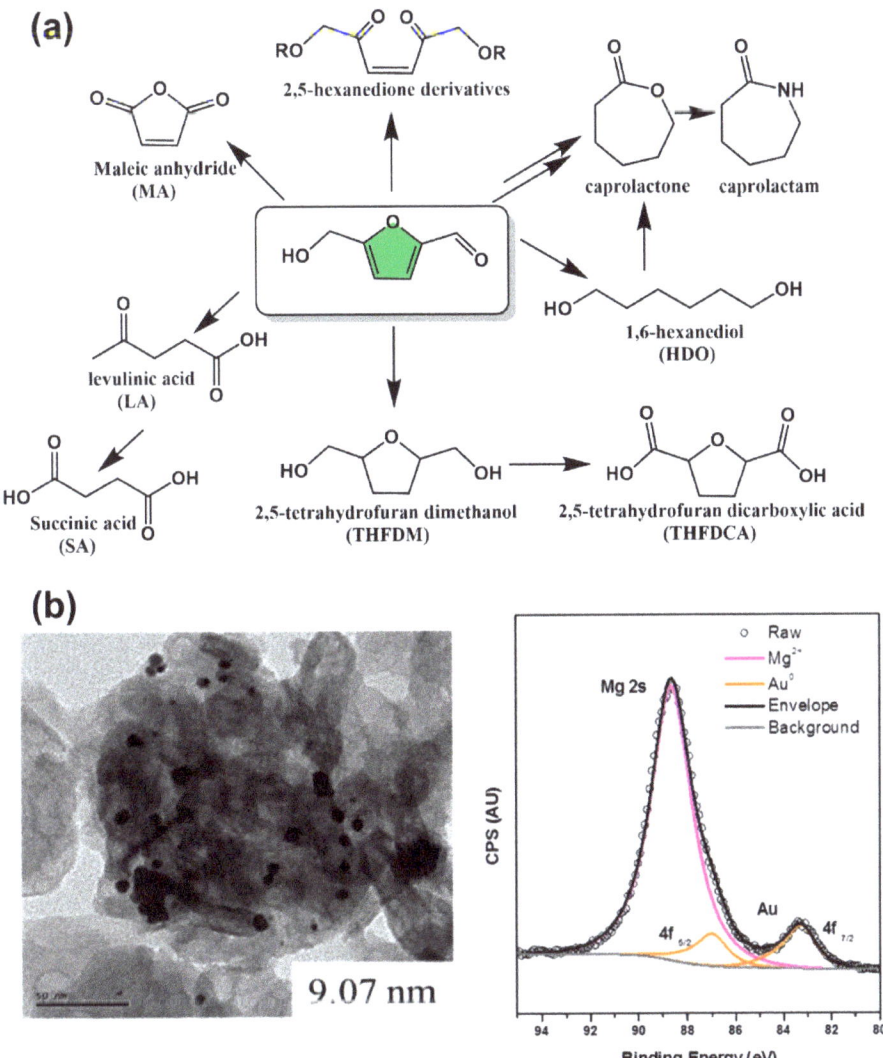

Figure 4. (a) 5-HMF-derived bio-based monomers without furan ring structures and (b) AuNP/HT catalyst for efficient conversion of THFDM to THFDCA [33]. Reprinted/adapted with permission from Ref. [33]. Copyright 2019, American Chemical Society.

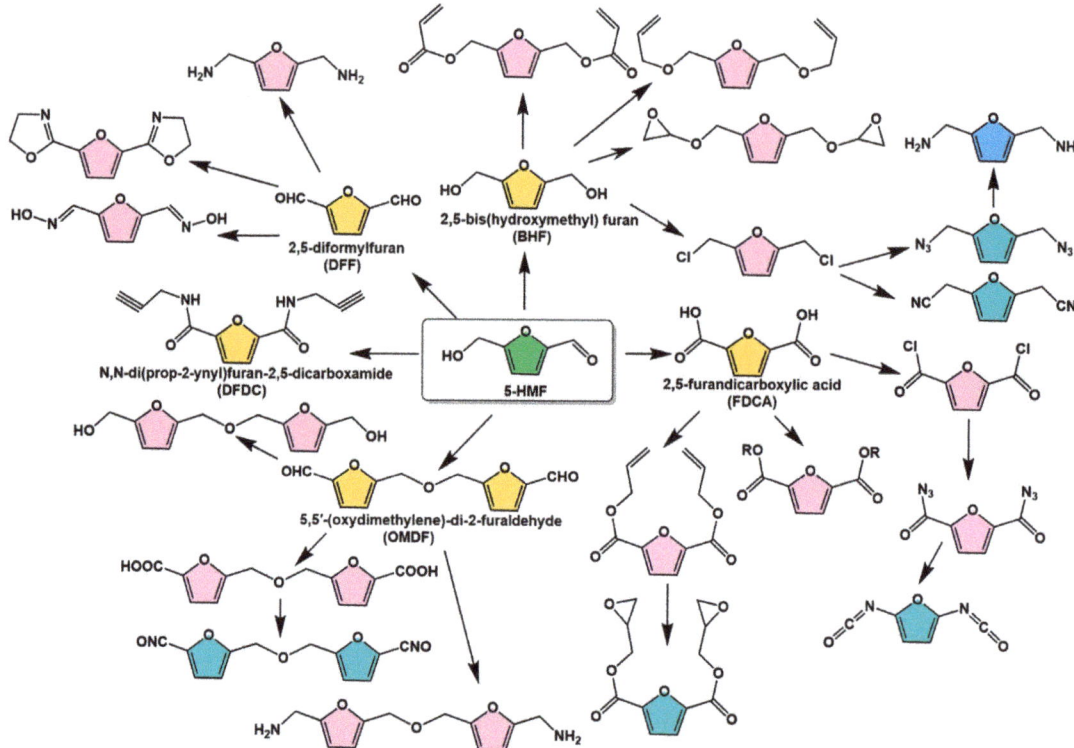

Figure 5. 5-HMF-derived bio-based monomers containing furan ring structures [25]. Reprinted/adapted with permission from Ref. [25]. Copyright 2017, John Wiley and Sons.

The production of FDCA from the oxidation of bio-based 5-HMF is of great interest and can be achieved by processes of aerobic catalytic oxidation, electrochemical catalytic oxidation, and biocatalytic reaction [40]. The most commonly used catalysts in the aerobic catalytic oxidation method are precious metal oxides (e.g., palladium catalysts: Pd-Bi-Te/C [41], platinum catalysts: CeCP@Pt [42], and gold nanoparticles loaded on metal oxides: Au/TiO$_2$ [43], Au/CeO$_2$ [44], Au-Cu/TiO$_2$ [45], and Au/Ce$_{1-x}$ Bi$_x$ O$_{2-\delta}$ [46]). However, precious metal catalysts suffer from high cost, harsh reaction conditions (high reaction temperature as well as high reaction pressure), lack of universality, and non-recyclability, thus hindering large-scale industrial applications. Recently, Li et al. have reported a mild solvent-free and alkali-free reaction system with a PdO/AlPO$_4$-5 catalyst under an oxygen atmosphere. [47] Using this reaction system, FDCA selectivity of up to 83.6% can be obtained at 80 °C for 5 h. Mechanistic studies showed that the hydroxyl group in the chemical structure of 5-HMF was oxidized first in this reaction system, while in the reaction that happened in the aqueous phase, the aldehyde group was oxidized first. Transition metal oxides are potential alternatives to noble metal catalysts. For instance, Hu et al. achieved near 100% conversion efficiency of 5-HMF to FDCA and up to 99.5% FDCA yield using two-dimensional Mn$_2$O$_3$ nanoflakes as catalysts [48].

The electrochemical oxidation of 5-HMF for the preparation of FDCA has a number of advantages, including (1) ambient reaction temperature and reaction pressure, (2) no need for high oxygen pressure, (3) no need for toxic oxidants, and (4) non-precious metals catalysts. However, the low surface area of typical catalysts results in low overall electro-catalytic activity, and the development of new catalysts with electrically active centers and high porosity is an effective way to improve reaction efficiency [49]. To address this issue,

Pila et al. loaded a thin layer of Co(OH)$_2$ on the surface of the porous metal–organic framework (MOF) material ZIF-67 and achieved excellent catalytic performance with conversions of up to 90.9%, FDCA yields of 81.8% and Faraday efficiencies of 83.6% [49]. Importantly, the applied potential for the reaction was reduced to 1.42 V (vs RHE), one of the lowest potentials reported.

The biocatalytic process is another synthetic method that allows the production of FDCA under mild conditions. 5-HMF oxidase and aryl alcohol oxidase can catalyze the complete oxidation of 5-HMF to FDCA [50,51]. Although a biocatalytic process generally needs a long reaction time, selecting suitable oxidases and optimizing experimental conditions could improve the reaction rate and achieve high conversion with a short reaction time. A tandem enzymatic one-pot reaction using galactose oxidase M3-5 and aldehyde oxidase PaoABC can also convert 5-HMF to pure FDCA with an isolated yield of 74% [52]. In particular, at 10 mM 5-HMF concentration, almost complete conversion (97%) of 5-HMF to FDCA after 1 h was observed. The galactose oxidase (GOase) variant prepared by W. Birmingham et al. was very active against 5-HMF, with high oxygen adsorption and very high production capacity [53]. The selective oxidation of 5-HMF to 2,5-diformylfuran was also possible. The biocatalyst and reaction conditions provide a blueprint for the further development of effective enzymatic catalysts and also facilitate the development of large-scale biocatalytic techniques for the preparation of furan-based compounds from sustainable feedstocks.

Similar to FDCA, DFF is one of the main products of the oxidative transformation of 5-HMF and is used as a chemical intermediate as a starting material for the synthesis of ligands, drugs, pesticide antifungals, fluorescent materials and new polymeric materials [54–56]. The oxidation of 5-HMF to DFF is always accompanied by a number of side reactions, such as the over-oxidation of DFF to FDCA, the oxidation of aldehydes to 5-hydroxymethyl-2-furancarboxylic acid (HMFCA), decarbonylation, and cross-polymerization to produce unwanted by-products. Therefore, selective oxidation of 5-HMF to DFF remains challengeable [56]. A number of catalysts have been reported for the selective oxidation of 5-HMF to DFF, such as manganese oxide-based catalysts [57], graphitized carbonitrides (g-C$_3$N$_4$) based catalysts [58], Ru-loaded catalysts [59], and vanadium-based catalysts [60], etc. Recently, a visible light catalyst developed by Huang et al. has achieved 95% DFF selectivity and good cycling stability [58]. The in situ oxidation using Co$_3$O$_4$ electrochemical catalyst developed by Zhang et al. achieved 100% conversion of 5-HMF to FDCA in 93.2% yield due to its unique defect structure and abundant electroactive sites [61]. More notably, the above reaction process simultaneously produces hydrogen, thus allowing the preparation of clean hydrogen energy as well as obtaining high-value chemicals, which can contribute to the realization of carbon neutrality from many aspects.

In addition, the catalytic hydrogenation of 5-HMF to prepare BHF is also a very important reaction for the high-value conversion of biomass. BHF can be used as a feedstock for renewable polymers such as bio-based polyesters and polycarbonates. Common catalysts are metal oxide-loaded platinum catalysts (MO/Pt), which vary greatly in their reaction selectivity depending on the type of metal oxide. Basic metal oxides such as MgO/Pt can achieve up to 99% BHF selectivity, while acidic loaded oxides such as TiO$_2$/Pt have a lower BHF selectivity [62].

2.4. Lignin-Based Platform Chemicals and Polymeric Monomers

The aromatic structure of lignin dictates that its degradation products are platform compounds with unique methoxyphenol structures [63]. However, due to its structural complexity and heterogeneity, its selective degradation of lignin to obtain chemicals with high selectivity and yield is very challenging. The main approaches to make lignin depolymerization include thermochemical (e.g., hydrolysis, gasification, hydrothermal liquefaction, and microwave methods), chemical (e.g., acid-base catalysis, ionic liquids/supercritical fluids, oxidation,) and biotechnological (e.g., bacterial, fungal, enzymatic) routes [64,65]. The resulting platform chemicals can be structurally simpler compounds by the selective breaking of C-O bonds; new fine chemicals and bio-based monomers can then be prepared by selective modification of specific functional groups [66–69]. 3-Methoxy-4-hydroxybenzaldehyde (vanillin) is one of the most important platform compounds derived from lignin conversion, containing both aldehyde and hydroxyl groups for further functionalization. It can be used as an intermediate compound for the preparation of a variety of bio-based monomers with aromatic structures (Figure 6a). Vanillin is the only commercially available bio-based aromatic compound because the depolymerization and purification of lignin are complex and difficult. Therefore, it has received a great deal of attention from the polymer community and can be used as a starting material for the preparation of various vanillin-based polymers, such as phenolic resins, benzoxazine resins, polyesters, acrylates, and methacrylate polymers (Figure 6b) [70,71]. A derivative, vanillin methacrylate was also reported as a bio-based monomer for the synthesis of aldehyde-containing porous materials [72]. The hydroxyl group in vanillin can also react with epichlorohydrin to synthesize compounds with both epoxy and aldehyde groups for the preparation of epoxy resins.

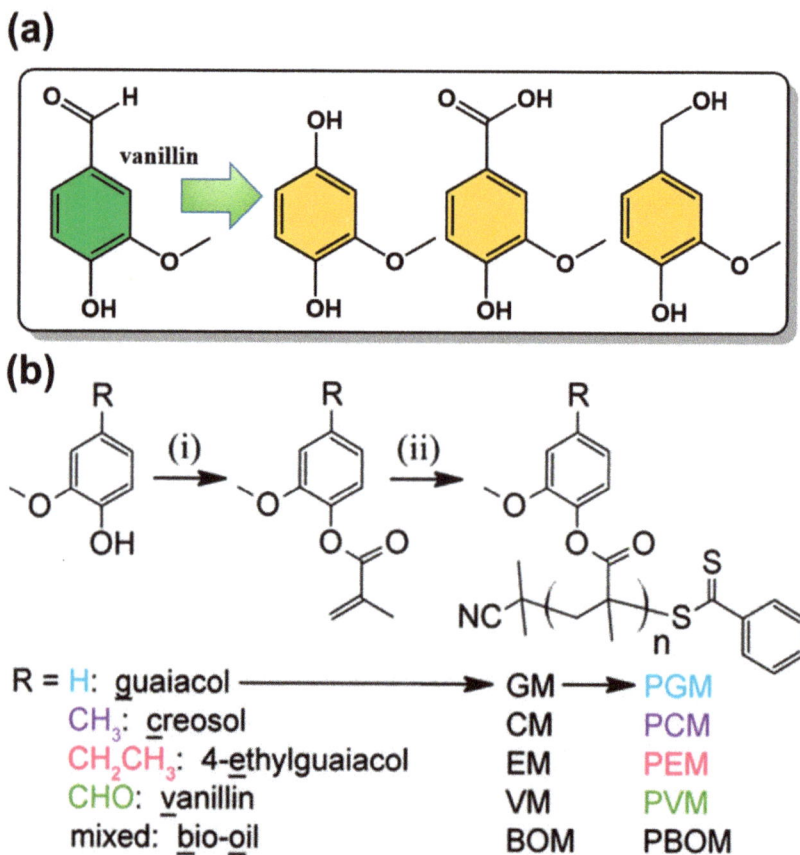

Figure 6. (a) Representative platform compounds derived from vanillin and (b) synthetic route of vanillin-based methacrylate monomers and the corresponding polymers. Reprinted/adapted with permission from Ref. [71]. Copyright 2016, American Chemical Society.

3. Common Polymerization Methods of Bio-Based Monomers

Since most polymers are synthesized depending on the petrochemical industry, which always accompanies a negative impact on the environment, there is a strong need to develop a facile and efficient synthesis of bio-based polymers using renewable platform chemicals as monomers [73]. Similar to petroleum-based monomers, the polymerization of bio-based monomers can be divided into chain-growth and step-growth polymerization depending on the reaction mechanism proposed by Flory in 1953 [74]. In chain-growth polymerization, the reaction starts from the active center generated by initiators and consists of chain initiation, chain propagation, and chain termination. The reaction rates and activation energies vary considerably between the various stages. Common free radical polymerization, ionic polymerization, coordination polymerization, living polymerization, and ring-opening polymerization belong to the category of chain-growth polymerization. Different from chain-growth polymerization which is proceeded by the addition of a monomer to a growing polymer with an active center, step-growth polymerization is gradually proceeded by reactions between any two compounds of monomers and growth chains with the rate and activation energy of each reaction step being approximately the same. Polymers including polyesters, polyamides, polyurethanes, etc., are produced through step-growth polymerization. Since most bio-based monomers contain functional groups such as hydroxyl, amine, and carboxyl groups, step-growth polymerization is the most common method for the preparation of bio-based synthetic polymers. Ring-opening polymerization is also used for the polymerization of cyclic molecules such as caprolactones. Although it is rare, there are some bio-based monomers with C=C double bonds by chemically modifying 5-HMF possible for free radical polymerization.

3.1. Step-Growth Polymerization by Condensation or Addition

In step-growth polymerization, condensation polymerization, and addition polymerization are referring to the formation of a polymer from bifunctional monomers with the release of small molecules or not. In the former, the molecular weight of the repeating units in the polymer is less than that of the monomer, e.g., the dehydration condensation polymerization of a dicarboxylate monomer with a diol. In the latter, the molecular weight of the repeating units in the polymer is same as that of the monomer, such as the reaction of diisocyanate with a diol to form a polyurethane. Figure 7 illustrates the types of polymers that can be prepared from 5-HMF derived bio-monomers, including polyesters (PEt), polyamides (PA), aromatic resins (AR), phenolic resins (PhR), polyimides (PI), and polyurethanes (PU).

Among all 5-HMF derived bio-monomers, FDCA is the most promising alternative to the petroleum based monomers, sharing the same carboxyl functional groups and similar aromatic center and reactivity with terephthalic acid (PTA), allowing the preparation of high-performance bio-based polymeric materials by condensation polymerization with diols or diamines [75–79]. However, before 2009, the development of FDCA-based polymers was limited by the purity and the price of the bio-monomer feedstocks. With the rapid development of biotechnology and the chemical industry, the economics and purity of FDCA feedstock have been greatly improved. In 2009, Gandini et al. prepared poly(ethylene glycol) 2,5-furandicarboxylate (PEF) by performing a transesterification and polycondensation between FDCA dimethyl ester with ethylene glycol using an antimony oxide catalyst (Sb_2O_3, 70–200 °C under high vacuum) and obtained high-molecular-weight products (degree of polymerization: 250–300) [76]. Since then, different diols such as 1,3-propanediol, 1,4-butanediol, 1,6-hexanediol, and 1,8-octanediol have been used as monomers in condensation reactions with FDCA to prepare bio-based polyesters containing different alkyl chain lengths [77].

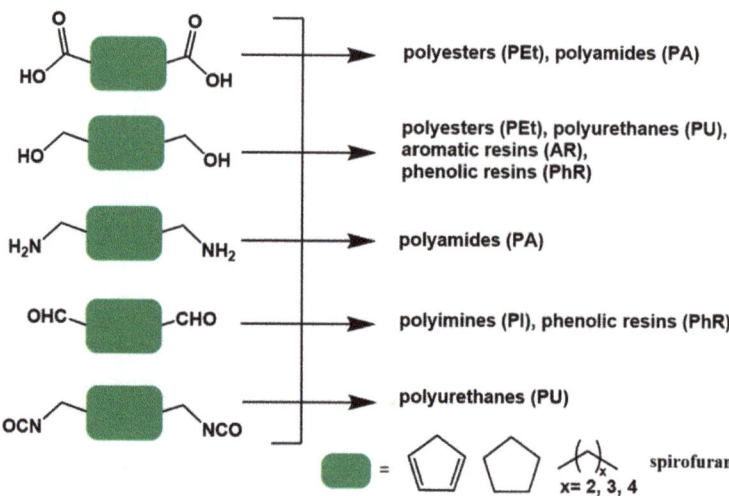

Figure 7. 5-HMF-derived bio-based monomers for polymer species. (The green block is used for structure simplification).

FDCA-derived bio-based polyesters are non-toxic, renewable, and have controlled degradation and similar properties to polymers of PTA and diols (e.g., polyethylene glycol terephthalate, PET) in terms of glass transition temperature (T_g), melting point (T_m), Young's modulus (η), and tensile strength (σ). Table 1 compares the thermomechanical properties of PEF and PET as well as polybutylene glycol 2,5-furandicarboxylate (PBF) and polybutylene glycol terephthalate (PBT) [77–79]. Although FDCA and TPA have very similar structures, there are differences in aromatic ring size, molecular polarity, and molecular symmetry, e.g., the distance between the two carboxyl groups in TPA is 5.731 Å, whereas in FDCA this value is less than 4.830 Å [78]. In addition, the angle between the two carboxylic acid groups is different, 180° in TPA and 129.4° in FDCA. These structural differences therefore lead to some specific properties of PEF, such as better gas barrier properties than PET, because of the polarity of the furan ring which could favorably interact with polar molecules. In particular, PEF has 6 times higher O_2 barrier than PET and 14 times higher CO_2 barrier than PET, thus offering great potential for applications such as high-performance food packaging materials. PET and PEF have similar chemical structures, which makes their recycling face similar problems. For instance, under a mechanical recycling process, incomplete separation and recovery of pure streams are general problems limited by technical difficulties. The degradation properties of the PEF in comparison with the PET were also well studied [80]. It was found that the PEF films degraded 1.7 times faster than PET ones under enzymatic hydrolysis. For instance, not only in the academic field, with the growth of global environmental issues and for achieving carbon neutrality by 2050, various large chemical companies such as Coca-Cola, DuPont, BASF, Mitsubishi, and Avantium are also vigorously pursuing the application of bio-based polymers instead of traditional petroleum-based materials. They are devoted to the commercialization of PEF and its copolymers. Among them, Avantium's 100% bio-based PEF, developed based on YXY technology, has been used in renewable plastic bottles, fibers as well as films, and other commodities [4].

Table 1. Comparison of various properties of PEF vs. PET and PBF vs. PBT.

Polymer	T_g (°C)	T_m (°C)	T_d (°C)	η (GPa)	σ (MPa)	M_n (kg/mol)
PEF	82–89	210–250	389	2.5–2.8	67–85	83–105
PET	71–79	246–260	407	2.0–2.5	65–72	6.4
PBF	36–44	169–172	373	1.8–2.0	55–62	11.8–17.8
PBT	24–48	220–227	384	1.4–1.6	51–56	17.7–44

3.2. Chain-Growth Polymerization

3.2.1. Ring-Opening Polymerization

Ring-opening polymerization (ROP) of cyclic bio-based monomers is a common method for the production of bio-based polymers, for instance, ROP of bio-based caprolactones [81]. The ROP products of caprolactones are aliphatic polyesters with biodegradability, good compatibility with other polymers, and easy processing. As U.S. Food and Drug Administration (FDA)-approved materials for human use, bio-based polycaprolactones (PCLs) have a wide range of applications in resin modification, coatings, drug-carrying materials, binders, etc. [82]. Depending on the type of initiator used, the ROP mechanism varies by anionic ROP, cationic ROP, and acid-activated monomer ROP mechanisms [83].

In the anionic ROP, nucleophilic reagents such as organometallics, alcohol salts and alcohols are used as initiators. The mechanism of polymerization is shown in Figure 8a, where an initiator such as an alcohol salt attacks the carbonyl carbon in the lactone monomer to open the ring structure of the lactone and then creates growing anionic chain ends. It is worth noting that the reaction produces a number of side reactions such as inter- and intra-molecular ester exchange reactions leading to the shortening of the polymer chain or lengthening of the chain, thereby expanding the molecular weight distribution. In addition, some reverse tail-bite reactions can also occur leading to the formation of oligomeric macrocyclic molecules. In contrast to the anionic mechanism, the cationic initiator in the cationic ROP first electrophilically attacks the carbonyl oxygen atom to produce an oxonium ion, which is a class of cation with an oxygen substituent attached to the central sp_2 hybridized carbon atom and capable of carrying a positive charge through π-bond dispersion between the central carbon and oxygen atoms. This then leads to alkyl oxygen cleavage and growth occurs through the continued reaction of this oxonium ion with the monomers (Figure 8b). Because the growing polymer chain end carries a positively charged terminus, the cationic mechanism is also known as the active chain end mechanism. Unlike the cationic mechanism where the active chain end is growing, a proton ion is added to the monomer side to continuously generate activated monomers for chain growth in the activated monomer mechanism (Figure 8c). First, the protonated lactone cation is approached by the hydroxyl nucleophilic reagent at the chain end, which opens the C=O double bond and leads to ring opening. The ring-opening reaction via the activation monomer mechanism is less susceptible to side reactions or molecular rearrangements because it does not have charged chain ends.

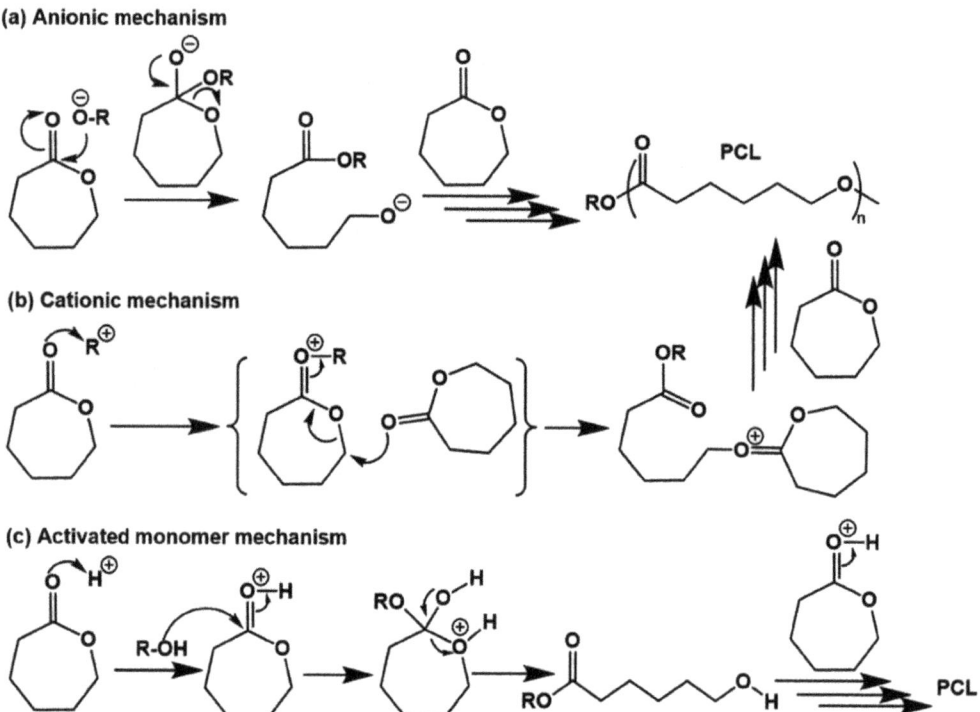

Figure 8. Reaction mechanism for ring-opening polymerization of caprolactones: (**a**) anionic polymerization mechanism, (**b**) cationic polymerization mechanism, (**c**) acid-activated monomer mechanism.

3.2.2. Free Radical Polymerization

Free radical polymerization is carried out in the mechanism of chain-growth polymerization initialized by an initiator. The application of 5-HMF as a direct feedstock in free radical polymerization is less common. However, the aldehyde group in 5-HMF can be selectively modified to other chemical structures suitable for free radical polymers, such as vinyl, acetylene-based monomers, etc. In 2008, Yoshida et al. transformed 5-HMF into 2-hydroxymethyl-5-vinylfuran (5-HMVF) and 2-methoxymethyl-5-vinylfuran (5-MMVF) by the Wittig reaction (Figure 9) [84]. 5-MMVF requires a methylation step to convert the hydroxyl group to methoxy group. Polymerization of these monomers is achieved at 70 °C under a nitrogen atmosphere using azodiisobutyronitrile (AIBN) as an initiator. The resulting polymers, PHMVF and PMMVF are in the form of oligomers and have number average molecular weights (M_n) of 2170 and 2890 g/mol, respectively.

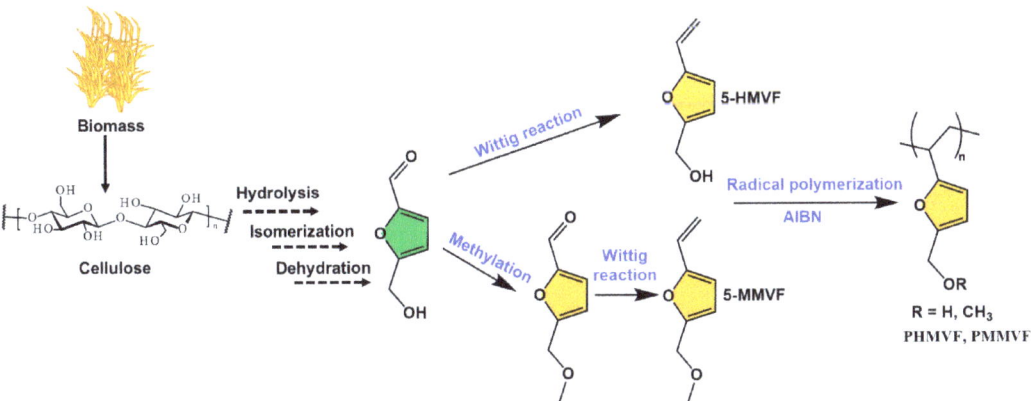

Figure 9. Synthetic routes of 5-HMVF and 5-MMVF bio-based monomers and their free radical polymerization.

4. Conclusions and Future Prospects of Bio-Based Polymers

For any new material, moving from concept to market launch is a challenge, with many hurdles to overcome, including the technology required to scale up the production and processing of the material. In recent years, the development of bio-based monomers and polymers as well as qualitative improvements in mechanistic research and commercialization have advanced by leaps and bounds, for example, the commercialization of PEF. However, the market share of bio-based polymers is still only about 1%, so there is still a long way to go in terms of their popularity and dissemination. This further requires increased research and development efforts in the areas of large-scale and efficient preparation of bio-based monomers, the achievement of improved properties of the bio-based polymers to approach those of petroleum-based polymers, and new synthetic technologies. In the production of bio-based monomers, conversion studies for different lignocellulosic feedstocks, the development of efficient production units, and the design of new catalytic systems may help to achieve more efficient and economical production of bio-based monomers than petroleum-based monomers [85].

In terms of polymer performance, bio-based polymers that have been commercialized or are close to commercialization are still lacking in terms of high performance and therefore require further optimization of molecular design and synthetic process. For petroleum-based polymer materials, hard monomers with aromatic structures characterized by high glass transition temperature (T_g) can provide the polymer with higher mechanical strength and heat resistance temperatures. In recent years, the design and synthesis of bio-based hard monomers have been exploited to develop high performance polymeric materials for replacement of engineering plastics. Some bio-based hard monomers were used to obtain polymers with high glass transition temperatures, such as spirofuran-based monomers (Figure 10) [86–93]. It is also found that the CO_2 emission for production of the bio-based rigid monomer 1 (Figure 10a,b) was much lower than that of fossil-based monomers, which gave rise to high performance polymers with a degradation temperature of up to 300 °C based on thermal gravimetric analysis (TGA) (Figure 10c) [86]. In addition to this, advanced computational methods can be used in the design of bio-based monomer structures to quickly and efficiently select suitable bio-based monomers among numerous molecular structures and to design rational polymerization routes, for example, material informatics methods using domain modeling to explore polymer structure–property relation [94].

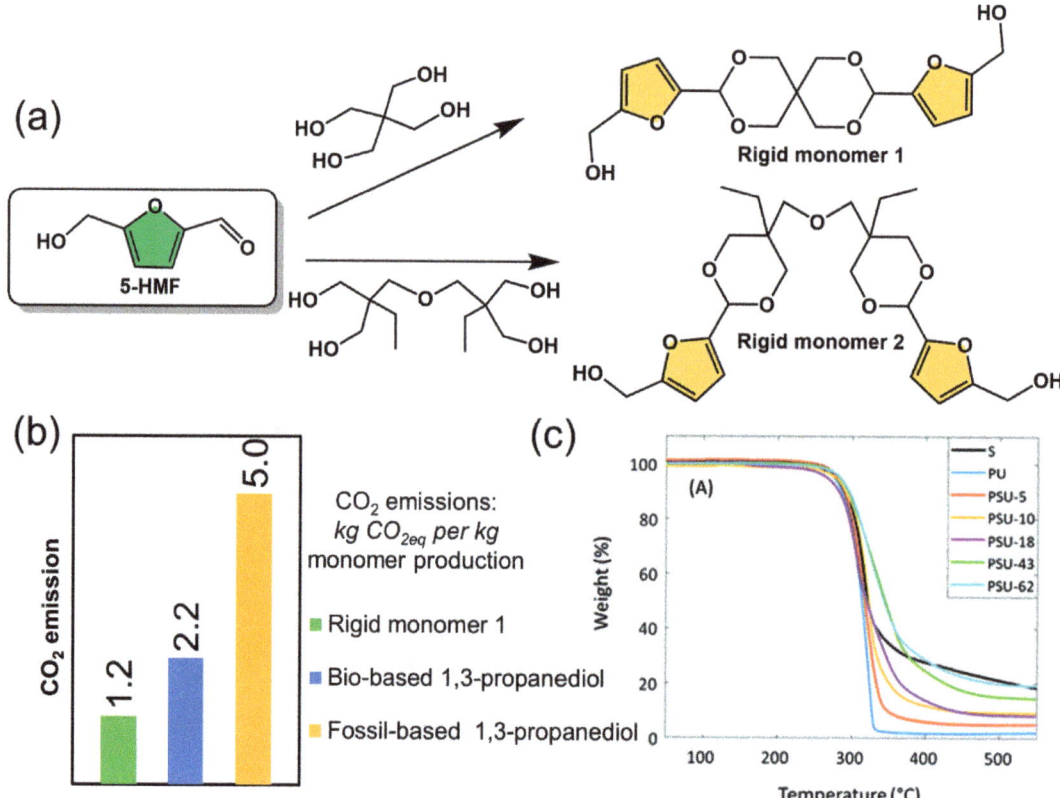

Figure 10. (a) Novel bio-based rigid monomers derived from 5-HMF in spirofuranyl, (b) CO_2 emission during monomer production and (c) TGA curved of rigid monomer 1 (black line) and its poly(urea-urethane) copolymers. Reprinted/adapted with permission from Ref. [86]. Copyright year, Royal Society of Chemistry.

Author Contributions: Conceptualization, H.G. and H.Z.; validation, F.P., H.G. and H.Z.; investigation, F.P. and L.L.; writing—original draft preparation, F.P., L.L., H.G. and H.Z.; writing—review and editing, F.P., L.L., H.G. and H.Z.; funding acquisition, F.P. and H.G. All authors have read and agreed to the published version of the manuscript.

Funding: The work is also partially supported by the National Key Research and Development Program of China (2020YFC1807804). H.G. is grateful for the support of the Elite Youth Program of the Chinese Academy of Agricultural Sciences, Basic Frontier Project of Agro-Environmental Protection Institute, Ministry of Agriculture and Rural Affairs of China (2022-jcqyrw-ghx).

Institutional Review Board Statement: Not applicable.

Informed Consent Statement: Not applicable.

Data Availability Statement: No new data were created.

Conflicts of Interest: The authors declare no conflict of interest.

References

1. Serrano-Ruiz, J.C.; Luque, R.; Sepúlveda-Escribano, A. Transformations of biomass-derived platform molecules: From high added-value chemicals to fuelsvia aqueous-phase processing. *Chem. Soc. Rev.* **2011**, *40*, 5266–5281. [CrossRef]
2. Alsabri, A.; Tahir, F.; Al-Ghamdi, S.G. Life-Cycle Assessment of Polypropylene Production in the Gulf Cooperation Council (GCC) Region. *Polymers* **2021**, *13*, 3793. [CrossRef] [PubMed]
3. MacArthur, E. Beyond plastic waste. *Science* **2017**, *358*, 843. [CrossRef]
4. Isikgor, F.H.; Becer, C.R. Lignocellulosic biomass: A sustainable platform for the production of bio-based chemicals and polymers. *Polym. Chem.* **2015**, *6*, 4497–4559. [CrossRef]
5. Cywar, R.M.; Rorrer, N.A.; Hoyt, C.B.; Beckham, G.T.; Chen, E.Y.X. Bio-based polymers with performance-advantaged properties. *Nat. Rev. Mater.* **2022**, *7*, 83–103. [CrossRef]
6. Yu, I.K.M.; Tsang, D.C.W. Conversion of biomass to hydroxymethylfurfural: A review of catalytic systems and underlying mechanisms. *Bioresour. Technol.* **2017**, *238*, 716–732. [CrossRef] [PubMed]
7. Cai, J.; He, Y.; Yu, X.; Banks, S.W.; Yang, Y.; Zhang, X.; Yu, Y.; Liu, R.; Bridgwater, A.V. Review of physicochemical properties and analytical characterization of lignocellulosic biomass. *Renew. Sustain. Energy Rev.* **2017**, *76*, 309–322. [CrossRef]
8. Loow, Y.-L.; Wu, T.Y.; Tan, K.A.; Lim, Y.S.; Siow, L.F.; Jahim, J.M.; Mohammad, A.W.; Teoh, W.H. Recent Advances in the Application of Inorganic Salt Pretreatment for Transforming Lignocellulosic Biomass into Reducing Sugars. *J. Agric. Food Chem.* **2015**, *63*, 8349–8363. [CrossRef] [PubMed]
9. Akhtar, N.; Gupta, K.; Goyal, D.; Goyal, A. Recent advances in pretreatment technologies for efficient hydrolysis of lignocellulosic biomass. *Environ. Prog. Sustain. Energy* **2016**, *35*, 489–511. [CrossRef]
10. Dahmen, N.; Lewandowski, I.; Zibek, S.; Weidtmann, A. Integrated lignocellulosic value chains in a growing bioeconomy: Status quo and perspectives. *Glob. Chang. Biol. Bioenergy* **2019**, *11*, 107–117. [CrossRef]
11. Bhatia, S.K.; Jagtap, S.S.; Bedekar, A.A.; Bhatia, R.K.; Patel, A.K.; Pant, D.; Rajesh Banu, J.; Rao, C.V.; Kim, Y.-G.; Yang, Y.-H. Recent developments in pretreatment technologies on lignocellulosic biomass: Effect of key parameters. *Bioresour. Technol.* **2020**, *300*, 122724. [CrossRef] [PubMed]
12. Baruah, J.; Nath, B.K.; Sharma, R.; Kumar, S.; Deka, R.C.; Baruah, D.C.; Kalita, E. Recent Trends in the Pretreatment of Lignocellulosic Biomass for Value-Added Products. *Front. Energy Res.* **2018**, *6*, 141. [CrossRef]
13. Ricciardi, L.; Verboom, W.; Lange, J.P.; Huskens, J. Production of furans from C5 and C6 sugars in the presence of polar organic solvents. *Sustain. Energy Fuels* **2022**, *6*, 11–28. [CrossRef]
14. Wu, X.; Luo, N.; Xie, S.; Zhang, H.; Zhang, Q.; Wang, F.; Wang, Y. Photocatalytic transformations of lignocellulosic biomass into chemicals. *Chem. Soc. Rev.* **2020**, *49*, 6198–6223. [CrossRef] [PubMed]
15. Delidovich, I.; Hausoul, P.J.C.; Deng, L.; Pfützenreuter, R.; Rose, M.; Palkovits, R. Alternative Monomers Based on Lignocellulose and Their Use for Polymer Production. *Chem. Rev.* **2016**, *116*, 1540–1599. [CrossRef] [PubMed]
16. Ahmed, S.F.; Mofijur, M.; Chowdhury, S.N.; Nahrin, M.; Rafa, N.; Chowdhury, A.T.; Nuzhat, S.; Ong, H.C. Pathways of lignocellulosic biomass deconstruction for biofuel and value-added products production. *Fuel* **2022**, *318*, 123618. [CrossRef]
17. Binder, J.B.; Raines, R.T. Simple Chemical Transformation of Lignocellulosic Biomass into Furans for Fuels and Chemicals. *Chem. Soc.* **2009**, *131*, 1979–1985. [CrossRef]
18. Román-Leshkov, Y.; Dumesic, J.A. Solvent Effects on Fructose Dehydration to 5-Hydroxymethylfurfural in Biphasic Systems Saturated with Inorganic. *Top. Catal.* **2009**, *52*, 297–303. [CrossRef]
19. Lee, Y.; Kwon, E.E.; Lee, J. Polymers derived from hemicellulosic parts of lignocellulosic biomass. *Rev. Environ. Sci. Biotechnol.* **2019**, *18*, 317–334. [CrossRef]
20. Corma, A.; Iborra, S.; Velty, A. Chemical Routes for the Transformation of Biomass into Chemicals. *Chem. Rev.* **2007**, *107*, 2411–2502. [CrossRef]
21. Nakagawa, Y.; Tomishige, K. Production of 1,5-pentanediol from biomass via furfural and tetrahydrofurfuryl alcohol. *Catal. Today* **2012**, *195*, 136–143. [CrossRef]
22. Iqbal, S.; Liu, X.; Aldosari, O.F.; Miedziak, P.J.; Edwards, J.K.; Brett, G.L.; Akram, A.; King, G.M.; Davies, T.E.; Morgan, D.J.; et al. Conversion of furfuryl alcohol into 2-methylfuran at room temperature using Pd/TiO$_2$ catalyst. *Catal. Sci. Technol.* **2014**, *4*, 2280–2286. [CrossRef]
23. Li, X.; Lan, X.; Wang, T. Selective oxidation of furfural in a bi-phasic system with homogeneous acid catalyst. *Catal. Today* **2016**, *276*, 97–104. [CrossRef]
24. Zhu, C.; Wang, H.; Liu, Q.; Wang, C.; Xu, Y.; Zhang, Q.; Ma, L. Chapter 3-5-Hydroxymethylfurfural—A C6 precursor for fuels and chemicals. In *Biomass, Biofuels, Biochemicals*; Saravanamurugan, S., Pandey, A., Li, H., Riisager, A., Eds.; Elsevier: Amsterdam, The Netherlands, 2020; pp. 61–94.
25. Zhang, D.; Dumont, M.-J. Advances in polymer precursors and bio-based polymers synthesized from 5-hydroxymethylfurfural. *J. Polym. Sci. A Polym. Chem.* **2017**, *55*, 1478–1492. [CrossRef]
26. Lichtenthaler, F.W.; Brust, A.; Cuny, E. Sugar-derived building blocks. *Green Chem.* **2001**, *3*, 201–209. [CrossRef]
27. Du, Z.; Ma, J.; Wang, F.; Liu, J.; Xu, J. Oxidation of 5-hydroxymethylfurfural to maleic anhydride with molecular oxygen. *Green Chem.* **2011**, *13*, 554–557. [CrossRef]
28. Tang, Y.; Fu, J.; Wang, Y.; Guo, H.; Qi, X. Bimetallic Ni-Zn@OMC catalyst for selective hydrogenation of levulinic acid to γ-valerolactone in water. *Fuel Process. Technol.* **2023**, *240*, 107559. [CrossRef]

29. Carnevali, D.; Rigamonti, M.G.; Tabanelli, T.; Patience, G.S.; Cavani, F. Levulinic acid upgrade to succinic acid with hydrogen peroxide. *Appl. Catal. A Gen.* **2018**, *563*, 98–104. [CrossRef]
30. Song, L.; Wang, R.; Che, L.; Jiang, Y.; Zhou, M.; Zhao, Y.; Pang, J.; Jiang, M.; Zhou, G.; Zheng, M.; et al. Catalytic Aerobic Oxidation of Lignocellulose-Derived Levulinic Acid in Aqueous Solution: A Novel Route to Synthesize. *ACS Catal.* **2021**, *11*, 11588–11596. [CrossRef]
31. Liu, S.; Cheng, X.; Sun, S.; Chen, Y.; Bian, B.; Liu, Y.; Tong, L.; Yu, H.; Ni, Y.; Yu, S. High-Yield and High-Efficiency Conversion of HMF to Levulinic Acid in a Green and Facile Catalytic Process by a Dual-Function Brønsted-Lewis Acid HScCl4 Catalyst. *ACS Omega* **2021**, *6*, 15940–15947. [CrossRef]
32. Kumalaputri, A.J.; Bottari, G.; Erne, P.M.; Heeres, H.J.; Barta, K. Tunable and Selective Conversion of 5-HMF to 2,5-Furandimethanol and 2,5-Dimethylfuran over Copper-Doped Porous Metal Oxides. *ChemSusChem* **2014**, *7*, 2266–2275. [CrossRef] [PubMed]
33. Yuan, Q.; Hiemstra, K.; Meinds, T.G.; Chaabane, I.; Tang, Z.; Rohrbach, L.; Vrijburg, W.; Verhoeven, T.; Hensen, E.J.M.; van der Veer, S.; et al. Bio-Based Chemicals: Selective Aerobic Oxidation of Tetrahydrofuran-2,5-dimethanol to Tetrahydrofuran-2,5-dicarboxylic Acid Using Hydrotalcite-Supported Gold Catalysts. *ACS Sustain. Chem. Eng.* **2019**, *7*, 4647–4656. [CrossRef]
34. Tuteja, J.; Choudhary, H.; Nishimura, S.; Ebitani, K. Direct Synthesis of 1,6-Hexanediol from HMF over a Heterogeneous Pd/ZrP Catalyst using Formic Acid as Hydrogen Source. *ChemSusChem* **2014**, *7*, 96–100. [CrossRef]
35. Xiao, B.; Zheng, M.; Li, X.; Pang, J.; Sun, R.; Wang, H.; Pang, X.; Wang, A.; Wang, X.; Zhang, T. Synthesis of 1,6-hexanediol from HMF over double-layered catalysts of Pd/SiO$_2$ + Ir-ReO$_x$/SiO$_2$ in a fixed-bed reactor. *Green Chem.* **2016**, *18*, 2175–2184. [CrossRef]
36. Buntara, T.; Noel, S.; Phua, P.H.; Melián-Cabrera, I.; de Vries, J.G.; Heeres, H.J. Caprolactam from Renewable Resources: Catalytic Conversion of 5-Hydroxymethylfurfural into Caprolactone. *Angew. Chem. Int. Ed.* **2011**, *50*, 7083–7087. [CrossRef]
37. Long, L.; Ye, B.; Wei, J.; Wu, B.; Li, Y.; Wang, Z. Structure and enhanced mechanical properties of bio-based poly(ethylene 2,5-furandicarboxylate) by incorporating with low loadings of talc platelets. *Polymer* **2021**, *237*, 124351. [CrossRef]
38. van Putten, R.-J.; van der Waal, J.C.; de Jong, E.; Rasrendra, C.B.; Heeres, H.J.; de Vries, J.G. Hydroxymethylfurfural, A Versatile Platform Chemical Made from Renewable Resources. *Chem. Rev.* **2013**, *113*, 1499–1597. [CrossRef]
39. Ardemani, L.; Cibin, G.; Dent, A.J.; Isaacs, M.A.; Kyriakou, G.; Lee, A.F.; Parlett, C.M.A.; Parry, S.A.; Wilson, K. Solid base catalysed 5-HMF oxidation to 2,5-FDCA over Au/hydrotalcites: Fact or fiction? *Chem. Sci.* **2015**, *6*, 4940–4945. [CrossRef]
40. Sajid, M.; Zhao, X.; Liu, D. Production of 2,5-furandicarboxylic acid (FDCA) from 5-hydroxymethylfurfural (HMF): Recent progress focusing on the chemical-catalytic routes. *Green Chem.* **2018**, *20*, 5427–5453. [CrossRef]
41. Ahmed, M.S.; Mannel, D.S.; Root, T.W.; Stahl, S.S. Aerobic Oxidation of Diverse Primary Alcohols to Carboxylic Acids with a Heterogeneous Pd-Bi-Te/C (PBT/C) Catalyst. *Org. Process Res. Dev.* **2017**, *21*, 1388–1393. [CrossRef]
42. Gong, W.; Zheng, K.; Ji, P. Platinum deposited on cerium coordination polymer for catalytic oxidation of hydroxymethylfurfural producing 2,5-furandicarboxylic acid. *RSC Adv.* **2017**, *7*, 34776–34782. [CrossRef]
43. Gorbanev, Y.Y.; Klitgaard, S.K.; Woodley, J.M.; Christensen, C.H.; Riisager, A. Gold-Catalyzed Aerobic Oxidation of 5-Hydroxymethylfurfural in Water at Ambient Temperature. *ChemSusChem* **2009**, *2*, 672–675. [CrossRef] [PubMed]
44. Casanova, O.; Iborra, S.; Corma, A. Biomass into Chemicals: Aerobic Oxidation of 5-Hydroxymethyl-2-furfural into 2,5-Furandicarboxylic Acid with Gold Nanoparticle Catalysts. *ChemSusChem* **2009**, *2*, 1138–1144. [CrossRef]
45. Pasini, T.; Piccinini, M.; Blosi, M.; Bonelli, R.; Albonetti, S.; Dimitratos, N.; Lopez-Sanchez, J.A.; Sankar, M.; He, Q.; Kiely, C.J.; et al. Selective oxidation of 5-hydroxymethyl-2-furfural using supported gold-copper nanoparticles. *Green Chem.* **2011**, *13*, 2091–2099. [CrossRef]
46. Miao, Z.; Zhang, Y.; Pan, X.; Wu, T.; Zhang, B.; Li, J.; Yi, T.; Zhang, Z.; Yang, X. Superior catalytic performance of Ce$_{1-x}$Bi$_x$O$_{2-\delta}$ solid solution and Au/Ce$_{1-x}$Bi$_x$O$_{2-\delta}$ for 5-hydroxymethylfurfural conversion in alkaline aqueous solution. *Catal. Sci. Technol.* **2015**, *5*, 1314–1322. [CrossRef]
47. Yu, L.; Chen, H.; Wen, Z.; Ma, X.; Li, Y.; Li, Y. Solvent- and Base-Free Oxidation of 5-Hydroxymethylfurfural over a PdO/AlPO$_4$-5 Catalyst under Mild Conditions. *Ind. Eng. Chem. Res.* **2021**, *60*, 13485–13491. [CrossRef]
48. Bao, L.; Sun, F.-Z.; Zhang, G.-Y.; Hu, T.-L. Aerobic Oxidation of 5-Hydroxymethylfurfural to 2,5-Furandicarboxylic Acid over Holey 2 D Mn$_2$O$_3$ Nanoflakes from a Mn-based MOF. *ChemSusChem* **2020**, *13*, 548–555. [CrossRef]
49. Pila, T.; Nueangnoraj, K.; Ketrat, S.; Somjit, V.; Kongpatpanich, K. Electrochemical Production of 2,5-Furandicarboxylic from 5-Hydroxymethylfurfural Using Ultrathin Co(OH)$_2$ on ZIF-67. *ACS Appl. Energy Mater.* **2021**, *4*, 12909–12916. [CrossRef]
50. Dijkman, W.P.; Fraaije, M.W. Discovery and Characterization of a 5-Hydroxymethylfurfural Oxidase from Methylovorus sp. Strain MP688. *Appl. Environ. Microbiol.* **2014**, *80*, 1082–1090. [CrossRef]
51. Carro, J.; Ferreira, P.; Rodríguez, L.; Prieto, A.; Serrano, A.; Balcells, B.; Ardá, A.; Jiménez-Barbero, J.; Gutiérrez, A.; Ullrich, R.; et al. 5-hydroxymethylfurfural conversion by fungal aryl-alcohol oxidase and unspecific peroxygenase. *FEBS J.* **2015**, *282*, 3218–3229. [CrossRef]
52. McKenna, S.M.; Leimkühler, S.; Herter, S.; Turner, N.J.; Carnell, A.J. Enzyme cascade reactions: Synthesis of furandicarboxylic acid (FDCA) and carboxylic acids using oxidases in tandem. *Green Chem.* **2015**, *17*, 3271–3275. [CrossRef]
53. Birmingham, W.R.; Toftgaard Pedersen, A.; Dias Gomes, M.; Bøje Madsen, M.; Breuer, M.; Woodley, J.M.; Turner, N.J. Toward scalable biocatalytic conversion of 5-hydroxymethylfurfural by galactose oxidase using coordinated reaction and enzyme engineering. *Nat. Commun.* **2021**, *12*, 4946. [CrossRef]

54. Liu, X.; Xiao, J.; Ding, H.; Zhong, W.; Xu, Q.; Su, S.; Yin, D. Catalytic aerobic oxidation of 5-hydroxymethylfurfural over VO^{2+} and Cu^{2+} immobilized on amino functionalized SBA-15. *Chem. Eng. J.* **2016**, *283*, 1315–1321. [CrossRef]
55. Liu, B.; Zhang, Z. One-Pot Conversion of Carbohydrates into Furan Derivatives via Furfural and 5-Hydroxylmethylfurfural as Intermediates. *ChemSusChem* **2016**, *9*, 2015–2036. [CrossRef]
56. Lai, J.; Liu, K.; Zhou, S.; Zhang, D.; Liu, X.; Xu, Q.; Yin, D. Selective oxidation of 5-hydroxymethylfurfural into 2,5-diformylfuran over VPO catalysts under atmospheric pressure. *RSC Adv.* **2019**, *9*, 14242–14246. [CrossRef]
57. Dhingra, S.; Chhabra, T.; Krishnan, V.; Nagaraja, C.M. Visible-Light-Driven Selective Oxidation of Biomass-Derived HMF to DFF Coupled with H_2 Generation by Noble Metal-Free $Zn_{0.5}Cd_{0.5}S/MnO_2$ Heterostructures. *ACS Appl. Energy Mater.* **2020**, *3*, 7138–7148. [CrossRef]
58. Bao, X.; Liu, M.; Wang, Z.; Dai, D.; Wang, P.; Cheng, H.; Liu, Y.; Zheng, Z.; Dai, Y.; Huang, B. Photocatalytic Selective Oxidation of HMF Coupled with H_2 Evolution on Flexible Ultrathin $g-C_3N_4$ Nanosheets with Enhanced N-H Interaction. *ACS Catal.* **2022**, *12*, 1919–1929. [CrossRef]
59. Antonyraj, C.A.; Jeong, J.; Kim, B.; Shin, S.; Kim, S.; Lee, K.-Y.; Cho, J.K. Selective oxidation of HMF to DFF using Ru/γ-alumina catalyst in moderate boiling solvents toward industrial production. *J. Ind. Eng. Chem.* **2013**, *19*, 1056–1059. [CrossRef]
60. Grasset, F.L.; Katryniok, B.; Paul, S.; Nardello-Rataj, V.; Pera-Titus, M.; Clacens, J.-M.; De Campo, F.; Dumeignil, F. Selective oxidation of 5-hydroxymethylfurfural to 2,5-diformylfuran over intercalated vanadium phosphate oxides. *RSC Adv.* **2013**, *3*, 9942–9948. [CrossRef]
61. Chen, C.; Zhou, Z.; Liu, J.; Zhu, B.; Hu, H.; Yang, Y.; Chen, G.; Gao, M.; Zhang, J. Sustainable biomass upgrading coupled with H_2 generation over in-situ oxidized Co_3O_4 electrocatalysts. *Catal. B Environ.* **2022**, *307*, 121209. [CrossRef]
62. Wang, J.; Zhao, J.; Fu, Y.; Miao, C.; Jia, S.; Yan, P.; Huang, J. Highly selective hydrogenation of 5-hydroxymethylfurfural to 2,5-bis(hydroxymethyl)furan over metal-oxide supported Pt catalysts: The role of basic sites. *Appl. Catal. A Gen.* **2022**, *643*, 118762. [CrossRef]
63. Sun, Z.; Fridrich, B.; de Santi, A.; Elangovan, S.; Barta, K. Bright Side of Lignin Depolymerization: Toward New Platform Chemicals. *Chem. Rev.* **2018**, *118*, 614–678. [CrossRef] [PubMed]
64. Khan, R.J.; Lau, C.Y.; Guan, J.; Lam, C.H.; Zhao, J.; Ji, Y.; Wang, H.; Xu, J.; Lee, D.-J.; Leu, S.-Y. Recent advances of lignin valorization techniques toward sustainable aromatics and potential benchmarks to fossil refinery products. *Bioresour. Technol.* **2022**, *346*, 126419. [CrossRef]
65. Ralph, J.; Lundquist, K.; Brunow, G.; Lu, F.; Kim, H.; Schatz, P.F.; Marita, J.M.; Hatfield, R.D.; Ralph, S.A.; Christensen, J.H.; et al. Lignins: Natural polymers from oxidative coupling of 4-hydroxyphenyl-propanoids. *Phytochem. Rev.* **2004**, *3*, 29–60. [CrossRef]
66. del Río, J.C.; Rencoret, J.; Gutiérrez, A.; Elder, T.; Kim, H.; Ralph, J. Lignin Monomers from beyond the Canonical Monolignol Biosynthetic Pathway: Another Brick in the Wall. *ACS Sustain. Chem. Eng.* **2020**, *8*, 4997–5012. [CrossRef]
67. Fache, M.; Darroman, E.; Besse, V.; Auvergne, R.; Caillol, S.; Boutevin, B. Vanillin, a promising bio-based building-block for monomer synthesis. *Green Chem.* **2014**, *16*, 1987–1998. [CrossRef]
68. Decostanzi, M.; Auvergne, R.; Boutevin, B.; Caillol, S. Bio-based phenol and furan derivative coupling for the synthesis of functional monomers. *Green Chem.* **2019**, *21*, 724–747. [CrossRef]
69. Harvey, B.G.; Guenthner, A.J.; Meylemans, H.A.; Haines, S.R.L.; Lamison, K.R.; Groshens, T.J.; Cambrea, L.R.; Davis, M.C.; Lai, W.W. Renewable thermosetting resins and thermoplastics from vanillin. *Green Chem.* **2015**, *17*, 1249–1258. [CrossRef]
70. Holmberg, A.L.; Reno, K.H.; Nguyen, N.A.; Wool, R.P.; Epps III, T.H. Syringyl Methacrylate, a Hardwood Lignin-Based Monomer for High-T_g Polymeric Materials. *ACS Macro Lett.* **2016**, *5*, 574–578. [CrossRef]
71. Holmberg, A.L.; Nguyen, N.A.; Karavolias, M.G.; Reno, K.H.; Wool, R.P.; Epps III, T.H. Softwood Lignin-Based Methacrylate Polymers with Tunable Thermal and Viscoelastic Properties. *Macromolecules* **2016**, *49*, 1286–1295. [CrossRef]
72. Zhang, H.; Yong, X.; Zhou, J.; Deng, J.; Wu, Y. Biomass Vanillin-Derived Polymeric Microspheres Containing Functional Aldehyde Groups: Preparation, Characterization, and Application as Adsorbent. *ACS Appl. Mater. Interfaces* **2016**, *8*, 2753–2763. [CrossRef] [PubMed]
73. Chacón-Huete, F.; Messina, C.; Cigana, B.; Forgione, P. Diverse Applications of Biomass-Derived 5-Hydroxymethylfurfural and Derivatives as Renewable Starting Materials. *ChemSusChem* **2022**, *15*, e202200328. [CrossRef]
74. Flory, P.J. *Principles of Polymer Chemistry*; Cornell University Press: Ithaca, NY, USA, 1953.
75. Maniar, D.; Hohmann, K.; Jiang, Y.; Woortman, A.; van Dijken, J.; Loos, K. Enzymatic Polymerization of Dimethyl 2,5-Furandicarboxylate and Heteroatom Diamines. *ACS Omega* **2018**, *3*, 7077–7085. [CrossRef] [PubMed]
76. Gandini, A.; Silvestre, A.J.D.; Neto, C.P.; Sousa, A.F.; Gomes, M. The furan counterpart of poly(ethylene terephthalate): An alternative material based on renewable resources. *J. Polym. Sci. A Polym. Chem.* **2009**, *47*, 295–298. [CrossRef]
77. Jiang, M.; Liu, Q.; Zhang, Q.; Ye, C.; Zhou, G. A series of furan-aromatic polyesters synthesized via direct esterification method based on renewable resources. *J. Polym. Sci. A Polym. Chem.* **2012**, *50*, 1026–1036. [CrossRef]
78. Burgess, S.K.; Leisen, J.E.; Kraftschik, B.E.; Mubarak, C.R.; Kriegel, R.M.; Koros, W.J. Chain Mobility, Thermal, and Mechanical Properties of Poly(ethylene furanoate) Compared to Poly(ethylene terephthalate). *Macromolecules* **2014**, *47*, 1383–1391. [CrossRef]
79. Fei, X.; Wang, J.; Zhu, J.; Wang, X.; Liu, X. Bio-based Poly(ethylene 2,5-furancoate): No Longer an Alternative, but an Irreplaceable Polyester in the Polymer Industry. *ACS Sustain. Chem. Eng.* **2020**, *8*, 8471–8485. [CrossRef]
80. Weinberger, S.; Haernvall, K.; Scaini, D.; Ghazaryan, G.; Zumstein, M.T.; Sander, M.; Pellis, A.; Guebitz, G.M. Enzymatic surface hydrolysis of poly(ethylene furanoate) thin films of various crystallinities. *Green Chem.* **2017**, *19*, 5381–5384. [CrossRef]

81. Li, C.; Wang, L.; Yan, Q.; Liu, F.; Shen, Y.; Li, Z. Rapid and Controlled Polymerization of Bio-Sourced δ-Caprolactone toward Fully Recyclable Polyesters and Thermoplastic Elastomers. *Ang. Chem. Int. Ed.* **2022**, *61*, e202201407.
82. Thakur, M.; Majid, I.; Hussain, S.; Nanda, V. Poly(ε-caprolactone): A potential polymer for biodegradable food packaging applications. *Technol. Sci.* **2021**, *34*, 449–461. [CrossRef]
83. Grobelny, Z.; Golba, S.; Jurek-Suliga, J. Mechanism of ε-caprolactone polymerization in the presence of alkali metal salts: Investigation of initiation course and determination of polymers structure by MALDI-TOF mass spectrometry. *Polym. Bull.* **2019**, *76*, 3501–3515. [CrossRef]
84. Yoshida, N.; Kasuya, N.; Haga, N.; Fukuda, K. Brand-new Biomass-based Vinyl Polymers from 5-Hydroxymethylfurfural. *Polym. J.* **2008**, *40*, 1164–1169. [CrossRef]
85. Davidson, M.G.; Elgie, S.; Parsons, S.; Young, T.J. Production of HMF, FDCA and their derived products: A review of life cycle assessment (LCA) and techno-economic analysis (TEA) studies. *Green Chem.* **2021**, *23*, 3154–3171. [CrossRef]
86. Warlin, N.; Garcia Gonzalez, M.N.; Mankar, S.; Valsange, N.G.; Sayed, M.; Pyo, S.-H.; Rehnberg, N.; Lundmark, S.; Hatti-Kaul, R.; Jannasch, P.; et al. A rigid spirocyclic diol from fructose-based 5-hydroxymethylfurfural: Synthesis, life-cycle assessment, and polymerization for renewable polyesters and poly(urethane-urea)s. *Green Chem.* **2019**, *21*, 6667–6684. [CrossRef]
87. Bonjour, O.; Liblikas, I.; Pehk, T.; Khai-Nghi, T.; Rissanen, K.; Vares, L.; Jannasch, P. Rigid bio-based polycarbonates with good processability based on a spirocyclic diol derived from citric acid. *Green Chem.* **2020**, *22*, 3940–3951. [CrossRef]
88. Warlin, N.; Nilsson, E.; Guo, Z.; Mankar, S.V.; Valsange, N.G.; Rehnberg, N.; Lundmark, S.; Jannasch, P.; Zhang, B. Synthesis and melt-spinning of partly bio-based thermoplastic poly(cycloacetal-urethane)s toward sustainable textiles. *Polym. Chem.* **2021**, *12*, 4942–4953. [CrossRef]
89. Valsange, N.G.; Garcia Gonzalez, M.N.; Warlin, N.; Mankar, S.V.; Rehnberg, N.; Lundmark, S.; Zhang, B.; Jannasch, P. Bio-based aliphatic polyesters from a spirocyclic dicarboxylate monomer derived from levulinic acid. *Green Chem.* **2021**, *23*, 5706–5723. [CrossRef]
90. Nguyen, H.T.H.; Qi, P.; Rostagno, M.; Feteha, A.; Miller, S.A. The quest for high glass transition temperature bioplastics. *J. Mater. Chem. A* **2018**, *6*, 9298–9331. [CrossRef]
91. Lingier, S.; Spiesschaert, Y.; Dhanis, B.; De Wildeman, S.; Du Prez, F.E. Rigid Polyurethanes, Polyesters, and Polycarbonates from Renewable Ketal Monomers. *Macromolecules* **2017**, *50*, 5346–5352. [CrossRef]
92. Choi, G.-H.; Hwang, D.Y.; Suh, D.H. High Thermal Stability of Bio-Based Polycarbonates Containing Cyclic Ketal Moieties. *Macromolecules* **2015**, *48*, 6839–6845. [CrossRef]
93. Arza, C.R.; Zhang, B. Synthesis, Thermal Properties, and Rheological Characteristics of Indole-Based Aromatic Polyesters. *ACS Omega* **2019**, *4*, 15012–15021. [CrossRef] [PubMed]
94. Hara, K.; Yamada, S.; Kurotani, A.; Chikayama, E.; Kikuchi, J. Materials informatics approach using domain modelling for exploring structure-property relationships of polymers. *Sci. Rep.* **2022**, *12*, 10558. [CrossRef] [PubMed]

Disclaimer/Publisher's Note: The statements, opinions and data contained in all publications are solely those of the individual author(s) and contributor(s) and not of MDPI and/or the editor(s). MDPI and/or the editor(s) disclaim responsibility for any injury to people or property resulting from any ideas, methods, instructions or products referred to in the content.

MDPI
St. Alban-Anlage 66
4052 Basel
Switzerland
www.mdpi.com

Polymers Editorial Office
E-mail: polymers@mdpi.com
www.mdpi.com/journal/polymers

Disclaimer/Publisher's Note: The statements, opinions and data contained in all publications are solely those of the individual author(s) and contributor(s) and not of MDPI and/or the editor(s). MDPI and/or the editor(s) disclaim responsibility for any injury to people or property resulting from any ideas, methods, instructions or products referred to in the content.

www.ingramcontent.com/pod-product-compliance
Lightning Source LLC
LaVergne TN
LVHW070156100526
838202LV00015B/1952